Lecture Notes in Computer Science 9140

Commenced Publication in 1973
Founding and Former Series Editors:
Gerhard Goos, Juris Hartmanis, and Jan van Leeuwen

More information about this series at http://www.springer.com/series/7407

Ying Tan · Yuhui Shi
Fernando Buarque · Alexander Gelbukh
Swagatam Das · Andries Engelbrecht (Eds.)

Advances in Swarm and Computational Intelligence

6th International Conference, ICSI 2015
held in conjunction with the Second BRICS
Congress, CCI 2015
Beijing, China, June 25–28, 2015
Proceedings, Part I

 Springer

Editors

Ying Tan
Peking University
Beijing
China

Yuhui Shi
Xi'an Jiaotong-Liverpool University
Suzhou
China

Fernando Buarque
Universidade de Pernambuco
Recife
Brazil

Alexander Gelbukh
Instituto Politécnico Nacional
Mexico
Mexico

Swagatam Das
Indian Statistical Institute
Kolkata
India

Andries Engelbrecht
University of Pretoria
Pretoria
South Africa

ISSN 0302-9743   ISSN 1611-3349 (electronic)
Lecture Notes in Computer Science
ISBN 978-3-319-20465-9   ISBN 978-3-319-20466-6 (eBook)
DOI 10.1007/978-3-319-20466-6

Library of Congress Control Number: 2015941139

LNCS Sublibrary: SL1 – Theoretical Computer Science and General Issues

Springer International Publishing AG Switzerland is part of Springer Science+Business Media
(www.springer.com)

Preface

This book and its companion volumes, LNCS vols. 9140, 9141, and 9142, constitute the proceedings of the 6th International Conference on Swarm Intelligence in conjunction with the Second BRICS Congress on Computational Intelligence (ICSI-CCI 2015) held during June 25–28, 2015, in Beijing, China.

The theme of ICSI-CCI 2015 was "Serving Our Society and Life with Intelligence." With the advent of big data analysis and intelligent computing techniques, we are facing new challenges to make the information transparent and understandable efficiently. ICSI-CCI 2015 provided an excellent opportunity for academics and practitioners to present and discuss the latest scientific results and methods as well as the innovative ideas and advantages in theories, technologies, and applications in both swarm intelligence and computational intelligence. The technical program covered all aspects of swarm intelligence, neural networks, evolutionary computation, and fuzzy systems applied to all fields of computer vision, signal processing, machine learning, data mining, robotics, scheduling, game theory, DB, parallel realization, etc.

The 6th International Conference on Swarm Intelligence (ICSI 2015) was the sixth international gathering for researchers working on all aspects of swarm intelligence, following successful and fruitful events in Hefei (ICSI 2014), Harbin (ICSI 2013), Shenzhen (ICSI 2012), Chongqing (ICSI 2011), and Beijing (ICSI 2010), which provided a high-level academic forum for the participants to disseminate their new research findings and discuss emerging areas of research. It also created a stimulating environment for the participants to interact and exchange information on future challenges and opportunities in the field of swarm intelligence research. The Second BRICS Congress on Computational Intelligence (BRICS-CCI 2015) was the second gathering for BRICS researchers who are interested in computational intelligence after the successful Recife event (BRICS-CCI 2013) in Brazil. These two prestigious conferences were held jointly in Beijing this year so as to share common mutual ideas, promote transverse fusion, and stimulate innovation.

Beijing is the capital of China and is now one of the largest cities in the world. As the cultural, educational, and high-tech center of the nation, Beijing possesses many world-class conference facilities, communication infrastructures, and hotels, and has successfully hosted many important international conferences and events such as the 2008 Beijing Olympic Games and the 2014 Asia-Pacific Economic Cooperation (APEC), among others. In addition, Beijing has rich cultural and historical attractions such as the Great Wall, the Forbidden City, the Summer Palace, and the Temple of Heaven. The participants of ICSI-CCI 2015 had the opportunity to enjoy Peking operas, beautiful landscapes, and the hospitality of the Chinese people, Chinese cuisine, and a modern China.

ICSI-CCI 2015 received 294 submissions from about 816 authors in 52 countries and regions (Algeria, Argentina, Australia, Austria, Bangladesh, Belgium, Brazil, Brunei Darussalam, Canada, Chile, China, Christmas Island, Croatia, Czech Republic,

Egypt, Finland, France, Georgia, Germany, Greece, Hong Kong, India, Ireland, Islamic Republic of Iran, Iraq, Italy, Japan, Republic of Korea, Macao, Malaysia, Mexico, Myanmar, New Zealand, Nigeria, Pakistan, Poland, Romania, Russian Federation, Saudi Arabia, Serbia, Singapore, South Africa, Spain, Sweden, Switzerland, Chinese Taiwan, Thailand, Tunisia, Turkey, UK, USA, Vietnam) across six continents (Asia, Europe, North America, South America, Africa, and Oceania). Each submission was reviewed by at least two reviewers, and on average 2.7 reviewers. Based on rigorous reviews by the Program Committee members and reviewers, 161 high-quality papers were selected for publication in this proceedings volume with an acceptance rate of 54.76 %. The papers are organized in 28 cohesive sections covering all major topics of swarm intelligence and computational intelligence research and development.

As organizers of ICSI-CCI 2015, we would like to express our sincere thanks to Peking University and Xian Jiaotong-Liverpool University for their sponsorship, as well as to the IEEE Computational Intelligence Society, World Federation on Soft Computing, and International Neural Network Society for their technical co-sponsorship. We appreciate the Natural Science Foundation of China and Beijing Xinhui Hi-tech Company for its financial and logistic support. We would also like to thank the members of the Advisory Committee for their guidance, the members of the international Program Committee and additional reviewers for reviewing the papers, and the members of the Publications Committee for checking the accepted papers in a short period of time. Particularly, we are grateful to Springer for publishing the proceedings in their prestigious series of *Lecture Notes in Computer Science*. Moreover, we wish to express our heartfelt appreciation to the plenary speakers, session chairs, and student helpers. In addition, there are still many more colleagues, associates, friends, and supporters who helped us in immeasurable ways; we express our sincere gratitude to them all. Last but not the least, we would like to thank all the speakers, authors, and participants for their great contributions that made ICSI-CCI 2015 successful and all the hard work worthwhile.

April 2015

Ying Tan
Yuhui Shi
Fernando Buarque
Alexander Gelbukh
Swagatam Das
Andries Engelbrecht

Yaochu Jin	University of Surrey, UK
Xiaodong Li	RMIT University, Australia
Gary G. Yen	Oklahoma University, USA
Rachid Chelouah	EISTI, Cergy, France
Kunqing Xie	Peking University, China

Special Sessions Chairs

Benlian Xu	Changshu Institute of Technology, China
Yujun Zheng	Zhejing University of Technology, China
Carmelo Bastos	University of Pernambuco, Brazil

Publications Co-chairs

| Radu-Emil Precup | Polytechnic University of Timisoara, Romania |
| Tom Hankes | Radboud University, Netherlands |

Competition Session Chair

| Jane Liang | Zhengzhou University, China |

Tutorial/Symposia Sessions Chairs

Jose Alfredo Ferreira Costa	Federal University of Rio Grande do Norte, Brazil
Jianhua Liu	Fujian University of Technology, China
Xing Bo	University of Limpopo, South Africa
Chao Zhang	Peking University, China

Publicity Co-chairs

Yew-Soon Ong	Nanyang Technological University, Singapore
Hussein Abbass	The University of New South Wales, ADFA, Australia
Carlos A. Coello Coello	CINVESTAV-IPN, Mexico
Eugene Semenkin	Siberian Aerospace University, Russia
Pramod Kumar Singh	Indian Institute of Information Technology and Management, India
Komla Folly	University of Cape Town, South Africa
Haibin Duan	Beihang University, China

Finance and Registration Chairs

| Chao Deng | Peking University, China |
| Suicheng Gu | Google Corporation, USA |

Organization

Honorary Chairs

Xingui He   Peking University, China
Xin Yao   University of Birmingham, UK

Joint General Chair

Ying Tan   Peking University, China

Joint General Co-chairs

Fernando Buarque   University of Pernambuco, Brazil
Alexander Gelbukh   Instituto Politécnico Nacional, Mexico, and Sholokhov
                                 Moscow State University for the Humanities, Russia
Swagatam Das   Indian Statistical Institute, India
Andries Engelbrecht   University of Pretoria, South Africa

Advisory Committee Chairs

Jun Wang   Chinese University of Hong Kong, HKSAR, China
Derong Liu   University of Chicago, USA, and Institute
                                 of Automation, Chinese Academy of Science, China

General Program Committee Chair

Yuhui Shi   Xi'an Jiaotong-Liverpool University, China

PC Track Chairs

Shi Cheng   Nottingham University Ningbo, China
Andreas Janecek   University of Vienna, Austria
Antonio de Padua Braga   Federal University of Minas Gerais, Brazil
Zhigang Zeng   Huazhong University of Science and Technology,
                                 China
Wenjian Luo   University of Science and Technology of China, China

Technical Committee Co-chairs

Kalyanmoy Deb   Indian Institute of Technology, India
Martin Middendorf   University of Leipzig, Germany

Conference Secretariat

Weiwei Hu   Peking University, China

ICSI 2015 Program Committee

Hafaifa Ahmed   University of Djelfa, Algeria
Peter Andras   Keele University, UK
Esther Andrés   INTA, Spain
Yukun Bao   Huazhong University of Science and Technology,
                                 China
Helio Barbosa   LNCC, Laboratório Nacional de Computação
                                 Científica, Brazil
Christian Blum   IKERBASQUE, Basque Foundation for Science, Spain
Salim Bouzerdoum   University of Wollongong, Australia
David Camacho   Universidad Autonoma de Madrid, Spain
Bin Cao   Tsinghua University, China
Kit Yan Chan   DEBII, Australia
Rachid Chelouah   EISTI, Cergy-Pontoise, France
Mu-Song Chen   Da-Yeh University, Taiwan
Walter Chen   National Taipei University of Technology, Taiwan
Shi Cheng   The University of Nottingham Ningbo, China
Chaohua Dai   Southwest Jiaotong University, China
Prithviraj Dasgupta   University of Nebraska, Omaha, USA
Mingcong Deng   Tokyo University of Agriculture and Technology,
                                 Japan
Yongsheng Ding   Donghua University, China
Yongsheng Dong   Henan University of Science and Technology, China
Madalina Drugan   Vrije Universiteit Brussel, Belgium
Mark Embrechts   RPI, USA
Andries Engelbrecht   University of Pretoria, South Africa
Zhun Fan   Technical University of Denmark, Denmark
Jianwu Fang   Xi'an Institute of Optics and Precision Mechanics
                                 of CAS, China
Carmelo-Bastos Filho   University of Pernambuco, Brazil
Shangce Gao   University of Toyama, Japan
Ying Gao   Guangzhou University, China
Suicheng Gu   University of Pittsburgh, USA
Ping Guo   Beijing Normal University, China
Fei Han   Jiangsu University, China
Guang-Bin Huang   Nanyang Technological University, Singapore
Amir Hussain   University of Stirling, UK
Changan Jiang   RIKEN-TRI Collaboration Center for
                                 Human- Interactive Robot Research, Japan
Liu Jianhua   Fujian University of Technology, China
Colin Johnson   University of Kent, UK

Zhenzhen Wang	Jinling Institute of Technology, China
Fang Wei	Southern Yangtze University, China
Ka-Chun Wong	University of Toronto, Canada
Zhou Wu	City University of Hong Kong, HKSAR, China
Shunren Xia	Zhejiang University, China
Bo Xing	University of Limpopo, South Africa
Benlian Xu	Changshu Institute of Technology, China
Rui Xu	Hohai University, China
Bing Xue	Victoria University of Wellington, New Zealand
Wu Yali	Xi'an University of Technology, China
Yingjie Yang	De Montfort University, UK
Guo Yi-Nan	China University of Mining and Technology, China
Peng-Yeng Yin	National Chi Nan University, Taiwan
Ling Yu	Jinan University, China
Zhi-Hui Zhan	Sun Yat-sen University, China
Defu Zhang	Xiamen University, China
Jie Zhang	Newcastle University, UK
Jun Zhang	Waseda University, Japan
Junqi Zhang	Tongji University, China
Lifeng Zhang	Renmin University, China
Mengjie Zhang	Victoria University of Wellington, New Zealand
Qieshi Zhang	Waseda University, Japan
Yong Zhang	China University of Mining and Technology, China
Wenming Zheng	Southeast University, China
Yujun Zheng	Zhejiang University of Technology, China
Zhongyang Zheng	Peking University, China
Guokang Zhu	Chinese Academy of Sciences, China
Zexuan Zhu	Shenzhen University, China
Xingquan Zuo	Beijing University of Posts and Telecommunications, China

BRICS CCI 2015 Program Committee

Hussein Abbass	The University of New South Wales, Australia
Mohd Helmy Abd Wahab	Universiti Tun Hussein Onn Malaysia
Aluizio Araujo	Federal University of Pernambuco, Brazil
Rosangela Ballini	State University of Campinas, Brazil
Gang Bao	Huazhong University of Science and Technology, China
Guilherme Barreto	Federal University of Ceará, Brazil
Carmelo J.A. Bastos Filho	University of Pernambuco, Brazil
Antonio De Padua Braga	Federal University of Minas Gerais, Brazil
Felipe Campelo	Federal University of Minas Gerais, Brazil
Cristiano Castro	Universidade Federal de Lavras, Brazil
Wei-Neng Chen	Sun Yat-Sen University, China
Shi Cheng	The University of Nottingham Ningbo, China

Ding Wang   Institute of Automation, Chinese Academy of Sciences,
                                     China
Qinglai Wei   Northeastern University, China
Benlian Xu   Changshu Institute of Technology, China
Takashi Yoneyama   Aeronautics Institute of Technology (ITA), Brazil
Yang Yu   Nanjing University, China
Xiao-Jun Zeng   University of Manchester, UK
Zhigang Zeng   Huazhong University of Science and Technology,
                                     China
Zhi-Hui Zhan   Sun Yat-sen University, China
Mengjie Zhang   Victoria University of Wellington, New Zealand
Yong Zhang   China University of Mining and Technology, China
Liang Zhao   University of São Paulo, Brazil
Zexuan Zhu   Shenzhen University, China
Xingquan Zuo   Beijing University of Posts and Telecommunications,
                                     China

Additional Reviewers for ICSI 2015

Bello Orgaz, Gema   Lin, Ying
Cai, Xinye   Liu, Jing
Chan, Tak Ming   Liu, Zhenbao
Cheng, Shi   Lu, Bingbing
Deanney, Dan   Manolessou, Marietta
Devin, Florent   Marshall, Linda
Ding, Ke   Menéndez, Héctor
Ding, Ming   Oliveira, Sergio Campello
Dong, Xianguang   Peng, Chengbin
Fan, Zhun   Qin, Quande
Geng, Na   Ramírez-Atencia, Cristian
Gonzalez-Pardo, Antonio   Ren, Xiaodong
Han, Fang   Rodríguez Fernández, Víctor
Helbig, Marde   Senoussi, Houcine
Jiang, Yunzhi   Shang, Ke
Jiang, Ziheng   Shen, Zhe-Ping
Jordan, Tobias   Sze-To, Antonio
Jun, Bo   Tao, Fazhan
Junfeng, Chen   Wang, Aihui
Keyu, Yan   Wen, Shengjun
Li, Jinlong   Wu, Yanfeng
Li, Junzhi   Xia, Changhong
Li, Wenye   Xu, Biao
Li, Xin   Xue, Yu
Li, Yanjun   Yan, Jingwei
Li, Yuanlong   Yang, Chun

Yaqi, Wu
Yassa, Sonia
Yu, Chao
Yu, Weijie
Yuan, Bo

Zhang, Jianhua
Zhang, Tong
Zhao, Minru
Zhao, Yunlong
Zong, Yuan

Additional Reviewers for BRICS-CCI 2015

Amaral, Jorge
Bertini, João
Cheng, Shi
Ding, Ke
Dong, Xianguang
Forero, Leonardo
Ge, Jing
Hu, Weiwei
Jayne, Chrisina
Lang, Liu
Li, Junzhi
Lin, Ying

Luo, Wenjian
Ma, Hongwen
Marques Da Silva, Alisson
Mi, Guyue
Panpan, Wang
Rativa Millan, Diego Jose
Xun, Li
Yan, Pengfei
Yang, Qiang
Yu, Chao
Zheng, Shaoqiu

Contents – Part I

Particle Swarm Optimization

Ant Colony Optimization

Artificial Bee Colony Algorithms

Evolutionary and Genetic Algorithms

Differential Evolution

Brain Storm Optimization Algorithm

Biogeography Based Optimization

Multi-Agent Systems and Swarm Robotics

Contents – Part II

Information Security

Automation Control

Combinatorial Optimization Algorithms

Constrained Optimization Algorithms

Scheduling and Path Planning

Machine Learning

Blind Source Separation

Swarm Interaction Behavior

Parameters and System Optimization

Contents – Part III

Simulation

Image and Texture Analysis

Dimension Reduction

System Optimization

Other Applications

Segmentation and Detection System

Machine Translation

Virtual Management and Disaster Analysis

Other Applications

Novel Swarm-based Optimization
Algorithms and Applications

Effects of Topological Variations on Opinion Dynamics Optimizer

Rishemjit Kaur[1,2(✉)], Ritesh Kumar[2], Amol P. Bhondekar[2], Reiji Suzuki[1], and Takaya Arita[1]

[1] Graduate School of Information Science, Nagoya University, Nagoya, Japan
rishemjit.kaur@gmail.com, {reiji,arita}@nagoya-u.jp
[2] CSIR-Central Scientific Instruments Organisation, Chandigarh, India
{riteshkr,amolbhondekar}@csio.res.in

Abstract. Continuous opinion dynamics optimizer (CODO) is an algorithm based on human collective opinion formation process for solving continuous optimization problems. In this paper, we have studied the impact of topology and introduction of leaders in the society on the optimization performance of CODO. We have introduced three new variants of CODO and studied the efficacy of algorithms on several benchmark functions. Experimentation demonstrates that scale free CODO performs significantly better than all algorithms. Also, the role played by individuals with different degrees during the optimization process is studied.

Keywords: Human opinion dynamics · Opinion dynamics optimizer · Scale free · Swarm intelligence · Topology

1 Introduction

Researchers have been using the multi-agent models to investigate collective phenomenon in human societies such as crowd behaviour [1], consensus formation [2, 3] and wisdom of crowds [4]. Recently, it has been shown that human group thinking or opinion dynamics can be used to solve complex mathematical optimization problems [5–8]. The optimizer for continuous problems is based on Durkheim theory of social integration [9] and referred to as Continuous Opinion Dynamics Optimizer (CODO) [8]. Unlike other population based meta-heuristics, the main advantage of this algorithm is the presence of few tuning parameters and an intuitive relation to collective human decision making. Also, it was proven to be efficient and better than a structurally similar algorithm, local best PSO on the CEC 2013 benchmark real parameter optimization problems [10]. This algorithm entails two important aspects i.e. updating dynamics and underlying societal structure

The updating dynamics, based on theory of social integration, has two components namely, integrating and disintegrating forces. The former represents the need of belongingness of an individual to the society and binds the individuals together in a society. The latter component represents the individualism which tends to threaten the

© Springer International Publishing Switzerland 2015
Y. Tan et al. (Eds.): ICSI-CCI 2015, Part I, LNCS 9140, pp. 3–13, 2015.
DOI: 10.1007/978-3-319-20466-6_1

social integration. Individualism, in general represents an individual's need to be distinct or unique from others, and it is incorporated as an 'adaptive noise' in the algorithm. The integrating forces tend to increase with increase in the number of individuals having the similar opinion in the society. Intuitively, the integrating forces are responsible for exploitation of the search space, whereas disintegrating forces contribute towards exploration of the search space. Together, these two components contribute towards finding an optimum solution of a problem.

Another aspect of this algorithm is the societal structure or topology. The topology may be regular, fully connected, random or scale free in nature. It is evident from the previous works on other optimization algorithms [11–14], that the topology can significantly improve the optimization performance. Such type of study has never been performed before in case of CODO and it is intuitive to assume that in the case of human opinion dynamics, that the societal structure and its formation strategies may improve the optimization performance.

In the previous work [8], a society was considered to be a 2D square grid structure with Moore neighbourhood. But in real world, the networks are not this simple. In this paper, an investigation of the impact of network topology on the performance of CODO has been conducted. Here, we propose two variants of CODO having scale free topology based on two different network generation mechanisms i.e. Barabasi Albert (BA) model and Bianconi-Barabasi (BB) model (also known as fitness model). These variants are referred to as BA-CODO and BB-CODO henceforth. Further, a variant of CODO with leaders (L-CODO) has also been developed, where leaders carry the best information and act as super spreaders in a society by connecting to all the individuals. The optimization performances of these three variants of CODO have been compared with fully connected CODO (F-CODO, where all individuals are connected to each other) as well as basic CODO algorithm (2D grid topology with Moore neighbourhood).

2   Continuous Opinion Dynamics Optimizer

The society has M number of individuals representing candidate solutions to the problem in D-dimensional space. Each individual i is characterized by its opinion vector $o_i = [o_i^1, o_i^2, ... o_i^D] \in R^D$, fitness f_i and social rank SR_i at particular time or iteration t. The opinions are real valued and initialized randomly using a uniform distribution. Individuals are placed on a 2D grid topology with Moore neighbourhood as shown in Fig. 1(a).

For every individual i, its fitness value f_i, which is the output of the function to be minimized, is calculated depending upon its opinion vector. Further, each individual is ranked by the fitness value. The individual with the minimum fitness value is assigned the highest Social Rank (SR). The individuals with the same fitness values have the same SR. The highest possible SR is equal to the total number of individuals in a society. At each iteration t, individuals update their opinion vector as follows.

$$\Delta o_i = \frac{\sum_{j=1}^{N}(o_j(t) - o_i(t))w_{ij}(t)}{\sum_{j=1}^{N}w_{ij}(t)} + \xi_i(t), j \neq i \qquad (1)$$

where N is the no. of neighbours of individual i, $o_j(t)$ is the opinion of the neighbour j, $w_{ij}(t)$ is the social influence exerted by neighbour j on individual i and $\xi_i(t)$ represents the '*adaptive noise*'.

The first component of this equation represents the integrative forces of a society, which bind the individuals together. It is the weighted average of the opinion differences of an individual from its neighbours, and the weights represent the social influence, which are defined as $w_{ij} = SR_j(t)/d_{ij}(t)$. Here, dij is the Euclidean distance between individual i and j in topological space. So the better-fit and nearby individuals will have more social influence on others.

The second component of (1) stands for the disintegrative forces in a society. It is the normally distributive random noise with zero mean and a standard deviation of σi(t) defined as follows

$$\sigma_i(t) = S \sum_{j=1}^{N} e^{-|f_{ij}(t)|} \qquad (2)$$

where S is the strength of the disintegrating forces and $|f_{ij}(t)|$ denotes the difference in fitness value of an individual i and its neighbour j. As the fitness values of individuals get closer to the fitness value of individual i, σi(t) increases and leads to more individualization in a society.

The pseudo code for CODO is as follows:

```
1.   society.opinion = GenerateInitialSociety(Xmin, Xmax);
2.   iter = 0;
3.  while (iter < max_iter && error >= min_error ) do
4.      society.fitness =EvaluateFitnessFcn(society.opinion);
5.      society.ranking = CalcRank(society.fitness); //It ranks
        the individuals based on society fitness
6.      iter = iter + 1;
7.     for each individual i and each dimension d do
8.         Calculate w_ij of neighbours j of individual i with re-
           spect to dimension d.
9.         Update opinion of individual i as defined in (1).
10.    end (for)
11. end (while)
```

3   CODO with Leaders (L-CODO)

Every society is characterized by some leaders, followers and agnostics [16]. We focus on the leaders and exploit its various characteristics to design a variant of CODO for mathematical optimization. Leaders are normally highly connected, socially active individuals [17], which leads them to play an important role in opinion dissemination in a society. A leader can be autocratic or democratic, an autocratic leader usually does not get influenced by others, whereas, a democratic leader is accommodating. In this study, we have considered only autocratic leaders, and have developed our algorithm based on the following assumptions:

1. A leader is fully connected
2. Leaders may change with time/iterations

The topology adopted for L-CODO is also 2D grid with Moore neighbourhood as shown in Fig 1(b). Each individual is connected to leaders. This may help in faster diffusion of solutions in a society resulting in faster convergence. Here, leader (l) is defined as an individual with a minimum fitness value or maximum SR. There can be more than one leader in the society depending upon the social ranking. With M individuals in the society, the SR of a leader at particular iteration t can be defined as

$$SR_l(t) = \max(SR_i(t)) \quad i = 1,2,\dots,M \tag{3}$$

The original opinion updating equation is modified as follows:

$$\Delta o_i = \frac{\sum_{j=1}^{N+L}(o_j(t)-o_i(t))w_{ij}(t)}{\sum_{j=1}^{N+L}w_{ij}(t)} + \xi_i(t), j \neq i \tag{4}$$

$$\sigma_i(t) = S\sum_{j=1}^{N+L} e^{-|f_{ij}(t)|} \tag{5}$$

where, L is the number of leaders in a society apart from the neighbours. It is to be noted that the leader can also be among the neighbours of the individual.

The pseudo code for L-CODO

```
1.   society.opinion = GenerateInitialSociety(Xmin, Xmax);
2.   iter = 0;
3.   while (iter < max_iter && error >= min_error ) do
4.      society.fitness = EvaluateFitnessFcn(society.opinion);
5.      society.ranking = CalcRank(society.fitness);
        l = FindLeaders(society.ranking)
6.      iter = iter+1;
7.      for each individual i (if not l) and each dimension d do
8.         Calculate wij of neighbours j of individual i and
           leaders l with respect to dimension d.
9.         Update opinion of individual i as defined in (4).
10.  end (for)
11. end (while)
```

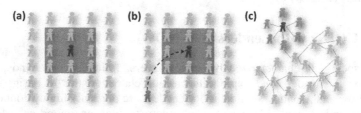

Fig. 1. Society structure for   a) CODO, b) L-CODO and c) BA and BB-CODO

4   Scale Free CODO

This section provides an overview of scale free networks, its generation mechanisms and finally discusses the optimizers based on those mechanisms.

4.1   Scale Free Networks

Their degree distribution, which is defined as the probability of finding a node with degree k, follows a power law i.e. $p(k) = k^{-\gamma}$ and $2 \leq \gamma \leq 3$. Following subsections describe the two different models proposed to simulate a scale free topology.

Barabasi and Albert Model. According to this model, the emergence of scale free nature of a network depends upon two co-existing mechanisms[18]:

Growth: Real world networks are not fixed but they grow with time. Starting with a small number of initial nodes m_0, network continuously expands by connecting a new node in every time step to a fixed number of existing nodes (m).

Preferential Attachment: The probability that a new node connects to existing nodes is proportional to their degree k and is given by following equation

$$p_i = k_i / \sum_j k_j \qquad (6)$$

New nodes are more likely to connect to existing nodes with more number of connections, i.e. "rich-gets-richer" phenomenon which gives rise to the formation of hubs in a network.

Bianconi and Barabasi Model. Another variant of BA model was proposed by Bianconi and Barabasi [19] where the probability of connection depends upon the product of node's degree k and as well as its fitness f. It follows "fit-gets-richer" phenomenon.

$$p_i = k_i f_i / \sum_j k_j f_j \qquad (7)$$

4.2   Scale Free Opinion Dynamics Optimizers (BA-CODO and BB-CODO)

The above mentioned models result into a scale free topology, but they follow different generation principles. In this work, these models have been used to develop two scale free opinion dynamics optimizers, namely, BA-CODO and BB-CODO. It may provide us interesting insights about how evolution of topology can affect the optimization performance. The optimization process of these algorithms involves two stages.

Firstly, the topology of a society is constructed, in which, initially there are m_0 number of individuals fully connected with each other. At each iteration, a new individual connects to m existing individuals depending upon the model (either BA or BB model). After creation of these m new links, the individuals collectively search in the space and update their opinions. This process keeps on repeating till the number of individuals has reached the maximum society size (M). In this stage, the topology generation and optimization take place simultaneously. Secondly, the individuals are connected to each other in a scale free topology and they collectively search in the opinion space for the optimum position. The total number of function evaluations is the summation of function evaluations of stage 1 and stage 2.

The pseudo code for BA-CODO and BB-CODO

```
1. society.opinion = GenerateInitialSociety(Xmin, Xmax)
2. Initialize society.network
3. // stage 1: Topology construction & search
   society.network = GenerateTopology(society.network, socie-
   ty.opinion, generation_mechanism);
4. //stage 2: Search
   iter =0
5. while (iter < max_iter && error >= min_error )  do
6. society.opinion =CODO(society.opinion,society.network)
7. end(while)


society.network = function GenerateTopology (society.network,
                society.opinion, generation_mechanism)

1. iter = 0
2. while ( iter < M-m₀) do
3.    society.opinion = CODO (society.opinion, society.network)
4.    if generation_mechanism == BA
5.       // perform equation (6)
6.    else
7.       // perform equation (7)
8.    end (else)
9.    iter = iter+1
10. end(while)
11. return society.network
end (function)


society.opinion    =    function   CODO(society.opinion,    socie-
                ty.network)
1. society.fitness = EvaluateFitnessFcn(society.opinions);
2. society.ranking = calcRank(society.fitness);
3. for each individual i and each dimension d do
4.    Calculate wᵢⱼ of neighbours j of individual with respect to
      dimension d.
5.    Update opinion of individual i as defined in equation (1).
6. end (for)
7. return society.opinion
end (function)
```

5   Experimental Results

The experiments were performed for five algorithms namely, CODO, L-CODO, BA-CODO, BB-CODO and F-CODO. The benchmark functions used to test the performance, parameters of the algorithms and finally the results are discussed in this section.

5.1   Benchmark Functions

The initialization range, optimum solution (x*) and the optimum fitness value f(x*) of the benchmark functions used have been listed in Table 1

Table 1. Benchmark Functions

Function Name	Initialization Space	x*	f(x*)
F1 Sphere	$[-100, 100]^D$	$[0, 0, ..., 0]^D$	0
F2 Rosenbrock	$[-2.048, 2.048]^D$	$[1, 1, ..., 1]^D$	0
F3 Rastrigin	$[-5.12, 5.12]^D$	$[0, 0, ..., 0]^D$	0
F4 Griewank	$[-100, 100]^D$	$[0, 0, ..., 0]^D$	0
F5 Ackley	$[-10, 10]^D$	$[0, 0, ..., 0]^D$	0

5.2   Parameterization

Table 2 summarizes the parameters used for the algorithms. For every function, all the algorithms were initialized with same society in each run for fair comparisons. The parameters of algorithms were not tuned specifically for each function. It has been ensured that the average degree of scale free society is same as 2D grid society for CODO and their comparison is independent of average degree number.

5.3   Results

For analyzing and comparing the performance of algorithms, the fitness values of the best individuals were recorded. Table 3 lists the mean, standard deviation and minimum fitness value attained by the best individual over 25 independent runs for 10 and 30 dimensions for all benchmark functions. Further for detailed statistical study, the unpaired Wilcoxon tests [21] were conducted between CODO and other algorithms for every function. It is two sided test with the null hypothesis that the results of two algorithms come from identical distribution with equal median. The significance level was set to be 5%. In Table 3 the statistically insignificant results are asterisk (*) marked.

It can be observed from Table 3 that BA-CODO and BB-CODO out-performed other algorithms for 10D problems in a statistically significant manner, except for function F2. The L-CODO has also performed significantly better than CODO and F-CODO. The F-CODO performed the worst in this case except for function F5 where its performance is statistically not different from CODO. In case of 30D problems, again BA-CODO and BB-CODO out-performed other algorithms except for F3, where L-CODO has performed the best. Surprisingly, for F3 the performance of BB-CODO and BA-CODO is similar to CODO. Also, F-CODO has maintained the previous trend and showed the worst performance for 30D problems. The poor performance of F-CODO can be attributed to the fact that since in such type of network, every node is connected with each other, hence the diversity of society reduces to the point that the individuals get stuck in local minima.

Fig. 3 shows the convergence rate i.e. average fitness value of the best individual achieved over 25 runs vs the number of function evaluations for 10D functions. It can be observed that the initial convergence rate for BA-CODO and BB-CODO is slower than other algorithms (prominently visible in F1), but eventually it achieves better fitness or higher quality solutions, whereas the other algorithms converge to poorer regions. This slow convergence is may be due to the additional stage where evolution of network takes place. F-CODO shows the fastest convergence in the initial stage, but then it gets trapped in local optimum.

Table 2. Parameters used in algorithms

Parameters	Value
Dimension	[10, 30]
Society Structure	2D cellular grid for CODO, CODO with leaders and fully connected CODO
	Scale free for CODO-BA and CODO-BB
Max number function evaluations (MaxFES)	10000*No of features
Strength of disintegrating forces (S)	8
No of runs	25
Stopping criterion	Max number function evaluations (MaxFES)
Initialization	Uniform random distribution
Society Size (M)	100
Moore Neighbourhood	1
m	4
m0	4

It can also be observed that in some cases, BA-CODO and BB-CODO show decreasing trend at function evaluation limit, whereas the other algorithms have converged/stabilized. It may mean that with more number of function evaluations, BA-CODO and BB-CODO can achieve better solutions. It may be noted that performance of BB-CODO and BA-CODO is similar, which leads us to conclude that generation mechanisms may not have any effect on the optimization performance, though, it requires further investigations.

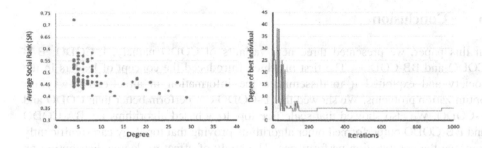

Fig. 2. a) Average social rank vs degree for all individuals during optimization process b) Degree vs iterations of the best individual

However, the results suggest the importance and dependence of topology on the optimization performance. Therefore, in order to investigate the effect of degree distribution of a scale free society during the optimization process, we calculated the average SR of each individual over all the iterations.   Fig. 2(a) shows the average SR vs the degree of all individuals for one of the runs of F1. It can be observed that average SR is higher for lower degree nodes as compared to the higher degree nodes. It may be due to the fact that the high degree individuals get affected by more number of individuals as compared to the lower degree ones, therefore they have access to information from diverse regions of solution space. As a result, they are involved in larger exploration of the solution space. The degree of the best indivdual vs iterations was also studied (Fig. 2(b)). The large fluctuations in the degree of the best individual can be observed during the initial iterations, whereas the degree has almost stabilized to a lower value in the later iterations. Similar results were obtained for other functions and with BB scale free network.

Table 3. Comparison of Fitness Values Obtained for 10D and 30D for all functions

10D	CODO		F-CODO		L-CODO		BA-CODO		BB-CODO	
	Mean	Std. Dev	Mean	Std. Dev	Mean	Std. Dev	Mean	Std. Dev	Mean	Std. Dev
F1	0.66	0.20	6.35	2.44	0.42	0.09	**0.16**	0.08	0.20	0.09
F2	15.41	1.30	19.10	4.52	**10.09**	0.79	11.83	1.26	11.71	1.55
F3	11.48	3.33	23.64	3.87	8.15	3.87	**4.34**	1.83	5.22	2.47
F4	0.83	0.10	0.93	0.08	0.77	0.11	0.63	0.08	**0.58**	0.09
F5	10.96	2.19	11.71*	0.77	10.32	3.07	7.98	1.39	**7.17**	1.32
30D	CODO		F-CODO		L-CODO		BA-CODO		BB-CODO	
	Mean	Std. Dev	Mean	Std. Dev	Mean	Std. Dev	Mean	Std. Dev	Mean	Std. Dev
F1	7.01	1.16	72.14	8.33	3.81	0.39	**1.86**	0.60	2.23	0.59
F2	46.1	2.22	64.83	4.43	39.1	1.33	**39.07**	2.15	40.37	2.64
F3	49.40	8.88	119.14	7.99	**42.39**	8.44	48.85*	16.89	58.42*	25.37
F4	1.21	0.03	1.20	0.02	1.12	0.04	1.05	0.02	**1.04**	0.02
F5	13.69	0.27	13.69*	0.276	13.69*	0.27	12.62	0.74	**12.04**	1.188

6   Conclusion

In this paper, we presented three new variants of CODO namely, L-CODO, BA-CODO and BB-CODO. The first of which introduced the concept of leaders in the society and exploited it to disseminate the information in an effective way for optimization problems. We showed that L-CODO can perform better than CODO and F-CODO. We also showed that scale free topology based algorithms i.e. BA-CODO and BB-CODO outperformed other algorithms proving that topology can significantly improve the optimization performance. The study of effect of degree distribution on optimization performance gave us insight into the role played by hubs and non-hubs for opinion dissemination in a society. Since in this study the algorithmic parameters such as strength of disintegrating forces, society size etc were not fine tuned, future research on the effects of these parameters on the optimization performance needs attention.

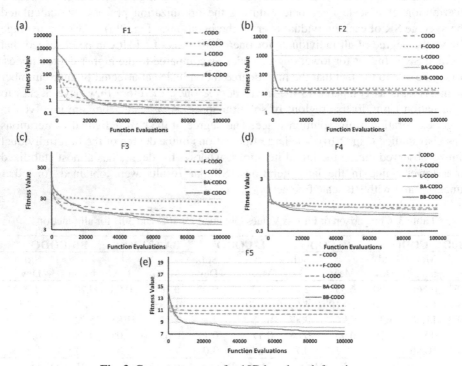

Fig. 3. Convergence rate for 10D benchmark functions

References

1. Castellano, C., Fortunato, S., Loreto, V.: Statistical physics of social dynamics. Rev. Mod. Phys. **81**, 591 (2009)
2. Deffuant, G., Neau, D., Amblard, F., Weisbuch, G.: Mixing beliefs among interacting agents. Adv. Complex Syst. **3**, 87–98 (2000)

3. Hegselmann, R., Krause, U.: Opinion dynamics driven by various ways of averaging. Comput. Econ. **25**, 381–405 (2005)
4. Lorenz, J., Rauhut, H., Schweitzer, F., Helbing, D.: How social influence can undermine the wisdom of crowd effect. Proc. Natl. Acad. Sci. **108**, 9020–9025 (2011)
5. Kaur, R., Kumar, R., Gulati, A., Ghanshyam, C., Kapur, P., Bhondekar, A.P.: Enhancing electronic nose performance: a novel feature selection approach using dynamic social impact theory and moving window time slicing for classification of Kangra orthodox black tea (Camellia sinensis(L.) O. Kuntze). Sensors Actuators B Chem. **166-167**, 309–319 (2012)
6. Bhondekar, A.P., Kaur, R., Kumar, R., Vig, R., Kapur, P.: A novel approach using Dynamic Social Impact Theory for optimization of impedance-Tongue (iTongue). Chemom. Intell. Lab. Syst. **109**, 65–76 (2011)
7. Macaš, M., Bhondekar, A.P., Kumar, R., Kaur, R., Kuzilek, J., Gerla, V., Lhotská, L., Kapur, P.: Binary social impact theory based optimization and its applications in pattern recognition. Neurocomputing **132**, 85–96 (2014)
8. Kaur, R., Kumar, R., Bhondekar, A.P., Kapur, P.: Human opinion dynamics: An inspiration to solve complex optimization problems. Sci. Rep. **3** (2013)
9. Mäs, M., Flache, A., Helbing, D.: Individualization as driving force of clustering phenomena in humans. PLoS Comput. Biol. **6**, e1000959 (2010)
10. Liang, J.J., Qu, B.Y., Suganthan, P.N.: Problem Definitions and Evaluation Criteria for the CEC 2013 Special Session on Real-Parameter Optimization (2013)
11. Liu, C., Du, W.-B., Wang, W.-X.: Particle Swarm Optimization with Scale-Free Interactions. PLoS One. **9**, e97822 (2014)
12. Zhang, C., Yi, Z.: Scale-free fully informed particle swarm optimization algorithm. Inf. Sci. (Ny) **181**, 4550–4568 (2011)
13. Kennedy, J., Mendes, R.: Population structure and particle swarm performance. In: Proc. IEEE Congr. Evol. Comput., pp. 1671–1676. IEEE computer Society, Honolulu
14. Mendes, R., Kennedy, J., Neves, J.: The fully informed particle swarm: simpler, maybe better. Evol. Comput. IEEE Trans. **8**, 204–210 (2004)
15. Toscano-Pulido, G., Reyes-Medina, A.J., Ramírez-Torres, J.G.: A statistical study of the effects of neighborhood topologies in particle swarm optimization. In: Madani, K., Correia, A.D., Rosa, A., Filipe, J. (eds.) Computational Intelligence. SCI, vol. 343, pp. 179–192. Springer, Heidelberg (2011)
16. Kurmyshev, E., Juárez, H.A.: What is a leader of opinion formation in bounded confidence models? arXiv Prepr. arXiv1305.4677 (2013)
17. Düring, B., Markowich, P., Pietschmann, J.F., Wolfram, M.T.: Boltzmann and Fokker-Planck equations modelling opinion formation in the presence of strong leaders. Proc. R. Soc. A Math. Phys. Eng. Sci. **465**, 3687–3707 (2009)
18. Barabási, A.-L., Albert, R.: Emergence of scaling in random networks. Science (80-.). **286**, 509–512 (1999)
19. Bianconi, G., Barabási, A.-L.: Competition and multiscaling in evolving networks. EPL (Europhysics Lett.) **54**, 436 (2001)
20. Bratton, D., Kennedy, J.: Defining a standard for particle swarm optimization. In: IEEE Swarm Intelligence Symposium, SIS 2007, pp. 120–127. IEEE (2007)
21. García, S., Molina, D., Lozano, M., Herrera, F.: A study on the use of non-parametric tests for analyzing the evolutionary algorithms' behaviour: a case study on the CEC 2005 special session on real parameter optimization. J. Heuristics. **15**, 617–644 (2009)

Utilizing Abstract Phase Spaces in Swarm Design and Validation

Sanza Kazadi[✉], Dexin Jin, and M. Li

Jisan Research Institute, 308 S. Palm Ave., Alhambra, CA
91803, USA
skazadi@jisan.org

Abstract. We introduce a swarm design methodology. The methodology uses a seven step process involving a high-level phase space to map the desired goal to a set of behaviors, castes, deployment schedules, and provably optimized strategies. We illustrate the method on the stick-pulling task.

1 Introduction

Swarms are continually interacting groups of autonomous agents whose interactions at least partially shape their behaviors. Different modalities of interaction range from direct communication to avoidance to stigmergy, or communication through environmental manipulation, to cooperation and competition. Swarms tend to amplify the impact that a single agent can have by leveraging the accumulated actions of a group. Swarms are, too, capable of accomplishing complex actions together that individuals cannot. What makes them interesting and exciting is the possibility of taking on complex tasks that the individual agents couldn't possibly do, using interaction and dynamics to extend individuals' capabilities.

As an engineering tool the application of swarms has been limited. While many swarms have been designed for a variety of different tasks, it is not yet clear that the tasks to which swarms have been applied cannot be done in other ways as well or better. Given a task that one wishes to accomplish, no provable methodology exists which indicates that the task is appropriate for and achievable by a swarm.

Swarm engineering is, loosely speaking, the process of taking a task together with a list of available technologies and constraints and using these to develop (1) a provable, minimalist plan to accomplish the task, (2) a list of requisite technologies, (3) a minimal set of agents, (4) the agent capabilities, and (5) the agent deployment schedule. We'd like to achieve the full design of the swarm aside from the design of the available technologies (which are assumed to already be in existence). This clarifies whether a swarm is an appropriate tool for the task as well as if and how it can be achieved.

In this paper we present a seven step swarm design framework. The framework is designed to help the swarm designer decide whether a swarm is the

Y. Tan et al. (Eds.): ICSI-CCI 2015, Part I, LNCS 9140, pp. 14–29, 2015.
DOI: 10.1007/978-3-319-20466-6_2

appropriate solution or not. Our framework begins with a problem description which describes the problem in terms of things that agents within the system can measure. An abstract phase space representation of the problem is generated. The initial and final points in the phase space are determined, and feasibility tests are carried out. Next pathways through phase space resulting from the use of available technologies and techniques are generated. The pathways are used to analyze the number of agents needed (if a swarm is needed), which technologies must be deployed and how they will be deployed, and how subtasks can be handled (agent number / technologies). Castes and swarm teams, or subsets of the swarm which comprise cooperating agents, can be developed. We show that the swarm's sufficiency is provable; any swarm enabling the movement in phase space along any generated path will achieve the task. The task requires a swarm only if the minimal number of agents required on any path exceeds one.

2 Seven Steps

We propose the following framework for the design of a swarm of agents aimed at accomplishing a specific task.

1. Define the set of measurables that the swarm is working on. These are $\{s_i\}_{i=1}^{N_s}$. These are quantities that describe the state of the system and can be measured by individual agents within the swarm.
2. Define the set of global measurables that one wants to achieve. These are $\{P_k\}_{k=1}^{N_g}$ as functions of the measurables $\{s_i\}$. That is, $P_k = f_k(s_1, s_2, \ldots, s_{N_s})$.
3. Define the initial point or set of initial points and the final point or set of final points for the $(P_1, P_2, \ldots, P_{N_g})$. That is, define $\left(P_1^i, P_2^i, \ldots, P_{N_g}^i\right)$ and $\left(P_1^f, P_2^f, \ldots, P_{N_g}^f\right)$.
4. Define the transitions one might go through in travelling through phase space. These are also called the requisite technologies
5. Define the path through phase space while avoiding the unfeasible areas of P-space.
6. Determine whether these technology levels and costs require a swarm. If so, determine the swarm requirement. If not, determine the agent requirement.
7. Design behaviors that satisfy the swarm requirements and agents that carry out these behaviors.

This methodology adds steps to the swarm design process that focus on the phase space describing the task. Most swarm design efforts use only the last two steps. This approach focuses attention on the technological levels available and their effect on the swarm's possible deployment. Differing technology will connect the phase space differently. The path through phase space must enable the application of the technological levels in a way that minimizes the path length.

3   System and Problem Description

As indicated in Steps 1 and 2, we must carry out a description of the system's agent-level measurables. Agents are individual autonomous actors in the system endowed with a location and the capability of independently interacting with one or more parts of the system. Let us define *local measurables* as the the set of things that can be measured by swarm agents as $\{s_i\}_{i=1}^{N_s}$. These are distinct from things that can be measured by external individuals, such as omnipotent observers or thinking users that employ other techniques than those available in the swarm's environment. For instance, minimalist puck clusterers [3] cannot know that a single cluster has been formed; an external observer can.

The functions need not be simple or have pre-determined dimensionality; they can have any form practically capable of being calculated. Let the set of these functions be denoted as $\{P_i (s_1, \ldots, s_{N_s})\}_{i=1}^{N_P}$. We define the *design goal of the swarm* as the creation of a dynamics that changes $\overrightarrow{P^{initial}} = (P_1^{initial}, P_2^{initial}, \ldots, P_{N_p}^{initial})$ to the desired final state $\overrightarrow{P^{final}} = \left(P_1^{final}, P_2^{final}, \ldots, P_{N_p}^{final} \right)$. The effect of the swarm is to create a change in the state of $\overrightarrow{P}$. The swarm must control $\left\{ \frac{\partial P_i}{\partial s_j} \frac{ds_j}{dt} \right\}_{i=1}^{N_P}$ through designed control of $\frac{\partial P_i}{\partial s_j}$. Yet, this defines any control algorithm, not specifically that accomplished by a swarm. In fact, many of the control algorithms designed in this way function equally well if a single agent is performing the task as if a group of agents is performing the task. Such a system does not indicate the need for a swarm-based solution; it cannot be a "killer ap".

4   Requisite Technologies (Phase Space Connectivity)

The phase space is defined as the set of all points that are definable by the global measurables. That is, $\Psi = \left\{ \overrightarrow{P} | \overrightarrow{P} = (P_1 (s_1, \ldots, s_{N_s}), \ldots, P_{N_P} (s_1, \ldots, s_{N_s})) \right\}$. The states of the elements $\overrightarrow{P}$ are constrained by their functional relationship to $(s_1, \ldots, s_{N_s})$. The points that might actually be in the phase space might be significantly smaller than the cross products of the ranges of the functions P_i. Those elements in the cross product space that are not in the phase space are said to be unfeasible.

4.1   Feasibility

The swarm task is feasible if we can "connect" the initial points to the final points using a path through phase space that is feasible at each point along the path. Different points within the phase space are connected to one-another by the behaviors of the swarm agents. These behaviors are generated by the technologies available and the method of using them. In order to determine the connectedness of the phase space, we need to understand the technologies that are being used and their method of use.

Any point in phase space is a representation of the state of the system. Each point in phase space can be represented as $\vec{P} = (P_1, P_2, \ldots, P_{N_P})$. We define a point as *feasible* if the point represents a system state that can be achieved given the constraints imposed upon the global properties $\{P_i\}_{i=1}^{N_P}$. Given a point $\left(P_1^\circ, \ldots, P_{N_P}^\circ\right)$, we define the inverse set $\iota\left(P_1^\circ, \ldots, P_{N_P}^\circ\right)$ as the set of all points $\vec{s} = (s_1, \ldots, s_{N_s})$ in the cross product space of the local measurables such that $P_i\left(\vec{s}\right) = P_i^\circ \ \forall i$. A point is feasible if $\iota\left(P_1^\circ, \ldots, P_{N_P}^\circ\right) \neq$.

First pass feasibility test. We define a *task* as a set of initial and final points for the set in phase space. We define a *first pass feasibility test* as a test to determine whether the starting and ending points of the phase space are feasible. Let the initial point in the phase space be $\vec{P_i}$ and the final point in the phase space be $\vec{P_f}$. Then the task is unfeasible if either $\iota\left(\vec{P_i}\right) =$ or $\iota\left(\vec{P_f}\right) =$. Otherwise, the task is feasible.

Path feasibility. We define a *path through phase space* as an ordered set of points Γ in the phase space. These points may be parametrized by a variable t with the property that $\Gamma(t=0) = \vec{P_i}$ and $\Gamma(t \to \infty) = \vec{P_f}$. Given the agent behaviors as defined by $\left\{\frac{\partial P_i}{\partial s_j}\frac{ds_j}{dt}\right\}_{i=1}^{N_P}$ in the continuous case and $\left\{\frac{\Delta P_i}{\Delta s_j}\frac{\Delta s_j}{\Delta t}\right\}_{i=1}^{N_P}$ in the discrete case, we have that

$$\Gamma(t+dt) = \Gamma(t) + \left(\frac{\partial P_1}{\partial s_j}\frac{ds_j}{dt}, \frac{\partial P_2}{\partial s_j}\frac{ds_j}{dt}, \ldots, \frac{\partial P_{N_P}}{\partial s_j}\frac{ds_j}{dt}\right) dt \qquad (1)$$

(continuous) and that

$$\Gamma(t+\Delta t) = \Gamma(t) + \left(\frac{\Delta P_1}{\Delta s_j}\frac{\Delta s_j}{\Delta t}, \frac{\Delta P_2}{\Delta s_j}\frac{\Delta s_j}{\Delta t}, \ldots, \frac{\Delta P_{N_P}}{\Delta s_j}\frac{\Delta s_j}{\Delta t}\right) \Delta t \qquad (2)$$

(discrete). Equations (1) and (2) illustrate the requirement that the path in phase space must be connected and linked by behaviors of the agents; it cannot be disjoint or occur by any means other than by the actions of the agents. The actions of the agents are, in turn, limited by the available technologies.

Let us suppose that we have a path Γ. The path Γ is said to be feasible if (1) $\iota(\Gamma_i) \neq$ and (2) for each Δs associated with the transition from Γ_i to Γ_{i+1}, a technology exists that can achieve this change. The path through phase space defines the required transitions and the transitions define what technologies the agents must be able to deploy.

Suppose that two points in phase space are $\Gamma_i = \left(P_1^i, \ldots, P_{N_P}^i\right)$ and $\Gamma_{i+1} = \left(P_1^{i+1}, \ldots, P_{N_P}^{i+1}\right)$. In order for this transition to be feasible, there must first be a nonempty set $\iota(\Gamma_i)$ and a second nonempty set $\iota(\Gamma_{i+1})$. A total set of transitions $N_{i,i+1} = |\iota(\Gamma_i)| \, |\iota(\Gamma_{i+1})|$ exist which make this transition possible. $N_{i,i+1}$ different technologies are therefore possible which might generate the phase space transition.

The importance of this lies in the identification of the *necessary and sufficient technology set* required to enable the task. Without having *a priori* knowledge of the technology set required, we can generate a technology list that must be in existence in order to achieve the task.

Task feasibility. A task is said to be feasible if it passes the first pass feasibility test, there exists at least one path that connects the initial and final points of the system through a connected feasible subset of the phase space, and the requisite technologies are in existence and can be deployed on existing agents.

4.2   Energy, Time, Agents, Swarm Teams, and Teams of Swarms

Carrying out any task requires consumption of resources, including energy and time. These can be included in the global measurables vector $\overrightarrow{P}$ by assigning the measures to two vector components. Making the assignment places restrictions on the energy consumption and the execution time of the tasks, as the beginning and ending points now have associated finite energy and time and each step consumes a finite amount of each.

Another resource that is used in any task is agents. This is a familiar concept among military planners; "spending men" is part of warfare. Let us define a number of states including "active", "fully functional", "executing behavior #k", "defective", "inactive", and "nonfunctional". If we limit the states to "active" and "inactive", agents transitioning between states is equivalent to "consumption" or "generation". The differences between transitions are defined by the behaviors Δs that the agents are able to execute. As an example, it is clear that "nonfunctional" is appropriately defined a $\Delta s = 0$. Agents using their bodies to form a static structure transition from a state of "moving" to "nonmoving". The measurables include the number of agents in each state, and the consumption of agents is then encoded as an appropriate definition of one or more components of the $\overrightarrow{P}$ vector.

In order to appropriately choose a path through phase space, we must correctly determine the costs of the use of individual technologies. For each technology T_i is defined as $T_i = \left(\overrightarrow{\Delta P}, \Delta E, \Delta t, \Delta A \right)$ where $\overrightarrow{\Delta P}$ represents the change in position in phase space, ΔE represents the energy consumption of the action, Δt represents the time required to accomplish the task, and ΔA represents the number of agents required for the task to be accomplished.

Once a path has been specified, we can define the running agent tally as $R_i\left(\Gamma\right) = \sum_{j=1}^{i} |\Delta A_j|$. This is the number of agents required to accomplish the task along this path up to the ith point along the path. In order to accomplish the task along this path, the minimal number of agents required $S\left(\Gamma\right)$ is given by $S\left(\Gamma\right) = \max_{i=1}^{|\Gamma|} R_i$. This is the minimal swarm size of the swarm required to accomplish the task along this path.

Let us define the steps of the path Γ as independent or dependent according to the value of R_i. If both R_i and R_{i-1} are zero then the step Γ_i is *independent* of all other steps (if i is 1, we need only consider R_i). Otherwise, the step is *dependent* on previous steps. Each set of steps R_i that are dependent are bordered by independent steps. These set of steps must be accomplished by multiple agents; these steps are appropriate for swarms. We call these steps *subtasks*. Note that these subtasks are defined by the path Γ, and can change according to the path utilized to move through the phase space. We can also define *swarm teams* as the minimal swarm required to accomplish a specific subtask of the overall phase space task along the given path. Let γ_k be the kth subtask in the path Γ. Then $\gamma_k \in \Gamma$. We can define the size of the swarm team required for the subtask γ_k as $r_k = \max_i \left| \sum_{j=1}^{i} \Delta A_j \right|$ where i runs over all the steps of the subtask and ΔA_j is inherited from the task Γ. Clearly,

$$S\left(\Gamma\right) = \max_{i=1}^{|\{\gamma_k\}|} r_k. \tag{3}$$

Each of the swarm teams is a smaller subset of the overall swarm that must be deployed in order to accomplish the task. Each agent must employ a technology that enables it to accomplish the transition indicated in the subtask step; these technologies need not be identical. If the technologies are different the swarm team can utilize different *castes* of agents having differing technologies or designs.

The same analysis may be performed within a subtask, identifying progressively smaller swarm teams and teams within teams. The examination of these cases is beyond the scope of this paper and will be addressed in a later paper.

In the case that multiple subtasks are independent, the swarm may be made up of multiple swarm teams that act independently. Such an organization might increase the speed with which a swarm accomplishes its task without interference.

4.3 Choosing Paths, Minimizing Swarm Size

The choice of the path through phase space indicates a swarm's strategy. The technologies deployed at each of the steps define what technologies we must use. The technology sets that must be employed by the individual agents during subtasks define the minimal capabilities, order of deployment, and number of agent castes. In most phase spaces multiple paths connect the start point to the end point. We define in this subsection a process adapted from [4] which enables us to choose a path according to a number of criterion of potential importance in the design of a swarm.

The algorithm proceeds as follows.

1. Create a linked list containing a single point which contains the following information:
   (a) Path in phase space up to this point (initially it will be the starting point)
   (b) Total energy expended (initialized to zero)
   (c) Total time expended (initialized to zero)

   (d) Total number of agents required (R_i) (initialized to zero)
   (e) Ordering measurable (energy, time, agent number, or some real function
       of the three)
2. Set the linked list pointer to the first element. Set a flag to -1 indicating that
   a feasible path has not yet been found.
3. Using the record pointed to by the list pointer to calculate new points in
   phase space by applying, independently, each transition enabled by the set
   of technologies within our technology set to the record. In each case, create
   another record for placement on the linked list recording the resulting point
   in phase space and the updated totals of energy, time, and agent number
   as well as the updated ordering measurable (energy, time, agent number, or
   some real function of the three). Store the new point in the stored path for
   this record. Place each record in the linked list if the new point is feasible,
   making sure to order them from least to greatest according to the value of
   the ordering measurable.
4. Move the linked list pointer forward.
5. If the linked list pointer is NULL, end.
6. If the linked list pointer is pointing to a record whose position in phase space
   is in the set of final points, go on to step 7. Otherwise, go to step 3.
7. If the flag is -1, go to step 8. Otherwise, go to step 9.
8. Assign the minimum ordering measurable to that of the current record. Set
   the flag to 1.
9. If the ordering measurable exceeds that of the minimum ordering measurable,
   end.
10. Output the path, energy, time, agent number and ordering measurable.
11. Go to step 3.

It can be demonstrated that one of two outcomes will happen. In one case, there
will be no output path. In this case, there is no feasible path from the initial
state to the final state using the technology currently available. In the second
case, one or more paths will be output. These will be degenerate in terms of the
minimal ordering measurable. Once these paths are determined (and there may
be only one), swarm teams can be designed according to the process indicated
in Section 4.2. These will indicate how many teams are required, what size they
should be, what their capabilities and castes will be, and how they are deployed.

5   Stick Pulling Problem as an Example Problem

In [5] Martinoli et. al. introduced the *stick-pulling problem*. In this problem, a
group of agents must remove sticks from holes. The sticks initially sit in holes
in the floor, partially protruding upward. The agents must grasp the sticks, pull
them up out of the hole, and return them to a collection location. The situation
is depicted in Fig. 1.
   Complications occur when the agents are assumed to be incapable of pulling a
stick out individually. In this case, they require assistance of other agents, adding
the need for communication and/or cooperation if the task is to be accomplished.

Fig. 1. A diagram of the stick-pulling problem. (Reprinted with permission of the author. [5])

5.1   Step 1

We first define the local measurables – things that can be measured by the agents within the system. In our system, the states correspond to those of the sticks. These sticks have four potential states: (0) in the hole, (1) pulled up, (2) pulled out, and (3) stored.

We can define the local measurable space by an $N_s - dimensional$ space $(s_1, s_2, ..., s_{N_s})$ where $s_i \in \{0, 1, 2, 3\}$. This is illustrated by Figure 2.

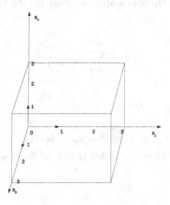

Fig. 2. A 3-dimensional representation of $S - space$

As a result of this, the feasible S-space is defined as $\{(s_1, s_2, ..., s_{N_s}) | s_i \in \{0, 1, 2, 3\}\}$.

5.2   Step 2

Now, we define the global functions – functions of the local measurables that will define the state of the system. We choose four natural measurables: P1 – sticks fully in their hole, P2 – sticks partially lifted, P3 – sticks fully pulled out, P4 – stored sticks. The four functions define the phase space of the system which defines its state. The phase space is a four-dimensional cartesian product space of integers between 0 and N_s, the number of sticks in the system. It is illustrated in Figure 3.

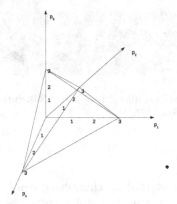

Fig. 3. The four-dimensional phase space describing the stick-pulling problem

Unlike the S space, the P-space is four-dimensional no matter the number of sticks in the system. Mathematically, we can relate the properties in S and P space by

$$P_j = \sum_{i=1}^{N_s} \delta_{S_i, j-1} \tag{4}$$

5.3   Step 3

It is clear that the initial point in S-space is $(0, 0, \ldots, 0)$ and in P-space is $(N_s, 0, 0, 0)$. It is also clear that the final point in S-space position is $(3, 3, \ldots, 3)$ and a P-space is $(0, 0, 0, N_s)$.

5.4   Step 4

Now, we examine the transitions needed. The transitions between the different states of P-space are limited only by the constraint that

$$\sum_{i=1}^{4} P_i = N_s \tag{5}$$

which communicates the conservation of the sticks. Therefore

$$\Delta P_1 = -\Delta P_2 - \Delta P_3 - \Delta P_4. \tag{6}$$

Some sample transitions one might create (and their associated agent costs) are

- $\Delta \vec{P} = (-1, 1, 0, 0)$
  - This transition indicates that the stick has been located and pulled halfway out of the ground. Required technological and strategic capabilities include the ability to locate the stick and the ability to extract it halfway and hold it there. $\Delta A = 1$
- $\Delta \vec{P} = (0, -1, 1, 0)$
  - This transition indicates that the stick being held halfway out of the ground has been located and that it has been fully extracted from the ground. Required technological and strategic capabilities include the ability to locate a held stick and fully pull it out of the ground. $\Delta A = 0$
- $\Delta \vec{P} = (0, 0, -1, 1)$
  - This transition indicates that the stick, having been removed from of the ground, is moved back to some storage location. Required technological and strategic capabilities include the ability to find and to directly transport the stick from its location to a storage location. $\Delta A = -1$
- $\Delta \vec{P} = (-1, 0, 1, 0)$
  - This transition indicates that the stick is pulled directly out of the ground and held out of the ground in one step. Required technological and strategic capabilities include the ability to find the stick and pull it completely out of the hole in one step. $\Delta A = 0$
- $\Delta \vec{P} = (-1, 0, 0, 1)$
  - This transition indicates that the stick is moved directly from its location in the hole to the storage facility without being moved halfway out of the hole and held or being lifted from halfway out of the hold and held. Required technological and strategic capabilities include the ability to move the stick directly from the hole to storage. It is not likely that this technological capability is in our repository of technologies. $\Delta A = 0$
- $\Delta \vec{P} = (-n, 0, n, 0)$
  - This transition indicates that multiple sticks are lifted directly out of their holes simultaneously in one move. Required technological and strategic capabilities include the ability to locate and move sticks from their holes simultaneously. This technological capability is significantly different from the strategies given in [5]. $\Delta A = 0$

These transitions correspond to a movement on the three-dimensional hyperplane defined by equation (5).

Notably, also, in these strategies is that location of the target, either a stick or a half-pulled stick, is part of the technological and strategic capability requirement of many of the steps. As a result, strategies for minimizing the cost of accomplishing this part of the task will ultimately reduce the cost of completing the overall task. This is an important part of the task optimization.

The costs of the steps of the task can be determined once the specific technology and strategy are identified. This too may require further optimization after the technologies are settled.

5.5 Step 5

After identifying the technological and strategic capabilities, a feasible path through phase space using them can be generated. The methodology developed in Section 4.3 will generate a minimal feasible path.

In the stick-pulling problem, moving from one feasible state to another requires that the constraint given in equation (6) is respected. Movements are thereby restricted to the three-dimensional hyperplane defined in (5).

We are asserting is that any point $\overrightarrow{P}$ on the hyperplane has the property that $\iota\left(\overrightarrow{P}\right) \neq$. We can prove this in the following way. Let $P_i = \sum_{j=1}^{n} \delta_{S_j i}$. If a point $\overrightarrow{P}$ is on the hyperplane, this means that $\sum_i P_i = N_s$. This also means that the phase space condition is satisfied if $\sum_i \delta_{s_i j-1} = P_j$. Since this is trivially true of the first point on the path $\overrightarrow{P^0}$ and each transition also maintains the same relation, each point on the hyperplane has the desired property. Therefore our path through phase space has to only require this property.

We applied the algorithm of Section 4.3 to stick pulling tasks involving one, two, and three sticks. The following paths through phase space resulted:

One stick

$(1,0,0,0)\ (0,1,0,0)\ (0,0,1,0)\ (0,0,0,1)\ \triangle A = 2$

Two stick

$(2,0,0,0)\ (1,1,0,0)\ (1,0,1,0)\ (0,1,1,0)\ (0,0,2,0)\ (0,0,1,1)\ (0,0,0,2)$
$\triangle A = 3$
$(2,0,0,0)\ (1,1,0,0)\ (1,0,1,0)\ (0,1,1,0)\ (0,1,0,1)\ (0,0,1,1)\ (0,0,0,2)$
$\triangle A = 2$
$(2,0,0,0)\ (1,1,0,0)\ (1,0,1,0)\ (1,0,0,1)\ (0,1,0,1)\ (0,0,1,1)\ (0,0,0,2)$
$\triangle A = 2$
$(2,0,0,0)\ (1,1,0,0)\ (0,2,0,0)\ (0,1,1,0)\ (0,0,2,0)\ (0,0,1,1)\ (0,0,0,2)$
$\triangle A = 3$
$(2,0,0,0)\ (1,1,0,0)\ (0,2,0,0)\ (0,1,1,0)\ (0,1,0,1)\ (0,0,1,1)\ (0,0,0,2)$
$\triangle A = 3$

Three stick

$(3,0,0,0) \to (2,1,0,0), (1,e,t) \to (2,0,1,0), (0,e,t) \to (2,0,0,1), (-1,e,t) \to$
$(1,1,0,1), (1,e,t,) \to (1,0,1,1), (0,e,t) \to (1,0,0,2), (-1,e,t) \to (0,1,0,2),$
$(1,e,t) \to (0,0,1,2), (0,e,t) \to (0,0,0,3), (-1,e,t)$
$\triangle A = 2$
$(3,0,0,0) \to (2,1,0,0), (1,e,t) \to (2,0,1,0), (1,e,t) \to (2,0,0,1), (-1,e,t) \to$
$(1,1,0,1), (1,e,t,) \to (1,0,1,1), (0,e,t) \to (0,1,1,1), (1,e,t) \to (0,1,0,2),$
$(-1,e,t) \to (0,0,1,2), (0,e,t) \to (0,0,0,3), (-1,e,t)$
$\triangle A = 3$
$(3,0,0,0) \to (2,1,0,0), (1,e,t) \to (2,0,1,0), (1,e,t) \to (2,0,0,1), (-1,e,t) \to$
$(1,1,0,1), (1,e,t,) \to (1,0,1,1), (0,e,t) \to (0,1,1,1), (1,e,t) \to (0,0,2,1),$
$(0,e,t) \to (0,0,1,2), (-1,e,t) \to (0,0,0,3), (-1,e,t)$

$\triangle A = 3$

$(3,0,0,0) \rightarrow (2,1,0,0), (1,e,t) \rightarrow (2,0,1,0), (1,e,t) \rightarrow (2,0,0,1), (-1,e,t) \rightarrow (1,1,0,1), (1,e,t) \rightarrow (0,2,0,1), (1,e,t) \rightarrow (0,1,1,1), (1,e,t) \rightarrow (0,1,0,2), (-1,e,t) \rightarrow (0,0,1,2), (0,e,t) \rightarrow (0,0,0,3), (-1,e,t)$

$\triangle A = 3$

$(3,0,0,0) \rightarrow (2,1,0,0), (1,e,t) \rightarrow (2,0,1,0), (1,e,t) \rightarrow (2,0,0,1), (-1,e,t) \rightarrow (1,1,0,1), (1,e,t) \rightarrow (0,2,0,1), (1,e,t) \rightarrow (0,1,1,1), (1,e,t) \rightarrow (0,0,2,1), (0,e,t) \rightarrow (0,0,1,2), (-1,e,t) \rightarrow (0,0,0,3), (-1,e,t)$

$\triangle A = 3$

$(3,0,0,0) \rightarrow (2,1,0,0), (1,e,t) \rightarrow (2,0,1,0), (0,e,t) \rightarrow (1,1,1,0), (1,e,t) \rightarrow (1,0,2,0), (0,e,t) \rightarrow (1,0,1,1), (-1,e,t) \rightarrow (1,0,0,2), (-1,e,t) \rightarrow (0,1,0,2), (1,e,t) \rightarrow (0,0,1,2), (0,e,t) \rightarrow (0,0,0,3), (-1,e,t)$

$\triangle A = 3$

$(3,0,0,0) \rightarrow (2,1,0,0), (1,e,t) \rightarrow (2,0,1,0), (0,e,t) \rightarrow (1,1,1,0), (1,e,t) \rightarrow (1,1,0,1), (-1,e,t) \rightarrow (1,0,1,1), (0,e,t,) \rightarrow (1,0,0,2), (-1,e,t) \rightarrow (0,1,0,2), (1,e,t) \rightarrow (0,0,1,2), (0,e,t) \rightarrow (0,0,0,3), (-1,e,t)$

$\triangle A = 3$

$(3,0,0,0) \rightarrow (2,1,0,0), (1,e,t) \rightarrow (1,2,0,0), (1,e,t) \rightarrow (1,1,1,0), (0,e,t) \rightarrow (1,0,2,0), (0,e,t) \rightarrow (1,0,1,1), (-1,e,t) \rightarrow (1,0,0,2), (-1,e,t) \rightarrow (0,1,0,2), (1,e,t) \rightarrow (0,0,1,2), (0,e,t) \rightarrow (0,0,0,3), (-1,e,t)$

$\triangle A = 3$

$(3,0,0,0) \rightarrow (2,1,0,0), (1,e,t) \rightarrow (1,2,0,0), (1,e,t) \rightarrow (1,1,1,0), (0,e,t) \rightarrow (1,1,0,1), (-1,e,t) \rightarrow (1,0,1,1), (0,e,t,) \rightarrow (1,0,0,2), (-1,e,t) \rightarrow (0,1,0,2), (1,e,t) \rightarrow (0,0,1,2), (0,e,t) \rightarrow (0,0,0,3), (-1,e,t)$

$\triangle A = 3$

$(3,0,0,0) \rightarrow (2,1,0,0), (1,e,t) \rightarrow (1,2,0,0), (1,e,t) \rightarrow (0,3,0,0), (1,e,t) \rightarrow (0,2,1,0), (o,e,t) \rightarrow (0,2,0,1), (-1,e,t) \rightarrow (0,1,1,1), (0,e,t) \rightarrow (0,0,2,1), (0,e,t) \rightarrow (0,0,1,2), (-1,e,t) \rightarrow (0,0,0,3), (-1,e,t)$

$\triangle A = 4$

$(3,0,0,0) \rightarrow (2,1,0,0), (1,e,t) \rightarrow (1,2,0,0), (1,e,t) \rightarrow (0,3,0,0), (1,e,t) \rightarrow (0,2,1,0), (o,e,t) \rightarrow (0,2,0,1), (-1,e,t) \rightarrow (0,1,1,1), (0,e,t) \rightarrow (0,1,0,2), (-1,e,t) \rightarrow (0,0,1,2), (0,e,t) \rightarrow (0,0,0,3), (-1,e,t)$

$\triangle A = 4$

$(3,0,0,0) \rightarrow (2,1,0,0), (1,e,t) \rightarrow (1,2,0,0), (1,e,t) \rightarrow (0,3,0,0), (1,e,t) \rightarrow (0,2,1,0), (o,e,t) \rightarrow (0,1,2,0), (0,e,t) \rightarrow (0,1,1,1), (0,e,t) \rightarrow (0,1,0,2), (-1,e,t) \rightarrow (0,0,1,2), (0,e,t) \rightarrow (0,0,0,3), (-1,e,t)$

$\triangle A = 4$

$(3,0,0,0) \rightarrow (2,1,0,0), (1,e,t) \rightarrow (1,2,0,0), (1,e,t) \rightarrow (0,3,0,0), (1,e,t) \rightarrow (0,2,1,0), (o,e,t) \rightarrow (0,1,2,0), (0,e,t) \rightarrow (0,1,1,1), (0,e,t) \rightarrow (0,0,2,1), (0,e,t) \rightarrow (0,0,1,2), (-1,e,t) \rightarrow (0,0,0,3), (-1,e,t)$

$\triangle A = 4$

$(3,0,0,0) \rightarrow (2,1,0,0), (1,e,t) \rightarrow (1,2,0,0), (1,e,t) \rightarrow (0,3,0,0), (1,e,t) \rightarrow (0,2,1,0), (o,e,t) \rightarrow (0,1,2,0), (0,e,t) \rightarrow (0,0,3,0), (0,e,t) \rightarrow (0,0,2,1), (0,e,t) \rightarrow (0,0,1,2), (-1,e,t) \rightarrow (0,0,0,3), (-1,e,t)$

$\triangle A = 4$

$(3,0,0,0) \rightarrow (2,1,0,0), (1,e,t) \rightarrow (1,2,0,0), (1,e,t) \rightarrow (1,1,1,0), (0,e,t) \rightarrow (1,1,0,1), (-1,e,t) \rightarrow (1,0,1,1), (0,e,t) \rightarrow (0,1,1,1), (1,e,t) \rightarrow (0,1,0,2), (-1,e,t) \rightarrow (0,0,1,2), (0,e,t) \rightarrow (0,0,0,3), (-1,e,t)$

$\triangle A = 4$

$(3,0,0,0) \rightarrow (2,1,0,0), (1,e,t) \rightarrow (1,2,0,0), (1,e,t) \rightarrow (1,1,1,0), (0,e,t) \rightarrow (1,0,2,0), (-1,e,t) \rightarrow (0,1,2,0), (0,e,t) \rightarrow (0,1,1,1), (1,e,t) \rightarrow (0,1,0,2), (-1,e,t) \rightarrow (0,0,1,2), (0,e,t) \rightarrow (0,0,0,3), (-1,e,t)$

$\triangle A = 4$

$(3,0,0,0) \rightarrow (2,1,0,0), (1,e,t) \rightarrow (2,0,1,0), (1,e,t) \rightarrow (1,1,1,0), (0,e,t) \rightarrow (1,0,2,0), (-1,e,t) \rightarrow (0,1,2,0), (0,e,t) \rightarrow (0,1,1,1), (1,e,t) \rightarrow (0,1,0,2), (-1,e,t) \rightarrow (0,0,1,2), (0,e,t) \rightarrow (0,0,0,3), (-1,e,t)$

$\triangle A = 4$

In each case, the minimum number of agents is two. It can be shown that in the n−stick case, the minimal number of agents is still two; the "swarm" must have at least two agents. In these cases each subtask also requires at least two agents. We can infer that swarm teams of size two may be utilized to accomplish each subtask individually, resulting in a superior overall performance if each team can act independently.

5.6   Step 6

In this step, we determine whether a swarm is required to accomplish the task. The paths identified in the Step 5 have a minimal swarm team size of 2. We conclude that a swarm containing at least two agents is required to accomplish this task. Moreover, a swarm made up of multiple swarm teams of size two can accomplish this task in parallel without interfering with one another.

As the task can only be carried out by pairs of agents, one of the agents must be able to locate the stick and execute the first transition. The second agent must be able to locate the half-raised stick/agent assembly and assist in removing the stick. It must also be able to return the stick to home. As a result, it must be able to locate home and carry the stick as well. It is not necessary for both agents to have both ability sets, enabling the task to be accomplished by two castes of agents.

It has been noted that, when a larger swarm is evolved with two castes of agents so a to determine an optimal proportion of one type to another, the swarm generates two different castes of equal numbers[5]. The considerations of this Section would seem to indicate a possible reason for this outcome.

5.7   Step 7

A variety of behaviors may be created in order to achieve the goals set out in Section 5.6. We illustrate these with the following behaviors:

- Caste 1
  - If a Caste 2 agent is in sight carrying a stick, follow it.
  - If a Caste 2 agent carrying a stick isn't in sight and a stick is in sight, lift the stick.
  - Otherwise, wander randomly.
- Caste 2
  - If at home and carrying a stick, drop stick in storage.
  - If not at home and carrying a stick, locate home and move toward it.
  - If not at home, not carrying a stick, and a Caste 1 agent is in view not holding a stick, follow it.
  - If not at home, not carrying a stick, and a Caste 1 agent is in view and holding a stick, move to the agent and lift the stick.
  - Otherwise wander randomly.

The basic capabilities that the agents must be able to are: identify other agents in a variety of states; identify home; determine the direction to home; lift the stick from the ground; lift the stick when held by another agent; and move unboundedly while holding or not holding the stick. These basic capabilities can be developed individually and when coupled with the behaviors that satisfy the swarm condition and a two-by-two deployment of agents of differing castes, the stick-pulling swarm will have been designed. It is easy to verify that these capabilities exist in stick-pulling studies in the literature.

6 Discussion

One of the early investigations of swarm engineering centers around a task called puck clustering[1,3]. This task involves pushing objects known as pucks[1] around an arena and eventually getting them to cluster in one big pile. The task is generally approached using a swarm of identical robots, but it can be shown using the analysis given above that the task requires only one agent. The situation is identical for the three-dimensional construction swarm described in [8]. While the task is accomplished by many agents in practice, it only requires one agent. As a result, the task itself is not a swarm task; it can be accomplished without a swarm.

Other tasks, such as the self-organization of the kilobot swarm described in [6,7], require the entire swarm of agents for completion; they are indeed swarm-based tasks. Moreover, they may be easily be modeled using the seven-step system. In both cases, every agent put through the system is "consumed", indicating the reason these are swarm-based tasks.

Using this analysis, subtasks and swarm teams become clearly elucidated and can be utilized to improve the swarm's effectiveness. The effect of adding new technology to the swarm design "soup" can be examined in a way that provides insight into how technology affects swarm-based strategies. The tool also enables

---

[1] The original experimental studies used hockey pucks; the name stuck.

the examination of constraints including money, time, and energy; some of these constraints can be overcome only by using multiple agents.

By analyzing the cost distribution and energy consumption throughout the pathway, one can determine the costliest parts of the task. This is helpful in determining focus areas of technological development. Improving technology deployed in a specific area of a task can ultimately reduce costs, complexity, and energy usage.

7   Conclusions

We have described a task and technology-based swarm design methodology. This methodology requires that the task be mapped to a phase space where the global measurables are captured as functions of local measurables based on agent sensory capability. The phase space is then traversed by paths reaching between the start and desired end point of the swarm. The paths are constructed via an algorithm that can be used to deliver paths with minimal agent counts, minimal energy usage, time consumption, cost, or any other constraint applied to the system. Analysis of the agent pathways enables determination of minimal agent counts required for the task, characterization of the technologies used, identification of the different agent castes, development of a deployment schedule, design of swarm teams which carry out subtasks, and tally of time, cost, energy usage, and other measurables associated with the task completion. The method was applied to the stick pulling problem. The application supported a conclusion of [5] that the optimal ratio of the two castes in the stick-pulling problem was 1:1.

Future work will investigate the algorithm's application to varied problems from the literature and others of interest to the broader swarm engineering community. A number of extensions will be applied enabling simultenaiety and continuity of action under perturbations to be modeled and designed.

Acknowledgments. This paper was made possible through private funding and help from: D. Jin, T. Li, C. Wong, A. An, A. Li, A. Pyon, R. Choi, C. Xu, and J. Park.

References

1. Beckers, R., Holland, O.E., Deneubourg, J.L.: From local actions to global tasks: stigmergy and collective robotics. In: Artificial Life IV, vol. 181, p. 189 (1994)
2. Kazadi, S.: Swarm Engineering. Caltech PhD Thesis (2000)
3. Kazadi, S., Abdul-Khaliq, A., Goodman, R.: On the convergence of puck clustering systems. Robotics and Autonomous Systems **38**(2), 93–117 (2002)
4. Lee, D., Kazadi, L., Goodman, R.: Swarm engineering for TSP. In: Proceedings of the ANTS 2000 From Ant Colonies to Artificial Ants: 2nd International Workshop on Ant Algorithms, Brussels, Belgium (2000)
5. Martinoli, A., Ijspeert, A.J., Mondada, F.: Understanding collective aggregation mechanisms: From probabilistic modelling to experiments with real robots. Robotics and Autonomous Systems **29**(1), 51–63 (1999)

6. Rubenstein, M., Cornejo, A., Nagpal, R.: Programmable self-assembly in a thousand-robot swarm. Science **345**(6198), 795–799 (2014)
7. Rubenstein, M., Cabrera, A., Werfel, J., Habibi, G., McLurkin, J., Nagpal, R.: Collective transport of complex objects by simple robots: theory and experiments. In: Proceedings of the 2013 International Conference on Autonomous Agents and Multi-Agent Systems. International Foundation for Autonomous Agents and Multiagent Systems, pp. 47–54 (2013)
8. Werfel, J., Petersen, K., Nagpal, R.: Designing collective behavior in a termite-inspired robot construction team. Science **343**(6172), 754–758 (2014)

Memetic Electromagnetism Algorithm for Finite Approximation with Rational Bézier Curves

Andrés Iglesias[1,2(✉)] and Akemi Gálvez[1]

[1] Department of Applied Mathematics and Computational Sciences,
University of Cantabria, Avda. de los Castros, s/n, E-39005 Santander, Spain
iglesias@unican.es
http://personales.unican.es/iglesias
[2] Department of Information Science, Faculty of Sciences,
Toho University, 2-2-1 Miyama, Funabashi 274-8510, Japan

Abstract. The problem of obtaining a discrete curve approximation to data points appears recurrently in several real-world fields, such as CAD/CAM (construction of car bodies, ship hulls, airplane fuselage), computer graphics and animation, medicine, and many others. Although polynomial blending functions are usually applied to solve this problem, some shapes cannot yet be adequately approximated by using this scheme. In this paper we address this issue by applying rational blending functions, particularly the rational Bernstein polynomials. Our methodology is based on a memetic approach combining a powerful metaheuristic method for global optimization (called the electromagnetism algorithm) with a local search method. The performance of our scheme is illustrated through its application to four examples of 2D and 3D synthetic shapes with very satisfactory results in all cases.

1 Introduction

This paper deals with the problem of obtaining a curve providing an accurate approximation to a finite set of data points. This problem, mathematically formulated as an optimization problem, arises in many theoretical and applied domains. Classical fields for the former case are numerical analysis and statistics, with computer aided-design and manufacturing (CAD/CAM) and medicine as good examples for the latter. In many cases (particularly, for real-world applications), data points are usually acquired through laser scanning and other digitizing devices and are, therefore, subjected to some measurement noise, irregular sampling, and other artifacts [2,36,37]. Consequently, a good fitting of data should be generally based on approximation schemes. In this case, the approximating curve is not required to pass through all input data points, but just near to them, according to some prescribed distance criteria.

A number of approximating families of functions have been applied to this problem. Among them, the free-form parametric curves such as Bézier, B-spline and NURBS, are widely applied in many industrial settings due to their great flexibility and the fact that they can represent smooth shapes with only a few

© Springer International Publishing Switzerland 2015
Y. Tan et al. (Eds.): ICSI-CCI 2015, Part I, LNCS 9140, pp. 30–40, 2015.
DOI: 10.1007/978-3-319-20466-6_3

parameters [2,30,31,34,35]. Some previous papers addressed this problem by using Bézier curves [15,18,32], which are given by a linear combination of polynomial basis functions (the Bernstein polynomials). Although they obtained good results for a number of shapes, this polynomial approach is still limited, as it cannot adequately describe some particular shapes (such as the conics). As a consequence, there is still a need for more powerful blending functions.

An interesting extension in this regard is given by the rational basis functions, which are mathematically described as the quotient of two polynomials. A remarkable advantage of this rational scheme is that the conics can be canonically described as rational functions. In this paper, we take advantage of this valuable feature to solve the finite curve approximation by using rational Bézier curves. Unfortunately, this rational approach becomes more difficult than the polynomial one, since new parameters are now introduced into the problem. Consequently, we are confronted with the challenge of obtaining optimal values for many (qualitatively different) parameters, namely, data parameters, poles, and weights. This leads to a difficult over-determined multivariate nonlinear optimization problem.

In this paper, we address this optimization problem by applying a memetic approach. It is based on the combination of a powerful physics-based algorithm, called electromagnetism algorithm and aimed at solving global optimization problems, and a local search procedure. This memetic approach can be effectively applied to obtain a very accurate approximation of a finite set of data points by using rational blending functions.

The structure of this paper is as follows: in Section 2 previous work in the field is briefly reported. Then, the fundamentals and main steps of the memetic electromagnetism algorithm are briefly explained in Section 3. Our proposed approach for curve approximation with rational Bézier curves is described in Section 4. To check the performance of our approach, it has been applied to four illustrative examples of 2D and 3D curves, as described in Section 5. Our experimental results show that the presented method performs very well, being able to replicate the underlying shape of data very accurately. The paper closes in Section 6 with the main conclusions of this contribution and our plans for future work in the field.

2   Previous Work

The problem of finite approximation with free-form parametric curves has been the subject of research for many years. First approaches in the field were mostly based on numerical procedures [5,6,38]. However, it has been shown that traditional mathematical optimization techniques fail to solve the problem in its generality. Consequently, there has been a great interest to explore other possible approaches to this problem. Some recent approaches in this line use error bounds [34], curvature-based squared distance minimization [40], or dominant points [35].

On the other hand, interesting research carried out during the last two decades has shown that the application of artificial intelligence and soft computing techniques can achieve remarkable results for this problem [1,23,24]. Most of these methods rely on some kind of neural networks, such as standard neural networks [23], RBS networks [28], and Kohonen's SOM (Self-Organizing Maps) nets [24]. In some cases, this neural approach is combined with partial differential equations [1] or other approaches [29]. The generalization of these methods to functional networks is also analyzed in [7,25–27]. The application of support vector machines to solve the least-squares B-spline curve fitting problem is reported in [30].

Other approaches are based on the application of nature-inspired metaheuristic techniques, which have been intensively applied to solve difficult optimization problems that cannot be tackled through traditional optimization algorithms. A previous paper in [21] describes the application of genetic algorithms and functional networks yielding pretty good results. Genetic algorithms have also been applied to this problem in both the discrete version [39] and the continuous version [22,41]. Other metaheuristic approaches applied to this problem include the use of the popular particle swarm optimization technique [8–10], artificial immune systems [18–20], firefly algorithm [12–14], cuckoo search [16], simulated annealing [32], estimation of distribution algorithms [42], memetic algorithms [17], and hybrid techniques [11,39].

3 Our Memetic Approach

During the last two decades, there has been an increasing interest upon the application of soft computing approaches (particularly, metaheuristic techniques) to solve hard optimization problems. Among them, the memetic algorithms - based on a metaheuristic strategy for global optimization coupled with a local search procedure - have shown a great potential for solving difficult nonlinear optimization problems such as that in this paper. Owing to these reasons, in this work we consider a memetic approach combining the electromagnetism algorithm and a local search method, as described in next paragraphs.

3.1 The Electromagnetism Algorithm

The electromagnetism algorithm (EMA) is a metaheuristic introduced by Birbil and Fang in [3] for optimization problems. This method utilizes an attraction-repulsion mechanism to move sample points towards optimality. Each point (called particle) is treated as a potential solution and an electric charge is assigned to each particle. Better solutions have stronger charges and each particle has an impact on others through charge. The exact value of the impact is given by a modification of original Coulomb's Law. In EMA, the power of the connection between two particles is proportional to the product of their charges and reciprocal to the distance between them. In other words, the particles with a higher charge will force the movement of other particles in their direction more strongly. Beside that, the best particle in this electromagnetic mechanism will stay unchanged. The charge of each particle relates to the objective function value, which is the subject of optimization. The reader is also referred to [4] for a comprehensive study about the convergence of the EMA approach.

The electromagnetism algorithm was originally proposed to study a special class of optimization problems with bounded variables in the form:

$$\min \psi(\Theta) \qquad \text{such that } \Theta \in [\mathbf{L}, \mathbf{U}] \tag{1}$$

where $[\mathbf{L}, \mathbf{U}] := \{\Theta \in \mathbb{R}^\nu / l_k \leq \Theta_k \leq u_k, k = 1, \ldots, \nu\}$, ν is the dimension of the problem, $\mathbf{L} = \{l_k\}_k$ and $\mathbf{U} = \{u_k\}_k$ represent respectively the lower bound and upper bound in $\mathbb{R}^\nu$, and $\psi(\Theta)$ is the function to be optimized. The algorithm consists of four main steps, which are summarized in next paragraphs. The corresponding pseudocode is depicted in Table 1. Note that in this paper vectors are denoted in bold.

Step 1: Initialization. In this step, μ sample points are selected at random from the feasible region, which is an ν-dimensional hypercube. To this purpose, each coordinate of the sampled point is assumed to be uniformly distributed between the corresponding lower and upper bound. Then, the objective function value of each sampled point is computed, and the point that has the best global value is stored in Θ^{best}.

Step 2: Local Search. In this step, a local search is carried out to gather the local information for each point Θ^i and exploit the local minima. To this aim, a *LocalSearch* procedure similar to that in [3] is applied. The procedure depends on a multiplier δ which is used to compute the maximum feasible step length for the local search. The search is performed for each coordinate and for a given number of iterations. In case a better point is obtained (according to the fitness function), the current point is replaced by this new (better) alternative. Note that this procedure does not require any gradient information. Note also that any other local search procedure might be alternatively used, opening the door for other hybridized schemes.

Step 3: Calculation of Total Force. In this step, the vector of the total force exerted on each particle from all other particles is computed. Firstly, a charged-like value ξ^i is assigned to each particle. The charge of a particle i determines its power of attraction or repulsion, and is evaluated as:

$$\xi^i = exp \left(-\nu \frac{\psi(\Theta^i) - \psi(\Theta^{best})}{\sum\limits_{k=1}^{\mu} [\psi(\Theta^k) - \psi(\Theta^{best})]} \right) \tag{2}$$

Then, the attraction/repulsion force between two particles is computed using a mechanism inspired in the electromagnetism theory for the charged particles. According to [3], the computation of this force is given by:

Table 1. General pseudocode of the electromagnetism algorithm

---

INPUT:
    μ: number of sampled points
    ν: dimension of the problem
    max_iter: maximum number of iterations for global loop
    max_lsiter: maximum number of iterations for local search
    δ: multiplier for local search

Step 1: Initialization
for i=1 **to** μ **do**
    for k=1 **to** ν **do**
        $\Theta_k^i \leftarrow l_k + \sigma(u_k - l_k)$ // $\sigma \sim U(0, 1)$
    end for
end for
$\Theta^{best} \leftarrow BestFitting(\{\Theta^i\}_{i=1,...,\mu})$ // initial best
iter $\leftarrow$ 1
while iter< *max_iter* **do**   // global loop
    Step 2: Local Search
    liter $\leftarrow$ 1
    for i=1 **to** μ **do**
        for k=1 **to** ν **do**
            while liter< *max_lsiter* **do**
                $\Theta_k^i \leftarrow LocalSearch(\delta)$ // local search improvement
                liter $\leftarrow$ liter +1
            end while
        end for
    end for
    Step 3: Total Force Computation
    for i=1 **to** μ **do**
        $\xi^i \leftarrow ChargeEvaluation()$ // given by Eq. (2)
    end for
    for i=1 **to** μ **do**
        $\Xi^i \leftarrow ForceEvaluation()$ // given by Eq. (3)
    end for
    Step 4: Movement According Total Force
    for i=1 **to** μ **do**
        if i$\neq$best **then**
            $\Theta^i \leftarrow Movement()$ // given by Eq. (4)
        end if
    end for
    iter $\leftarrow$ iter +1
end while

OUTPUT:
    Θ^{best}: best global solution

$$\Xi^i = \sum_{j=1, j \neq i}^{\mu} \frac{\xi^i \, \xi^j}{||\Theta^j - \Theta^i||^2} \begin{cases} (\Theta^j - \Theta^i) \text{ if } \psi(\Theta^j) < \psi(\Theta^i) \\ (\Theta^i - \Theta^j) \text{ if } \psi(\Theta^j) \geq \psi(\Theta^i) \end{cases} \quad (3)$$

Note, however, that the force computed in this way does not follow exactly Coulomb's law, where the force is inversely proportional to the square of the distance. Note also that, unlike electrical charges, there is no sign on the charge of individual particles in Eq. (2). Instead, the direction of a particular force between two particles is determined by comparing the objective function values at such particles. Then, the particle with a better fitness value attracts the other one, while the particle with a worse fitness value repels the other, as indicated by Eq. (3).

Step 4: Movement According to the Total Force. The force vector computed in previous step determines the direction of movement for the corresponding particle according to Eq. (4):

$$\Theta^{i+1} = \Theta^i + \lambda \frac{\Xi^i}{||\Xi^i||} \circ \Psi \quad (4)$$

where Ψ is the vector of the feasible movement toward the upper/lower bound for the corresponding dimension, λ is a random variable following the uniform distribution, and $\circ$ denotes the Hadamard product.

3.2   Local Optimization Method

The EMA is improved by its hybridization with a local search procedure. We apply the Luus-Jaakola local search method, a heuristic for optimization of real-valued functions [33]. This method starts with an initialization step, where random uniform values are chosen within the search space. Then, a random uniform value in-between boundary values is sampled for each component. This value is added to the current position of the potential solution to generate a new candidate solution, which replaces the current one only if the value of the fitness is improved. Otherwise, the sampling space is multiplicatively decreased by a self-adaptive size of a factor whose strength depends on the difference between consecutive parameters, with the effect of speeding up the convergence to the steady state. This process is repeated iteratively. With each iteration, the neighborhood of the point decreases, so the procedure eventually collapses to a point.

4   The Proposed Method

We assume that the reader is familiar with the main concepts of free-form parametric curves [6]. A *free-form rational Bézier curve* $\Phi(\tau)$ *of degree* η is defined as:

$$\Phi(\tau) = \frac{\displaystyle\sum_{j=0}^{\eta} \omega_j \Lambda_j \phi_j^{\eta}(\tau)}{\displaystyle\sum_{j=0}^{\eta} \omega_j \phi_j^{\eta}(\tau)} \tag{5}$$

where Λ_j are vector coefficients called the *poles*, ω_j are their scalar weights, $\phi_j^{\eta}(\tau)$ are the *Bernstein polynomials of index j and degree η*, given by:

$$\phi_j^{\eta}(\tau) = \binom{\eta}{j} \tau^j (1 - \tau)^{\eta-j}$$

and τ is the *curve parameter*, defined on a finite interval $[0,1]$. By convention, $0! = 1$.

Suppose now that we are given a set of data points $\{\Delta_i\}_{i=1,\ldots,\kappa}$ in $\mathbb{R}^{\nu}$ (usually $\nu = 2$ or $\nu = 3$). Our goal is to obtain the rational Bézier curve $\Phi(\tau)$ performing finite approximation of the data points $\{\Delta_i\}_i$. To do so, we have to compute all parameters (i.e. poles Λ_j, weights ω_j, and parameters τ_i associated with data points Δ_i for $i = 1,\ldots,\kappa$, $j = 0,\ldots,\eta$) of the approximating curve $\Phi(\tau)$ by minimizing the least-squares error, Υ, defined as the sum of squares of the residuals:

$$\Upsilon = \underset{\substack{\{\tau_i\}_i \\ \{\Lambda_j\}_j \\ \{\omega_j\}_j}}{\text{minimize}} \left[\sum_{i=1}^{\kappa} \left(\Delta_i - \frac{\displaystyle\sum_{j=0}^{\eta} \omega_j \Lambda_j \phi_j^{\eta}(\tau_i)}{\displaystyle\sum_{j=0}^{\eta} \omega_j \phi_j^{\eta}(\tau_i)} \right)^2 \right]. \tag{6}$$

Our strategy for solving this problem consists of applying the memetic electromagnetism method described in the previous section to determine suitable values for the unknowns of the least-squares minimization of functional Υ according to (6). All these parameters are initialized with random values within their respective domains. Application of our method yields new positions and charges of the particles representing the potential solutions. The process is performed iteratively until the convergence of the minimization of the error is achieved.

5   Experimental Results

The method described in previous section has been applied to several examples. Unfortunately, the lack of a standardized benchmark in the field forced us to choose the examples by ourselves. It also prevented us from making a comparative analysis with other metaheuristic methods in the literature. We think, however, that the examples reported here will be useful to determine the good applicability of our method to this problem. To keep the paper in manageable size, in this section we describe only four of them, corresponding to 2D and 3D curves and shown in Figure 1.

First example corresponds to a set of 150 data points, approximated with a rational Bézier curve. Our results are depicted in Fig. 1(top-left), where the original data points are displayed as red × symbols whereas the reconstructed curve

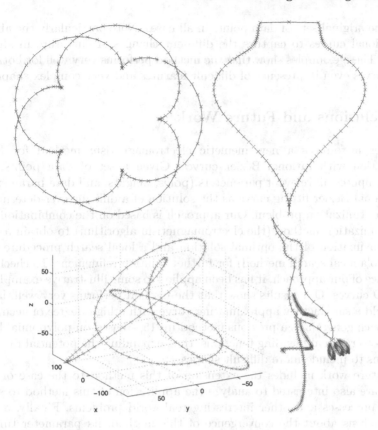

Fig. 1. Application of our memetic electromagnetism algorithm to four illustrative examples of finite approximation with rational Bézier curves: (top-left) epicycloid; (top-right) a spinning top; (bottom-left) a 3D curve; (bottom-right) a collection of curves representing a crane bird. In all cases, the original points are displayed as red × symbols and the approximating curve as a blue solid line.

appears displayed as a blue solid line. This example is particularly challenging because it contains a number of difficult features such as several self-intersections and turning points, where the curve is continuous but not differentiable. Note however the good visual matching between the original data points and the approximating curve. Despite of all these difficult features, the method performs very well, being able to replicate the original shape with high accuracy.

Second example corresponds to a set of 100 data points from the free-form shape of a spinning top, depicted in Fig. 1(top-right). Once again, we obtained a very good matching of data points. Third example corresponds to a 3D curve obtained from 200 data points (see Figure 1(bottom-left)), while the fourth example (Figure 1 (bottom-right)) corresponds to a collection of 7 different curves (with 50 data points each) from a free-form shape representing a crane bird. As the reader can see, our method obtained a very good approximating curve of the original sets of data points in all cases. Note particularly the ability of the rational curves to capture the different changes of curvature in all our examples. These examples show that the method performs very well for both 2D and 3D curves even in presence of difficult features and very complex shapes.

6   Conclusions and Future Work

This paper introduces a new memetic electromagnetism method for finite approximation with rational Bézier curves. Given a set of data points, the method computes all relevant parameters (poles, weights, and data parameters) of the rational Bézier fitting curve as the solution of a difficult over-determined nonlinear optimization problem. Our approach is based on the combination of a global optimization method (the electromagnetism algorithm) to obtain a very good approximation of the optimal solution and a local search procedure (the Luus-Jaakola local search method) for further solution refinement. To check the performance of our approach, it has been applied to some illustrative examples of 2D and 3D curves. Our results show that the method performs very well, being able to yield a satisfactory approximating curve with a high degree of accuracy. Our approach generalizes a previous method in [15] - based on polynomial basis functions - to rational blending functions, thus expanding the potential range of applications to include more difficult shapes.

Our future work includes the extension of this method to the case of surfaces. We are also interested to analyze the application of this method to some industrial processes and other interesting real-world problems. Finally, a theoretical analysis about the convergence of this method, its parameter tuning, and a comparative analysis with other alternative approaches on a standardized benchmark (when available) are also part of our future goals.

Acknowledgments. This research has been kindly supported by the Computer Science National Program of the Spanish Ministry of Economy and Competitiveness, Project Ref. #TIN2012-30768, Toho University, and the University of Cantabria.

References

1. Barhak, J., Fischer, A.: Parameterization and reconstruction from 3D scattered points based on neural network and PDE techniques. IEEE Trans. on Visualization and Computer Graphics **7**(1), 1–16 (2001)
2. Barnhill, R.E.: Geometric Processing for Design and Manufacturing. SIAM, Philadelphia (1992)
3. Birbil, S.I., Fang, S.C.: An electromagnetism-like mechanism for global optimization. Journal of Global Optimization **25**, 263–282 (2003)
4. Birbil, S.I., Fang, S.C., Sheu, R.L.: On the convergence of a population-based global optimization algorithm. Journal of Global Optimization **30**, 301–318 (2004)
5. Dierckx, P.: Curve and Surface Fitting with Splines. Oxford University Press, Oxford (1993)

6. Farin, G.: Curves and surfaces for CAGD, 5th edn. Morgan Kaufmann, San Francisco (2002)
7. Echevarría, G., Iglesias, A., Gálvez, A.: Extending neural networks for b-spline surface reconstruction. In: Sloot, P.M.A., Tan, C.J.K., Dongarra, J., Hoekstra, A.G. (eds.) ICCS-ComputSci 2002, Part II. LNCS, vol. 2330, pp. 305–314. Springer, Heidelberg (2002)
8. Gálvez, A., Cobo, A., Puig-Pey, J., Iglesias, A.: Particle Swarm optimization for Bézier surface reconstruction. In: Bubak, M., van Albada, G.D., Dongarra, J., Sloot, P.M.A. (eds.) ICCS 2008, Part II. LNCS, vol. 5102, pp. 116–125. Springer, Heidelberg (2008)
9. Gálvez, A., Iglesias, A.: Efficient particle swarm optimization approach for data fitting with free knot B-splines. Computer-Aided Design 43(12), 1683–1692 (2011)
10. Gálvez, A., Iglesias, A.: Particle swarm optimization for non-uniform rational B-spline surface reconstruction from clouds of 3D data points. Information Sciences 192(1), 174–192 (2012)
11. Gálvez, A., Iglesias, A.: A new iterative mutually-coupled hybrid GA-PSO approach for curve fitting in manufacturing. Applied Soft Computing 13(3), 1491–1504 (2013)
12. Gálvez, A., Iglesias, A.: Firefly algorithm for polynomial Bzier surface parameterization. Journal of Applied Mathematics, Article ID 237984, 9 (2013)
13. Gálvez, A., Iglesias, A.: From nonlinear optimization to convex optimization through firefly algorithm and indirect approach with applications to CAD/CAM. The Scientific World Journal, Article ID 283919, 10 (2013)
14. Gálvez, A., Iglesias, A.: Firefly algorithm for explicit B-Spline curve fitting to data points. Mathematical Problems in Engineering, Article ID 528215, 12 (2013)
15. Gálvez, A., Iglesias, A.: An electromagnetism-based global optimization approach for polynomial Bezier curve parameterization of noisy data points. In: Proc. of Cyberworlds 2013. IEEE Computer Society Press, Los Alamitos, pp. 259–266 (2013)
16. Gálvez, A., Iglesias, A.: Cuckoo search with Lévy flights for weighted Bayesian energy functional optimization in global-support curve data fitting. The Scientific World Journal, Article ID 138760, 11 (2014)
17. Gálvez, A., Iglesias, A.: New memetic self-adaptive firefly algorithm for continuous optimization. International Journal of Bio-Inspired Computation (in press)
18. Iglesias, A., Gálvez, A., Avila, A.: Discrete Bézier curve fitting with artificial immune systems. In: Plemenos, D., Miaoulis, G. (eds.) Intelligent Computer Graphics 2012. SCI, vol. 441, pp. 59–75. Springer, Heidelberg (2013)
19. Gálvez, A., Iglesias, A., Avila, A.: Immunological-based approach for accurate fitting of 3d noisy data points with Bézier surfaces. In: Proc. of Int. Conference on Comp. Science-ICCS 2013. Procedia Computer Science, vol. 18, pp. 50–59 (2013)
20. Gálvez, A., Iglesias, A., Avila, A., Otero, C., Arias, R., Manchado, C.: Elitist clonal selection algorithm for optimal choice of free knots in B-spline data fitting. Applied Soft Computing 26, 90–106 (2015)
21. Gálvez, A., Iglesias, A., Cobo, A., Puig-Pey, J., Espinola, J.: Bézier curve and surface fitting of 3d point clouds through genetic algorithms, functional networks and least-squares approximation. In: Gervasi, O., Gavrilova, M.L. (eds.) ICCSA 2007, Part II. LNCS, vol. 4706, pp. 680–693. Springer, Heidelberg (2007)
22. Gálvez, A., Iglesias, A., Puig-Pey, J.: Iterative two-step genetic-algorithm method for efficient polynomial B-spline surface reconstruction. Information Sciences 182(1), 56–76 (2012)

23. Gu, P., Yan, X.: Neural network approach to the reconstruction of free-form surfaces for reverse engineering. Computer-Aided Design **27**(1), 59–64 (1995)
24. Hoffmann, M.: Numerical control of Kohonen neural network for scattered data approximation. Numerical Algorithms **39**, 175–186 (2005)
25. Iglesias, A., Echevarría, G., Gálvez, A.: Functional networks for B-spline surface reconstruction. Future Generation Computer Systems **20**(8), 1337–1353 (2004)
26. Iglesias, A., Gálvez, A.: A new artificial intelligence paradigm for computer-aided geometric design. In: Campbell, J., Roanes-Lozano, E. (eds.) AISC 2000. LNCS (LNAI), vol. 1930, pp. 200–213. Springer, Heidelberg (2001)
27. Iglesias, A., Gálvez, A.: Applying functional networks to fit data points rom B-spline surfaces. In: Proc. of Computer Graphics International, CGI 2001, Hong-Kong (China). IEEE Computer Society Press, Los Alamitos, pp. 329–332 (2001)
28. Iglesias, A., Gálvez, A.: Curve fitting with RBS functional networks. In: Proc. of Int. Conference on Convergence Information Technology-ICCIT 2008 - Busan (Korea). IEEE Computer Society Press, Los Alamitos, vol. 1, pp. 299–306 (2008)
29. Iglesias, A., Gálvez, A.: Hybrid functional-neural approach for surface reconstruction. Mathematical Problems in Engineering, Article ID 351648, 13 (2014)
30. Jing, L., Sun, L.: Fitting B-spline curves by least squares support vector machines. In: Proc. of the 2nd. Int. Conf. on Neural Networks & Brain, Beijing (China). IEEE Press, pp. 905–909 (2005)
31. Li, W., Xu, S., Zhao, G., Goh, L.P.: Adaptive knot placement in B-spline curve approximation. Computer-Aided Design **37**, 791–797 (2005)
32. Loucera, C., Gálvez, A., Iglesias, A.: Simulated annealing algorithm for Bezier curve approximation. In: Proc. of Cyberworlds 2014. IEEE Computer Society Press, Los Alamitos, pp. 182–189 (2014)
33. Luus, R., Jaakola, T.H.I.: Optimization by direct search and systematic reduction of the size of search region. American Inst. of Chemical Engineers Journal **19**(4), 760–766 (1973)
34. Park, H.: An error-bounded approximate method for representing planar curves in B-splines. Computer Aided Geometric Design **21**, 479–497 (2004)
35. Park, H., Lee, J.H.: B-spline curve fitting based on adaptive curve refinement using dominant points. Computer-Aided Design **39**, 439–451 (2007)
36. Patrikalakis, N.M., Maekawa, T.: Shape Interrogation for Computer Aided Design and Manufacturing. Springer Verlag, Heidelberg (2002)
37. Pottmann, H., Leopoldseder, S., Hofer, M., Steiner, T., Wang, W.: Industrial geometry: recent advances and applications in CAD. Computer-Aided Design **37**, 751–766 (2005)
38. Powell, M.J.D.: Curve fitting by splines in one variable. In: Hayes, J.G. (ed.) Numerical approximation to functions and data. Athlone Press, London (1970)
39. Sarfraz, M., Raza, S.A.: Capturing outline of fonts using genetic algorithms and splines. In: Proc. of Fifth International Conference on Information Visualization IV 2001. IEEE Computer Society Press, pp. 738–743 (2001)
40. Wang, W.P., Pottmann, H., Liu, Y.: Fitting B-spline curves to point clouds by curvature-based squared distance minimization. ACM Transactions on Graphics **25**(2), 214–238 (2006)
41. Yoshimoto, F., Harada, T., Yoshimoto, Y.: Data fitting with a spline using a real-coded algorithm. Computer-Aided Design **35**, 751–760 (2003)
42. Zhao, X., Zhang, C., Yang, B., Li, P.: Adaptive knot adjustment using a GMM-based continuous optimization algorithm in B-spline curve approximation. Computer-Aided Design **43**, 598–604 (2011)

Solving the Set Covering Problem with Binary Cat Swarm Optimization

Broderick Crawford[1,4,5]($\boxtimes$), Ricardo Soto[1,2,3], Natalia Berríos[1],
Franklin Johnson[1,6], and Fernando Paredes[7]

[1] Pontificia Universidad Católica de Valparaíso, Valparaíso, Chile
{broderick.crawford,ricardo.soto}@ucv.cl,
natalia.berrios.p@mail.pucv.cl, franklin.johnson@upla.cl
[2] Universidad Autónoma de Chile, Santiago, Chile
[3] Universidad Científica del Sur, Lima, Perú
[4] Universidad Central de Chile, Santiago, Chile
[5] Universidad San Sebastián, Santiago, Chile
[6] Universidad de Playa Ancha, Valparaíso, Chile
[7] Universidad Diego Portales, Santiago, Chile
fernando.paredes@udp.cl

Abstract. The Set Covering Problem is a formal model for many practical optimization problems. It consists in finding a subset of columns in a zero–one matrix such that they cover all the rows of the matrix at a minimum cost. To solve the Set Covering Problem we use a metaheuristic called Binary Cat Swarm Optimization. This metaheuristic is a binary version of Cat Swarm Optimization generated by observing cat behavior. Cats have two modes of behavior: seeking mode and tracing mode. We are the first ones to use this metaheuristic to solve the Set Covering Problem, for this the proposed algorithm has been tested on 65 benchmarks instances.

Keywords: Binary Cat Swarm Optimization · Set Covering Problem · Metaheuristic

1 Introduction

The Set Covering Problem (SCP) [5,8,23] is a classic problem that consists in finding a set of solutions which allow to cover a set of needs at the lowest cost possible. In the field of optimization, many algorithms have been developed to solve the SCP. Examples of these optimization algorithms include: Genetic Algorithm (GA) [30], Ant Colony Optimization (ACO) [25] and Particle Swarm Optimization (PSO) [15,28]. Our proposal of algorithm uses cat behavior to solve optimization problems, it is called Binary Cat Swarm Optimization (BCSO) [27].

BCSO refers to a serie of heuristic optimization methods and algorithms based on cat behavior in nature. Cats behave in two ways: seeking mode and tracing mode. BCSO is based in CSO [12] algorithm, proposed by Chu and Tsai recently [24]. The difference is that in BCSO the vector position consists of ones and zeros, instead the real numbers of CSO.

© Springer International Publishing Switzerland 2015
Y. Tan et al. (Eds.): ICSI-CCI 2015, Part I, LNCS 9140, pp. 41–48, 2015.
DOI: 10.1007/978-3-319-20466-6_4

2   Set Covering Problem

The SCP [13,14,21] can be formally defined as follows. Let $A = (a_{ij})$ be an m-row, n-column, zero-one matrix. We say that a column j can cover a row if $a_{ij} = 1$. Each column j is associated with a nonnegative real cost c_j. Let $I = \{1,...,m\}$ and $J = \{1,...,n\}$ be the row set and column set, respectively. The SCP calls for a minimum cost subset $S \subseteq J$, such that each row $i \in I$ is covered by at least one column $j \in S$. A mathematical model for the SCP is

$$v(\text{SCP}) = \min \sum_{j \in J} c_j x_j \tag{1}$$

subject to

$$\sum_{j \in J} a_{ij} x_j \geq 1, \quad \forall\, i \in I, \tag{2}$$

$$x_j \in \{0,1\}, \forall\, j \in J \tag{3}$$

The objective is to minimize the sum of the costs of the selected columns, where $x_j = 1$ if column j is in the solution, 0 otherwise. The constraints ensure that each row i is covered by at least one column.

The SCP has been applied to many real world problems such as crew scheduling [3], location of emergency facilities [31], production planning in industry [32], ship scheduling [18], network attack or defense [6], assembly line balancing [19], traffic assignment in satellite communication systems [9], simplifying boolean expressions [7], the calculation of bounds in integer programs [10], information retrieval, political districting [20], stock cutting, crew scheduling problems in airlines [22] and other important real life situations. Because it has wide applicability, we deposit our interest in solving the SCP.

3   Binary Cat Swarm Optimization

Among the known felines, there are about thirty different species, e.g., lion, tiger, leopard, cat, among others [2]. Though many have different living environments, cats share similar behavior patterns. For wild cats, this hunting skill ensures their food supply and survival of their species [17]. Feral cats are groups with a mission to hunt their food, are very wild feline colonies, ranging from 2-15 individuals. Domestic cats also show the same ability to hunt, and are curious about moving objects [16]. Watching the cats, you would think that most of the time is spent resting, even when awake [1]. This alertness they do not never leave, they may be listening or with wide eyes to look around [26]. Based on these behaviors we known BCSO.

Binary Cat Swarm Optimization [27] is an optimization algorithm that imitates the natural behavior of cats [11,29]. Each cat is represented by cat_k, where $k \in [1, C]$, has its own position consisting of M dimensions, which are composed by ones and zeros. Besides, they have speed for each dimension d, Mixture Ratio (MR), a percentage for indicating if the cat is on seeking mode or tracing mode and finally a fitness value that is calculated based on the SCP (Eq. 1). The BCSO keeps the best solution until the end of iterations. Next is described the BCSO general diagram (Fig. 1):

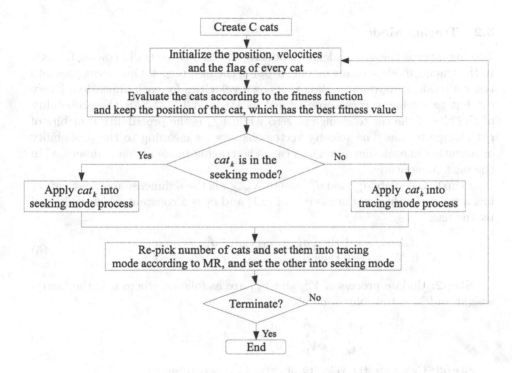

Fig. 1. Cat Swarm Optimization

3.1 Seeking Mode

This sub-model is used to model the situation of the cat, which is resting, looking around and seeking the next position to move to. Seeking mode has essential factors: PMO, Probability of Mutation Operation; CDC, Counts of Dimensions to Change, it indicates how many of the dimensions varied; SMP, Seeking Memory Pool, it is used to define the size of seeking memory for each cat.

The following pseudocode describe cat behavior seeking mode. In wish $FS_b = FS_{max}$ for finding the minimum solution and $FS_b = FS_{min}$ for finding the maximum solution.

Step1: Create SMP copies of cat_k

Step2: Based on CDC update the position of each copy by randomly according to PMO

Step3: Evaluate the fitness of all copies

Step4: Calculate the selecting probability of each copy according to

$$P_i = \frac{FS_i - FS_b}{FS_{max} - FS_{min}} \tag{4}$$

Step5: Apply roulette wheel to the candidate points and select one
Step6: replace the current position with the selected candidate

3.2   Tracing Mode

Tracing mode is the sub-model for modeling the case of the cat in tracing targets. In the tracing mode, cats are moving towards the best target. Once a cat goes into tracing mode, it moves according to its own velocities for each dimension. Every cat has two velocity vector are defined as V_{kd}^1 and V_{kd}^0. V_{kd}^0 is the probability of the bits of the cat to change to zero while V_{kd}^1 is the probability that bits of cat change to one. The velocity vector changes its meaning to the probability of mutation in each dimension of a cat. The tracing mode action is described in the next pseudocode.

Step1: Calculate d_{kd}^1 and d_{kd}^0 where $X_{best,d}$ is the d dimension of best cat, r_1 has a random values in the interval of [0,1] and c_1 is a constant which is defined by the user

$$\begin{aligned} &\text{if } X_{best,d} = 1 \text{ then } d_{kd}^1 = r_1 c_1 \text{ and } d_{kd}^0 = -r_1 c_1 \\ &\text{if } X_{best,d} = 0 \text{ then } d_{kd}^1 = -r_1 c_1 \text{ and } d_{kd}^0 = r_1 c_1 \end{aligned} \tag{5}$$

Step2: Update process of V_{kd}^1 and V_{kd}^0 are as follows, where w is the inertia weight and M is the column numbers.

$$\begin{aligned} V_{kd}^1 &= wV_{kd}^1 + d_{kd}^1 \\ V_{kd}^0 &= wV_{kd}^0 + d_{kd}^0 \end{aligned} \qquad d = 1,...,M \tag{6}$$

Step3: Calculate the velocity of cat_k, $V_{kd}^{'}$, according to

$$V_{kd}^{'} = \begin{cases} V_{kd}^1 \text{ if } X_{kd} = 0 \\ V_{kd}^0 \text{ if } X_{kd} = 1 \end{cases} \tag{7}$$

Step4: Calculate the probability of mutation in each dimension, this is defined by parameter t_{kd}, t_{kd} takes a value in the inverval of [0,1]

$$t_{kd} = \frac{1}{1 + e^{-V_{kd}^{'}}} \tag{8}$$

Step5: Based on the value of t_{kd} the new value of each dimension of cat is update as follows

$$X_{kd} = \begin{cases} X_{best,d} \text{ if } rand < t_{kd} \\ X_{kd} \text{ if } t_{kd} < rand \end{cases} \qquad d = 1,...,M \tag{9}$$

The maximun velocity vector of $V_{kd}^{'}$ should be bounded to a value V_{max}. If the value of $V_{kd}^{'}$ becomes larger than V_{max}. V_{max} should be selected for velocity in the corresponding dimension.

4 Solving the Set Covering Problem

Next is described the Solving SCP pseudocode:

Algorithm 1. *Solving SCP()*

1: Initialize parameters in cats;
2: Initialization of cat positions, randomly, with values between 0 and 1;
3: Evaluation of the fitness of the population (Eq. 1);
4: Change of the position of the cat based in seeking mode or tracing mode;
5: If solution is not feasible then repaired. Rows have not yet been covered and choose the needed columns for coverage must be determined to make a solution that is feasible. The search for these columns is based on: (the cost of a column)/(number of not covered row that can cover column j). Once the solution has become feasible applies a step of optimization to eliminate those redundant columns. A redundant column is one that if removed, the solution remains feasible;
6: Memorizes the best found solution. Increases the number of iterations;
7: Stop the process if the completion criteria are met. Completion criteria used in this work are the number specified maximum of iterations. Otherwise, go to step 3;

---

5 Results

The BCSO performance was evaluated experimentally using 65 SCP test instances from the OR-Library of Beasley [4]. The algorithm was coded in Java in NetBeans IDE 7.1 and executed on a Computer with 2.53 GHz and 3.0 GB of RAM under Windows 7 Operating System. In all experiments the BCSO was executed with 200 iterations and 30 times each instance. The parameters in the Table 1 were selected empirically so after a large number of tests. This number was determined by the rapid convergence to a near local optimum to the global optimum.

Table 1. Parameter values

Name	Parameter	Value
Number of Cats	C	20
Mixture Ratio	MR	0.5
Counts of Dimensions to Change	CDC	0.001
Seeking Memory Pool	SMP	20
Probability of Mutation Operation	PMO	1
Aleatory Variable	$rand$	$\in [0,1]$
Inertia weight	w	1
Factor c_1	c_1	1

The table 2 shows the results of the 65 instances. The Z_{opt} reports the best known solution for each instance. The Z_{min}, Z_{max} and Z_{avg} report the lowest, highest and the average cost of the best solutions obtained in 30 runs respectively. The quality of a solution is evaluated in terms of the percentage deviation relative

(RPD) of the solution reached Z_{min} and Z_{opt} (which can be either the optimal or the best known objective value).

$$RPD = \left(\frac{Z_{min} - Z_{opt}}{Z_{opt}}\right) * 100 \tag{10}$$

Table 2. Computational results on 65 instances of SCP

Instance	Z_{opt}	Z_{min}	Z_{max}	Z_{avg}	RPD							
4.1	429	459	485	479.6	11.79	B.3	80	85	87	85.4	6.5
4.2	512	570	599	594.2	16.05	B.4	79	89	89	89	12.66
4.3	516	590	614	606.8	17.6	B.5	72	73	73	73	1.53
4.4	494	547	585	578.3	17.06	C.1	227	242	243	242.4	6.7
4.5	512	545	558	554.2	8.24	C.2	219	240	244	240.8	9.86
4.6	560	637	655	649.9	16.05	C.3	243	277	279	278	14.40
4.7	430	462	469	467.4	8.7	C.4	219	250	250	250	13.24
4.8	492	546	571	566.9	15.22	C.5	215	243	247	244.3	13.63
4.9	641	711	741	725.0	13.10	D.1	60	65	66	65.7	9.50
4.10	514	537	556	552.1	7.41	D.2	66	70	71	70.1	6.21
5.1	253	279	283	281.6	11.30	D.3	72	79	81	80.8	12.22
5.2	302	339	340	339.9	12.55	D.4	62	64	67	66.6	7.42
5.3	226	247	252	250.5	10.84	D.5	61	65	66	65.6	7.54
5.4	242	251	254	253.2	4.63	NRE.1	29	29	30	29.9	3.1
5.5	211	230	231	230.4	9.19	NRE.2	30	34	35	34.2	14
5.6	213	232	244	242.7	13.94	NRE.3	27	31	32	31.5	16.67
5.7	293	332	343	338.0	15.36	NRE.4	28	32	33	32.9	17.5
5.8	288	320	331	329.9	14.55	NRE.5	28	30	31	30.3	8.21
5.9	279	295	299	298.6	7.03	NRF.1	14	17	18	17.1	22.14
5.10	265	285	288	286.9	8.26	NRF.2	15	18	19	18.2	21.33
6.1	138	151	166	159.9	15.87	NRF.3	14	17	18	17.2	22.86
6.2	146	152	160	157.4	7.81	NRF.4	14	17	18	17.1	22.14
6.3	145	160	166	164.3	13.31	NRF.5	13	15	16	15.9	22.31
6.4	131	138	143	141.7	8.17	NRG.1	176	190	194	192.7	9.49
6.5	161	169	176	172.8	7.33	NRG.2	154	165	167	166	7.79
A.1	253	286	287	286.9	13.40	NRG.3	166	187	191	187.7	21.1
A.2	252	274	280	276.3	9.64	NRG.4	168	179	185	183.2	9.05
A.3	232	257	264	263.1	13.41	NRG.5	168	181	186	184.3	9.7
A.4	234	248	252	251.3	7.16	NRH.1	63	70	74	71.2	13.02
A.5	236	244	244	244	3.31	NRH.2	63	67	67	67	6.35
B.1	69	79	79	79	14.49	NRH.3	59	68	74	69.6	17.97
B.2	76	86	90	88.5	16.18	NRH.4	58	66	68	66.6	14.83
						NRH.5	55	61	66	61.5	11.82

6   Conclusions

In this paper we use BCSO to solve SCP using its column based representation (binary solutions). In binary discrete optimization problems the position vector is binary, this causes significant change in BCSO with respect to CSO with real numbers. In fact in BCSO in the seeking mode the slight change in the position takes place by introducing the mutation operation. The interpretation of velocity vector in tracing mode also changes to probability of change in each dimension of position of the cats. As can be seen from the results, metaheuristic performs well in most all cases. This paper has shown that the BCSO is a valid alternative to solve the SCP, with its primary use is for continuous domains. The algorithm performs well regardless of the scale of the problem.

We can see the premature convergence, a typical problem in metaheuristics, which occurs when the cats quickly attain to dominate the population, constraining it to converge to a local optimum. For future works the objective will be make them highly immune to be trapped in local optima and thus less vulnerable to premature convergence problem. Thus, we could propose an algorithm that shows improved results in terms of both computational time and quality of solution.

Acknowledgments. Broderick Crawford is supported by Grant CONICYT/FONDECYT/REGU-LAR/1140897 Ricardo Soto is supported by Grant CONICY/-FONDECYT/ INICIACION/11130459 and Fernando Paredes is supported by Grant CONICYT/FONDECYT/REGULAR/1130455.

References

1. Adler, H.: Some factors of observation learning in cats. Journal of Genetic Psychology 159–177 (1995)
2. Aspinall, V.: Complete Textbook of Veterinary Nursing. Butterworth-Heinemann, Oxford, UK (2006)
3. Bartholdi, J.J.: A guaranteed-accuracy round-off algorithm for cyclic scheduling and set covering. Operations Research **29**(3), 501–510 (1981)
4. Beasley, J.: A lagrangian heuristic for set covering problems. Naval Research Logistics **37**, 151–164 (1990)
5. Beasley, J., Jornsten, K.: Enhancing an algorithm for set covering problems. European Journal of Operational Research **58**(2), 293–300 (1992)
6. Bellmore, M., Ratliff, H.D.: Optimal defense of multi-commodity networks. Management Science **18**(4–part-i), 174–185 (1971)
7. Breuer, M.A.: Simplification of the covering problem with application to boolean expressions. J. ACM **17**(1), 166–181 (1970)
8. Caprara, A., Fischetti, M., Toth, P.: Algorithms for the set covering problem. Annals of Operations Research **98**, 353–371 (2000)
9. Ceria, S., Nobili, P., Sassano, A.: A lagrangian-based heuristic for large-scale set covering problems. Mathematical Programming **81**(2), 215–228 (1998)
10. Christofides, N.: Zero-one programming using non-binary tree-search. Comput. J. **14**(4), 418–421 (1971)
11. Chu, S., Tsai, P.: Computational intelligence based on the behavior of cats. International Journal of Innovative Computing, Information and Control 163–173 (2007)
12. Chu, Shu-Chuan, Tsai, Pei-wei, Pan, Jeng-Shyang: Cat swarm optimization. In: Yang, Qiang, Webb, Geoff (eds.) PRICAI 2006. LNCS (LNAI), vol. 4099, pp. 854–858. Springer, Heidelberg (2006)

13. Crawford, B., Soto, R., Cuesta, R., Paredes, F.: Application of the artificial bee colony algorithm for solving the set covering problem. The Scientific World Journal **2014**(189164), 1–8 (2014)

14. Crawford, Broderick, Soto, Ricardo, Monfroy, Eric: Cultural algorithms for the set covering problem. In: Tan, Ying, Shi, Yuhui, Mo, Hongwei (eds.) ICSI 2013, Part II. LNCS, vol. 7929, pp. 27–34. Springer, Heidelberg (2013)

15. Crawford, B., Soto, R., Monfroy, E., Palma, W., Castro, C., Paredes, F.: Parameter tuning of a choice-a function based hyperheuristic using particle swarm optimization. In: Expert Systems with Applications, pp. 1690–1695 (2013)

16. Crowell-Davis, S.: Cat behaviour: social organization, communication and development. In: Rochlitz, I. (ed.) The Welfare Of Cats, pp. 1–22. Springer, Netherlands (2005)

17. Dards, J.: Feral cat behaviour and ecology. Bulletin of the Feline Advisory Bureau **15**, (1976)

18. Fisher, M.L., Rosenwein, M.B.: An interactive optimization system for bulk-cargo ship scheduling. Naval Research Logistics (NRL) **36**(1), 27–42 (1989)

19. Freeman, B., Jucker, J.: The line balancing problem. Journal of Industrial Engineering **18**, 361–364 (1967)

20. Garfinkel, R.S., Nemhauser, G.L.: Optimal political districting by implicit enumeration techniques. Management Science **16**(8), B495–B508 (1970)

21. Gouwanda, D., Ponnambalam, S.: Evolutionary search techniques to solve set covering problems. World Academy of Science, Engineering and Technology **39**, 20–25 (2008)

22. Housos, E., Elmroth, T.: Automatic optimization of subproblems in scheduling airline crews. Interfaces **27**(5), 68–77 (1997)

23. Lessing, Lucas, Dumitrescu, Irina, Stützle, Thomas: A comparison between ACO algorithms for the set covering problem. In: Dorigo, Marco, Birattari, Mauro, Blum, Christian, Gambardella, Luca Maria, Mondada, Francesco, Stützle, Thomas (eds.) ANTS 2004. LNCS, vol. 3172, pp. 1–12. Springer, Heidelberg (2004)

24. Panda, G., Pradhan, P., Majhi, B.: Iir system identification using cat swarm optimization. Expert Systems with Applications **38**, 12671–12683 (2011)

25. Ren, Z., Feng, Z., Ke, L., Zhang, Z.: New ideas for applying ant colony optimization to the set covering problem. Computers & Industrial Engineering 774–784 (2010)

26. Santosa, B., Ningrum, M.: Cat swarm optimization for clustering. In: International Conference of Soft Computing and Pattern Recognition, pp. 54–59 (2009)

27. Sharafi, Y., Khanesar, M., Teshnehlab, M.: Discrete binary cat swarm optimization algorithm. Computer, Control and Communication 1–6. (2013)

28. Shi, Y., Eberhart, R.: Empirical study of particle swarm optimization. In: Proc. of the Congress on Evolutionary Computation, pp. 1945–1950 (1999)

29. Tsai, P., Pan, J., Chen, S., Liao, B.: Enhanced parallel cat swarm optimization based on the taguchi method. Expert Systems with Applications **39**, 6309–6319 (2012)

30. Aickelin, U.: An indirect genetic algorithm for set covering problems. Journal of the Operational Research Society pp. 1118–1126 (2002)

31. Vasko, F.J., Wilson, G.R.: Using a facility location algorithm to solve large set covering problems. Operations Research Letters **3**(2), 85–90 (1984)

32. Vasko, F.J., Wolf, F.E., Stott, K.L.: A set covering approach to metallurgical grade assignment. European Journal of Operational Research **38**(1), 27–34 (1989)

Bird Mating Optimizer in Structural Damage Identification

H. Li, J.K. Liu, and Z.R. Lu(⊠)

Department of Applied Mechanics, Sun Yat-Sen University,
Guangzhou 510006, Guangdong Province, People's Republic of China
lvzhr@mail.sysu.edu.cn

Abstract. In this paper, a structural damage detection approach based on bird mating optimizer (BMO) is proposed. Local damage is represented by a perturbation in the elemental stiffness parameter of the structural finite element model. The damage parameters are determined by minimizing the error derived from modal data, and natural frequency and modal assurance criteria (MAC) of mode shape is employed to formulate the objective function. The BMO algorithm is adopted to optimize the objective and optimum set of stiffness reduction parameters are predicted. The results show that the BMO can identify the perturbation of the stiffness parameters effectively even under measurement noise.

Keywords: Damage identification · Bird mating optimizer · Frequency domain

1   Introduction

The identification of structural damage is a vital part of structural health monitoring during the period of construction and service. The objective of structural damage identification is to localize and quantify the deterioration in a physical structural system from the measured response or modal parameters. In the past few decades, more and more researchers have applied global optimization techniques to the problem of structural damage identification [1,2,3]. Swarm intelligence methods, such as particle swarm optimization (PSO) and ant colony optimization (ACO), are highly adaptive methods originated from the laws of nature and biology. The usual swarm intelligence methods minimize an objective function, which is defined in terms of the discrepancies between the vibration data identified by modal testing and those computed from the analytical model. In recent years, PSO [4,5,6] is a novel population-based global optimization technique developed. Mohan et al. [4] evaluated the use of frequency response function with the help of particle swarm optimization technique, for structural damage detection and quantification. An immunity enhanced particle swarm optimization algorithm with the artificial immune system is proposed for damage detection of structures by Kang et al. [5]. A two-stage method, including modal strain energy based index (MSEBI) and PSO, is proposed to properly identify the site and extent of multiple damage cases by Seyedpoor [6]. In recent literature, several applications of ACO [7,8,9] are observed to solve problems successfully such as in traveling salesman problem, vehicle routing

© Springer International Publishing Switzerland 2015
Y. Tan et al. (Eds.): ICSI-CCI 2015, Part I, LNCS 9140, pp. 49–56, 2015.
DOI: 10.1007/978-3-319-20466-6_5

problems etc. Kaveh et al. [7] proposed ACO for topology optimization of 2D and 3D structures, which was to find the stiffest structure with a certain amount of material. Yu and Xu [8] proposed an ant colony optimization based algorithm for continuous optimization problems on structural damage detection in the SHM field. Majumdar et al. [9] presented ant colony optimization algorithm to detect and assess structural damages from changes in natural frequencies.

Recently, a novel heuristic algorithm named bird mating optimizer (BMO) was proposed by Askarzadeh and Rezazadeh [10]. It is a population-based optimization algorithm which employs mating process of birds as a framework. Detail of the BMO algorithm is presented in [11]. BMO algorithm has been utilized to extract maximum power of solar cells by Askarzadeh and Rezazadeh [12]. The researches show that BMO is an efficient algorithm for multimodal optimization. And it can be applied on different fields.

In this paper, a structural damage identification method based on BMO algorithm is proposed. The damage parameters (i.e. stiffness reduction parameters) are determined by minimizing a global error derived from modal data. In order to employ the BMO algorithm, an objective function in frequency domain is introduced. Two numerical examples are utilized to verify the effectiveness of BMO, including a simply supported beam and a planar truss. The identified results indicate that the BMO algorithm is practical and efficient. The multiple local damages can be identified accuracy even under measurement noise.

2   Methodology for Damage Detection

2.1   Damage Model in Frequency Domain

Neglecting the damping of the structure, the eigenvalue equation for an n degree-of-freedom structural system can be expressed as

$$(\mathbf{K} - \omega_j^2 \mathbf{M})\phi_j = 0 \tag{1}$$

where $\mathbf{M}$, $\mathbf{K}$ are the mass, stiffness matrices, respectively. ω_j is the jth natural frequency and ϕ_j is the corresponding mode shapes.

When a structure is damaged, the reduction of the stiffness can be evaluated by a set of damage parameters $\alpha_i (i = 1, 2, \cdots, nel)$ and the loss in mass is ignored. The damaged stiffness matrix can be written as

$$\mathbf{K_d} = \sum_{i=1}^{nel} \alpha_i \mathbf{k_i^e} \tag{2}$$

where $\mathbf{K_d}$ is the stiffness matrix of the damaged system, and $\mathbf{k_i^e}$ presents the ith elemental stiffness matrix in the global form. The parameter α_i ranges between 0 and 1. $\alpha_i = 0$ represents the complete damaged status. Base idea of damages identification in a structure is equal to identify the values of the damage parameter vector $\{\alpha\}$.

2.2   Objective Function in Frequency Domain

It is known that the changes of stiffness will lead to changes of the structural properties, such as vibration frequencies and mode shapes. For damage detection our task is to minimize the difference between the measured data and the calculated one.

Taking account of mode shape data and natural frequency data, the objective function used for damage detection can be defined as

$$f = \sum_{j=1}^{NF} w_{\omega j}^2 \Delta \omega_j^2 + \sum_{j=1}^{NM} w_{\phi j}^2 (1 - MAC_j^R) \qquad (3)$$

where $w_{\omega j}$ is a weight factor of the output error of the jth natural frequency, $\Delta \omega_j$ is the differences of natural frequencies. $w_{\phi j}$ is a weight factor of the output error of the jth mode shape. And MAC_j^R represents the jth MAC obtained by incomplete mode shape data. NF and NM are the numbers of frequencies and mode shapes.

3   Brief of BMO Algorithm

Bird mating optimizer (BMO) imitates the behavior of bird species metaphorically to breed broods with superior genes for designing optimum searching techniques. The concept of BMO is very easy and it can be effectively employed in damage detection.

In BMO algorithm, a feasible solution of the problem is called bird. There are male and female birds in the society. And the females represent the better solutions. Based on the way by which birds breed broods, females are categorized into two groups (i.e. polyandrous and promiscuous) while males are categorized into three groups (i.e. monogamous, polygynous and promiscuous).

Monogamous birds are those males that tend to mate with one female. If a monogamous bird $\vec{x}_M$ wants to mate with his interesting female $\vec{x}_i$. The resultant brood is produced by

$$\vec{x}_b = \vec{x}_M + w \times \vec{r}. \times (\vec{x}_i - \vec{x}_M)$$

$$if\ r_1 > mcf$$

$$x_b(c) = l(c) + r_2 \times (u(c) - l(c)); \qquad (4)$$

$$end$$

where $\vec{x}_b$ is the resultant brood, w is a time-varying weight factor from 1.9 to 0.1., $\vec{r}$ is a vector with elements distributed randomly in [0,1], mcf is a mutation control factor varying, which is set to be 0.9 in this paper. u and l are the upper and lower bounds of the elements, respectively.

Polygynous birds are those males that have a tendency to couple with multiple females. The resultant brood is given as follows:

$$\vec{x}_b = \vec{x}_{Pg} + w \times \sum_{j=1}^{n_i} \vec{r}.\times(\vec{x}_{ij} - \vec{x}_{Pg})$$

$$if \ r_1 > mcf \tag{5}$$

$$x_b(c) = l(c) + r_2 \times (u(c) - l(c));$$

$$end$$

where n_i is the number of interesting birds, and $\vec{x}_{ij}$ represents the jth elite female. r_i is the random number between 0 and 1.

Polyandrous birds are those females seek for superior males to breed a brood with high-quality genes. And the resultant brood is produced as same as Eq. (5).

Parthenogenesis birds represent the best solutions. Each parthenogenetic bird produces a brood by the following process

$$for \ i = 1:n$$

$$if \ r_1 > mcf_p$$

$$x_b(i) = x(i) + \mu \times (r_2 - r_3) \times x(i);$$

$$else \tag{6}$$

$$x_b(i) = x(i);$$

$$end$$

$$end$$

where mcf_p is the parthenogenetic mutation control factor, which changes from 0.15 to 0.8 and μ is the step size taken as 9.0×10^{-3}.

Promiscuous birds are produced by a chaotic sequence. At the initial generation, each promiscuous bird is produced using Eq. (7), where z is chaos variable and its initial value is a random number between 0 and 1. And the way by which they breed is same as that of monogamous birds, i.e. Eq. (4).

$$for \ i = 1:n$$

$$x_{Pro}(i) = l(i) + z_{gen} \times (u(i) - l(i));$$

$$end \tag{7}$$

$$z_{gen+1} = 4 z_{gen}(1 - z_{gen})$$

Fig. 1 shows the flowchart of BMO algorithm using in this paper. More details can be consulted in Askarzadeh's studies [10,11]. In BMO, the percentage of each type is determined manually.

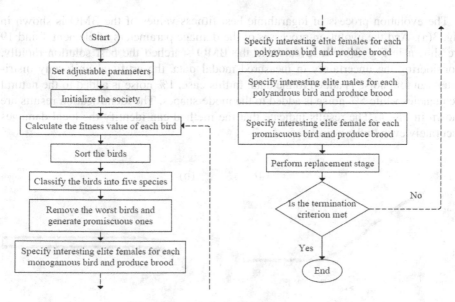

Fig. 1. Flow chart of the detection process

4 Numerical Simulation

4.1 A Simply Supported Beam

A simply supported beam shown in Fig. 2 is used as the numerical example to illustrate the effectiveness of the proposed method. The geometrical parameters of the simply supported beam are Length $L = 1.2\,\text{m}$, Cross section $b \times h = 0.05\,\text{m} \times 0.006\,\text{m}$, Young's modulus $E = 70\,\text{GPa}$, and Density $\rho = 2.70 \times 10^3\,\text{kg/m}^3$.

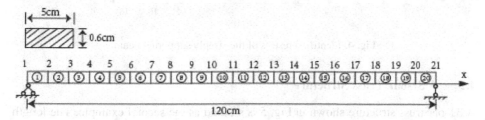

Fig. 2. A simply supported beam

It is assumed two damages located at elements 4 and 19 with reduction of 15% and 20% in each stiffness, i.e. $\alpha_4 = 0.15$, $\alpha_{19} = 0.20$, respectively. The first four natural frequencies and modes shapes are used in the identification. The society scale is 100. And the number of monogamous, polygynous, promiscuous, polyandrous, and parthenogenetic birds is respectively set at 50, 30, 10, 5 and 5. The maximum generation in this case is 500.

The evolution process of logarithmic best fitness values of the BMO is shown in Fig. 3(a). And the evolution processes of the damage parameters of element 4 and 19 are shown in Fig. 3(b). It reveals that the BMO searched the best solution rapidly. Considering the uncertainty in measured modal data, the artificial uniformly distributed random noise is added to the data. In this case, 1% noise is added to the natural frequencies while 5% noise is added to the mode shapes. The identification results are shown in Fig. 4. The results indicate that the method can identify the local damages accurately even under noise.

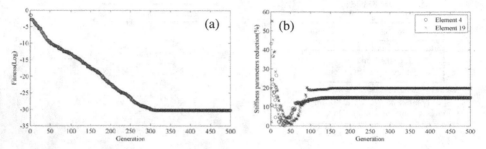

Fig. 3. Evolution process of BMO

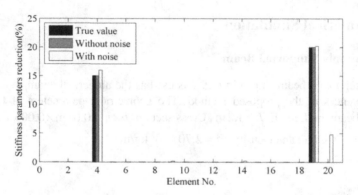

Fig. 4. Identified results of the simply supported beam

4.2   A 31-bar Truss Structure

A 31-bar truss structure shown in Fig. 5 is studied as the second example. The length of each bar is $L = 1\,\text{m}$, and the section area of each bar is $A = 0.004\,\text{m}^2$. Young's modulus $E = 200\,\text{GPa}$, and Density $\rho = 7800\,\text{kg/m}^3$.

The first example showed that the BMO can identify the severe damages successfully. In this example, relatively small damages are added to the structure. In this case, 1% and 5% noise is added to the natural frequencies and mode shapes, respectively. The first five natural frequencies and mode shape displacements are used for damage detection. The solution still converged real quickly. Fig. 6 presents the identified results of the damaged truss. It can be found that the method can identify all local damages even under measurement noise. However, false alarms exist and the damage extents are not accurate enough under noise.

In order to find a way to improve the accuracy of identified result under noisy circumstance, detection results using different number of modal data under measurement noise are shown in Fig. 7. The comparison indicates that the accuracy of the identification increase as the number of adopted modal data increase. Apparently, the first ten frequencies and mode shapes are sufficient under the presupposed noise level in this case.

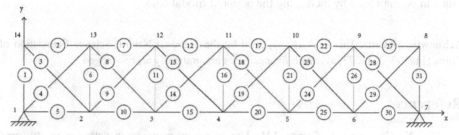

Fig. 5. A 31-bar truss structure

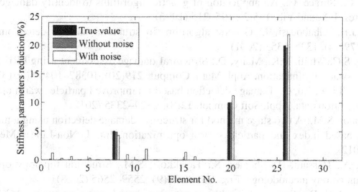

Fig. 6. Identified results of the 31-bar truss structure

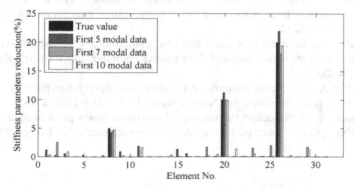

Fig. 7. Comparison of identified results obtained from different number of modal data

5   Conclusions

Making use of the frequency and mode shape data, a damage identification method based on BMO algorithm is proposed. The effectiveness of the proposed method is verified by a simply supported beam and a 31-bar truss structure. The evolution processes shows that the proposed procedure can search the solution rapidly. And the simulations indicates the proposed method can identify the damage parameters successfully even under measurement noise. It can be seen that the accuracy of the detection can be improved by increasing the adopted modal data.

Acknowledgements. This work is supported by the National Natural Science Foundation of China (11172333, 11272361). Such financial aids are gratefully acknowledged.

References

1. Wu, X., Ghaboussi, J., Garrett, J.H.: Use of neural networks in detection of structural damge. Comput. Struct. **42**(4), 649–659 (1992)
2. Mares, C., Surace, C.: An application of genetic algorithms to identify damage in elastic structures. J. Sound Vib. **195**(2), 195–215 (1996)
3. Chou, J.H., Ghaboussi, J.: Genetic algorithm in structural damage detection. Comput. Struct. **79**(14), 1335–1353 (2001)
4. Mohan, S.C., Maiti, D.K., Maity, D.: Structural damage assessment using FRF employing particle swarm optimization. Appl. Math. Comput. **219**(20), 10387–10400 (2013)
5. Kang, F., Li, J., Xu, Q.: Damage detection based on improved particle swarm optimization using vibration data. Appl. Soft Comput. **12**(8), 2329–2335 (2012)
6. Seyedpoor, S.M.: A two stage method for structural damage detection using a modal strain energy based index and particle swarm optimization. Int. J. Non-Linear Mech. **47**(1), 1–8 (2012)
7. Kaveh, A., Hassani, B., Shojaee, S., Tavakkoli, S.M.: Structural topology optimization using ant colony methodology. Eng. Struct. **30**(9), 2559–2565 (2008)
8. Yu, L., Xu, P.: Structural health monitoring based on continuous ACO method. Microelectr. Reliab. **51**(2), 270–278 (2011)
9. Majumdar, A., Maiti, D.K., Maity, D.: Damage assessment of truss structures from changes in natural frequencies using ant colony optimization. Appl. Math. Comput. **218**(19), 9759–9772 (2012)
10. Askarzadeh, A., Rezazadeh, A.: new heuristic optimization algorithm for modeling of proton exchange membrane fuel cell: bird mating optimizer. Int. J. Energy Res. **37**(10), 1196–1204 (2013)
11. Askarzadeh, A.: Bird mating optimizer: An optimization algorithm inspired by bird mating strategies. Commun. Nonlinear Sci. Numer. Simul. **19**(4), 1213–1228 (2014)
12. Askarzadeh, A., Rezazadeh, A.: Extraction of maximum power point in solar cells using bird mating optimizer-based parameters identification approach. Sol. Energy. **90**, 123–133 (2013)

Structural Damage Detection and Moving Force Identification Based on Firefly Algorithm

Chudong Pan[1] and Ling Yu[1,2(✉)]

[1] Department of Mechanics and Civil Engineering, Jinan University,
Guangzhou 510632, China
[2] MOE Key Lab of Disaster Forecast and Control in Engineering,
Jinan University, Guangzhou 510632, China
lyu1997@163.com

Abstract. Based on firefly algorithm (FA), a structural damage detection (SDD) and moving force identification (MFI) method is proposed in this paper. The basic principle of FA is introduced, some key parameters, such as light intensity, attractiveness, and rules of attraction are defined. The moving forces-induced responses of damage structures are defined as a function of both damage factors and moving forces. By minimizing the difference between the real and calculated responses with given damage factors and moving forces, the identified problem is transformed into a constrained optimization problem and then it can be hopefully solved by the FA. In order to assess the accuracy and the feasibility of the proposed method, a three-span continuous beam subjected to moving forces is taken as an example for numerical simulations. The illustrated results show that the method can simultaneously identify the structural damages and moving forces with a good accuracy and better robustness to noise.

Keywords: Firefly algorithm (FA) · Structural damage detection (SDD) · Moving force identification (MFI) · Structural health monitoring (SHM)

1 Introduction

Both structural damage detection (SDD) and moving force identification (MFI) are the crucial problems in the field of bridge structural health monitoring. Many scholars have paid attention and proposed effective methods on either one of them. Most of the proposed methods are established when either structural damage or moving force is known. However, in practice, the unknown damages and unknown moving forces are usually coexisted, and the studies on coexistent identification of damage and moving forces seems not to be so much. Based on the virtual distortion method, Zhang et al. [1] have carried out systematic study of identifying both the coexistent unknown damages and unknown loads. Naseralavi et al. [2] presented a technique for damage detection in structures under unknown periodic excitations using the transient displacement responses. Similar to SDD and MFI, identification of coexistent structural damage and moving forces is a typical inverse problem, therefore, it is inevitable to

© Springer International Publishing Switzerland 2015
Y. Tan et al. (Eds.): ICSI-CCI 2015, Part I, LNCS 9140, pp. 57–64, 2015.
DOI: 10.1007/978-3-319-20466-6_6

solve the ill-posed problem on them. If the inverse problem is considered as a constrained optimization problem, the above problem can be solved to some extent.

Nature-inspired algorithms are among the most powerful algorithms for optimazation [3]. Some of them have been successfully applied to the field of SDD [4-6]. As an important member of nature-inspired algorithm, although the firefly algorithm (FA) has successfully solved many problems from various areas, such as the application in economic dispatch [7], structural optimization [8], software testing [9], MFI [10] and so on, unfortunately, the FA has not been applied to the SDD yet. This study will introduce the FA into the SDD and MFI field and explore the applicability of FA. A FA-based method is proposed for both SDD and MFI, the feasibility and robustness of new method are also studied in this paper.

2   Firefly Algorithm

Firefly algorithm (FA) is a new swarm intelligence (SI) optimization algorithms inspired by nature fireflies flashing behavior. For simplicity in describing the new firefly-inspired algorithms, three idealized rules are allowed to use as follow: 1) All fireflies are unisex so that one firefly will be attracted to other fireflies regardless of their sex; 2) Attractiveness is proportional to their brightness, thus for any two flashing fireflies, the less brighter one will move towards the brighter one. The attractiveness is proportional to the brightness and they both decrease as their distance increases; 3) For a specific problem, the brightness of firefly is associated with the objective function.

The light intensity of a firefly will decrease with the increasing distance of viewer. In addition, light is also absorption by the media. Therefore, it can be defined as:

$$I(r) = I_0 e^{-\gamma r^2} \tag{1}$$

where I_0 is the original light intensity ($r = 0$) and its value is determined by the objective function. γ is the absorption coefficient for simulating the environmental characteristics of the weakening light. Generally, a firefly's attractiveness is proportional to the light intensity seen by adjacent fireflies, therefore, it can be defined as:

$$\beta(r) = \beta_0 e^{-\gamma r^2} \tag{2}$$

where β_0 is the attractiveness at $r = 0$ and it can take $\beta_0 = 1$ for most cases. Because the differences of light intensity and attractiveness existed, the fireflies will try to move themselves to the best position. It means that the lower light intensity one will be attracted by the higher one, and this step can be expressed as:

$$x_j(t+1) = x_j(t) + \beta(r_{ij})(x_i(t) - x_j(t)) + \alpha \cdot (rand - 0.5) \tag{3}$$

where $x_j(t+1)$ is the j-th firefly position after $(t+1)$ times generations. α is a step factor and usually taken a constant. **rand** is a random number generator uniformly distributed in [0, 1]. r_{ij} is the distance between firefly i and firefly j, and it can be defined as the Cartesian distance as: $r_{ij} = \|x_i - x_j\|$.

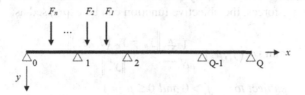

Fig. 1. A continuous beam subjected to moving forces

3   FA-Based SDD and MFI

In practice, the bridge-vehicle system is a very complicated system. Therefore, an appropriate simplified model is always assumed for analysis [11]. A continuous beam subjected to moving forces $F_i\,(i=1,2,\cdots,n)$ is shown in Fig. 1. The beam is regarded as the assembly of the BEAM element and the magnitude of each moving force is assumed to be a constant during the duration when the moving forces pass across the bridge. The structural damage is simulated by element stiffness reduction and quantified by the damage factor μ_i at i-th element of finite element model as:

$$\tilde{K}_i = (1-\mu_i)K_i \tag{4}$$

where K_i, $\tilde{K}_i$ are the stiffness matrix of i-th element for the intact and damage structures, respectively. Therefore, the motion equation of damage structure subjected to moving forces can be expressed as:

$$M\ddot{x} + C\dot{x} + \sum_i (1-\mu_i)K_i x = \sum_i D_i F_i \tag{5}$$

where M, C are the mass and damping matrix of the structure, respectively. F_i denotes the i-th moving force. D_i is a transfer matrix corresponding to i-th moving force, it can transfer the moving forces to the related nodes. Eq. (5) can be solved by Wilson-θ method. For a linear system, the responses of a damage structure subjected to moving forces can be expressed as:

$$y_j = \sum_i H(\mu)_{ji} f_i \tag{6}$$

where y_j denotes the structure responses measured by sensor j. $H(\mu)_{ji}$ is the structure responses of j-th sensor when the structure is only subjected to i-th moving force with an unit load. f_i denotes the magnitude of i-th moving force. It can be seen from Eq. (6) that the responses of the damage structure can be expressed as a function of both damage factors and moving force magnitudes.

For an optimization problem, the essence of SDD and MFI is to find group of damage factors and moving forces, which can minimize the difference between real and calculated responses. Therefore, by minimizing the difference with given damage factors and moving forces, the objective function can be expressed as:

$$\min \, val\left(f_i, \mu_j\right) = \frac{1}{N} \sum_n^N \frac{\left\| y_n^R - y_n^C\left(f_i, \mu_j\right)\right\|}{\left\| y_n^R \right\|};$$

$$subject \, to: \quad f_i > 0 \, and \, 0 \le \mu_j < 1 \tag{7}$$

where y_n^R is the real responses at sensor n. $y_n^C\left(f_i, \mu_j\right)$ is the calculated responses at sensor n under given moving force f_i and damage factor μ_j. N is the number of sensors. If the problem is considered as a maximization problem, then Eq. (7) can be rewrote as:

$$\max \, val\left(f_i, \mu_j\right) = 1 - \frac{1}{N} \sum_n^N \frac{\left\| y_n^R - y_n^C\left(f_i, \mu_j\right)\right\|}{\left\| y_n^R \right\|};$$

$$subject \, to: \quad f_i > 0 \, and \, 0 \le \mu_j < 1 \tag{8}$$

Eq. (8) can be solved by the FA method. Actually, the magnitude of moving force f_i is considerably larger than damage factor μ_j, therefore, the searching space will be very 'long-narrow'. In order to improve the computational efficiency, a reference moving force f_{ref} is introduced into the objective function. The ratio $\lambda_i = f_i / f_{ref}$ is used as the optimization variable, thus, the objective function can be rewrote as:

$$\max \, val\left(\lambda_i, \mu_j\right) = 1 - \frac{1}{N} \sum_n^N \frac{\left\| y_n^R - y_n^C\left(\lambda_i, \mu_j\right)\right\|}{\left\| y_n^R \right\|};$$

$$subject \, to: \quad \lambda_i > 0 \, and \, 0 \le \mu_j < 1 \tag{9}$$

Some special aspects should be noted in implementation. 1) The displacement responses of structures contain system information mainly in the lower frequency bandwidth, while the acceleration responses contain system information in the higher frequency bandwidth. It should be noted that the magnitude of moving force does not vary with the time, therefore, the responses caused by the moving force are mainly in the lower frequency bandwidth. However, the damage factors are more sensitive to the higher frequency bandwidth than the lower one. It would be beneficial to use both the displacement and acceleration measurements. 2) The locations of damages are known before SDD and MFI, therefore, both levels of damage elements and magnitudes of moving forces will be identified by the proposed method. 3) The reference moving force f_{ref} should be estimated before SDD and MFI. For simplicity in implementation, one assumption can be used as: the displacement responses are caused by one moving force passing across the intact bridge. It means that the responses can be expressed as:

$$y^R = H(0) f_{ref} \qquad (10)$$

where y^R is real responses and $H(0)$ is system matrix. The reference moving force f_{ref} can be estimated as:

$$f_{ref} = H(0)^+ y^R \qquad (11)$$

Here, $H(0)^+$ is the Moore–Penrose pseudo inverse of $H(0)$. Generally, the optimization value λ_i can be taken as: $0 < \lambda_i \leq 2$.

4   Numerical Simulations

In order to check the correctness and effectiveness of the proposed method, a three-span continuous beam subjected to moving forces, as shown in Fig. 2(a), is taken an example for numerical simulations. The system is represented by a three-span continuous bridge with the following parameters: $L = 30\text{m} + 30\text{m} + 30\text{m}$, $EI = 1.274916 \times 10^{11} \text{N} \cdot \text{m}^2$, $\rho A = 12,000 \text{ kg} \cdot \text{m}^{-1}$. The bridge is equally divided into 18 elements. The first two damping ratio are both set to be 0.002. Rayleigh damping $C = \alpha M + \beta K$ is used, and assuming that α, β will not change for the damage structure.

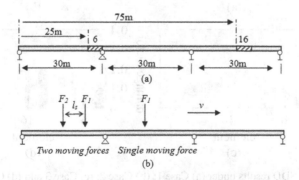

Fig. 2. A three-span continuous beam model

The real responses of structures can be calculated by the *Wilson-θ* method. Both displacement and acceleration responses at the middle of each span are used. The sampling frequency is 500Hz and 0%, 5%, 10% noise level are studied with: $y = y_c + E_p \times \sigma(y_c) \times N_{oise}$, where y_c, y are the responses with no noise and noise, respectively. E_p denotes noise level. $\sigma(y_c)$ is the standard deviations of y_c. N_{oise} is a standard normal distribution noise vector with zero mean value and unit standard deviation. Four cases are studied as follows (Fig. 2):

Case 1: single moving force with single damage.
  $\mu_6 = 10\%$; $F_1 = 35000\text{N}$; $v = 40\,\text{m/s}$;

Case 2: single moving force with two damages.
  $\mu_6 = 10\%$; $\mu_{16} = 20\%$; $F_1 = 35000\text{N}$; $v = 40\,\text{m/s}$;

Case 3: two moving forces with single damage.
  $\mu_6 = 10\%$; $F_1 = 15000\text{N}$; $F_2 = 20000\text{N}$; $v = 40\,\text{m/s}$; $l_s = 4\,\text{m}$;

Case 4: two moving forces with two damages.
  $\mu_6 = 10\%$; $\mu_{16} = 20\%$; $F_1 = 15000\text{N}$; $F_2 = 20000\text{N}$; $v = 40\,\text{m/s}$; $l_s = 4\text{m}$;

For both SDD and MFI, the FA is used with the parameters: number of fireflies $n = 10$, absorption coefficient $\gamma = 1$, original attractiveness $\beta_0 = 1$, step factor $\alpha = 0.05$ and max generation $\max G = 50$. The initial positions of fireflies are distributed randomly. The final identified result is taken by the mean value of 5-times running results.

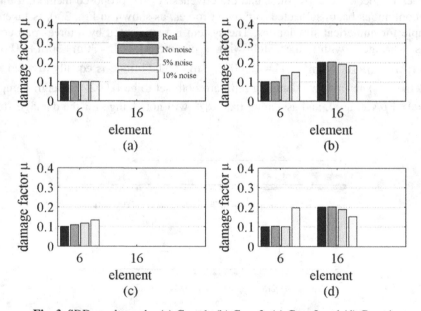

Fig. 3. SDD results under (a) Case 1; (b) Case 2; (c) Case 3 and (d) Case 4

The SDD results are shown in Fig. 3. It can be seen that: 1) Under two single moving force cases, i.e. Cases 1 and 2, the SDD results are accurate for single damage in Fig. 3(a) and two damage cases in Fig. 3(b), respectively. 2) Under two two-moving force cases, i.e. Cases 3 and 4, the SDD results from Figs. 3(c) and 3(d) are accurate except the white spike in Fig. 3(d), which is corresponding to the 10% noise level under Case 4. This mainly because that the searching space is too large, while the swarm size or the generations are not enough to enhance the convergence. Further, Fig. 4 shows the results when the swarm size and max generation are taken as: $n = 20$, $\max G = 80$, respectively. It clearly shows that the SDD results are accurate. 3) The identified damages will be affected by the noise level.

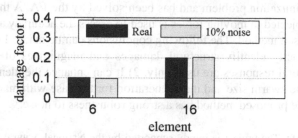

Fig. 4. SDD results under Case 4 with 10% noise level

The MFI results, as shown in Table 1, show that the forces identified by the proposed method from both the displacement and acceleration are accurate, except under Case 3 with noise level 5%. This mainly because that the searching space is too large, while the swarm size or the generations are not enough to enhance the convergence. If the max generation is set to be 80, the MFI results will be more accurate. The identified results are shown in Table 1 with bracket.

Table 1. MFI results under four cases

Case	Noise level	Moving forces	Real f_r (kN)	Identified f_i (kN)	Error $|f_r - f_i| / f_r (\%)$
1	0%	F1	35	35.01	0.03
	5%	F1	35	34.96	0.11
	10%	F1	35	35.08	0.23
2	0%	F1	35	34.97	0.09
	5%	F1	35	36.01	2.89
	10%	F1	35	33.98	2.91
3	0%	F1	15	14.76	1.60
		F2	20	19.98	0.10
	5%	**F1**	**15**	**18.37 (15.06)**	**22.47 (0.40)**
		F2	**20**	**24.89 (20.04)**	**24.45 (0.20)**
	10%	F1	15	13.42	10.53
		F2	20	19.38	3.10
4	0%	F1	15	15.71	4.73
		F2	20	20.65	3.25
	5%	F1	15	15.76	5.07
		F2	20	19.30	3.50
	10%	F1	15	15.24	1.60
		F2	20	19.69	1.55

5   Conclusions

A novel firefly algorithm (FA)-based method is proposed for structural damage detection (SDD) and moving force identification (MFI) in this paper. The basic principle of FA is introduced. The inverse problem on SDD and MFI is transformed into a constrained optimization problem and has been solved by the FA. A three-span continuous beam subjected to moving forces is used to assess the accuracy and the feasibility of the proposed method. The following conclusions can be made. 1) The FA-based method can exactly identify structural damages and magnitudes of moving forces when the structural responses are used only. 2) It can enhance the identified accuracy by increasing the swarm size and max generation for the case with many optimization variables. 3) The proposed method has a strong robustness to noise.

Acknowledgments. The project is jointly supported by the National Natural Science Foundation of China with Grant Numbers 51278226 and 11032005 respectively.

References

1. Zhang, Q., Jankowski, L., Duan, Z.: Simultaneous identification of excitation time histories and parameterized structural damages. Mech. Syst. Signal Process. **33**, 56–58 (2012)
2. Naseralavi, S.S., Salajegheh, E., Fadaee, M.J., Salajegheh, J.: A novel sensitivity-based method for damage detection of structures under unknown periodic excitations. J Sound Vib. **333**, 2776–2803 (2014)
3. Yang, X.-S.: Firefly algorithms for multimodal optimization. In: Watanabe, O., Zeugmann, T. (eds.) SAGA 2009. LNCS, vol. 5792, pp. 169–178. Springer, Heidelberg (2009)
4. Yu, L., Xu, P., Chen, X.: A SI-based algorithm for structural damage detection. In: Tan, Y., Shi, Y., Ji, Z. (eds.) ICSI 2012, Part I. LNCS, vol. 7331, pp. 21–28. Springer, Heidelberg (2012)
5. Yu, L., Li, C.: A global artificial fish swarm algorithm for structural damage detection. Adv. Struct. Eng. **17**(3), 331–346 (2014)
6. Yu, L., Xu, P.: Structural health monitoring based on continuous ACO method. Microelectron. Reliab. **51**(2), 270–278 (2011)
7. Yang, X.S., Hosseini, S.S.S., Gandomi, A.H.: Firefly algorithm for solving non-convex economic dispatch problems with valve loading effect. Appl. Soft Comput. J. **12**, 1180–1186 (2012)
8. Gandomi, A.H., Yang, X.S., Alavi, A.H.: Mixed variable structural optimization using firefly algorithm. Comput. Struct. **89**, 2325–2336 (2011)
9. Srivatsava, P.R., Mallikarjun, B., Yang, X.S.: Optimal test sequence generation using firefly algorithm. Swarm Evol. Comput. **8**, 44–53 (2013)
10. Pan, C.D., Yu, L.: Moving force identification based on firefly algorithm. Adv. Mater. Res. **919–921**, 329–333 (2014)
11. Yu, L., Chan, T.H.T.: Recent research on identification of moving loads on bridges. J Sound Vib. **305**, 3–21 (2007)

Unit Commitment with Electric Vehicles Based on an Improved Harmony Search Algorithm

Qun Niu[✉], Chao Wang, and Letian Zhang

Shanghai Key Laboratory of Power Station Automation Technology,
School of Mechatronic Engineering and Automation,
Shanghai University, Shanghai 200072, China
comelycc@hotmail.com

Abstract. This paper presents an intelligent economic operation of unit commitment (UC) incorporated with electric vehicles (EV) problem. The model of UC with EV is formulated, which includes constraints of power balance, spinning reserve, minimum up-down time, generation limits and EV limits. An improved harmony search, namely NPAHS-M, is proposed for UC problem with vehicle-to-grid (V2G) technology. This method contains a new pitch adjustment which can enhance the diversity of newly generated harmony and provide a better searching guidance. Simulation results show that EVs can reduce the running cost effectively and NPAHS-M can achieve comparable results compared with the methods in literatures.

Keywords: Unit commitment · Electric vehicles · Vehicle-to-grid · Harmony search

1 Introduction

Unit Commitment (UC), which is a key step for power system operation, refers to the optimization problem of determining the on/off states and real power outputs of thermal units that minimize the operation cost over a time horizon, however, energy crisis and environment problems arises at the same time [1]. Therefore, electric vehicle (EV), which is economical and eco-friendly, attracts more and more countries to develop it. Vehicle-to-grid (V2G), as an important technology in this field, makes EV become a portable distributed power facility to give feedback to grid when needed and increases the flexibility for the power gird to better utilize the renewable energy sources. Incorporating EV to UC lightens the load of thermal units so that the operation cost and emission is reduced, however, the introduction of EV also brings more constraints such as minimum and maximum number of charging/discharging vehicle limit, state of charge and inverter efficiency. This makes the scheduling of UC-EV become more complex. If a reasonable scheduling can reduce the cost by little as 0.5%, millions of dollars can be saved per year for large utilities [2].Therefore, it is meaningful to investigate UC incorporated with EV problem.

In the past few decades, various approaches were proposed to solve unit commitment problems. Unit commitment, as a NP-hard, large scale and complex mixed

© Springer International Publishing Switzerland 2015
Y. Tan et al. (Eds.): ICSI-CCI 2015, Part I, LNCS 9140, pp. 65–73, 2015.
DOI: 10.1007/978-3-319-20466-6_7

integer combinatorial optimization problem, is addressed mostly by classic methods [3-5] and intelligent evolutionary techniques [6-8]. The nature and biologically inspired techniques always provide a better way for reaching the optimal solutions than traditional methods. With economic and environmental problems concentrating more and more attention, some researches about unit commitment incorporated with EV were presented. A model of unit commitment with vehicle-to-gird for minimizing total running cost in addition to environmental emissions was formulated in [9]. Ahmed et al. [10] presented a cyber-physical energy system containing EVs, renewable energy and conventional thermal units. In [11], load-leveling model and smart grid model for gridable-vehicles were optimized with an improved PSO. Shantanu et al. [12] showed a fuzzy-logic-based intelligent operational economic strategy for smart grid. In [13], a UC model with integrating EVs as flexible load was designed to analyse the effects of EVs on generation side. Ehsan et al. [14] proposed a charging and discharging schedule of EVs and a combination of GA and LR was used to solve this problem. Although these methods were proposed, some new optimization approaches should be explored due to the complexity of the UC-EV problem.

In this paper, an improved harmony search (HS) with a new pitch adjustment called NAPHS-M is proposed for solving UC-EV problem, where it is assumed that EVs can be discharged via electric smart grid. UC-EV is a non-convex, non-linear and mixed integer optimization problem that has more variables and constraints than UC problem. NPAHS-M extends NPAHS [16] to the area of mixed integer programming problems. In NPAHS-M, the simplicity of HS is kept and exploration ability is improved. Considering the characteristic of UC-EV problem, integer handling mechanism is introduced in NPAHS-M. Analysis in two cases and comparisons between different algorithms are also presented.

2   UC-EV Problem Formulation

2.1   Objective Function

Unit commitment with electric vehicles is a cost and emission minimization problem that determines the on/off state and power outputs of all units and the number of discharging vehicles during each hour of the planning period T while considering a set of equality constraints and inequality constraints. The objective function consists of fuel costs, start-up costs and emission and it is formulated as follows [9]:

$$\min Obj = \sum_{t=1}^{T}\sum_{i=1}^{N}[\lambda_c(C_{fuel,i}+C_{start,i}(1-S_{i,t-1}))+\lambda_e epf_i E_{em,i}]S_{i,t} \tag{1}$$

where T is the number of hours and N is quantity of units. λ_c and λ_e are weight factors between cost and emission. epf_i is unit's emission plenty factor. $S_{i,t}$ represents the on-off state of ith unit in t hour. Similar to literature [14], emission is not considered in this paper so the value of λ_e is zero and the value of λ_c is one.

(a) Fuel costs: It is a quadratic function related to the output power of unit at each hour. a_i, b_i and c_i are the fuel cost coefficients of units. $P_{i,t}$ is the output power of ith unit in t hour.

$$C_{fuel,i} = a_i + b_i P_{i,t} + c_i (P_{i,t})^2 \tag{2}$$

(b) Start-up costs: Due to the temperature of boiler, hot-start cost and cold-start cost should all be considered. So the start-up cost is given as equation (3). $T_{off,i}^{\min}$ is the minimum shutdown time and $X_{off,i}$ is the continuous shutdown time. $cshour_i$ is the cold-start hour of unit.

$$C_{start,i} = \begin{cases} h\cos t_i & T_{off,i}^{\min} \le X_{off,i} \le H_{off,i} \\ c\cos t_i & X_{off,i} > H_{off,i} \end{cases} \tag{3}$$

$$H_{off,i} = T_{off,i}^{\min} + cshour_i$$

(c) Emission: α_i, β_i and γ_i are emission coefficients of units.

$$E_{em,i} = \alpha_i + \beta_i P_{i,t} + \gamma_i (P_{i,t})^2 \tag{4}$$

2.2   Constraints

Main constraints in this paper are power balance, spinning reserves, minimum up-down time, generation limits and EV limits. Prohibited operating zones and ramp rate limits are not considered. Constraints which should be met are given as follows:

(a) Power balance: At each hour, output power of on-state units with the sum of V2G power must satisfy the load demand named as PD_t. P_{EV} is the vehicle's average output power and N_{EV} is the number of discharging vehicles in period t.

$$\sum_{i=1}^{N} P_{i,t} S_{i,t} + P_{EV} N_{EV,t} = PD_t \tag{5}$$

(b) Spinning reserves: To maintain the system reliability, sufficient spinning reserves must be satisfied before considering power balance. $P_{i,t}^{\max}$ is the maximum output of unit i and $P_{EV}^{\max}$ is the vehicle's maximum output power. SR is spinning reserves rate and is 10% in this paper [15].

$$\sum_{i=1}^{N} P_{i,t}^{\max} S_{i,t} + P_{EV}^{\max} N_{EV,t} \ge PD_t + PR_t \tag{6}$$

$$PR_t = SR * PD_t$$

(c) Minimum up-down time: Estimating whether there are enough continuous on/off hours before changing an unit's state is important.

$$\begin{cases} X_{on,i} \geq T_{on,i}^{\min}(1 - S_{i,t+1}) & S_{i,t} = 1 \\ X_{off,i} \geq T_{off,i}^{\min} S_{i,t+1} & S_{i,t} = 0 \end{cases} \tag{7}$$

(d) Generation limits: Every unit should generate between its maximum output and minimum output, which is represented as equation (8).

$$P_{i,\min} \leq P_{i,t} \leq P_{i,\max} \tag{8}$$

(e) EV limits: Number of discharging electric vehicles has a range and sum of them in all hours must satisfy total vehicles as displayed in (9) and (10).

$$N_{EV,\min} \leq N_{EV,t} \leq N_{EV,\max} \tag{9}$$

$$\sum_{t=1}^{T} N_{EV,t} = N_{EV,total} \tag{10}$$

3 NPAHS-M for UC-EV

Harmony search (HS) originally proposed by Geem et al. imitates the music pitch adjustment process employed by musicians [15]. The structure of HS is simple and it has fewer control parameters. Although HS shows high performance on many optimization problems, its applications in mixed integer programming are fewer. Therefore, an improved HS is proposed for solving UC-EV problem in this paper.

NPAHS in [16] is a novel harmony search algorithm with new pitch adjustment rule, which is only used to solve continuous problem. In this paper, NPAHS-M is proposed to solve mixed integer programming problem based on NPAHS. For integer variables, a 0-1 variation mechanism [17] and integer processing are introduced. Continuous, 0-1 and integer variables are separated when updating. The new pitch adjustment rule combines random perturbation and mean value of HM in it, which can give better diversity to the new produced harmony and make the search direction go for better. NPAHS-M remains the simplicity and easy implementation with respect to the conventional HS. Meanwhile, no new parameters are introduced. The computational steps of the proposed NPAHS-M for UC-EV are as follows:

- **Step 1:** Input system data and parameters of algorithm HMS, HMCR, PAR, BW and Max iteration.
- **Step 2:** Initialize the state of units, output power of units, and number of discharging vehicles in harmony memory.

$$HM - state_{s,i*t} = S_{i,t} = \begin{cases} 1 & if \ rand < 0.5 \\ 0 & if \ rand \geq 0.5 \end{cases}, \ s \in HMS, i \in N, t \in T \tag{11}$$

$$HM - power_{s,i*t} = P_{i,t} = (P_{i,\min} + rand*(P_{i,\max} - P_{i,\min}))*S_{i,t} \tag{12}$$

$$HM - EV_{s,t} = N_{EV,t} = N_{EV,\min} + round((N_{EV,\max} - N_{EV,\min})*rand) \tag{13}$$

- **Step 3:** Adjust the $HM-EV$, $HM-state$ and $HM-power$ in turn to make them satisfy the electric vehicles limits, spinning reserves constraints, power balance constraints.
- **Step 4:** Calculate the objective value for $HM-EV$, $HM-state$ and $HM-power$ and pick out the global optimal solution.
- **Step 5:** Execute iteration and update the solution of UC-EV. The detailed process is presented as follows:

```
for j = 1 to N*T
  if rand ≤ HMCR
    r1 = ceil(rand * HMS);
```
$$state_{new}(1,j) = HM-state(r1,j);$$
```
    if rand ≤ PAR
```
$$state_{new}(1,j) = state_{new}(1,j) + BW*(2*rand-1)*(mean(HM\_$$
$$state(:,j)) - rand\ int)\ ;\qquad\qquad /\text{*new pitch adjustment*}/$$
$$EX = 1/(1+e^{-state_{new}(1,j)});$$
$$state_{new}(1,j) = \begin{cases} 1 & EX \geq 0.5 \\ 0 & EX < 0.5 \end{cases};\quad /\text{*0-1 variation mechanism*}/$$
```
    endif
  else
```
$$state_{new}(1,j) = rand\ int;$$
```
  endif
endfor
```

Generate $power_{new}$ within generation limits according to $state_{new}$ and then update $power_{new}$ by the same way of $state_{new}$ while the 0-1 variation mechanism is not used. Update EV_{new} by the same method of $power_{new}$. Then adjust EV_{new}, $power_{new}$ and $state_{new}$ for satisfying all constraints. Calculate the objective value of HM_{new} and update the harmony memory.

```
if obj(HM_new) < obj(HM_worst)
```
$$HM_{worst} = HM_{new};$$
```
else
```
$$HM = HM;$$
```
endif
```

- **Step 6:** If iteration number is within the max iteration number, return to step 5 for continuing.

4   Simulation and Discussion

A standard IEEE 10-unit system with 50000 EVs [14] was simulated on a personal computer with Intel Core i5-3470 CPU and 8G RAM. All the codes were compiled on MTLAB R2008b. The load demand, spinning reserve rate, characteristics of EV and the data of 10-unit system were got from [14]. It was assumed that electric vehicles were charged from solar, wind or other renewable energy, so the cost of vehicle to grid is ignored in this paper. The control parameters of NPAHS-M are HMS=5, HMCR=0.99, PAR=0.01 and BW=0.001, which are same with [16]. The number of maximum fitness evaluations FES is 12000 in this paper. All the experiments in this paper were executed 10 independent runs for an overall presentation.

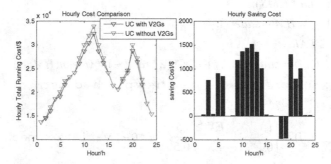

Fig. 1. Hourly cost comparison between UC with and without V2G

In this section, Results of unit commitment with and without V2G are all analyzed and the consequences of NPAHS-M are compared with other algorithms. Solution of 10-unit system with V2G is shown in Table 1. Due to the restriction of space, solution of 10-unit system without V2G is not listed. As shown in Fig. 1, UC with V2G can save considerable cost from an overall perspective especially in peak period. Total running cost of UC with V2G is \$551,715.4, and UC without V2G's is \$563,977.3. Therefore, \$12,261.9 is saved one day with the introduction of V2G. It follows that electric vehicles share a portion of power outputting and reduce heavily money every day.

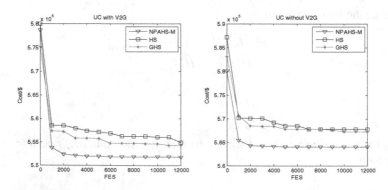

Fig. 2. Convergence curves of different methods in UC with and without V2G

Table 2 shows different results of UC with and without V2G obtained by traditional and intelligent optimization methods. ' - ' represents that results are not reported in the reference. As is shown in Table 2, performance of the proposed NPAHS-M is superior to PSO, GA-LR and LR while the FES of NPAHS-M is a little less than others'. According to compare the variance of these methods, it can be seen that NPAHS-M has a better stability for multiple runs. Basic HS [15] and a Global best harmony search (GHS) [18] are also compared and it presents that NPAHS-M can produce better results compared with different HS versions. Moreover, NPAHS-M also shows better results in dealing with UC without V2G problem.

Table 1. Best schedule and dispatch of generating units and V2G (Total Cost=$551,715.4)

T (h)	P1	P2	P3	P4	P5 (MW)	P6	P7	P8	P9	P10	No. of EVs	Hourly Cost ($)
1	455	245	0	0	0	0	0	0	0	0	18	13681.1
2	455	293	0	0	0	0	0	0	0	0	253	14526.4
3	455	381	0	0	0	0	0	0	0	0	2237	16052.3
4	455	455	0	0	38	0	0	0	0	0	293	18560.3
5	455	455	0	0	66	0	0	0	0	0	3802	19116.5
6	455	455	130	0	43	0	0	0	0	0	2745	21539.7
7	455	411	129	130	25	0	0	0	0	0	5	23261.9
8	455	455	130	130	30	0	0	0	0	0	77	24140.5
9	455	455	130	130	85	20	0	0	0	0	3955	26073.0
10	455	455	130	130	155	20	25	0	0	0	4785	28686.4
11	455	455	130	130	162	51	25	10	0	0	5000	30472.0
12	455	455	130	130	162	80	25	21	10	0	4983	32372.3
13	455	455	130	130	155	20	25	0	0	0	4666	28702.4
14	455	455	130	130	76	0	25	0	0	0	4569	26247.5
15	455	455	130	130	28	0	0	0	0	0	251	24118.4
16	455	310	130	130	25	0	0	0	0	0	123	21500.1
17	455	260	130	129	26	0	0	0	0	0	15	20644.0
18	455	360	130	110	25	20	0	0	0	0	21	22862.6
19	440	455	130	130	25	20	0	0	0	0	53	24613.8
20	455	455	130	130	158	20	25	0	0	0	4279	28754.2
21	455	455	130	130	86	0	25	0	0	0	2932	26459.7
22	455	455	0	0	136	0	25	0	0	0	4591	21725.8
23	455	419	0	0	25	0	0	0	0	0	151	17667.8
24	455	344	0	0	0	0	0	0	0	0	196	15405.5

Table 2. Comparison of total cost between proposed method and others

Total cost/$	Methods	PSO [10]	GA-LR [15]	LR	HS	GHS	NPAHS-M
(with EV)	Best	554,509	552,427	557,920	554,852	554,254	551,715
	Worst	559,987	553,765	560,208	556,747	555,514	552,024
	Ave	557,584	552,965	558,751	555,658	555,042	551,925
(without EV)	Best	563,741	-	565,825	567,797	566,723	563,977
	Worst	565,443	-	-	569,286	567,776	564,029
	Ave	564,743	564,703	-	568,459	567,286	564,007

5   Conclusion

This paper proposed NPAHS-M to solve unit commitment with vehicle-to-grid problem. Comparisons between UC with and without V2G show that EVs could bring enormous economic benefits for power system operation. By comparing the results of different intelligent and traditional methods, NPAHS-M shows the better global searching ability and effectiveness in solving this problem. In the future, NPAHS-M would be explored to solve more complex real world MILP problems.

Acknowledgements. This work is supported by the National Natural Science Foundation of China (61273040), and Shanghai Rising-Star Program (12QA1401100).

References

1. Yilmaz, M., Krein, P.T.: Review of the Impact of Vehicle-to-Grid Technologies on Distribution Systems and Utility Interfaces. IEEE Trans. Power Elec. **28**, 5637–5689 (2013)
2. Saber, A.Y., Venayagamoorthy, G.K.: Resource Scheduling Under Uncertainty in a Smart Grid with Renewables and Plug-in Vehicles. IEEE Systems Journal **6**, 103–109 (2012)
3. Ongsakul, W., Petcharaks, N.: Unit Commitment by Enchanced Adaptive Lagrangian Relaxation. IEEE Trans. Power Syst. **19**, 620–628 (2004)
4. Lowery, P.G.: Generating Unit Commitment by Dynamic Programming. IEEE Trans. Power Appa. Syst. **85**, 422–426 (1966)
5. Senjyu, T., Shimabukuro, K., Uezato, K., Funabashi, T.: A fast technique for unit commitment problem by extended priority list. In: T&D Conference and Exhibition 2002, pp. 244–249. IEEE press (2002)
6. Damousis, I.G., Bakirtzis, A.G., Dokopulos, P.S.: A Solution to the Unit-commitment Problem Using Integer-code Genetic Algorithm. IEEE Trans. Power Syst. **19**, 577–585 (2004)
7. Yuan, X., Su, A., Nie, H., Yuan, Y., Wang, L.: Unit Commitment Problem Using Enhanced Particle Swarm Optimization Algorithm. Soft Comput. **15**, 139–148 (2011)
8. Yuan, X., Su, A., Nie, H., Yuan, Y., Wang, L.: Application of Enhanced Discrete Differential Evolution Approach to Unit Commitment Problem. Energy Convers. Mange. **50**, 2449–2456 (2009)
9. Saber, A.Y., Venayagamoorthy, G.K.: Intelligent Unit Commitment with Vehicle-to-grid: A Cost-emission optimization. Journal of Power Sources **195**, 898–911 (2010)
10. Saber, A.Y., Venayagamoorthy, G.K.: Efficient Utilization of Renewable Energy Sources by Gridable Vehicles in Cyber-Physical Energy Systems. IEEE Systems Journal **4**, 285–294 (2010)
11. Saber, A.Y., Venayagamoorthy, G.K.: Plug-in Vehicles and Renewable Energy Sources for Cost and Emission Reductions. IEEE Trans. Industr. Electr. **58**, 1229–1238 (2011)
12. Chakraborty, S., Ito, T., Senjyu, T., Saber, A.Y.: Intelligent Economic Operation of Smart-Grid Facilitating Fuzzy Advanced Quantum Evolutionary Method. IEEE Trans. Sustain. Energy **4**, 905–916 (2013)
13. Madzharov, D., Delarue, E., D'haeseleer W.: Integrating Electric Vehicles as Flexible Load in Unit Commitment Modeling **66**, 285–294 (2014)

14. Talebizadeh, E., Rashidinejad, M., Abdollahi, A.: Evaluation of Plug-in Electric Vehicles Impact on Cost-based Unit Commitment. Journal of Power Sources **248**, 545–552 (2014)
15. Geem, Z.W., Kim, J.H., Loganathan, G.V.: A New Heuristic Optimization Algorithm: Harmony Search. Simulation **76**, 60–68 (2001)
16. Niu, Q., Zhang, H.Y., Li, K., Irwin, G.W.: An Efficient Harmony Search with New Pitch Adjustment for Dynamic Economic Dispatch. Energy **10**, 085 (2013)
17. Modiri, A., Kiasaleh, K.: Modification of Real-Number and Binary PSO Algorithms for Accelecrated Convergence. IEEE Trans. Antennas and Propag. **59**, 214–224 (2011)
18. Omran, M.G.H., Mahdavi, M.: Global-best Harmony Search. Appl. Math. Comput. **198**, 643–656 (2008)

A *Physarum*-Inspired Vacant-Particle Model with Shrinkage for Transport Network Design

Yuxin Liu[1], Chao Gao[1], Mingxin Liang[1], Li Tao[1], and Zili Zhang[1,2(✉)]

[1] School of Computer and Information Science, Southwest University,
Chongqing 400715, China
[2] School of Information Technology, Deakin University,
Geelong, VIC 3217, Australia
zhangzl@swu.edu.cn

Abstract. *Physarum* can form a higher efficient and stronger robust network in the processing of foraging. The vacant-particle model with shrinkage (VP-S model), which captures the relationship between the movement of *Physarum* and the process of network formation, can construct a network with a good balance between exploration and exploitation. In this paper, the VP-S model is applied to design a transport network. We compare the performance of the network designed based on the VP-S model with the real-world transport network in terms of average path length, network efficiency and topology robustness. Experimental results show that the network designed based on the VP-S model has better performance than the real-world transport network in all measurements. Our study indicates that the *Physarum*-inspired model can provide useful suggestions to the real-world transport network design.

Keywords: *Physarum polycephalum* · *Physarum*-inspired model · Transport network design · Network analysis

1 Introduction

Physarum polycephalum is a unicellular slime mold with many diploid nuclei. It has a complex life cycle. The vegetative phase of *Physarum* is the plasmodium, which is the most common form in nature. The plasmodium lives in a dark and moist environment and feeds on the bacteria or fungal spores [1]. The intelligence of the plasmodium lies that it can form a high efficient, self-adaptive and robust network connecting distributed food sources without central consciousness [2]. Nakagaki et al. [3] have evaluated the characteristics of networks generated by the plasmodium based on some measurements. Statistical analyses in response to several different arrangements of food sources show that networks generated by the plasmodium have short total length, close connections among any two food sources and high robustness of accidental disconnection of tubes [4–7].

Inspired by the self-adaptive and self-organization abilities of *Physarum*, researchers have proposed lots of effective computing models [8–13]. In particular, Gunji et al. [14] have proposed a vacant-particle model with shrinkage

© Springer International Publishing Switzerland 2015
Y. Tan et al. (Eds.): ICSI-CCI 2015, Part I, LNCS 9140, pp. 74–81, 2015.
DOI: 10.1007/978-3-319-20466-6_8

(shorted as VP-S model) to explore the relationship between the process of network formation and the movement of *Physarum* for the first time. The VP-S model evolves by updating the grid state and can be thought of as a generalized cellular automaton. It focuses on the construction of complex networks with a good trade-off between exploration and exploitation. In this paper, we utilize the VP-S model to solve network design problems. We compare the network designed based on the VP-S model (shorted as VP-S network) with the real-world transport network (shorted as real network) in terms of average path length, network efficiency and topology robustness.

The rest of this paper is organized as follows. Sect. 2 presents the working mechanism of the VP-S model. Sect. 3 compares the performance of the VP-S network with the real network based on some measurements. Sect. 4 summarizes the main results and contributions.

2 The Vacant-Particle Model with Shrinkage

2.1 Basic Idea of the VP-S Model

The VP-S model uses an agent-based system to simulate the evolution mechanism of *Physarum* (Fig. 1), which consists of three parts: environment, agent and its behaviors. An environment represents a culture dish with $M \times N$ planar grids. Each grid has four neighbors. The regions corresponding to the positions of food sources are defined as active zones, which are some grids surrounded by dotted lines in Fig. 1(a). During the evolving process of the VP-S model, grids are divided into two groups: internal grids and external grids. An agent has three behaviors: generation, movement and adaptation. Fig. 1 shows a life-cycle of an agent in an environment, which can be described as follows.

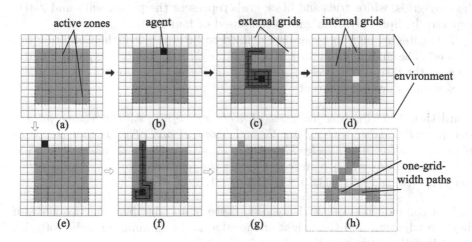

Fig. 1. The environment of the VP-S model and the life cycle of an agent in the environment

Step 1 Generation behavior: An agent is randomly generated from active zones (Fig. 1(b)) with probability Q ($0<Q<1$) or it is randomly generated from external grids that are adjacent to one-grid-width paths with probability $1-Q$. One-grid-width paths are consisted of internal grids whose north-south or east-west neighbors are external grids. Especially, if an environment doesn't have one-grid-width paths, an agent is randomly generated from external grids that are adjacent to internal girds (Fig. 1(e)) with probability $1-Q$.

Step 2 Movement behavior: The agent randomly chooses an internal grid from its neighbors to reside, which has not been resided by itself (Fig. 1(c) and Fig. 1(f)). Step 2 is iterated until there is no internal grid for the agent to reside.

Step 3 Adaptation behavior: If the agent has been generated from external grids, it updates the state of the initial grid from external to internal (Fig. 1(g)). Otherwise the state of the initial grid is internal, which will not be changed (Fig. 1(d)). If the agent does not stop in active zones, it adapts the state of the current grid to external (Fig. 1(d)). Otherwise the state of the current grid will not be changed (Fig. 1(g)). After that, a new agent is randomly generated again.

Adaptation behavior makes sure that the internal grids in active zones will never lost, which implies that there is highly concentrated protoplasm of *Physarum* at food sources [14]. These three steps are repeated until a route connecting active zones appears (e.g., Fig. 2(h)). The final route is consisted of internal grids and approximates an efficient network produced by *Physarum*.

However, it's hard to measure the characteristics of a VP-S network since it is composed of grids. In Sect. 2.2, we present how to transfer a disordered VP-S network to a common network with weights.

2.2   Transformation Strategy

The VP-S network can be regarded as a maze. Taking Fig. 2(a) as an example, the gray grids, white grids and black grids represent the paths, walls and gates, respectively. Specially, each gate is composed of four grids.

Two gates x and y in a maze are defined to be adjacent if the Condition 1 is satisfied. The minimum number of grids x_1, x_2,..., x_n connecting gates x and y is set as the distance between gates x and y. If the Condition 1 is not satified, gates x and y are not adjacent.

Condition 1 *There is a finite set of grids x_1, x_2,..., x_n connecting gates x and y, and grids x_1 and x_n are the neighbors of one of grids that composed of gates x and y, respectively. Meanwhile, $x_1 \neq x_2 \neq ... \neq x_n$ and $x_i = x_{i-1} + f(x_{i-1}, k_{i-1})$, where k_{i-1} represents whether the four neighbors of x_{i-1} are paths, and $f(x_{i-1}, k_{i-1}) \rightarrow \{(0,1), (0,-1), (1,0), (-1,0)\}$.*

In a common network with weights, there is an edge between two nodes if they are adjacent, and the weight of the edge is the minimum number of grids connecting these two nodes, as shown in Fig. 2(b).

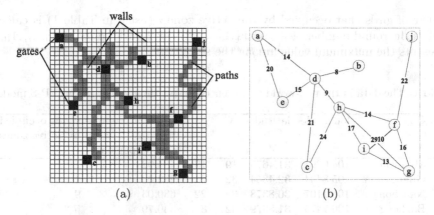

Fig. 2. (a) The VP-S network, which can be seemed as a maze. (b) The common network with weights based on (a).

3   Empirical Study

3.1   The Experimental Setup

In this section, the VP-S model is used to design the Sichuan transport network. The internal grids of the VP-S model is initialized as a square aggregation consisting of 53×53 grids. Each grid has the size of 10×10 square-millimeters. Active zones in the VP-S model are corresponding to cities in the real-world map. Hence, the positions of active zones in the environment are setup according to the geographical locations of cities, i.e., we first get the longitude and latitude of each city in the real-world from Google Maps. Then, we convert the longitudes and latitudes to screen coordinates (X, Y) according to (1) and (2),

$$X = \frac{(lon - minLon) \times w}{maxLon - minLon} \tag{1}$$

$$Y = \frac{(maxLat - lat) \times h}{maxLat - minLat} \tag{2}$$

where *lon* is a city's lonitude, and *lat* is a city's latitude. *minLon*, *maxLon*, *minLat* and *maxLat* are the minimum longitude, the maximum longitude, the minimum latitude and the maximum latitude among the set, respectively. w is the width of the screen, and h is the height of the screen, they are both set as 50 in the experiments. Lists from 3 to 6 in Table 1 show the longitudes, latitudes and screen coordinates (X, Y) of 17 cities in Sichuan province, respectively.

How many grids that an active zones occupied are set based on the resident population in the city. We acquire the resident population of each city in Sichuan province from the website *Sichuan Statistics Yearbook*[1], as shown in the list 7 in Table 1. Let A_i represent the resident population in a city i ($i \in [0, 16]$). The

---

[1] http://www.sc.stats.gov.cn/tjcbw/tjnj/2013/index.htm

number of girds that occupied by the active zone i (list 8 in Table 1) is calculated as the round number of $A_i/max\{A_0, ..., A_{16}\} \times 20$, where $max\{A_0, ..., A_{16}\}$ represents the maximum value among the set.

Table 1. The data in the real-world and their corresponding values in the VP-S model

NO.	City	longitudes	latitudes	X	Y	Resident Population (2012 year) (10000 persons)	Grids occupied by active zones
0	Mianyang	104.6791	31.4675	19	13	464.02	7
1	Guangyuan	105.8434	32.4354	32	0	253.00	4
2	Nanchong	106.1107	30.8378	35	22	630.03	9
3	Bazhong	106.7475	31.8679	42	8	330.79	5
4	Dazhou	107.4680	31.2096	50	17	549.27	8
5	Ya'an	103.0133	29.9805	0	33	152.65	2
6	Luzhou	105.4423	28.8718	27	48	425.00	6
7	Deyang	104.3979	31.1269	16	18	353.13	5
8	Suining	105.5929	30.5328	29	26	326.77	5
9	Guangan	106.6332	30.4560	41	27	321.64	5
10	Meishan	103.8485	30.0754	9	32	296.64	4
11	Ziyang	104.6276	30.1289	18	31	358.85	5
12	Leshan	103.7656	29.5521	8	39	325.44	5
13	Neijiang	105.0584	29.5802	23	39	371.81	5
14	Zigong	104.7784	29.3390	20	42	271.82	4
15	Yibin	104.6434	28.7518	18	50	446.00	6
16	Chengdu	104.0665	30.5723	12	25	1417.78	20

The distance of each path in the real network is measured from Google maps. Then, a traditional data normalization method is used to unify the dimension. That is, $y(k) = \frac{x(k) - min(x(n))}{x(k)}$, where $k = 1, 2, ..., n$, and $min(x(n))$ means the minimum value among the data set that to be normalized. $x(k)$ is the original data and $y(k)$ is the value of $x(k)$ after being normalized.

3.2 Evaluation Measurements

For a comlete weighted network $G = (V, L)$, let $V = \{1, 2, ..., n\}$ represent the set of N nodes, $L = \{(i, j)|i, j \in V, i \neq j\}$ represent the set of edges. Then, we introduce three measurements to evaluate the characteristics of the network G.

(i) Average path length (APL). APL is calculated based on (3). The distance d_{ij} between two nodes i and j in a network is defined as the shortest path length connecting two nodes.

$$\text{APL} = \frac{1}{N(N-1)} \sum\nolimits_{i \neq j} d_{ij}, (i, j \in V) \tag{3}$$

(ii) Network efficiency (NE). NE defines the average proximity between all nodes in a network, which is defined as (4).

$$NE = \frac{1}{N(N-1)} \sum_{i \neq j} \frac{1}{d_{ij}}, (i, j \in V) \tag{4}$$

(iii) Topological robustness (TR). TR is defined as (5):

$$TR = \sum_{k=1}^{N} s(k) \tag{5}$$

where $s(k)$ is the fraction of nodes in the largest connected subgraph after removing k nodes. This measure of robustness considers the size of the largest component during all possible attacks.

3.3   Experimental Comparison and Analysis

The VP-S model has been run 30 times. In each run, we construct a VP-S network. What's more, we have built the VP-S network with probability p. For example, the VP-S network with probability 0.6 means that edges which occurred in over 60% of experiments be reserved.

Firstly, Fig. 3(a) plots the average path length (APL) against the network efficiency (NE). We select 10 VP-S networks for the clear presentation. As shown in Fig. 3(a), APL of VP-S networks are all shorter than that of the real network, and NE of VP-S networks are no less than that of the real network. These mean that VP-S networks will cost less time if traverse each pair of cities than that of the real network.

Secondly, Fig. 3(b) plots the average path length (APL) against the topological robustness (TR) under malicious attacks. To simulate a malicious attack, we first remove the most connected nodes, and continue selecting and removing nodes in decreasing order of their connectivity [15]. As shown in Fig. 3(b), APL

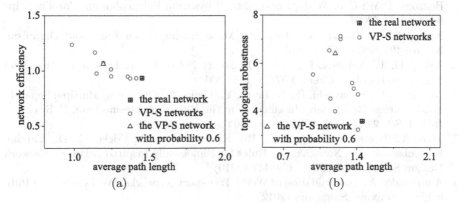

Fig. 3. The performance comparison among the real network, 10 VP-S networks and the VP-S network with probability 0.6: (a) APL against NE and (b) APL against TR

and TR of the real network are 1.45 and 3.59, respectively. APL and TR of the VP-S network with probability 0.6 are 1.19 and 6.41, respectively. That is, the VP-S model can construct network whose APL is shorter than that of the real network. In particular, TR of the VP network is more than 1.78 times higher than that of the real network.

Based on the above experiments, we can conclude that the performance of the VP-S model network is better than the real network by comparing APL, NE and TR. The results show that the construction of high performance transport network based on the VP-S model is a working approach.

4 Conclusion

In this paper, we proposed to use the *Physarum*-inspired model, i.e., the VP-S model, for the transport network design. With regard to the disorder of a VP-S network, we presented how to transfer it to a common network with weights. The experimental results showed that the network designed based on the VP-S model has shorter average length, higher efficiency and stronger robustness than the real network. Our study showed that it might be worth to consider the *Physarum*-inspired model for the problem of designing high performance transport networks.

Acknowledgments. This work was supported by National High Technology Research and Development Program of China (No. 2013AA013801), National Science and Technology Support Program (No. 2012BAD35B08), National Natural Science Foundation of China (Nos. 61402379, 61403315), Natural Science Foundation of Chongqing (Nos. cstc2012jjA40013, cstc2013jcyjA40022), and Research Fund for the Doctoral Program of Higher Education (20120182120016).

References

1. Bozzone, D.M.: Cells With "Personality" Physarum Polycephalum. Carolina Tips **64**(3), 9–11 (2001)
2. Nakagaki, T., Yamada, H., Tóth, A.: Maze-Solving by an Amoeboid Organism. Nature **407**(6803), 470 (2000)
3. Nakagaki, T., Yamada, H., Hara, M.: Smart Network Solutions in an Amoeboid Organism. Biophys. Chem. **107**(1), 1–5 (2004)
4. Nakagaki, T., Kobayashi, R., Nishiura, Y., Ueda, T.: Obtaining Multiple Separate Food Sources: Behavioural Intelligence in the Physarum plasmodium. P. Roy. Soc. B-Biol. Sci. **271**(1554), 2305–2310 (2004)
5. Tero, A., Takagi, S., Saigusa, T., Ito, K., Bebber, D.P., Fricker, M.D., Yumiki, K., Kobayashi, R., Nakagaki, T.: Rules for Biologically Inspired Adaptive Network Design. Science **327**(5964), 439–442 (2010)
6. Adamatzky, A.: Bioevaluation of World Transport Networks. World Scientific Publishing Company, Singapore (2012)
7. Reid, C.R., Beekman, M.: Solving the Towers of HanoiHow an Amoeboid Organism Efficiently Constructs Transport Networks. J. Exp. Biol. **216**(9), 1546–1551 (2013)

8. Tero, A., Kobayashi, R., Nakagaki, T.: A Mathematical Model for Adaptive Transport Network in Path Finding by True Slime Mold. J. Theor. Biol. **244**(4), 553–564 (2007)
9. Adamatzky, A.: Physarum Machines: Encapsulating Reaction Diffusion to Compute Spanning Tree. Naturwissenschaften **94**(12), 975–980 (2007)
10. Jones, J.: The Emergence and Dynamical Evolution of Complex Transport Networks from Simple Low-Level Behaviours. Int. J. Unconv. Comput. **6**(2), 125–144 (2010)
11. Gunji, Y.P., Shirakawa, T., Niizato, T., Haruna, T.: Minimal Model of a Cell Connecting Amoebic Motion and Adaptive Transport Networks. J. Theor. Biol. **253**(4), 659–667 (2008)
12. Tsompanas, M.-A.I., Sirakoulis, G.C., Adamatzky, A.I.: Evolving Transport Networks with Cellular Automata Models Inspired by Slime Mould. IEEE Trans. Cybern. **PP**(99), 1–13 (2014)
13. Aono, M., Hara, M.: Spontaneous Deadlock Breaking on Amoeba-Based Neurocomputer. Biosystems **91**(1), 83–93 (2008)
14. Gunji, Y.P., Shirakawa, T., Niizato, T., Yamachiyo, M., Tani, I.: An Adaptive and Robust Biological Network Based on the Vacant-particle Transportation Model. J. Theor. Biol. **272**(1), 187–200 (2011)
15. Schneider, C.M., Moreira, A.A., Andrade, J.S., Havlin, S., Herrmann, H.J.: Mitigation of Malicious Attacks on Networks. PNAS **108**(10), 3838–3841 (2011)

Bean Optimization Algorithm Based on Negative Binomial Distribution

Tinghao Feng[1], Qilian Xie[2], Haiying Hu[1], Liangtu Song[1],
Chaoyuan Cui[1], and Xiaoming Zhang[1(✉)]

[1] Institute of Intelligent Machines,
Chinese Academy of Sciences, Hefei 230031, China
xmzhang@iim.ac.cn
[2] Electronic and Information Engineering School,
Anhui University, Hefei 230031, China

Abstract. Many complex self-adaptive phenomena in the nature often give us inspirations. Some scholars are inspired from these natural bio-based phenomena and proposed many nature-inspired optimization algorithms. When solving some complex problems which cannot be solved by the traditional optimization algorithms easily, the nature-inspired optimization algorithms have their unique advantages. Inspired by the transmission mode of seeds, a novel evolutionary algorithm named Bean Optimization Algorithm (BOA) is proposed, which can be used to solve complex optimization problems by simulating the adaptive phenomenon of plants in the nature. BOA is the combination of nature evolutionary tactic and limited random search. It has stable robust behavior on explored tests and stands out as a promising alternative to existing optimization methods for engineering designs or applications. Through research and study on the relevant research results of biostatistics, a novel distribution model of population evolution for BOA is built. This model is based on the negative binomial distribution. Then a kind of novel BOA algorithm is presented based on the distribution models. In order to verify the validity of the Bean Optimization Algorithm based on negative binomial distribution model (NBOA), function optimization experiments are carried out, which include four typical benchmark functions. The results of the experiments are made a comparative analysis with that of particle swarm optimization (PSO) and BOA. From the results analysis, we can see that the performance of NBOA is better than that of PSO and BOA. We also conduct a research on the characters of NBOA. A contrast analysis is carried out to verify the research conclusions about the relations between the algorithm parameters and its performance.

Keywords: Swarm intelligence · Bean optimization algorithm · Negative binomial distribution · Function optimization

1 Introduction

Many complex self-adaptive phenomena in nature often give us inspirations. For example, organisms and natural ecosystems can solve many highly complex

© Springer International Publishing Switzerland 2015
Y. Tan et al. (Eds.): ICSI-CCI 2015, Part I, LNCS 9140, pp. 82–88, 2015.
DOI: 10.1007/978-3-319-20466-6_9

optimization problems through their own evolutions. Some scholars have been inspired from these natural phenomena and many nature-inspired optimization algorithms have been proposed to solve complex optimization problems. The idea of the novel optimization algorithms which simulated the natural ecosystem mechanisms is different from the idea of the classic optimization algorithms. Their appearances greatly enriched the optimization technology and brought new life and hope for the solution of complex optimization problems which traditional optimization methods are difficult to deal with.

Nature-inspired optimization algorithm refers to the computing technology and algorithms which based on the functions, characteristics and mechanism of the nature to solve the optimization problems, such as genetic algorithms (GA) [1], particle swarm optimization (PSO) [2], ant colony optimization (ACO) [16], artificial fish-swarm algorithm [3], free search algorithm [4], human evolution model algorithm [5], group search optimization algorithm [6], Bees Algorithm [17] etc. Because the structure of nature biology is complex and sophisticated, they have a high degree of adaptive capacity and strong collaborative capabilities both in the evolutions and behaviors. Through collaboration, they can get the best environment for survival. Therefore, most of the nature-inspired algorithms have the character of self-organizing, self-adaptive and self-learning. When solving some complex problems which the traditional optimization algorithm cannot solve easily, the nature-inspired optimization algorithms have its own unique advantages. At present, the nature-inspired optimization algorithms have been used to solve complex optimization problems in many fields successfully, for example in task assignment [7], classification [8], and gene selection [9].

In 2008, inspired by the transmission mode of seeds, a novel evolutionary algorithm named Bean Optimization Algorithm (BOA) is proposed [10] [18] [19], which can be used to solve complex optimization problems by simulating the adaptive phenomenon of plants in nature. BOA is the combination of nature evolutionary tactic and limited random search. It has a stable robust behavior on explored tests and stands out as a promising alternative to existing optimization methods for engineering designs or applications. At present, two algorithm models have been constructed for BOA, including piecewise function model and normal distribution model. BOA has been successfully applied in solving TSP [11] [12], materials scheduling [13], earthquake recovery and reconstruction planning of China [14].

2   Overview of the Negative Binomial Distribution

In probability theory and statistics, the negative binomial distribution is a discrete probability distribution of the times of successes in a sequence of Bernoulli trials before a specified (non-random) number of failures occur. Now we give the definition of negative binomial distribution. Suppose there is a sequence of independent Bernoulli trials, each trial having two potential outcomes called "success" and "failure". In each trial the probability of success is p and of failure is $(1 - p)$. We are observing this sequence until a predefined number r of successes has occurred. Then the random number of successes we have seen, f, will have the negative binomial distribution:

$$f(k,r,p) = \binom{k+r-1}{r-1} \cdot p^r \cdot (1-p)^k \tag{1}$$

It is possible to extend the definition of the negative binomial distribution to the case of a positive real parameter r. Although it is impossible to visualize a non-integer number of failures, we can still formally define the distribution through its probability mass function.

The main characteristics of negative binomial distribution are listed in table 1.

Table 1. Characteristics of Negative Binomial Distribution

Parameters	$r > 0$
	$0 \le p \le 1$
Support	$k \in \{0, 1, 2, \dots\}$
Mean	$r \cdot \dfrac{(1-p)}{p}$
Variance	$r \cdot \dfrac{(1-p)}{p^2}$

Negative binomial distribution is relatively common in the biological cluster distribution. It can be used to reflect the aggregations (like groups, clusters, or the plaque aggregations) of individuals in biological populations. In the nature, due to the environmental heterogeneity and biological clustering, the spatial distributions of most of the animals (especially insects) and a large proportion of plants are often negative binomial distributions [15].

3   Population Distribution Model Based on the Negative Binomial Distribution

The main idea of the population distribution model based on the negative binomial distribution is that each offspring individual distributes around the father bean according to the negative binomial distribution. The number of successes of the negative binomial distribution which the distribution pattern of individuals uses is according to the value of the current father bean's position coordinate. The setting of corresponding probability of successes changes dynamically according to the domain of target problem and the solving progress of the problem. If the number of success times is set to be the value of the current father bean's position coordinate, the corresponding probability of successes can be set 0.5. Similarly, the worst result of the father bean get, the smaller population size of the corresponding offspring is.

In the actual design of the model, we combine the negative binomial distribution model with the normal distribution model. The threshold of population reproduction should be set first. Before reaching the threshold value, bean population will multiply in line with the normal distribution. Otherwise, they will multiply in line with the negative binomial distribution.

The flowchart of parent beans generating offsprings is shown as follows :

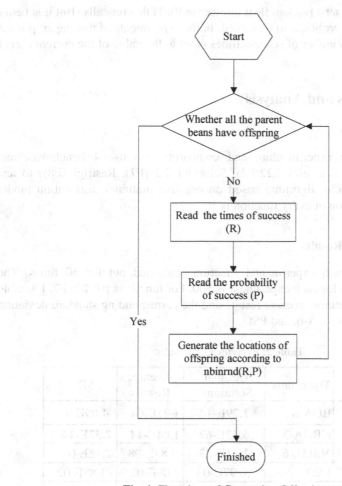

Fig. 1. Flowchart of Generating Offsprings

Because negative binomial distribution is a kind of discrete distribution and the problems to be solved by bean optimization algorithm are always continuous, so when constructing the population distribution patterns which are based on the negative binomial distribution, we need to do the following process:

Based on nbinrnd (R, P) in MATLAB (nbinrnd (R, P) is a matrix of random number generated by negative binomial distribution in line with the parameters R and P), parameter R is the times of successes, and P is the probability of success in a single experiment. Both R and P may be vectors, matrix, or multidimensional arrays with the same dimension.

To generate each parent bean's offsprings according with negative binominal distribution, it should be processed as follows:

$$R = abs(FthPop(1,:)*(100/FthPop(1))); \tag{2}$$

Parameter P can be set a random float number in [0, 1] theoretically. But it is better to set according to the problems to be solved. In the experiments of this paper, p is set to be 0.5, because the number of success times is set to the value of the current parent bean's position.

4   Experiments and Analysis

4.1   Test Functions

In order to facilitate experimenting and comparing, we use 4 benchmark test functions(Sphere(F1), Schwefel 2.22(F2), Schwefel 2.26(F7), Rastrigin(F8)) to test the performance of BOA algorithm based on negative binomial distribution model (NBOA). The dimension of every function is 30.

4.2   Experimental Results

Each algorithm for each experimental function is carried out for 50 times. The number of generations for each experiment is 500. For functions F1, F2, F7, F8, table 2 lists the optimal solutions, average results and the corresponding standard deviation of BOA-6, NBOA-3, NBOA-6, and PSO.

Table 2. Experimental Results

Experimental Functions	Algorithms	Optimal Solutions	Average Results	SD
F_1	BOA-6	1.29E-14	1.21E-04	4.09E-4
	NBOA-3	**3.42E-67**	**1.04E-14**	**7.37E-14**
	NBOA-6	1.31E-33	3.84E-08	2.18E-07
	PSO	6.97E-03	3.09E-02	1.88E-02
F_2	BOA-6	2.42E-12	3.45E-04	2.01E-3
	NBOA-3	**5.21E-68**	**8.44E-17**	**5.97E-16**
	NBOA-6	8.28E-21	1.81E-07	9.12E-07
	PSO	5.99E-01	1.99	1.04
F_7	BOA-6	-9.091E3	**-7.94E+03**	6.41E2
	NBOA-3	-8.99E3	-7.28E+03	7.04E2
	NBOA-6	**-1.11E4**	-7.80E+03	7.55E2
	PSO	-7.69E03	-6.09E03	**6.39E02**
F_8	BOA-6	1.61E-08	1.62E-03	3.07E-3
	NBOA-3	2.11E-07	8.04E-04	1.97E-3
	NBOA-6	**1.36E-09**	**5.27E-04**	**1.41E-3**
	PSO	9.63	2.79E01	8.74

4.3   Experiment Analysis

We will analyze the experimental results from two aspects as follows.
(1) Optimal Results
   Optimal result is the best result of the algorithm running 50 times. SD in Table 7 is standard deviation. It measures the amount of variation from the average result. We can see from the above results in table 2 that NBOA-3 obtains the relatively optimal values of the two functions: F1 and F2. NBOA6 obtains relatively optimal value of two functions: F7 and F8. The performance of BOA-6 is relatively poor. But compared with PSO and BOA-3 [15], there is an evident progress.
(2) Average Results
   Average result is the average result of the algorithm running 50 times. Considering the average results of the experiments, NBOA-3 obtains the best average results of the two functions: F1, F2. NBOA-6 obtains the best average results of function F8. BOA-6 obtains the best average results of function F7.

5   Summaries

Inspired by the transmission mode of seeds in nature and the adaptive phenomenon of plants, a novel evolutionary algorithm named BOA is proposed. This paper firstly introduces a discrete distribution named negative binomial distribution. By processing the distribution variable, the distribution can be converted into a model of population distribution for solving the continuous optimization problems. Combined with the normal distribution, a novel kind of BOA named NBOA is proposed. Four benchmark function optimization experiments are carried out to test its performance. By comparing and analysis of experimental results of four algorithms, we can see that the negative binomial distribution model is effective.

   In the research of BOA, three kinds of population distribution models have been designed to improve the performance of BOA algorithm. But in the area of Biostatistics research, this is only a very small part of numerous natural biological population distributions. There are lots of distribution functions which can be used to construct population distribution models of BOA, such as Neyman distribution, Polya-Eggenberger distribution. In addition, at present, the study of combining with other intelligent algorithm is very popular. But it is not carried out in BOA, such as individual cross thought based on GA, chaos theory. In the future research, we will continue to explore BOA. We also hope researchers to pay more attention to BOA and carry out related research work.

Acknowledgment. This work was supported by the National Science Foundation of China under Grant 61203373.

References

1. Holland, J.H.: Adaption in Natural and Artificial Systems. The MIT Press, Cambridge (1992)
2. Kennedy, J., Eberhart R.C.: Particle swarm optimisation. In: Proceedings of IEEE International Conference on Neural Networks, pp. 1942–1948 (1995)
3. Xiaolei, L., Zhijiang, S., Jixin, Q.: An Optimizing Method Based on Autonomous Animats: Fish-swarm Algorithm. Systems Engineering Theory & Practice 22(11), 32–38 (2002)
4. Kalin, P., Guy, L.: Free Search-a comparative analysis. Information Sciences 172(1–2), 173–193 (2005)
5. Montiela, O., Castillob, O., Melinb, P., Díazc, A.R., Sepúlvedaa, R.: Human evolutionary model: A new approach to optimization. Information Sciences 177(10), 2075–2098 (2007)
6. He, S., Wu, Q.H., Saunders, J.R.: Group Search Optimizer: An Optimization Algorithm Inspired by Animal Searching Behavior. IEEE Transactions on Evolutionary Computation 13(5), 973–990 (2009)
7. Ho, S.-Y., Lin, H.-S., Liauh, W.-H., Ho, S.-J.: OPSO: Orthogonal Particle Swarm Optimization and Its Application to Task Assignment Problems. IEEE Transactions on Systems, Man and Cybernetics, Part A 38(2), 288–298 (2008)
8. Souda, T., Silva, A., Neves, A.: Particle Swarm based Data Mining Algorithms for classification task. Parallel Computing 30(5), 767–783 (2004)
9. Li, S., Xixian, W., Tan, M.: Gene selection using hybrid particle swarm optimization and genetic algorithm. Soft Computing 12(11), 1039–1048 (2008)
10. Zhang, X., Wang, R., Song, L.: A Novel Evolutionary Algorithm—Seed Optimization Algorithm. Pattern Recognition And Artificial Intelligence 21(5), 677–681 (2008)
11. Li, Y.: Solving TSP by an ACO-and-BOA-based hybrid algorithm. In: Proceedings of 2010 International Conference on Computer Application and System Modeling, pp. 189–192 (2010)
12. Zhang, X., Jiang, K., Wang, H., Li, W., Sun, B.: An improved bean optimization algorithm for solving TSP. In: Tan, Y., Shi, Y., Ji, Z. (eds.) ICSI 2012, Part I. LNCS, vol. 7331, pp. 261–267. Springer, Heidelberg (2012)
13. Wang, P., Cheng, Y.: Relief Supplies Scheduling Based on Bean Optimization Algorithm. Economic Research Guide 8, 252–253 (2010)
14. Zhang, X., Sun, B., Mei, T., Wang, R.: Post-disaster restoration based on fuzzy preference relation and bean optimization algorithm. In: IEEE YC-ICT 2010, pp. 253–256 (2010)
15. Zhang, X., Sun, B., Mei, T., Wang, R.: A Novel Evolutionary Algorithm Inspired by Beans Dispersal. International Journal of Computational Intelligence Systems 6(1), 79–86 (2013)
16. Dorigo, M., Birattari, M., Stützle, T.: Ant Colony Optimization– Artificial Ants as a Computational Intelligence Technique. IEEE Computational Intelligence Magazine 11(4), 28–39 (2006)
17. Pham, D.T., Ghanbarzadeh A., Koc, E., Otri, S., Rahim, S., Zaidi, M.: The bees algorithm – a novel tool for complex optimisation problems. In: IPROMS 2006, pp. 454–461 (2006)
18. Zhang, X., Wang, H., Sun, B., Li, W., Wang, R.: The Markov Model of Bean Optimization Algorithm and Its Convergence Analysis. International Journal of Computational Intelligence Systems 6(4), 609–615 (2013)
19. Zhang, X., Sun, B., Mei, T., Wang, R.: A Novel Evolutionary Algorithm Inspired by Beans Dispersal. International Journal of Computational Intelligence Systems 6(1), 79–86 (2013)

On the Application of Co-operative Swarm Optimization in the Solution of Crystal Structures from X-Ray Diffraction Data

Alexandr Zaloga[1], Sergey Burakov[2], Eugene Semenkin[2], Igor Yakimov[1], Shakhnaz Akhmedova[2], Maria Semenkina[2], and Evgenii Sopov[2(✉)]

[1] Siberian Federal University, Krasnoyarsk, Russia
{zaloga,i-s-yakimov}@yandex.ru
[2] Siberian State Aerospace University, Krasnoyarsk, Russia
{s.v.burakov,eugenesemenkin}@yandex.ru, shahnaz@inbox.ru,
semenkina88@mail.ru, evgenysopov@gmail.com

Abstract. A co-operative method based on five biology-related optimization algorithms is used in solving crystal structures from X-ray diffraction data. This method does not need essential effort for its adjustment to the problem in hand but demonstrates high performance. This algorithm is compared with a sequential two-level genetic algorithm, a multi-population parallel genetic algorithm and a self-configuring genetic algorithm as well as with two problem specific approaches. It is demonstrated on a special crystal structure with 7 atoms and 21 degrees of freedom on which the co-operative swarm optimization algorithm exhibits comparative reliability but works faster than other used algorithms. Perspective directions for improving the approach are discussed.

Keywords: Co-operation of biologically inspired algorithms · Genetic algorithms · Crystal structure solution · X-ray diffraction data

1 Introduction

One of the most important tasks in the material sciences is the explanation and prediction of the physical and chemical properties of materials that can be done on the background of information about crystal structure. This information is accumulated in the crystal structure databases (DB) [1, 2] and includes the coordinates of atoms in the symmetrically independent part of the crystal cell and some additional parameters. The structure of polycrystalline materials is studied using different methods of X-Ray powder diffraction [3]. Structural investigation includes the definition of an approximate model of the atomic crystal structure and its optimization. The initial data are the chemical formula, unit cell parameters, space group symmetry and the X-Ray powder diffraction pattern of the material. Typically, the optimization of the crystal structure model is performed using the method of Rietveld [4], which consists in the full-profile fitting of the calculated diffraction pattern to an experimental one using a nonlinear least squares method. The main problem here is the finding of a crystal structure as an appropriate model.

© Springer International Publishing Switzerland 2015
Y. Tan et al. (Eds.): ICSI-CCI 2015, Part I, LNCS 9140, pp. 89–96, 2015.
DOI: 10.1007/978-3-319-20466-6_10

Today, in order to achieve this objective, global optimization methods are used as an effective and useful tool. With computing power increasing, the number of structures solved by these methods is rising [5]. The main options are variations of a simulated annealing method [6]. Their disadvantage is the sequential approach, based on a large number ($\sim 10^7$-10^8) of structural configuration evaluations that allows the efficient full-profile structure refinement with the Rietveld method at the final stage only. Methods of the crystal structure model search based on genetic algorithms [7] immediately generate a lot of trial structural models (population) and carry out its evolution, but the Rietveld refinement is also used only at the final stage. The disadvantage of global optimization methods, including GA, is the stagnation of the convergence process in the local minima. An additional problem is the complexity of the adjustment of genetic algorithm operators that requires significant computational effort or a high level human expert in computational intelligence. Besides, genetic algorithms work with a binary representation that makes necessary the binarization of real-valued problem variables (atom coordinates) and bring additional troubles. In this study we use the self-configuring genetic algorithm [8] that allows us to avoid the problem of genetic algorithm adjustment and compare its performance with a co-operative bionic algorithm [9] that works with real variables and is also a self-tuning tool.

The remainder of this paper is organized as follows. Section 2 explains the ideas of conventional genetic algorithms used for solving crystal structures. Further, in Section 3, a self-configuring genetic algorithm is described that does not need human effort to be adjusted to the problem in hand. In Section 4, the meta-heuristic of the co-operative method based on swarm optimization algorithms is described. Results of numerical experiments and a comparison of algorithm performance is given in Section 5. Finally, the conclusion follows in Section 6.

2   Conventional Genetic Algorithms for Crystal Structure Solution

In [10], a two-level hybrid evolutionary method for determining the crystal structure has been described. The algorithm was based on the iterative use of two genetic algorithms (GA): one for the structural model search and another one for their local full profile optimization with the Derivative Difference Minimization method - DDM [11] (analogue of the Rietveld method). Level 1 and level 2 GAs are executed cyclically until a convergence of the R-factor to the expected value (the R-factor is a relative difference between the calculated and experimental X-ray diffraction profiles). The resultant structure can be finally refined by the DDM. The two-level GA evolutionary mechanism provides the accelerated convergence described in [12]. However, this algorithm (2LGA) was not reliable enough and worked very slowly.

The work [13] was dedicated to the multi-population parallel evolutionary modelling of atomic crystal structures based on a two-level GA (MPGA) with the organization of the evolution of different populations performed on the different units of the computing cluster with the following exchange of the best structural models. Populations of trial structures are evolving on the computational units P_1 - P_n (where n>4) during several cycles of the 1st level GA and the 2nd level GA. Then, the trial structures with a low R-factor value from these units are transmitted and accumulated at the main unit P_{main}. Further, these best structures from P_{main} are selectively transmitted to those P_k units, which have the GA convergence slowed down, in an amount inversely proportional to the rate of convergence. The application of MPGA needed significant time and effort for the choice of appropriate settings and tuning parameters. This work involved human experts in crystallography as well as in evolutionary computations [13].

3 Self-configuring Genetic Algorithm

For the aim of genetic algorithm self-configuration in [8] the dynamic adaptation on the level of population with centralized control techniques was applied to the operator probabilistic rates. Setting variants, e.g. types of selection, crossover, population control and level of mutation (medium, low, high) were used to avoid real parameter precise tuning. Each of these has its own probability distribution. During the initialization phase all probabilities are equal and they will be changed according to a special rule through the execution of the algorithm. No probability could be less than a predetermined minimum balance.

When the algorithm has to create the next offspring from the current population, it firstly must configure settings, i.e. form the list of operators with the use of probability operator distributions. Then the algorithm selects parents with the chosen selection operator, produces an offspring with the chosen crossover operator, mutates this offspring with the chosen mutation probability and puts it into an intermediate population. When the intermediate population is complete, the fitness evaluation is executed and the operator rates (probabilities to be chosen) are updated according to the performance of the operator. Then the next parents' population is formed with the chosen survival selection operator. The algorithm stops after a given number of generations or if a termination criterion (e.g., the given error minimum) is met.

As was reported in [8], this algorithm (SelfCGA) exhibits a high level performance without the necessity of its adjustment by a human expert and can be used in black-box like problems, i.e. directly without incorporating any problem-specific knowledge. The applicability of the algorithm to complicated real-world problems was later demonstrated [14].

4 Co-operation of Biology Related Algorithms

The meta-heuristic of Co-Operation of Biology Related Algorithms (COBRA) was developed in [9] on the base of five well-known optimization methods such as Particle Swarm Optimization Algorithm (PSO) [15], Wolf Pack Search Algorithm (WPS) [16], the Firefly Algorithm (FFA) [17], the Cuckoo Search Algorithm (CSA) [18] and the Bat Algorithm (BA) [19]. These algorithms are biology related optimization approaches originally developed for continuous variable space. They mimic the collective behaviour of the corresponding animal groups which allows the global optima of real-valued functions to be found. However, these algorithms are very similar in their ideology and behaviour. This brings end users to the inability to decide in advance which of the above-listed algorithms is the best or which algorithm should be used for solving any given optimization problem [9]. The idea of a developed meta-heuristic was the use of the cooperation of these algorithms instead of any attempts to understand which one is the best for the current problem in hand.

The proposed approach consists in generating five populations (one population for each algorithm) which are then executed in parallel cooperating with each other. The proposed algorithm is a self-tuning meta-heuristic that allows the user to avoid an arbitral (i.e., badly justified) choice of the population size for each algorithm. The number of individuals in the population of each algorithm can increase or decrease depending on whether the fitness value is increased or decreased. If the fitness value was not improved during a given number of generations, then the size of all populations increases. And vice versa, if the fitness value was constantly improved, then the size of all populations decreases. Besides, if the average fitness of some population is better than the average fitness of all other populations then this population "grows" by adding a number of randomly generated individuals. The corresponding number of individuals are then removed from another population by rejecting an equal number of the worst members of the population.

The result of this kind of competition allows the biggest resource (population size) to be presented to the most appropriate (in the current generation) algorithm. This property can be very useful in the case of a hard optimization problem when, as is known, there is no single best algorithm on all stages of the optimization process execution.

Another important driving force of the meta-heuristic is the migration operator that creates a cooperation environment for component algorithms. All populations communicate with each other: they exchange individuals in such a way that a part of the worst individuals of each population is replaced by the best individuals of other populations. It brings up to date information on the best achievements to all component algorithms and prevents their preliminary convergence to its own local optimum which improves the group performance of all algorithms.

The performance of the proposed algorithm was evaluated on the set of benchmark problems from the CEC'2013 competition [9]. This set of benchmark functions included 28 unconstrained real-parameter optimization problems. The validation of COBRA was carried out for functions with 10, 30 and 50 variables. The experiments showed that COBRA works successfully and is reliable on this benchmark. The results also showed that COBRA outperforms its component algorithms when the dimension grows and more complicated problems are solved. It means that COBRA can be used instead of any component algorithm.

The applicability of the algorithm to complicated real-world problems was also demonstrated [20].

5   Modelling Crystal Structure from X-Ray Diffraction Data with Adaptive Search Algorithms

We demonstrate the applicability and compare the performance of the described adaptive search algorithm to the solution of crystal structures with the example of orthorombic Ba_2CrO_4 with 7 independent atoms in the general positions. This choice could be explained as follows. This structure is not so easy to solve as it has 21 degrees of freedom of atomic coordinates, the cell parameters are a=7.67, b=5.89, c=10.39 angstroms and the cell angles are α=90, β=90, γ=90 degrees. However, it is also not so complicated and can be solved within a reasonable time, which is important for our approach as we use randomized search algorithms and must apply multiple runs for statistical evaluations of the algorithms performance. Additionally, this structure was solved many times by other techniques, the results of which can serve us as reference points for better evaluation.

First of all we fulfil the solving of the crystal structure by means of the latest available versions of specialized software FOX [21] and DASH [22]. This software is specially designed for the problems being solved and incorporates human experts' knowledge that gives the possibility, in particular, to evaluate very quickly the current model. This is not the case for direct search algorithms operating with a problem in black-box mode.

The comparison of results is fulfilled with three criteria. One criterion is the algorithm reliability which is the proportion of successful runs when the right solution has been found. Another criterion is the average number of structure models generated by the algorithm before the right structure was found. The last criterion is the average amount of computing time needed to find the solution. Here we should say that MPGA should be considered separately as it is executed on a computing cluster with a number of used computing units unknown in advance that was between 4 and 8. One should realize that the average time here has not the same meaning as in other cases. All parameters of algorithms were adjusted in order to have reliability not less than 50 %. The results averaged after 20 runs are given in Table 1 below. Statistical significance was evaluated with the t-test.

Table 1. Summary of results for Ba_2CrO_4

Index\Tool	FOX	DASH	2LGA	MPGA	SelfCGA	COBRA
Reliability	0.80	0.50	0.70	1.00	0.80	0.80
Number of structures	10^6	10^7	39000	30000	75000	29500
Average time (min)	14	5	30	1.5	1.6	1.15

As one can see, the simulations show that the cooperation of bionic algorithms is the fastest tool with a high enough reliability. A reliability higher then COBRA was demonstrated only by MPGA which uses much more computational resource. One should mention as well that both SelfCGA and COBRA outperform both kinds of specialized software. This could be explained through the relatively simple crystal structure used for evaluation. At the same time one can see that specialized software works significantly faster when estimating the single current model but far more of these models are required to find the solution. This can be a way of further improving the current tools through the hybridization of a problem-specific way of model estimation and direct search optimization algorithms.

An illustration of the SelfCGA and COBRA convergence is presented on Figure 1 and Figure 2, correspondingly. On both figures, horizontal axis counts the number of generation, vertical axis shows the fitness value (to be minimized), bold line depicts the fitness current best value on each generation, thin line depicts the average fitness value on current generation, and dashed line shows a portion of the penalty in the best fitness value. It is penalized the excessive convergence of atoms in the best structure found.

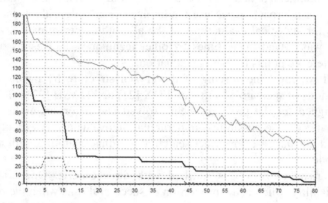

Fig. 1. SelfCGA convergence illustration when solving Ba_2CrO_4 structure

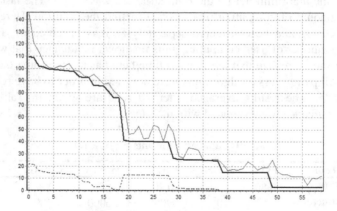

Fig. 2. COBRA convergence illustration when solving Ba_2CrO_4 structure

We can see in the Figure 1 that SelfCGA improves the fitness very fast on the initial stages but then it converges relatively slowly to the solution. Big enough difference between the best and average fitness values on each generation, including the last of them when solution has been found, demonstrates that the variation of individuals is supported on the sufficiently high level. This allows us to hope that SelfCGA can be workable on more complicated structures.

Figure 2 shows that COBRA works faster, constantly improves the fitness and finds solution much earlier. At the same time, it does not support the individuals' variation that can explain the fast convergence but also could be a problem in case of more complex structures to be solved. This question should attract developers' attention in further investigations.

Nevertheless, results of numerical experiments allow us to conclude that the most perspective algorithm here is COBRA which works fast, does not require so many models for estimation and does not need any human expert level knowledge in computational intelligence for its adjustment. COBRA is also a very suitable approach for parallelization on cluster computing systems.

6 Conclusions

The stochastic genetic algorithm provides an ab initio search of the atomic crystal structure of chemical compounds from powder diffraction patterns. It can be easily automated and does not require high skills in the field of structural analysis. This makes possible the performing of structural studies by investigators who have synthesized new chemical compounds. The structure search reliability depends on the complexity of the crystal structure, i.e. on the number of degrees of freedom of the atomic coordinates.

In this work, it is suggested to use the co-operative method based on swarm optimization algorithms. It resolves usual troubles of binary string based GA as the suggested algorithm works with strings of real numbers and uses self-adjustment to the given problem. The usefulness of the suggested approach is demonstrated on the special problem of the solution of Ba_2CrO_4 crystal structure that contains 21 real variables within an unconstrained optimization task. The co-operative method demonstrates high reliability and speed and outperforms alternative approaches. Additional observation gives a hint for the further development of the approach that should be directed to a hybridization of the suggested optimization techniques with the way of structure model estimation used in specialized software as it could give a higher speed of problem solving.

Acknowledgements. The study was fulfilled with the financial support of the Ministry of Science and Education of the Russian Federation within the state assignment 2014/71 to the Siberian Federal University, project 3098.

References

1. Inorganic Crystal Structure Database. FIZ Karlsruhe. http://www.fiz-karlsruhe.de/icsd. html
2. Cambridge Structural Database. Cambridge Crystallographic Data Centre. http://www.ccdc. cam.ac.uk/products/csd/
3. Dinnebier, R.E., Billinge, S.J.L.: Powder diffraction theory and practice. Royal Society of Chemistry, 507 (2008)

4. Young, R.A.: The rietveld method, p. 298. Oxford University Press (1995)
5. David, W.I.F., Shankland, K.: Structure determination from powder diffraction data. Acta Cryst. **A64**, 52–64 (2008)
6. Cerny, R., Favre-Nicolin, V.: Direct space methods of structure determination from powder diffraction: principles, guidelines, perspectives. Z. Kristallogr. **222**, 105–113 (2007)
7. Kenneth, D., Harris M.: Fundamentals and applications of genetic algorithms for structure solution from powder X-ray diffraction data. Computational Materials Science **45**(1) (2009)
8. Semenkin, E., Semenkina, M.: Self-configuring genetic algorithm with modified uniform crossover operator. In: Tan, Y., Shi, Y., Ji, Z. (eds.) ICSI 2012, Part I. LNCS, vol. 7331, pp. 414–421. Springer, Heidelberg (2012)
9. Akhmedova, Sh., Semenkin, E.: Co-operation of biology related algorithms. In: IEEE Congress on Evolutionary Computation (CEC 2013), pp. 2207–2214 (2013)
10. Yakimov, Y.I., Semenkin, E.S., Yakimov, I.S.: Two-level genetic algorithm for a fullprofile fitting of X-ray powder patterns. Z. Kristallogr. (30), 21–26 (2009)
11. Solovyov, L.A.: Full-profile refinement by derivative difference minimization. J. Appl. Cryst. **3**, 743–749 (2004)
12. Yakimov, Y.I., Kirik, S.D., Semenkin, E.S., Solovyov, L.A., Yakimov, I.S.: The evolutionary method of modelling a crystal structure by powder diffraction data. Journal of Siberian Federal University (Chemistry) **6**(2), 180–191 (2013)
13. Burakov, S.V., et al.: Multi-population two-level genetic algorithm for determination of crystal structure from X-ray powder diffraction data. In: Proceedings of Conference APMSIT-2014, Shanghai, China, pp. 13–14 (2014)
14. Semenkin, E., Semenkina, M.: Stochastic models and optimization algorithms for decision support in spacecraft control systems preliminary design. In: Ferrier, J.-L., Bernard, A., Gusikhin, O., Madani, K. (eds.) ICINCO 2012. LNEE, vol. 283, pp. 51–65. Springer, Heidelberg (2014)
15. Kennedy, J., Eberhart, R.: Particle swarm optimization. In: IEEE International Conference on Neural networks, IV, pp. 1942–1948 (1995)
16. Yang, Ch., Tu, X., Chen, J.: Algorithm of marriage in honey bees optimization based on the wolf pack search. In: International Conference on Intelligent Pervasive Computing, pp. 462–467 (2007)
17. Yang, X.-S.: Firefly algorithms for multimodal optimization. In: Watanabe, O., Zeugmann, T. (eds.) SAGA 2009. LNCS, vol. 5792, pp. 169–178. Springer, Heidelberg (2009)
18. Yang, X.S., Deb, S.: Cuckoo search via levy flights. In: World Congress on Nature & Biologically Inspired Computing, pp. 210–214. IEEE Publications (2009)
19. Yang, X.-S.: A new metaheuristic bat-inspired algorithm. In: González, J.R., Pelta, D.A., Cruz, C., Terrazas, G., Krasnogor, N. (eds.) NICSO 2010. SCI, vol. 284, pp. 65–74. Springer, Heidelberg (2010)
20. Akhmedova, S., Semenkin, E.: Data mining tools design with co-operation of biology related algorithms. In: Tan, Y., Shi, Y., Coello, C.A. (eds.) ICSI 2014, Part I. LNCS, vol. 8794, pp. 499–506. Springer, Heidelberg (2014)
21. Favre-Nicolin, V., Cerny, R.: FOX "free objects for crystallography": a modular approach to ab initio structure determination from powder diffraction. J. Appl. Cryst. **35**, 734–743 (2002)
22. Griffin, T.A.N., Shankland, K., Van de Streek, J., Cole, J.: GDASH: a grid-enabled program for structure solution from powder diffraction data. J. Appl. Cryst. **42**, 356–359 (2009)

Particle Swarm Optimization

Swarm Diversity Analysis of Particle Swarm Optimization

Yuanxia Shen[✉], Linna Wei, and Chuanhua Zeng

School of Computer Science and Technology,
Anhui University of Technology, Maanshan, Anhui, China
chulisyx@163.com

Abstract. When particle swarm optimization (PSO) solves multimodal problems, the loss of swarm diversity may bring about the premature convergence. This paper analyses the reasons leading to the loss of swarm diversity by computing and analyzing of the probabilistic characteristics of the learning factors in PSO. It also provides the relationship between the loss of swarm diversity and the probabilistic distribution and dependence of learning parameters. Experimental results show that the swarm diversity analysis is reasonable and the proposed strategies for maintaining swarm diversity are effective. The conclusions of the swarm diversity of PSO can be used to design PSO algorithm and improve its effectiveness. It is also helpful for understanding the working mechanism of PSO theoretically.

Keywords: Particle swarm optimization · Premature convergence · Swarm diversity

1 Introduction

Particle swarm optimization (PSO) algorithm is originally proposed by Kennedy and Eberhart [1] in 1995 as a member of the wide category of swarm intelligence methods for solving global optimization problems. PSO algorithm performs well on many optimization problems [2]. However, when solving complex multimodal tasks, PSO may easily get trapped in premature convergence. As particles share information in PSO, a single *gbest* spreading among them usually leads to particle clustering which results in swarm diversity declining quickly in the search prophase [3].

Maintaining swarm diversity can decrease the possibility of PSO getting into the premature convergence. Therefore, many schemes have been proposed to keep the swarm diversity during the evolution. Ratnaweera [4] developed PSO with linearly time-varying acceleration coefficients to adjust the local and global search ability. He also introduced mutation operation to increase the swarm

This work was supported by National Natural Science Foundation of China under Grant Nos.61300059. Provincial Project of Natural Science Research for Anhui Colleges of China (KJ2012Z031, KJ2012Z024).

© Springer International Publishing Switzerland 2015
Y. Tan et al. (Eds.): ICSI-CCI 2015, Part I, LNCS 9140, pp. 99–106, 2015.
DOI: 10.1007/978-3-319-20466-6_11

diversity. Mutation operation is helpful in increasing the swarm diversity and is often used to improve PSO variants [5]. Besides the mutation operation, other auxiliary operations are also introduced, such as selection [6], crossover [7], perturbation [8], and collision operator [9]. Zhan [10] proposed an elitist learning strategy. Jie [3] developed a knowledge-based cooperative particle swarm optimization by multi-swarm to maintain the diversity and a knowledge billboard to control the search process. With the aim to improve the performance in multimodal problems, Liang [11] presented a comprehensive learning strategy to preserve the diversity.

Though empirical results show that the auxiliary operations and the design of topological structures are effective for preserving swarm diversity, these techniques increase the complexity of PSO framework. This paper analyzes reasons leading to the loss of swarm diversity from the view of information processing. By the way of computing and analyzing the probabilistic characteristics of learning factors in information processing mechanism, we obtain relationships between the loss of swarm diversity and two characteristics of the learning parameters which are the probabilistic distribution and dependence. In order to verify the swarm diversity analysis of PSO, two experiments are designed in which several learning strategies without additional operation are proposed to keep swarm diversity. The experimental results show that the strategies proposed in this paper are effective. The conclusions of the swarm diversity on PSO can be used to design and improve PSO algorithm.

The rest of this paper is organized as follows. The original PSO is introduced in Section 2. The swarm diversity analysis of PSO is described in Section 3. Section 4 presents experimental design, results and discussions. Conclusions are given in Section 5.

2   The Original PSO

In PSO, each particle has a position and a velocity in a variable space. Assuming a D-dimensional search space and a swarm consisting of N particles, the current position X_i and the velocity V_i of the i-th particle are D-dimensional vectors, i.e. $X_i = (x_{i1}, x_{i2}, ..., x_{iD})$ and $V_i = (v_{i1}, v_{i2}, ..., v_{iD})$. $Pb_i = (Pb_{i1}, Pb_{i2}, ..., Pb_{iD})$ is the best previous position yielding the best fitness value for the i-th particle and $Gb = (Gb_1, Gb_2, ..., Gb_D)$ is the best position discovered by the whole population. Each particle updates its velocity and position with the following equations.

$$V_{id}(t+1) = wV_{id}(t) + c_1 r_{1,id}(t)(Pb_{id}(t) - X_{id}(t)) + c_2 r_{2,id}(t)(Gb_d(t) - X_{id}(t)) \quad (1)$$

$$X_{id}(t+1) = X_{id}(t) + V_{id}(t+1) \quad (2)$$

Where w is an inertia weight; $c1$ and $c2$ are acceleration coefficients reflecting the weighting of stochastic acceleration terms that pull each particle toward *pbest* and *gbest*, respectively. Random factors r_1 and r_2 are independent and uniformly distributed random numbers in $[0, 1]$.

3   The Swarm Diversity Analysis of PSO

Swarm diversity can reflect the exploration ability of particles and higher swarm diversity can decrease the possibility of PSO suffering from premature convergence. In order to improve the exploration ability of the particles of the original PSO, it is necessary to analyze the reasons leading to the loss of swarm diversity. From the view of information-processing of the particles, this paper analyzes reasons leading to the loss of swarm diversity.

3.1   Transformation of Updating Equation

As particle i is chosen arbitrarily, the result can be applied to all the other particles. Since it appears from Eq. (1) and Eq. (2) that each dimension is updated independently, we can reduce the algorithm description into the one-dimensional case without loss of generality. By omitting the notations for the particle and its dimension, update equations (3) and (4) can be stated as follows:

$$V(t + 1) = wV(t) + c_1 r_1(t)(Pb(t) - X(t)) + c_2 r_2(t)(Gb(t) - X(t)) \qquad (3)$$

$$X(t + 1) = X(t) + V(t + 1) \qquad (4)$$

By substituting Eq. (3) into Eq. (4), the following relation is obtained:

$$X(t + 1) = X(t) + c_1 r_1(Pb(t) - X(t)) + c_2 r_2(Gb(t) - X(t)) + wV(t) \qquad (5)$$

In order to analyze conveniently, we need to transform the velocity updating equation. According to Eq. (5), the following equation can be obtained.

$$X(t + 1) = X(t) + wV(t) + (c_1 + c_2) * \left[\frac{c_1 r_1}{(c_1 + c_2)} + \frac{c_2 r_2}{(c_1 + c_2)} \right]$$
$$* \left[\frac{c_1 r_1 Pb(t)}{(c_1 r_1 + c_2 r_2)} + \frac{c_2 r_2 Gb(t)}{(c_1 r_1 + c_2 r_2)} - X(t) \right] \qquad (6)$$

Let $k = c_1/(c_1 + c_2)$. By substituting k into Eq. (6), the following equation is obtained:

$$X(t + 1) = X(t) + wV(t) + (c_1 + c_2) * [kr_1 + (1 - k)r_2]$$
$$* \left[\frac{kr_1 Pb(t)}{(kr_1 + (1 - k)r_2)} + \frac{(1 - k)r_2 Gb(t)}{(kr_1 + (1 - k)r_2)} - X(t) \right] \qquad (7)$$

Let $Y = kr_1 + (1 - k)r_2$, $Z = kr_1/(kr_1 + (1 - k)r_2)$, then from eq.(7), we gets

$$X(t + 1) = X(t) + wV(t) + (c_1 + c_2)Y[ZPb(t) + (1 - Z)Gb(t) - X(t)] \qquad (8)$$

where Y and Z are functions with respect to random factors r_1, r_2 and undetermined parameter k. Y and Z are correlative random variables in $[0, 1]$. Given Pb, Gb and X in Eq.(8) and Y and Z traverse an interval value of $[0, 1]$, the second part of Eq. (8) can be viewed as a 2-dimension search space. For the convenience of elaboration, this search space is called one-step search space. The distribution of searching points in this space dependents on the probabilistic characteristics of Z and Y. Z is the weighting coefficient for the synthetic of *pbest* and *gbest*, its value can reflect the differences of exploitation for *pbest* and *gbest*. The learning attitude of the particle drifts toward *pbest* when the value of Z approaching 1 and it drifts toward *gbest* when the value of Z approaching 0. The value of Y can influences the size of the one-step searching space.

3.2   Computation and Analysis of the Probabilistic Characteristics of Learning Factors

We calculate the probability density of Z ($f_Z(z)$) and joint density of Z and Y ($f(y,z)$) under general situations that $c_1=c_2$ (i.e. $k=0.5$)

$$f_Z(z) = \begin{cases} 1/(2(1-z)^2), & 0 \leq z < 0.5 \\ 1/(2z^2), & 0.5 \leq z \leq 1 \\ 0, & other \end{cases} \tag{9}$$

$$f(y,z) = \begin{cases} 4y, 0 \leq y \leq 1, \max(0, 1-\frac{1}{2y}) \leq z \leq \min(1, \frac{1}{2y}) \\ 0, \qquad\qquad other \end{cases} \tag{10}$$

From Eq. (9) and Eq. (10), we can calculate $f_{Y|Z}(y|z)$ the conditional probability density of Y given Z

$$f_{Y|Z}(y|z) = \begin{cases} 8y(1-z)^2, 0 \leq z < \frac{1}{2}, 0 \leq y < \frac{1}{2(1-z)} \\ 8yz^2, \qquad \frac{1}{2} \leq z < 1, 0 \leq y < \frac{1}{2z} \end{cases} \tag{11}$$

Eq. (9) shows that $f(z)$ is a unimodal function and is symmetrical at z equals to 0.5. The function $f(z)$ obtains its maximum value at this point. In Eq. (11), $f(y|z)$ shows that the range of Y depends on the value of Z. Moreover, $f(y|z)$ is symmetric at $z=0.5$. When $z=0.5$, the value of Y falls into $[0, 1]$ and it decreases gradually with the changing of the value of Z. When Z equals to 0 or 1, the value of Y falls into $[0, 0.5]$.

The probability characteristics of the learning parameters Z and Y can be used to analyze the change of swarm diversity. There are two reasons that may lead to the loss of population diversity. On one hand, each particle tends to learn from *gbest* with the probability that $P(0 < z < 0.25) = 1/6$ in every iteration. When $z \in [0, 0.25)$ the maximum range of Y is restricted to $[0, 2/3]$ and when $z=0$ the range of Y decreases to $[0, 0.5]$.

On the other hand, the probability of the random variable z lies in the range of $[0.5-\delta, 0.5+\delta]$ (δ is a small value) is equal to 0.182, where $\delta=0.05$. When $z=0.5$, the value of Y should fall into its maximum range. But the probability that z lies in the range of $[0.5-\delta, 0.5+\delta]$ ($\delta=0.05$) is equal to 0.182, which means that the probability that Y cannot fall into its maximum range is about 0.818 and then the one-step search space of a particle is constrained.

4 Experimental Design and Simulation Results

4.1 Experimental Design

In order to verify the analysis of swarm diversity of PSO, two kinds of experiments are designed. In these experiments, particles update their velocity according to Eq.(6) in which the learning parameters adopt different strategies. The purpose of the first kind is to test the effect of the probability distribution of Z on swarm diversity. Two learning strategies are introduced, named strategy 1 and strategy 2.

Strategy 1: $z \sim U(0.4, 0.6)$, and $y \sim f_{Y|Z}(y|z)$,where z subjects to the uniform distribution on [0.4, 0.6] and y follows the conditional density $f_{Y|Z}(y|z)$ according to eq. (11) in which z and y are correlated.

Strategy 2: $z \sim U(0.4, 0.6)$, and $y=0.5*y1,(y1 \sim f_{Y|Z}(y|z))$, where learning strategy 2 is similar to strategy 1 and the value range of y is compressed. The second kind is to test the effect of the value range of Y on swarm diversity. Two learning strategies are designed, named strategy 3 and strategy 4.

Strategy 3: $z \sim f_Z(z)$, and $y \sim f_{Y|Z}(y|z = 0.5)$, where z follows the density function of eq. (9)and y subjects to the conditional density $f_{Y|Z}(y|z)$ given z=0.5 in which z and y are independent.

Strategy 4: $z \sim f_Z(z)$, and $y \sim U(0, 1)$. Where z follows the density function according to eq. (9) and y is subject to the uniform distribution on [0, 1] in which z and y are independent.

4.2 Measurement of Swarm Diversity

We measure the swarm diversity according to the average distance around the swarm center. A small value indicates swarm convergence around the swarm center while a large value indicates particles dispersion from the center. The measure of the swarm diversity (Div) is taken as the equation in [12]:

$$Div(S(t)) = \frac{1}{N_s} \sum_{i=1}^{N_s} \sqrt{\sum_{j=1}^{D_x} (x_{ij} - \overline{x_j(t)})^2} \tag{12}$$

where S is the swarm, $N_s = |S|$ is the size of the swarm, D_x is the dimensionality of the problem, x_{ij} is the j-th value of the i-th particle and $\overline{x_j(t)}$ is the average of the j-th dimension over all particles, i.e. $\overline{x_j(t)} = (\sum_{i=1}^{N_s} x_{ij}(t))/N_s$.

4.3 Experiment Settings

We test four benchmark functions to illustrate the variation of swarm diversity in PSO algorithms with different learning strategies during the search process.The properties and the formulas of these functions are presented below.We set the parameters as follows: inertia weights equals to 0.7298 and $c_1=c_2 = 1.49618$; 20

particles are used in all the experiments; the reported results are averages over 30 simulations; all functions are tested on 30 dimensions. All the test functions are minimized.

Sphere function $f_1(x) = \sum\limits_{i=1}^{D} x_i^2, x \in [-100, 100]$

Griewanks function $f_3(x) = \sum\limits_{i=1}^{D} \frac{x_i^2}{4000} - \prod\limits_{i=1}^{D} \cos(\frac{x_i}{\sqrt{i}}) + 1, x \in [-600, 600]$

Rastrigins function $f_4(x) = \sum\limits_{i=1}^{D} (x_i^2 - 10\cos(2\pi x_i) + 10), x \in [-5.12, 5.12]$

Schwefels function $f_6(x) = 418.9829 \times D - \sum\limits_{i=1}^{D} x_i \sin(|x_i|^{0.5}), x \in [-500, 500]$

4.4   Experimental Results

Experiment 1 The evolutionary trends of the swarm diversity indicator in PSO with learning strategy 1 and 2 for the four functions are shown in Figure 1.

As we can see from Figure 1, the swarm diversity indicator of the original PSO dropped faster than the PSO with learning strategy 1 and strategy 2 in all the four functions, which means that the value of z in the interval [0.4, 0.6] is beneficial for keeping swarm diversity. This simulation results show that the value range of z has effect on swarm diversity. In addition, from Figure 1 (a), (b), and (c), the swarm diversity indicator of PSO with learning strategy 1 dropped

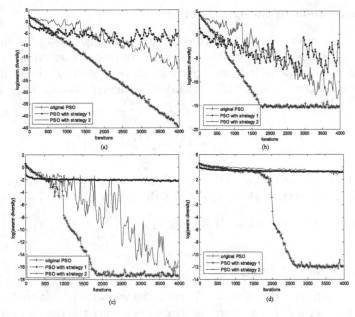

Fig. 1. The variation of swarm diversity with time in experiment 1 (a) Sphere function. (b) Griewank function (c) Rastrigin function (d) Schwefel function

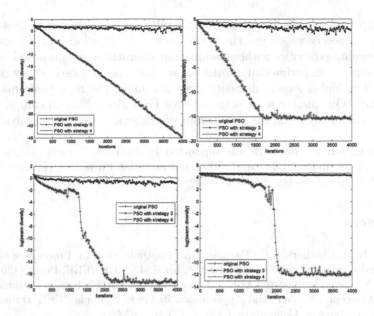

Fig. 2. The variation of swarm diversity with time in experiment 2 (a) Sphere function. (b) Griewank function (c) Rastigrin function (d) Schwefel function

slower than the PSO with learning strategy 2. In learning strategy 1, the value range of y is form 0 to 1, while the value range of y is form 0 to 0.5 in learning strategy 2. Simulation results show that under same probability distribution of z, the size of value range of y has effect on swarm diversity and the bigger value range of y is helpful for keeping swarm diversity.

Experiment 2 For four functions, the evolutionary trends of the swarm diversity indicator in PSO with learning strategy 3 and 4 are show in Figure 2.

In the learning strategy 3 and 4, z and y are independent, and z follows the density function of original PSO and the value range of y is form 0 to 1. In the original PSO, the value rang of y is dependent on the value of z. As can be seen from Figure 2, original PSO drops fast in swarm diversity indicator, and the two PSO algorithms with learning strategy 3 and 4 keep higher swarm diversity. This simulation results show that the value range of y has effect on swarm diversity. The swarm diversity indicator in PSO with learning strategy 3 is slightly higher than that of PSO with learning strategy 4, which means the uniform density distribution of y is helpful for maintaining swarm diversity.

5   Conclusion

This paper analyzes the reasons leading to the loss of swarm diversity by the way of the probabilistic characteristics of learning factors in PSO. It provides relationship between the loss of swarm diversity and two characteristics of the learning parameters which are the probabilistic distribution and dependence. Several learning strategies without additional operations are proposed to keep swarm diversity. Experimental results show that the proposed strategies are effective. The higher swarm diversity does not mean the high performance of optimization. Our further work is to improve PSO algorithms on complex multimodal problems by using the conclusions of the swarm diversity analysis.

Acknowledgments. This work was supported by National Natural Science Foundation of China under Grant Nos.61300059. Provincial Project of Natural Science Research for Anhui Colleges of China (KJ2012Z031, KJ2012Z024).

References

1. Kennedy, J., Eberhart, R.C.: Particle swarm optimization. In: Proceeding of International Conference on Neural Networks, pp. 1942–1948. IEEE, Perth (1995)
2. Valle, Y., Venayagamoorthy, G.K., Mohagheghi, S.: Particle Swarm Optimization: Basic Concepts, Variants and Applications in Power Systems. IEEE Transactions on Evolutionary on Computation **12**(2), 171–195 (2008)
3. Jing, J., Jianchao, Z., Chongzhao, H., Qinghua, W.: Knowledge-based cooperative particle swarm optimization. Applied Mathematics and Computation **205**(2), 861–873 (2008)
4. Ratnaweera, A., Halgamuge, S.K., Watson, H.C.: Self-Organizing Hierarchical Particle Swarm Optimizer with Time-Varying Acceleration Coefficients. IEEE Transaction on Evolutionary on Computation **8**(3), 240–255 (2004)
5. Andrews, P.S.: An investigation into mutation operators for particle swarm optimization. In: Proceedings of the IEEE Congress on Evolutionary Computation, pp. 1044–1051. IEEE, Vancouver (2006)
6. Angeline, P.J.: Using selection to improve particle swarm optimization. In: Proceedings of the IEEE Congress on Evolutionary Computation, pp. 84–89. IEEE, Anchorage (1998)
7. Yingping, C., Wenchih, P., Mingchung, J.: Particle swarm optimization with recombination and dynamic linkage discovery. IEEE Trans. Syst. Man, Cybern. B, Cybern. IEEE **37**, 1460–1470 (2007)
8. Xinchao, Z.: A perturbed particle swarm algorithm for numerical optimization. Applied soft compute. **10**, 119–124 (2010)
9. Arani, B.O., Mirzabeygi, P., Panahi, M.S.: An improved PSO algorithm with a territorial diversity-preserving scheme and enhanced explorationCexploitation balance. Swarm and Evolutionary Computation **11**, 1–15 (2013)
10. Zhihui, Z., Jun, Z., Yun, L., Chung, H.S.-H.: Adaptive Particle Swarm Optimization. IEEE Trans. Syst. Man, Cybern. B, cybernetics **39**, 1362–1382 (2009)
11. Liang, J.J., Qin, A.K., Suganthan, P.N., Baskar, S.: Comprehensive learning particle swarm optimizer for global optimization of multimodal functions. IEEE Trans. Evolutionary Computation **10**, 281–295 (2006)
12. Shi, Y.H., Eberhart, R.C.: Population diversity of particle swarm. In: Proceeding of International Conference on Evolutionary computation, pp. 1063–1068. IEEE, Sofia (2008)

A Self-learning Bare-Bones Particle Swarms Optimization Algorithm

Jian Chen, Yuanxia Shen$^{(\boxtimes)}$, and Xiaoyan Wang

School of Computer Science and Technology,
Anhui University of Technology, Maanshan, Anhui, China
{jianchen_ah,chulisyx}@163.com

Abstract. In order to solve the premature convergence of BBPSO, this paper proposes a self-learning BBPSO (SLBBPSO) to improve the exploration ability of BBPSO. First, the expectation of Gaussian distribution in the updating equation is controlled by an adaptive factor, which makes particles emphasize on the exploration in earlier stage and the convergence in later stage. Second, SLBBPSO adopts a novel mutation to the personal best position (*Pbest*) and the global best position (*Gbest*), which helps the algorithm jump out of the local optimum. Finally, when particles are in the stagnant status, the variance of Gaussian distribution is assigned an adaptive value. Simulations show that SLBBPSO has excellent optimization ability in the classical benchmark functions.

Keywords: BBPSO · Mutation · Premature convergence

1 Introduction

Bare-bones particle swarms optimization (BBPSO) was first proposed by Kennedy in 2003 [1]. In BBPSO, the item of speed is removed and Gaussian sampling with the information of personal and global optimal position is used to update particles position. Pan proved that BBPSO can be mathematically deduced from the standardized PSO [11]. Compared with traditional PSO [2], BBPSO is simpler because it does not involve some parameters, including inertia weight, acceleration factor, velocity threshold and so on. Because of its simplicity and efficiency, BBPSO is applied to some application areas such as image feature selection [3] and gene selection and classification [4].

Although BBPSO has showed its potential to solve practical problems but there exists the premature convergence. In order to improve search performance, Krohling and Mendel proposed a jump strategy for BBPSO to prevent the premature convergence in 2009 [5]; Blackwell and Masjid presented another jump strategy [6]; Orman and Haibo adopted the mutation and crossover operations

---

This work was supported by National Natural Science Foundation of China under Grant Nos.61300059. Provincial Project of Natural Science Research for Anhui Colleges of China (KJ2012Z031, KJ2012Z024).

© Springer International Publishing Switzerland 2015
Y. Tan et al. (Eds.): ICSI-CCI 2015, Part I, LNCS 9140, pp. 107–114, 2015.
DOI: 10.1007/978-3-319-20466-6_12

of differential algorithm to enhance BBPSO [7]; Zhang proposed an adaptive BBPSO which is based on cloud model in 2011 [8]. Meanwhile, Zhang developed a disturbance to keep swarm diversity in each dimension of each particle [9].

In order to maintain the relative balance between swarm diversity and convergence speed, this paper proposes a self-learning BBPSO (SLBBPSO). In SLBBPSO, the expectation of Gaussian distribution is not the arithmetic mean of the personal best position (*Pbest*) and the global best position (*Gbest*), but is the weighted average of *Pbest* and *Gbest* in which an adaptive factor is used to harmonize exploration and exploitation. In order to jump out of the local optima, the mutation operation is introduced to *Pbest* and *Gbest*. If the position of a particle is the position of *Gbest*, and then this particle is in the stagnant status in the next iteration as the variance of the Gaussian sampling is zero. In order to solve this problem, the variance of Gaussian sampling of this particle is assigned an adaptive value.

The structure of the remaining part of this paper is as follows: the second section briefly introduces the bare-bones particle swarms optimization (BBPSO). The self-learning bare-bones particle swarms optimization (SLBBPSO) presents in the third section. The fourth section focuses on the specific performance of SLBBPSO by contrast with other improved algorithms in BBPSO. Finally, the conclusion is given in the fifth section.

2   Bare-Bones PSO

Bare-bones particle swarms optimization (BBPSO) does not consider the item of speed, but use the information of *Pbest* and *Gbest* to update the position of the particles. The specific formula is as follows:

$$X_{i,j}(t+1) = N(\mu, \delta)$$
$$\mu = (Pbest_{i,j}(t) + Gbest_{i,j}(t))/2 \qquad (1)$$
$$\delta = |(Pbest_{i,j}(t) - Gbest_{i,j}(t)|$$

where the particles position subjects to a Gaussian distribution in which the mean is $\mu = (Pbest_{i,j}(t) + Gbest_{i,j}(t))/2$ and the standard deviation is $\delta = |(Pbest_{i,j}(t) - Gbest_{i,j}(t)|$.

Furthermore, Kennedy proposed a alternative BBPSO, named exploiting bare-bones PSO (BBPSO-E) [10]. The formula 1 is replaced by following formula 2:

$$X_{i,j}(t+1) = \begin{cases} N(\mu, \delta) & rand < 0.5 \\ Pbest_{i,j}(t) & otherwise \end{cases}$$
$$\mu = (Pbest_{i,j}(t) + Gbest_{i,j}(t))/2 \qquad (2)$$
$$\delta = |(Pbest_{i,j}(t) - Gbest_{i,j}(t)|$$

where the each dimension of particle changes to corresponding *Pbest* with 50% chance.

3   Self-learning Bare-Bones PSO

According to formula 1 and 2, if the *Pbest* of the particle is equal to *Gbest*, and then the particle should be in the stagnant status due to the variance of Gaussian distribution is zero. As the fast convergence of BBPSO, the swarm is more likely to be trapped into premature convergence. To solve this problem, SLBBPSO develops three strategies.

First, the expectation of Gaussian distribution is an adaptive value rather than the mean value of *Gbest* and *Pbest*, so algorithm enhances the global searching ability in the earlier stage and ensures the timely convergence in the later stage; Moreover, it will give particle an adaptive standard deviation to allow the particle continue to search when the particles *Pbest* is the same as the *Gbest*.

Taking the position updating equations of particle i as an example, new updated equation is as follows:

$$X_{i,j}(t+1) = \begin{cases} N(\mu, \delta) & rand < 0.5 \\ Pbest_{i,j}(t) & otherwise \end{cases}$$
$$\mu = l * Pbest_{i,j}(t) + (1 - l) * Gbest_{i,j}(t) \tag{3}$$
$$\delta = \begin{cases} R & |Pbest_{i,j}(t) - Gbest_{i,j}(t)| = 0 \\ |Pbest_{i,j}(t) - Gbest_{i,j}(t)| & otherwise \end{cases}$$

$Pbest_{i,j}(t)$ is the personal optimal position of particle i in j-th dimension at t-th iteration. $Gbest_{i,j}(t)$ is the global optimal position of particle i in j-th dimension at t-th iteration. l linearly decreases from 1 to 0 with the evolution iterations increasing, the variable l ensures particles' early exploration and later convergence, this is what we hope. The equation is as follows:

$$l = 1 - iter/iterMax \tag{4}$$

R is an adaptive factor, the equation is as follows:

$$R = R\max - (R\max - R\min) * iter/iterMax \tag{5}$$

where *iter* is the current number of particles iterations. *iterMax* is the maximum number of iterations allowed. *Rmax*, *Rmin* are default parameters, the *Rmax* is set to 1 and the *Rmin* is set to 0.1 in this paper according to the reference [12]. Variance will linearly decrease from *Rmax* to *Rmin* with evolution generations increasing when the *Pbest* is the same as the *Gbest*.

The positions updating of particles are dependent on the information of *Gbest* and *Pbest*. Hence, the mutation operation is employed to *Gbest* and *Pbest* with a certain probability, which can help the swarm jump out of local optimum effectively.

Taking particle i as an example, the mutation operation to *Pbest* is as follows:

$$Pbest'_{i,j} = \begin{cases} Pbest_{i,j} + sgn(r1) * r * (\overline{X}_{i,j} - \underline{X}_{i,j}) & rand < Ppb \\ Pbest_{i,j}(t) & otherwise \end{cases} \tag{6}$$

where r is a random number generated by a normal distribution whose mean is 0 and variance is R; $r1$ is a random number generated by a uniform distribution whose range is [-1, 1]. The sgn() is sign function. $\overline{X}_{i,j}$ and $\underline{X}_{i,j}$ are the upper and lower bounds of the search space in j-th dimension. The mutation operation to $Gbest$ is as follows:

$$Gbest'_{i,j} = \begin{cases} Gbest_{i,j} + sgn(r1) * r * (\overline{X}_{i,j} - \underline{X}_{i,j}) & rand < Pgb \\ Gbest_{i,j}(t) & otherwise \end{cases} \quad (7)$$

where Ppb and Pgb are the mutational probabilities of $Pbest$ and $Gbest$ respectively.

When the mutation value is better than before, SLBBPSO accept the new value, otherwise it still retain the old value.

The concrete steps of SLBBPSO are as follows:

Step1: Set up main parameters: $popSize$ (number of swarms), $dimSize$ (dimension of search space), $threshold$, Ppb (the personal optimal mutating probability), Pgb (the global optimal mutating probability) and so on;

Step2: Initialize swarms: Set up the initial position for each particle, in the initial time, set personal optimal position $Pbest$ as the particle current position, and set the global optimal position $Gbest$ as the current position optimum $Pbest$;

Step3: Update the position of particles according to the formula 3;

Step4: Mutate swarms according to the algorithm in the formula 5, 6;

Step5: If the number reaches the maximum of evolution iterations, progress will jump to the Step6, otherwise it will jump to the Step3;

Step6: Output the searched optimum solution.

4   Simulation Experiments

In order to test the performance, SLBBPSO is compared with 6 BBPSO-based algorithms. These algorithms are listed as follows:

- BBEXP: alternative BBPSO proposed by Kennedy (2003) [1];
- ABPSO: an adaptive bare-bones particle swarm optimization algorithm (Zhang et al.2014) [8];
- BBDE: the bare-bones differential evolution (Mahamed, Omran, Andries, Salman, Ayed 2009) [7];
- BBPSO-MC: BBPSO with mutation and crossover (Haibo, Kennedy, Rangaiah et al.2011) [10];
- BBPSO-GJ: BBPSO with Gaussian jump (Krohling et al.2009) [6];
- BBPSO-CJ: BBPSO with Cauchy jump (Krohling et al.2009) [6].

These selected algorithms are implemented to 10 benchmark functions. Table 1 shows the specific details of these functions.

For the 10 functions, this paper sets 50000 iterations and 30 running times; The size of swarm is 20; Assuming that the optimal position of a problem is

Table 1. Name, dimension, global optimum, range, optimal and accuracy of the test functions

Functions	Name	Dimension	Range	Optimal	Accuracy
f1	Sphere	30	[-100,100]	0	0
f2	Rosenbrock	30	[-30,30]	0	0
f3	Quadric	30	[-1.28,1.28]	0	0
f4	Quadric with noise	30	[-1.28,1.28]	0	0
f5	Ackley	30	[-32,32]	0	0
f6	Griewank	30	[-600,600]	0	0
f7	Schewfel problem 2.26	30	[-500,500]	-12569.5	0
f8	Shifted Ackley	30	[-32,32]	-140	0
f9	Expanded Extended Griewank's plus Rosenbrock's Function (F8F2)	30	[-5,5]	-130	0
f10	Shifted Rotated Ackley's Function with Global Optimum on Bounds	30	[-5,5]	-140	0

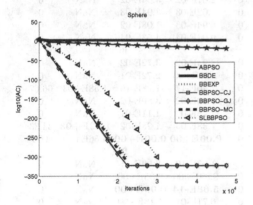

Fig. 1. Average optimum value with time for Sphere

Fig. 2. Average optimum value with time for Griewank

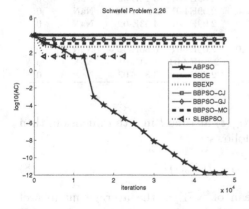

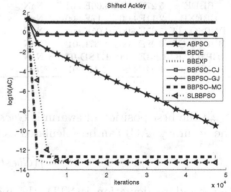

Fig. 3. Average optimum value with time for Schwefel Problem 2.26

Fig. 4. Average optimum value with time for Shifted Ackley

Table 2. Experimental result

Algorithms	AC	Std	NFE	SR(%)	AC	Std	NFE	SR(%)
Functions		f1				f2		
ABPSO	5.40E-21	5.28E-21	NaN	0	3.25E+01	3.49E+01	NaN	0
BBDE	3.54E+02	2.38E+02	NaN	0	4.47E+05	4.17E+05	NaN	0
BBEXP	**0.00E+00**	**0.00E+00**	2.49E+04	100	1.40E+00	2.25E+00	NaN	0
BBPSO-GJ	4.94E-324	**0.00E+00**	NaN	0	**1.05E-04**	**1.19E-04**	NaN	0
BBPSO-CJ	1.98E-323	**0.00E+00**	NaN	0	6.28E-04	1.04E-03	NaN	0
BBPSO-MC	**0.00E+00**	**0.00E+00**	**2.30E+04**	100	8.68E-04	1.33E-03	NaN	0
SLBBPSO	**0.00E+00**	**0.00E+00**	3.14E+04	100	2.76E+00	4.49E+00	NaN	0
Functions		f3				f4		
ABPSO	5.45E-03	2.59E-04	NaN	0	5.95E-03	6.76E-04	NaN	0
BBDE	5.10E-01	1.81E-01	NaN	0	6.08E-02	4.90E-02	NaN	0
BBEXP	6.95E-32	7.47E-32	NaN	0	**8.41E-04**	**1.96E-04**	NaN	0
BBPSO-GJ	**8.67E-49**	**1.45E-48**	NaN	0	2.07E-02	3.04E-02	NaN	0
BBPSO-CJ	5.88E-48	7.99E-48	NaN	0	4.02E-03	3.94E-03	NaN	0
BBPSO-MC	8.62E-33	1.11E-32	NaN	0	6.94E-03	1.08E-02	NaN	0
SLBBPSO	8.08E-30	4.58E-30	NaN	0	9.04E-04	4.53E-04	NaN	0
Functions		f5				f6		
ABPSO	3.70E-11	2.76E-11	NaN	0	3.94E-02	3.72E-02	NaN	0
BBDE	3.98E+00	5.30E-01	NaN	0	1.27E+00	2.72E-01	NaN	0
BBEXP	**7.11E-15**	**0.00E+00**	NaN	0	7.40E-17	1.28E-16	1.58E+03	66.67
BBPSO-GJ	2.45E+00	1.37E+00	NaN	0	2.19E-02	3.80E-02	NaN	0
BBPSO-CJ	2.61E-14	3.28E-14	NaN	0	7.75E-02	1.11E-01	NaN	0
BBPSO-MC	3.20E-14	6.15E-15	NaN	0	7.38E-03	1.28E-02	**1.42E+03**	33.33
SLBBPSO	**7.11E-15**	**0.00E+00**	NaN	0	**0.00E+00**	**0.00E+00**	4.56E+03	**100**
Functions		f7				f8		
ABPSO	1.82E-12	**0.00E+00**	NaN	0	1.03E-10	8.59E-11	NaN	0
BBDE	1.08E+04	3.45E+01	NaN	0	9.89E+00	3.35E+00	NaN	0
BBEXP	4.87E+02	2.68E+02	NaN	0	**5.68E-14**	**0.00E+00**	NaN	0
BBPSO-GJ	3.12E+03	8.00E+02	NaN	0	6.71E-01	1.16E+00	NaN	0
BBPSO-CJ	3.40E+03	7.21E+02	NaN	0	5.49E-01	9.50E-01	NaN	0
BBPSO-MC	1.20E+03	5.73E+02	NaN	0	2.18E-13	1.64E-14	NaN	0
SLBBPSO	**0.00E+00**	**0.00E+00**	1.42E+04	100	**5.68E-14**	**0.00E+00**	NaN	0
Functions		f9				f10		
ABPSO	3.00E+00	7.58E-01	NaN	0	2.09E+01	3.13E-02	NaN	0
BBDE	5.27E+01	6.66E+01	NaN	0	2.09E+01	**1.11E-02**	NaN	0
BBEXP	2.11E+00	1.17E+00	NaN	0	2.09E+01	4.53E-02	NaN	0
BBPSO-GJ	5.65E+00	4.05E+00	NaN	0	2.09E+01	5.51E-02	NaN	0
BBPSO-CJ	3.28E+00	8.37E-01	NaN	0	2.09E+01	6.50E-02	NaN	0
BBPSO-MC	**1.53E+00**	**1.18E-01**	NaN	0	**2.08E+01**	1.34E-01	NaN	0
SLBBPSO	1.94E+00	2.41E-01	NaN	0	2.09E+01	3.39E-02	NaN	0

X_{opt}, the best position of swarm is $Gbest_{i,j}(t)$ at time t in j-th dimension, then the accuracy (AC) can be calculated as follows:

$$AC = f(Gbest_{i,j}) - f(X_{opt}) \tag{8}$$

the standard deviation is STD, the mean of NFE is the average number of function evaluations required to find the global optima are considered, the SR is successful ratio. The Ppb is set to 0.3 and the Pgb is set to 0.7 in SLBBPSO. The results of the experiments are shown in Table 2.

From Table 2, we can know that SLBBPSO is able to find the optimal solution search space in less iterations for simple unimodal function f1, only BBPSO-MC is better than it; For the more complex functions f2 and f3, the improved BBPSOs (BBPSO-CJ, BBPSO-GJ, BBPSO-MC) performs better than SLBBPSO. But for the unimodal function f4, SLBBPSO still has strong competitiveness. Thus SLBBPSO is more excellent in classic unimodal function than others.

For multimodal function f5 and f8, BBEXP and SLBBPSO obtain the better value than other algorithms. For the f6 and f7, only SLBBPSO can find the global best solution with the 100% successful ratio and less iterations. For f9 and f10, SLBBPSO also can find the promising solution. Overall, SLBBPSO is competitive in multimodal functions.

5   Conclusion

This paper proposes a self-learning BBPSO (SLBBPSO) which adopts three strategies in order to solve the problem of premature convergence. First, the expectation of Gaussian distribution is controlled by an adaptive factor, so algorithm enhances the global searching ability in the earlier stage and ensures the timely convergence in the later stage; Then it will give particle an adaptive standard deviation to allow the particle continue to search when the particles *Pbest* is the same as the *Gbest*; Finally, the *Pbest* and *Gbest* mutate at a certain probability, and the novel mutation operation makes the swarm jump out of local optimum effectively. The experimental results of 10 benchmark functions show that SLBBPSO has strong competitiveness compared with other BBPSO-based algorithms.

Acknowledgments. This work was supported by National Natural Science Foundation of China under Grant Nos.61300059. Provincial Project of Natural Science Research for Anhui Colleges of China (KJ2012Z031, KJ2012Z024).

References

1. Kennedy, J.: Bare bones particle swarms. In: Proceedings of the IEEE Swarm Intelligence Symposium, pp. 80–87. IEEE (2003)
2. Kennedy, J., Eberhart, R.C.: Particle swarm optimization. In: Proceeding of International Conference on Neural Networks, pp. 1942–1948. IEEE, Perth (1995)
3. Yong, Z., Dunwei, G., Ying, H., Wanqiu, Z.: Feature Selection Algorithm Based on Bare Bones Particle Swarm Optimization. Neurocomputing **148**, 150–157 (2015)
4. Liyeh, C., Chengsan, Y., Kuochuan, W., Chenghong, Y.: Gene selection and classfication using Taguchi chaotic binary particle swarm optimization. Expert Systems with Applications **38**, 13367–13377 (2011)
5. Krohling, R.A., Mendel, E.: Bare bones particle swarm optimization with Gaussian or Cauchy jumps. In: Proceedings of the IEEE Congress on Evolutionary Computation, pp. 3285–3291. IEEE, Trondheim (2009)

6. al-Rifaie, M.M., Blackwell, T.: Bare bones particle swarms with jumps. In: Dorigo, M., Birattari, M., Blum, C., Christensen, A.L., Engelbrecht, A.P., Groß, R., Stützle, T. (eds.) ANTS 2012. LNCS, vol. 7461, pp. 49–60. Springer, Heidelberg (2012)

7. Omran, M.G.H., Andries, P., Salman, A.: Bare bones differential evolution. European Journal of Operational Research **196**, 128–139 (2009)

8. Junqi, Z., Lina, N., Jing, Y., Wei, W., Zheng, T.: Adaptive bare bones particle swarm inspired by cloud model. IEICE Transactions on Information and Systems **E94D**(8), 1527–1538 (2011)

9. Yong, Z., Dunwei, G., Xiaoyan, S., Na, G.: Adaptive bare-bones particle swarm optimization algorithm and its convergence analysis. Soft Computing **18**, 1337–1352 (2014)

10. Haibo, Z., Kennedy, D.D., Rangaiah, G.P.: Bonilla-Petriciolet: A novel bare-bones particle swarm optimization and its performancefor modeling vapor-liquid equilibrium data. Fluid Phase Equilib. **301**, 33–45 (2011)

11. Feng, P., Xiaohui, H., Eberhart R.C., Yaobin, C.: An analysis of bare bones particle swarm. In: Proceeding of the 2008 IEEE Swarm Intelligence Symposium, pp. 21–23 (2008)

12. Lim, W.H., Isa, N.A.M.: Teaching and peer-learning particle swarm optimization. Applied Soft Computing **18**, 39–58 (2014)

Improved DPSO Algorithm with Dynamically Changing Inertia Weight

Jing Xin$^{(\boxtimes)}$, Cuicui Yan, and Xiangshuai Han

Xi'an University of Technology, Xi'an, China
xinj@xaut.edu.cn, {yancuic,hanxiangshuai}@163.com

Abstract. Population Diversity in Particle Swarm Optimization (DPSO) algorithm can effectively balance the "exploration" and "exploitation" ability of the PSO optimization algorithm and improve the optimization accuracy and stability of standard PSO algorithm. However, the accuracy of DPSO for solving the multi peak function will be obviously decreased. To solve the problem, we introduce the linearly decreasing inertia weight strategy and the adaptively changing inertia weight strategy to dynamically change inertia weight of the DPSO algorithms and propose two kinds of the improved DPSO algorithms: linearly decreasing inertia weight of DPSO (Linearly-Weight- Diversity-PSO, LWDPSO) and adaptively changing inertia weight of DPSO (Adaptively-Weight-Diversity-PSO, AWDPSO). Three representative benchmark test functions are used to test and compare proposed methods, which are LWDPSO and AWDPSO, with state-of-the-art approaches. Experimental results show that proposed methods can provide the higher optimization accuracy and much faster convergence speed.

Keywords: Particle swarm optimization · Population diversity · Dynamically change · Inertia weight

1 Introduction

Particle Swarm Optimization (PSO) is proposed by Eberhart and Kennedy in 1995 [1], it is a stochastic and population based global optimization algorithm for simulating such social behavior of groups as bird flocking. It has been widely used to solve a large number of non-linear, non-differentiable and complex multi-peak function optimization problems due to its better computational efficiency and ability to quickly converge to a reasonably good solution [2]. And it has also been widely used in the fields of science and engineering, such as neural network training [3], economic dispatch [4], pattern recognition [5], structure design [6], electromagnetic field [7] and so on. Compared with other stochastic optimization algorithms, PSO algorithm has a higher convergence speed in the earlier search phase and is easy to fall into local optimum in the later search phase, especially for solving complex non-linear problems. To improve the performance of PSO, many improved PSO algorithms have been proposed. Generally, these improved algorithms could be classified into two categories as follows:

© Springer International Publishing Switzerland 2015
Y. Tan et al. (Eds.): ICSI-CCI 2015, Part I, LNCS 9140, pp. 115–123, 2015.
DOI: 10.1007/978-3-319-20466-6_13

1) *Parameters-based algorithms.* Most of the parameters-based improved PSO algorithm focuses on the change of the inertia weight [8, 9, 10, 11, 12, 13], the learning factors [14] and the size of the population [15, 16]. These improvements are efficient to improve the accuracy and convergence speed of the standard PSO.

2) *Diversity-based algorithms.* Two representative diversity-based improved PSO algorithms are quantum-behaved PSO (QPSO) algorithm proposed by Sun [17] and DPSO proposed by Shi [18, 19, 20]. These improved algorithms focus on enhancing the diversity of the population and find the optimal solution efficiently through introducing the difference between the current position and the mean velocity of all particles to the particle's position update formula. However, the optimization accuracy and convergence speed of the above algorithms would decrease for the non-linear optimization problem.

In order to further improve the optimization accuracy and convergence speed of DPSO, we propose two kinds of improved DPSO algorithm: LWDPSO and AWDPSO. Experimental results on the benchmark test functions show the effectiveness of the proposed LWDPSO and AWDPSO algorithm.

2   Original DPSO Algorithm

In PSO system, each particle represents a potential feasible solution to an optimization problem. PSO algorithm will find the best solution by adjusting the particle's velocity and position according to the velocity and position update strategy. Different PSO algorithms adopt the different particle velocity and/or position update strategy. In DPSO algorithm [20], the particle velocity and position update strategy can be described as follows:

$$V_i(t+1) = wV_i(t) + c_1 \text{rand}(\cdot)\left(P_{\text{best}_i} - X_i(t)\right) + c_2 \text{rand}(\cdot)\left(G_{\text{best}} - X_i(t)\right). \tag{1}$$

$$X_i(t+1) = X_i(t) + V_i(t+1) + c_3 \text{rand}(\cdot)\left(X_i(t) - \overline{V}(t)\right). \tag{2}$$

Where, $V_i(t)$ and $X_i(t)$ represent the current velocity and position of the i-th particle respectively; for D-dimensional space, $V_i = [V_{i1}, V_{i2}, \cdots, V_{id}]'$, $X_i = [X_{i1}, X_{i2}, \cdots, X_{id}]'$; $P_{\text{best}_i} = \left[P_{\text{best}_{i1}}, P_{\text{best}_{i2}}, \cdots, P_{\text{best}_{id}}\right]'$ represents the local best position (or local best solution) obtained so far by the i-th particle; $G_{\text{best}} = [G_{\text{best}1}, G_{\text{best}2}, \cdots, G_{\text{bestn}}]'$ represents the best position (or global best solution) obtained so far by all particles; t represents the current iteration number for total n iterations; w, which is originally proposed by Shi and Eberhart [10], represents velocity inertia weight coefficient that control the exploration and exploitation; c_1 and c_2 are nonnegative learning factors which represent the influence of social and cognitive components, in other words, the parameters of the c_1 and c_2 control particle towards individual best solution and global best solution; rand($\cdot$) is a uniformly distributed random number between 0 and 1 and it is used to introduce a stochastic element in the search process; c_3 represents diversity inertia weight. $\overline{V}(t)$ represents the mean velocity of all particles at iteration t.

Compared to standard PSO, DPSO adds diversity control item "c_3" to the position update formula of standard PSO shown in formula (2). Then the particle position relates not only to the particles themselves, but also to the mean velocity of all particles. The improvement enhances the information interaction between the particles and population diversity.

3   Proposed Improved DPSO Algorithms

A good optimization algorithm could balance the local and global search abilities. For DPSO algorithms, we need a better global search in starting phase to make the optimization algorithm converge to a proper area quickly and then we need a stronger local search to find high precision value. The inertia weight w determines the contribution rate of a particle's previous velocity to its velocity at the current iteration t and that control the exploration (global search) and exploitation (local search). At the same time many researches indicate that the bigger value of inertia weight w encourages the exploration and smaller value of inertia weight w encourages for the exploitation.

Therefore we need to set w as a variable value and not constant. A linearly decreasing inertia weight w can efficiently balance local and global search abilities and make the optimization algorithms to obtain the higher precision value [10]. Researches also indicate that adaptively changing inertia weight w according to the real-time state of all particles can improve the convergence speed of the algorithm [11]. So, we introduces linearly decreasing w strategy and adaptively changing w strategy to the DPSO algorithm respectively, then propose two kinds of improved DPSO algorithms: Linearly decreasing Inertia weight of DPSO (LWDPSO) and adaptively changing inertia weight of DPSO (AWDPSO). The detail introduction on proposed algorithm is as follows:

3.1   The Principle of Proposed LWDPSO Algorithm

The main idea of proposed linearly decreasing inertia weight of DPSO (LWDPSO) algorithm is to introduce a linearly decreasing strategy to adjust the value of diversity inertia weight c3 in the particles position update equation (shown in formula (2)) and velocity inertia weight w in the particles velocity update equation (shown in formula (1)) of DPSO. Compared to the original DPSO algorithm, at the premise of diversity assurance, LWDPSO algorithm can search the best solution in the global area in the earlier search phase to prevent premature convergence effectively. With the increase in the number of iterations, the global search gradually turn into refine local search, in the later search phase, the particles are concentrated in local search, which can assure to find the best solution with larger probability and higher accuracy. So the improved DPSO algorithm can balance the global search ability and the local search ability effectively. The velocity and position update formula used in LWDPSO algorithm are respectively as follow:

$$V_i(t+1) = \left[\frac{t_{max}-t}{t_{max}}(w_{max}-w_{min}) + w_{min} \right] V_i(t) + c_1 \text{rand}(\cdot)(P_{best_i} - X_i(t)) + c_2 \text{rand}(\cdot)(G_{best} - X_i(t)). \quad (3)$$

$$X_i(t+1) = X_i(t) + V_i(t+1) + \left[\frac{t_{max}-t}{t_{max}}(c_{3max} - c_{3min}) + c_{3min} \right] \text{rand}(\cdot)(X_i(t) - \overline{V}(t)). \quad (4)$$

Where, t represents the current iteration number for total t_{max} iterations; w_{max} is the maximum inertia weight and w_{min} is minimum inertia weight; It can been seen from the first item of the velocity update formula (3) that the value of inertia weight w is linearly decreased from an initial value (w_{max}) to a final value (w_{min}). It improves global search capability of the DPSO and make DPSO algorithm find the fit area quickly. As the value of w gradually reduces, the velocities of the particles slow down and DPSO algorithm start refine local search to obtain high precision solution; c_{3max} and c_{3min} represent the maximum and diversity inertia weight respectively. It can be seen from the third item of the position update formula (4) that the value of diversity inertia weight c_3 is also linearly decreased from an initial value (c_{3max}) to a final value (c_{3min}). Research results in [20] indicate that if c_3 is positive, the population diversity increases, and the search range is larger than the standard PSO; if c_3 is negative, the population diversity decreases, and the convergence speed is faster than the standard PSO. Therefore, we initialize c_3 a positive number, and gradually decrease c_3 to a negative number. Thus, performance of DPSO will be further improved. Our preliminary simulation results about the selection of the value of c_3 show that the best performance of DPSO will be achieved by linearly decreasing inertia weight c_3 from 0.1 to -0.1. The implementation process of proposed LWDPSO is shown in Table 1.

Table 1. The implementation process of proposed LWDPSO algorithm

Step1:	Initialization, generate N particles;
Step2:	Calculate the fitness value of each particle, determine the individual optimum P_{best_i} and the global optimum G_{best};
Step3:	Update the particle velocity according to the formula (3);
Step4:	Update the particle position according to the formula (4);
Step5:	Determine whether the termination condition are met, and if so, outputting the best solution, otherwise, back to step2.

3.2   The Principle of Proposed AWDPSO Algorithm

The main idea of proposed adaptively changing inertia weight of DPSO (AWDPSO) algorithm is

1) to introduce a linearly decreasing strategy to adjust the value of diversity inertia weight c_3 of DPSO algorithm in the particles position update equation (shown in formula (2))
2) to introduce an adaptively changing inertia weight strategy to adjust the value of velocity inertia weight w of DPSO algorithm in the particles velocity update equation (shown in formula (1)).

Compared to the original DPSO algorithm, at the premise of diversity assurance, AWDPSO algorithm can effectively improve the algorithm convergence speed and achieve a relatively optimal accuracy and speed. The velocity and position update formula used in AWDPSO algorithm are respectively as follows:

$$V_i(t+1) = \left[w_{max} - e \cdot w_e + s \cdot w_s \right] V_i(t) + c_1 \, rand(\cdot)(P_{best_i} - X_i(t)) + c_2 \, rand(\cdot)(G_{best} - X_i(t)). \quad (5)$$

$$X_i(t+1) = X_i(t) + V_i(t+1) + \left[\frac{t_{max} - t}{t_{max}}(c_{3max} - c_{3min}) + c_{3min} \right] rand(\cdot)(X_i(t) - \bar{V}(t)). \quad (6)$$

$$e = \frac{\min\left(G_{best}(t), G_{best}(t-1)\right)}{\max\left(G_{best}(t), G_{best}(t-1)\right)}. \quad (7)$$

$$s = \frac{\min\left(G_{best}(t), \bar{P}_{best}(t)\right)}{\max\left(G_{best}(t), \bar{P}_{best}(t)\right)}. \quad (8)$$

$$\bar{P}_{best}(t) = \frac{1}{N}\sum_{i=1}^{N} P_{best_i}(t). \quad (9)$$

Where, e (shown in formula.7) is the evolution speed factor ($0 < e \leq 1$) which represents that the current global optimal value $G_{best}(t)$ is always better than or at least equal to the previous global optimal value $G_{best}(t-1)$. w_e is the weight coefficient of the evolution speed factor e. s (shown in formula.8) is the aggregation degree factor which depicts that the current global optimal value $G_{best}(t)$ always better than the current mean of all individual optimal value, $\bar{P}_{best}(t)$, w_s is the weight coefficient of the aggregation factor s. Researches indicate that the algorithm would have a good adaptive performance [13] if we select the value of w_e from 0.4 to 0.6 and the value of w_s from 0.05 to 0.20. The implementation process of proposed AWDPSO is shown in Table 2.

Table 2. The implementation process of proposed AWDPSO algorithm

Step1: Initialization, generate N particles;
Step2: Calculate the fitness value of each particle, determine the individual optimum P_{best_i} and the global optimum G_{best};
Step3: Update the particle velocity according to the formula (5);
Step4: Update the particle position according to the formula (6);
Step5: Determine whether the termination condition are met, and if so, outputting the best solution, otherwise, back to step2.

4   Experimental Research

Three representative benchmark test functions are used to test proposed methods (LWDPSO and AWDPSO) with state-of-the-art approaches and the results are compared based on the final accuracy and convergence speed.

4.1   Experimental Setup

Table 3 provides a detailed description of these functions. Among them, Sphere function is a unimodal quadratic function; Rastrigrin function is a simple multimodal function with a large amount of local minimum points; Griewank function is a typical non-linear pathological multi-mode function with a wide range of search space and a large number of local minimum points. The dimensions D of the three test functions are taken as 2, 3, 5 and 10 respectively. For each test function, x^* represents the best solution to the optimization problem and $f(x^*)$ represents the best value or best achievable fitness for that function. All functions have symmetric search spaces.

Table 3. Benchmark test functions used in the experiments

Function Name	Function expression	Search Space & Global optimal
Sphere	$$f_1(x) = \sum_{i=1}^{D} x_i^2$$	$-5.12 \le x_i \le 5.12,$ $x^* = (0,\cdots,0), f(x^*) = 0$
Rastrigrin	$$f_2(x) = 10 + \sum_{i=1}^{n}(x_i^2 - 10\cos(2\pi x_i))$$	$-5.12 \le x_i \le 5.12,$ $x^* = (0,\cdots,0), f(x^*) = 0$
Griewank	$$f_3(x) = \sum_{i=1}^{n}\frac{x_i^2}{4000} - \prod_{i=1}^{n}\cos(x_i/\sqrt{i}) + 1$$	$-600 \le x_i \le 600,$ $x^* = (0,\cdots,0), f(x^*) = 0$

In all the experiments carried out in this paper, the number of particles in the swarm N is 100 and the maximum number of iteration t_{max} is 1000. The value of learning factors c_1, c_2 is 2 and the value of w_e (weight coefficient of the evolution speed factor) and w_s are 0.5 and 0.1, respectively.

4.2   Experimental Results and Discussion

Standard PSO[2] and original DPSO [20] are applied to the above three representative benchmark test functions and the results are compared with LWDPSO and AWDPSO proposed in this paper. Table 4 ~ Table 6 list the mean , variance and average running time (T, unit: s) of the best solutions found by each algorithm in 30 independent runs for three test functions in 4 different dimensions . The results indicate that proposed LWDPSO and AWDPSO can provide the higher accuracy in all test functions in 4 different dimensions.

Convergence curves of different PSO algorithms for test functions can provide more insight into their searching behavior. Fig. 1~ fig.3 show the average convergence curve of the best fitness value found by four algorithms on three test functions in 30 runs, respectively. For the same function, the average convergence curve of different dimensions show a similar varying tendency, so, here we just give the one of dimension D=10. It can be seen from the fig. 1~ fig.3 that proposed LWDPSO and AWDPSO can provide the much faster convergence speed in all test functions, and AWDPSO is much faster than LWDPSO in convergence to the optimum due to the adaptive nature of this algorithm.

Table 4. The best solutions of four algorithms on Sphere function in 1000 function iteration

Sphere		PSO[3]	DPSO[20]	LWDPSO(proposed)	AWDPSO(proposed)
	mean	0	0	0	0
D=2	var	0	0	0	0
	T	81.8330	82.4030	83.6310	78.9310
	mean	0.0217	0.0064	8.1294e-70	3.9177e-110
D=3	var	0.0022	0.0004	1.1998e-137	3.0292e-183
	T	105.9770	112.7070	115.2170	108.1860
	mean	0.2327	0.1142	3.1326e-53	1.883719e-93
D=5	var	6.8342	5.1533	6.5872e-105	3.1322e-142
	T	163.6530	175.5960	186.9470	169.3020
	mean	1.1459	0.7881	8.3276e-33	3.3653e-67
D=10	var	6.3427	5.2331	7.3226e-58	3.41755e-94
	T	297.0110	331.9380	337.9110	321.1240

Table 5. The best solutions of four algorithms on Rastrigin function in 1000 function iteration

Rastrigin		PSO[3]	DPSO[20]	LWDPSO(proposed)	AWDPSO(proposed)
	mean	0.0013	0.0011	0	0
D=2	var	4.4929e-006	1.5533e-06	0	0
	T	86.3730	89.8340	87.7020	84.8370
	mean	0.2139	0.2677	0	0
D=3	var	0.0305	0.0601	0	0
	T	117.0390	120.0390	116.2920	115.4870
	mean	5.7616	5.4461	0	0.0332
D=5	var	1.9117	1.8800	0	0.0330
	T	176.5150	176.2770	176.1460	169.1100
	mean	37.5536	38.1118	1.4293	2.2221
D=10	var	49.1731	31.7218	1.2113	1.5224
	T	330.2700	329.9310	330.7020	313.7790

Table 6. The best solutions of four algorithms on Griewank function in 1000 function iteration

Griewank		PSO[3]	DPSO[20]	LWDPSO(proposed)	AWDPSO(proposed)
	mean	0.0046	0.0036	3.9988e-06	0
D=2	var	1.4126e-005	1.0471e-05	4.7971e-10	0
	T	75.1920	83.4480	83.3310	80.2150
	mean	0.0454	0.0435	0.0036	0.0023
D=3	var	3.3118e-004	0.0002	1.2310e-05	1.5892e-05
	T	103.1950	114.2090	114.1250	109.0790
	mean	0.3005	0.2950	0.0155	0.0145
D=5	var	0.0049	0.0052	5.2333e-05	6.8795e-05
	T	166.3530	175.5040	175.3370	172.9830
	mean	1.1347	1.1551	0.0645	0.0521
D=10	var	0.0050	0.0013	0.0008	0.0003
	T	294.6530	328.5480	329.4310	323.4050

5   Conclusion

In order to further improve the optimization accuracy and convergence speed of the original DPSO [20] algorithm, two kinds of improved DPSO algorithms are proposed in this paper. Experimental results on the three representative benchmark test function show that proposed algorithms can provide the higher accuracy and much faster convergence speed.

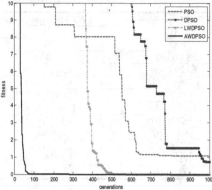

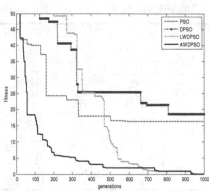

Fig. 1. The average convergence curve of Sphere function in 30 runs (D=10)

Fig. 2. The average convergence curve of Rastrigin function in 30 runs (D=10)

The proposed algorithms in this paper mainly focus on the dynamically adjusting the value of velocity inertia weight w of DPSO algorithm. In the first adjusting strategy, linearly decreasing strategy, the updating of the w is only associated with to the number of iterations. However, in the second adjusting strategy, adaptively changing strategy, the updating of the w is not only related to the number of iterations, but also closely related to the algorithm dynamic performance of all particles. Therefore, compared to the linearly decreasing strategy, adaptively changing strategy could further enhance the population diversity. Experimental results of the section 4 show that AWDPSO outper-

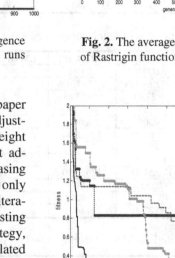

Fig. 3. The average convergence curve of Griewank function in 30 runs (D=10)

form the LWDPSO, and is in accord with the above theoretical analysis. In the future, we will further combine the proposed algorithm with non-rigid point clouds registration [21] and implement high-precision non-rigid point clouds registration.

Acknowledgment. This work is supported by the National Natural Science Foundation of China under Grant No. 61203345 and No. 61174101.

References

1. Eberhart, R., Kennedy, J.: A new optimizer using particle swarm theory. In: Proceedings of the Sixth International Symposium on Micro Machine and Human Science, pp. 39–43 (1995)
2. Shi, Y., Eberhart, R.: Empirical study of particle swarm optimization. In: 1999 Congress on Evolutionary Computing, vol. III, pp. 1945–1950 (1999)
3. Engelbrecht, A.: A Training product unit neural networks. Stability and Control: Theory and Applications **2**, 5972–5974 (1999)
4. Victoire, T., Jeyakumar, A.: Reserve constrained dynamic dispatch of units with valve-point effects. IEEE Trans. Power. Syst. **20**, 1273–1282 (2005)
5. Sousa, T., Silva, A., Neves, A.: Particle swarm based data mining algorithms for classification tasks. Parallel Computing, 767–783 (2004)
6. Elegbede, C.: Strutural reliability assessment based on particles swarm optimization. Structral Safety **27**, 171–186 (2005)
7. Pobinson, J., Rahmat-Samii, Y.: Particles swarm optimization in electromagnetics. IEEE Transaction on Antennas and Propagation **52**, 397–406 (2004)
8. Shi, Y.H., Eberahrt, R.C.: Parameter selection in particle swarm optimization. In: Porto, V., Waagen, D. (eds.) EP 1998. LNCS, vol. 1447, pp. 591–600. Springer, Heidelberg (1998)
9. Zhang, X., Du, Y., Qin, G., Qin, Z.: A dynamic adaptive inertia weight particle swarm optimization. Journal of Xi'an Jiaotong University **39** (2005)
10. Chen, G., Jia, J., Han, Q.: Study on the Strategy of Decreasing Inertia Weight in Particle Swarm Optimization Algorithm. Journal of Xi'an Jiaotong University **40** (2006)
11. Zhu, X., Xiong, W., Xu, B.: A Particle Swarm Optimization Algorithm Based on Dynamic Intertia Weight. Computer Simulation **24** (2007)
12. Bai, J., Yi, G., Sun, Z.: Random Weighted Hybrid Particle Swarm Optimization Algorithm Based on Second Order Oscillation and Natural Selection. Control and Decision **27** (2012)
13. Zhang, B.: Improved Particle Swarm Optimization algorithm and its application. Chongqing University, Chongqing (2007)
14. Mao, K., Bao, G., Xu, Z.: Particle Swarm Optimization Algorithm Based on Non-symmetric Learning Factor Adjusting. Computer Engineering **36** (2010)
15. Shi, Y., Eberhart, R.: Empirical study of particle swarm optimization. In: International Conference on Evolutionary Computation, Washington, USA (1999)
16. Zhang, W., Wang, G., Zhu, Z., Xiao, J.: Swarm optimization algorithm for population size selection. Computer Systems & Applications **9** (2010)
17. Sun, J., Feng, B., Xu, W.: Particle swarm optimization with particles having quantum behavior. In: IEEE Proceedings of Congress on Evolutionary Computation (2004)
18. Cheng, S., Shi, Y.: Diversity control in particle swarm optimization. In: Proceedings of 2011 IEEE Symposium on Swarm Intelligence, Paris, France, pp. 110–118 (2011)
19. Cheng, S., Shi, Y., Qin, Q.: Population diversity based study on search information propagation in particle swarm optimization. In: IEEE World Congress on Computational Intelligence, pp. 1272–1279. IEEE, Brisbane (2012)
20. Cheng, S.: Population Diversity in Particle Swarm Optimization: Definition, Observation, Control, and Application. Master thesis of University of Liverpool (2013)
21. Chui, H., Rnagarajna, A.: A new point matching algorithm for non-rigid registration. Computer Vision and Image Understanding **89**, 114–141 (2003)

An Improved PSO-NM Algorithm
for Structural Damage Detection

Zepeng Chen[1] and Ling Yu[1,2(✉)]

[1] Department of Mechanics and Civil Engineering,
Jinan University, Guangzhou 510632, China
[2] MOE Key Lab of Disaster Forecast and Control in Engineering,
Jinan University, Guangzhou 510632, China
lyu1997@163.com

Abstract. A hybrid particle swarm optimization (PSO) combined with an improved Nelder-Mead algorithm (NMA) is proposed and introduced into the field of structural damage detection (SDD). The improved NMA chooses parts of subplanes of the n-simplex for optimization, a two-step method uses modal strain energy based index (MSEBI) to locate damage firstly, and both of them can reduce the computational cost of the basic PSO-Nelder-Mead (PSO-NM). An index of solution assurance criteria (SAC) is defined to describe the correlation between the identified and actual damage of structures. Numerical simulations on a 2-storey frame model is adopted to assess the performance of the proposed hybrid method. The illustrated results show that the improved PSO-NM can provide a reliable tool for accurate SDD in both single and multiple damage cases. Meanwhile, the improved PSO-NM algorithm has a good robustness to noises contaminated in mode shapes.

Keywords: Structural damage detection (SDD) · Particle swarm optimization (PSO) · Nelder-Mead algorithm (NMA) · Particle swarm optimization - Nelder-Mead (PSO-NM) · Global superiority

1 Introduction

Structural damage detection (SDD) is one branch of structural health monitoring (SHM). Adopting optimization algorithm by converting the SDD problem into a mathematical optimal problem is a hot direction of SDD [1]. The particle swarm optimization (PSO) is one of the common algorithm used for optimization problem and has been proved to be an effective algorithm for SDD [2].

In 1995, Kennedy and Eberhart [3] proposed the PSO, a population-based, self-adaptive searching technique. Recently, the PSO has been applied to civil engineering problems and achieved good results. To improve the computational efficiency, reduce computational cost and avoid local optimum, scholars try efforts to make it suitable for such problems. Vakil Baghmisheh et al. [4] proposed a hybrid PSO-NM algorithm for a cantilever beam simulation with two optimized parameters. The experimental results of a cantilever beam indicate its capability for detecting small crack location

© Springer International Publishing Switzerland 2015
Y. Tan et al. (Eds.): ICSI-CCI 2015, Part I, LNCS 9140, pp. 124–132, 2015.
DOI: 10.1007/978-3-319-20466-6_14

and depth with a small error. Seyedpoor [5] presented a two-step method by using modal strain energy based index (MSEBI) to locate the damage and using PSO to determine the actual damage extent based on MSEBI results. Two numerical simulation examples of a cantilever beam and a truss have proved that it is a reliable tool to accurately identify multiple damages of structures. Begambre and Laier [6] proposed a new PSOS-model based damage identification procedure using frequency domain data, which makes the convergence of the PSOS independent of the heuristic constants, meanwhile its stability and confidence are also enhanced. Yildiz [7] proposed an idea of selection of cutting parameters in machining operations, which reduces cost of the products and increases quality at the same time.

Based on the original PSO-NM, the Nelder-Mead algorithm (NMA) is improved in this study for reducing the computational cost. The numerical simulation results indicate that the local optimum can be avoided more efficiently by using the new method. By introducing a two-step procedure based on the MSEBI, the damage elements are first located, the computational efficiency of the hybrid optimization can be further enhanced. Measurement noises of 5% and 10% added to the mode shapes are considered respectively. Some numerical simulation results of a 2-storey frame model shows that the new method has a great accuracy, the efficiency and global superiority are enhanced as well.

2   Theoretical Background

In this section, the basic theories are presented for the improved PSO-NM method. It is composed of three parts. The first one is an overview of the PSO-NM. The second one describes an improved processor of PSO-NM. Finally, the modal strain energy based index (MSEBI) is introduced, which will be used to provide a result of damage location for PSO-NM.

PSO-NM. The PSO-NM is a hybrid intelligent algorithm. For a n-dimensional optimization problem, the optimum solution x^* is firstly calculated by PSO, and NMA is then used to optimize x^* by constructing a n-dimensional simplex around x^*, finally the optimization result of NMA is deemed as one particle of PSO replacing the worst particle of the swarm.

PSO is a population-based, self-adaptive search technique. The PSO starts with a random population (particle) and finds the global best solutions based on the updating formula as follows,

$$V_i^{k+1} = \omega \times V_i^k + c_1 \times rand(\ ) \times (X_{Pb} - X_i^k) + c_2 \times rand(\ ) \times (X_{Gb} - X_i^k) \tag{1}$$

$$X_i^{k+1} = X_i^k + V_i^{k+1} \tag{2}$$

where V is velocity, X is particle location, ω is inertia weight, the superscript k means the k-th iteration and subscript i means i-th particle. In this study, the strategy of inertia weight reduction [8] shown below is employed,

$$\omega = \omega_{max} - \frac{CurCount}{LoopCount}(\omega_{max} - \omega_{min}) \tag{3}$$

where ω_{max}=0.95, ω_{min}=0.4, c_1, c_2 are cognitive and social coefficients respectively with a value of 2 both. *rand*( ) is the random number in the range [0, 1].

The NMA, introduced by Nelder and Mead in 1965 [9], is a non-derivative searching method for multidimensional unconstrained minimization. According to the process of the method, the worst point of n-simplex, with maximum cost values, is replaced by a better cost value calculated by four basic conversions (reflection, expansion, contraction and shrink). Basic conversions of original NMA are performed at every subplane of the n-simplex for *n*-dimensional problems [10]. Because the number of subplane of *n*-simplex is proportional to the cubic of n ($n>2$, the total subplane number is C_n^3), the great number of element of finite element model (FEM) will greatly increase the calculation cost on SDD of real structures.

Improved PSO-NM. In order to improve the calculation efficiency of PSO-NM for SDD, the PSO-NM procedure is improved as following:

(1) PSO process: calculating the current best solution; (2) Group: setting vertices of every subplane of the *n*-simplex as a group; (3) Choosing: choosing p subplanes ($p \leq C_n^3$) for conversions; (4) Conversions: considering only reflection, contraction and local shrink; (5) Replacing: replacing the worst particle of PSO with the vertices found by NMA.

For the five steps as shown above, the third and forth step improves the calculation efficiency by using p subplanes instead of C_n^3 and reducing the type of conversions. The last step is to improve the global superiority of PSO. Supposing that the vertices of subplane are a, b and c respectively, which satisfies $f(a) \leq f(b) \leq f(c)$, the reflection, contraction and local shrink can be formulized as,

$$X_r = X_a + X_b - X_c \tag{4}$$

$$X_c = (X_a + X_b + X_c)/3 \tag{5}$$

$$X_s = X_c + 0.618(X_a - X_c) \tag{6}$$

Modal Strain Energy. A SDD method based on modal strain energy (MSE) change before and after damage is presented by Shi *et al.* [11], and has been certified theoretically. The damage elements of structures can be identified by comparing the MSE between different elements because the damage element keeps a high value of MSE. So, it can be used to locate the damage efficiently.

For a linear vibration system without damping, the modal strain energy of *e-th* element in *i-th* vibration mode of the structure can be expressed as

$$mse_i^e = \frac{1}{2}\varphi_i^{eT}\mathbf{K}^e\varphi_i^e, \quad (i=1,\cdots,nm, \ e=1,\cdots,ne) \tag{7}$$

where ne is the number of elements, nm is the number of mode shapes used for calculation.

To indicate the state of the structure, an efficient parameter is defined as follows,

$$MSEBI^e = \frac{1}{nm} \sum_{i=1}^{nm} \frac{\varphi_i^{eT} \mathbf{K}^e \varphi_i^e}{\sum_{e=1}^{ne} \varphi_i^{eT} \mathbf{K}^e \varphi_i^e} \quad , \quad (e = 1, \cdots, ne) \tag{8}$$

As the stiffness of damage element is unknown, the undamaged element stiffness is used to calculate MSE before and after damage. Therefore, the damage occurrence is led to increasing the MSE and consequently the efficient parameter MSEBI. If the value of Δ_e defined as in Eq. (9) is much bigger than zero in $e\text{-}th$ element comparing to other elements, the $e\text{-}th$ element can be deemed as the damage element.

$$\Delta_e = \max\left[0, \frac{\left(MSEBI^d\right)_e - \left(MSEBI^h\right)_e}{\left(MSEBI^h\right)_e}\right], \quad (e = 1, \cdots, ne) \tag{9}$$

3   Application in Structural Damage Detection

3.1   Objective Function

The objective function of SDD problem can be defined as follows,

$$f(x) = \sum_{i=1}^{nm} \left(1 - MAC\left(\varphi_i^t, \varphi_i^a\right)\right) \tag{10}$$

where $MAC\left(\varphi_i^t, \varphi_i^a\right) = \left(\varphi_i^{tT} \varphi_i^a\right)^2 / \left[\left(\varphi_i^{aT} \varphi_i^a\right)\left(\varphi_i^{tT} \varphi_i^t\right)\right]$, φ_i^a, φ_i^t represent the analytical and test mode shapes respectively, x is the vector of damage factor ranged from 0 to 1. So the SDD problem can be changed into a typical constrained optimization problems as shown in Eq. (11). It is obviously that the minimum value of objective function is zero [12] because MAC=1 means the test results fits the analytical results completely.

$$\min f(x) = \sum_{i=1}^{nmode} \left(1 - MAC\left(\varphi_i^t, \varphi_i^a\right)\right) \tag{11}$$

$$subject \ to \ x \in [0,1] \ , \ (i = 1, \cdots nmode)$$

3.2   Numerical Simulations

The numerical simulations model used for SDD is shown in Fig. 1. The finite element model of frame is modeled by eighteen 2-dimensional beam elements with equal length. The properties of the materials are as follows: $E_c = E_b = 2.0 \times 10e+11$ N/m^2, $I_c = 0.0000126$ m^4, $I_b = 0.0000236$ m^4, $A_c = 0.00298$ m^2, $A_b = 0.0032$ m^2, material density $d_c = 8590$ kg/m^3 and $d_b = 7593$ kg/m^3, where the subscripts c and b represent the column and the beam respectively.

In the SDD process, the first five analytical mode shapes are used. Four damage cases are assumed to occur in the structure respectively. Cases 1 and 2 have a single damage that occurred in the 17th element with a loss of stiffness of 20% and 40% respectively, and Case 3 has multiple damages that occurred in the 17th and 8th elements with a loss of stiffness of 20% in both of them while the 20% and 40% stiffness loss occur in the 8th and 17th elements respectively for Case 4. The mode shapes contaminated with 5% and 10% random noise respectively are considered. The contaminated signal is represented as

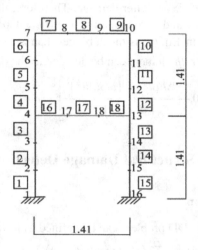

Fig. 1. 2-storey frame model

$$\varphi_n = \varphi_a + \psi\varepsilon R \tag{12}$$

where $\psi = \sqrt{\sum_i \sum_j \varphi_{ij}^2 / (row(\varphi_n) \times col(\varphi_n))}$, φ_a and φ_n are the mode shapes matrix

with no noise and with noise respectively. $row(\varphi)$ and $col(\varphi)$ represent the number of rows and columns of matrix φ_n respectively. R is a random matrix and its elements obeys the distribution $N(0,1)$. ε is the level of noise.

Here, the particle population was set to be 20 for each case. An average value will be calculated after 10 runs in each case and the subplane selected for NMA is 10. The SDD results under four cases are shown in Fig. 2 to Fig. 5 and the relative percentage errors are listed in Table 1. It can be seen that the damage coefficients close to the true value under four cases, the improved method can successful identify the damage even if the heavy damage cases. As the SDD error results listed in Table 1, it can be seen that the improved method can provide a good robustness to noise even under 10% noise although the SDD accuracy is not so high for two damage cases with noise.

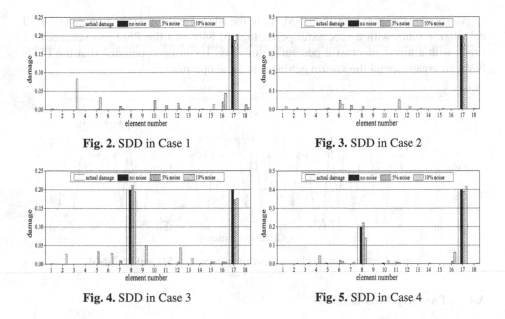

Fig. 2. SDD in Case 1   **Fig. 3.** SDD in Case 2

Fig. 4. SDD in Case 3   **Fig. 5.** SDD in Case 4

Table 1. Relative percentage errors between true and identified damage coefficients

Element	Case 1			Case 2		
	no noise	5% noise	10% noise	no noise	5% noise	10% noise
17	0.00%	7.13%	1.06%	0.2%	2.12%	0.47%
Element	Case 3			Case 4		
	no noise	5% noise	10% noise	no noise	5% noise	10% noise
8	0.06%	5.51%	2.63%	0.1%	10.8%	30.09%
17	0.10%	13.64%	11.89%	0.14%	3.02%	4.10%

3.3   Global Superiority of the Improved PSO-NM

In order to compare the performance of PSO-NM with the conventional PSO [8] , experiments have been done under the same parameters. The population size is 20. The maximum iteration number equals to 50. The algorithms have been executed 50 times in Case 4. To describe the correlation of the identified results with the actual ones, an index solution assurance criteria (SAC) is defined as follows,

$$SAC = \frac{\left(sol\_i \times sol\_a^T\right)^2}{\left(sol\_i \times sol\_i^T\right) \times \left(sol\_a \times sol\_a^T\right)} \tag{13}$$

where sol_a and sol_i are a row vector of actual and identified damage coefficients respectively. The identified results are accurate if the value of SAC is equal to 1, otherwise, the identified results can be deemed as a local optimum.

As shown in Fig. 6, the PSO-NM finds the global minimum 42 times within 50 calculation times with a successful percentage of 84% and the PSO just succeeds 12 times with a lower successful percentage of 24%. It can be concluded that the PSO-NM can greatly avoid the swarm getting into local optimum.

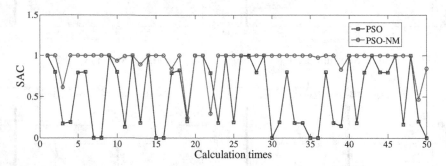

Fig. 6. Comparison on global superiority between PSO and PSO-NM

3.4   Two-Step PSO-NM

In order to make the algorithm more efficient, a two-step method is adopted here. The MSEBI is used for damage localization first, the PSO-NM is then quantifies the damage severity based on the results at the first step as shown in Fig. 7. Relatively higher values of MSEBI are found in the damage elements for both single and multiple damage cases in the structure, so it can be found that the damage element is element 17 in single damage cases and the damage elements are elements 8 and 17 in multiple damage cases, respectively.

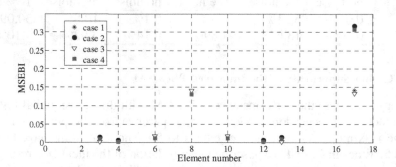

Fig. 7. Damage localization based on MSEBI

In order to study the computational efficiency improvement of two-step PSO-NM, some experiments are performed below. The population size is set to be 20 and the maximum iteration number is 50. When the fitness value of the best particle is less than 10e-8, the iteration is stopped and the times of iteration is recorded. 50 times of calculation were executed to identify multiple damages in Case 4. The iteration times for one-step and two-step PSO-NM is compared in Fig. 8. The average iteration times for one-step PSO is 46.34 while it is 36.8 for two-step method. The average damages of elements 8 and 17 identified by one-step PSO are 0.2029 and 0.3976 respectively, while that of two-step PSO-NM are 0.2000 and 0.4000 respectively. Obviously the two-step method is more accurate and less computational cost.

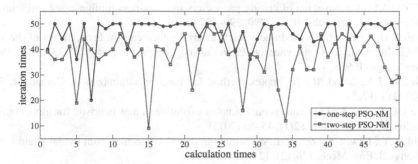

Fig. 8. Comparison on convergence rate between one-step and two-step methods

4   Conclusions

An improved particle swarm optimization - Nelder-Mead (PSO-NM) algorithm is proposed for structural damage detection (SDD) by using an improved NMA and a two-step method in this study. Based on the numerical simulation results of 2-storey rigid frame model with 18 elements, some conclusions can be made as below: (1) The proposed method provide a reliable tool for accurate SDD in both single and multiple damage cases. (2) The improved PSO-NM algorithm has a good robustness to noises contaminated in mode shapes. (3) The improved PSO-NM algorithm is more efficient and accurate in comparison with the original PSO-NM. (4) The improved PSO-NM can be applied to multi-parameter (more than 2) optimization and can give a good SDD.

Acknowledgments. The project is jointly supported by the National Natural Science Foundation of China with Grant Numbers 51278226 and 11032005 respectively.

References

1. Yu, L., Xu, P.: Structural health monitoring based on continuous ACO method. Microelectron. Reliab. **51**(2), 270–278 (2011)
2. Gokdag, H., Yidiz, A.R.: Structural damage detection using modal parameters and Particle Swarm Optimization. Mater Test **54**(6), 416–420 (2012)
3. Kennedy, J., Eberhart, R.: Particle swarm optimization. In: Proceedings of the 1995 IEEE International Conference on Neural Networks, pp. 1942–1948. IEEE Press, Perth (1995)
4. Vakil, B., Peimani, M., Sadeghi, M.H., Ettefagh, M.M., Tabrizi, A.F.: A hybrid particle swarm-Nelder-Mead optimization method for crack detection in cantilever beams. Appl. Soft Comput. J. **12**(8), 2217–2226 (2012)

5. Seyedpoor, S.M.: A two stage method for structural damage detection using a modal strain energy based index and particle swarm optimization. Int. J. Non Linear Mech. **47**(1), 1–8 (2012)
6. Begambre, O., Laier, J.E.: A hybrid particle swarm optimization–simplex algorithm (PSOS) for structural damage identification. Adv. Eng. Softw. **40**(9), 883–891 (2009)
7. Yildz, A.R.: Optimization of cutting parameters in multi-pass turning using artificial bee colony-based approach. Inf. Sci. **200**, 399–407 (2013)
8. Shi, Y.H., Eberhart, R.: Modified particle swarm optimizers. In: Proceedings of the 1998 IEEE International Conference on Evolutionary Computation, pp. 69–73. IEEE Press, Anchorage (1998)
9. Nelder, J.A., Mead, R.: A simplex method for function minimization. Comput. J. **7**(4), 308–313 (1965)
10. Luo, C.T., Yu, B.: Low dimensional simplex evolution: a new heuristic for global optimization. J. Global Optim. **52**(1), 45–55 (2012)
11. Shi, Z.Y., Law, S.S., Zhang, L.M.: Structural damage detection from modal strain energy change. J. Eng. Mech. **126**(12), 1216–1223 (2000)
12. Yu, L., Xu, P., Chen, X.: A SI-based algorithm for structural damage detection. In: Tan, Y., Shi, Y., Ji, Z. (eds.) ICSI 2012, Part I. LNCS, vol. 7331, pp. 21–28. Springer, Heidelberg (2012)

A Fully-Connected Micro-extended Analog Computers Array Optimized by Particle Swarm Optimizer

Yilin Zhu$^{(\boxtimes)}$, Feng Pan, and Xuemei Ren

School of Automation, Beijing Institute of Technology, Beijing 100081, China
{zhuyilin77,andropanfeng}@126.com,
xmren@bit.edu.cn

Abstract. The micro-Extended Analog Computer(uEAC) is a novel hardware implementation of Rubel's EAC model. In this study, we first analyse the basic uEAC mathematical model and two uEAC extensions with minus-feedback and multiplication-feedback, respectively. Then a fully-connected uEACs array is proposed to enhance the computational capability, and to get an optimal uEACs array structure for specific problems, a comprehensive optimization strategy based on Particle Swarm Optimizer(PSO) is designed. We apply the proposed uEACs array to Iris pattern classification database, the simulation results verify that all the uEACs array parameters can be optimized simultaneously, and the classification accuracy is relatively high.

Keywords: Micro-extended analog computer · Fully-connected topology · Particle swarm optimizer · Pattern classification

1 Introduction

Analog computers appear much earlier than digital computers and have been widely used [1–3], for example, the Dumaresq [4] invented in 1902 to relate variables of the fire control problems to the movement of one's own ship and that of a target ship, the FERMIAC [5] invented by Fermi in 1947 to aid in his studies of neutron transport, and the DeSTIN architecture [6,7] created by Arel and his colleagues to address the problem of general intelligence. However, there is not such a general model for analog computers so far as the Turing machine [8] in the digital computer area, which is a crucial problem that limit the development of analog computers.

Researchers never stopped proposing various models for analog computers. Shannon proposed the General Purpose Analog Computer(GPAC) [9] as a mathematical model of an analog device, the Differential Analyzer [10], and Rubel defined the Extended Analog Computer(EAC) [11] as an extension of the GPAC.

This work is supported by National Natural Science Foundation of China(61433003, 61273150), and Beijing Higher Education Young Elite Teacher Project(YETP1192).

© Springer International Publishing Switzerland 2015
Y. Tan et al. (Eds.): ICSI-CCI 2015, Part I, LNCS 9140, pp. 133–144, 2015.
DOI: 10.1007/978-3-319-20466-6_15

Rubel proved that EAC was able to compute partial differential equations, solve the inverse of functions and implement spatial continuity, and moreover, Mycka pointed out that the set of GPAC-computable functions was a proper subset of EAC-computable functions [12]. Thus we may assert that EAC is more powerful than GPAC. For a long period of time, researchers believe that EAC can not be physical implemented because it is a conceptual computer and no single physical, chemical or biological technique is enough to build such a computer. Fortunately, Mills and his colleagues designed and built an electronic implementation of Rubel's EAC model, the micro-Extended Analog Computer (uEAC), after a decade's research [13–15]. To make a detailed comparison between Rubel's EAC model and the uEAC, Mills introduced a Δ-digraph [16] and he related the EAC model to uEAC by dividing the "black boxes" of the EAC model into explicit functions and implicit functions. The up to date prototype of uEAC was designed at Indiana University in 2005 [16,17], and had been applied to letter recognition [18,19], exclusive-OR(XOR) problem [20], stock prediction [21], etc. All the aforementioned researches prove that uEAC is an efficient and powerful computational model, but additional significant improvements are still needed. Based on the basic uEAC model, in this study, we propose a fully-connected uEACs array that lots of uEAC units are integrated, aiming to design an efficient computational model that is able to solve much more complex nonlinear problems.

The rest of this paper is outlined as follows. Section 2 provides some detailed analyses of the uEAC mathematical model and its two feedback extensions, minus-feedback and multiplication-feedback. A fully-connected uEACs array is proposed in Section 3 to overcome uEAC's limitation on nonlinear mapping, moreover, PSO is employed to optimize the uEAC array to get the optimal array structure for specific problems. In Section 4, we apply the proposed fully-connected uEACs array to the famous pattern classification database, Iris data set. A relatively high classification accuracy is obtained and the simulation results verify that the uEACs array is able to solve nonlinear problems efficiently. Some conclusions are given in Section 5.

2   The Basic uEAC Model

The basic uEAC hardware mainly contains three computing modules, analog field computation unit, which complete the main analog computation, digital interface circuit, which is used to connect the host computer and the analog computer, and digital computer aided design unit, which complete the operation of optimization algorithm, data storage and analyses. Currents are placed at various locations on a conductive sheet of silicon foam, and can be read from different locations. Every current form a electrical field on the conductive sheet, and the output voltage is the sum of all the voltages generated by every current, it can either be read directly or mapped by a Lukasiewicz Logic Array(LLA). Specifically, LLA is an analog nonlinear function that takes some inputs and computes the corresponding output. There're totally 27 Lukasiewicz basic functions that form a piecewise-linear covering of the continuous [0,1] range and

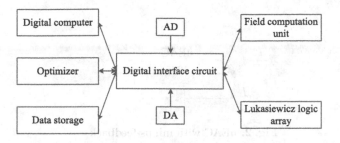

Fig. 1. Structure of a typical uEAC model

domain dissected by $\{0,0.5,1\}$. Readers who are interested in the hardware of uEAC are referred to [16, 17, 22] for more details.

2.1 uEAC Mathematical Model

Supposing that a current I is injected at location P and the electrical resistivity of the conductive sheet is ρ, without lose of generality, we can discretize the conductive sheet into n directions. Every tiny region of the conductive sheet can be represented by a resistance, then the resistance value R of a particular region is

$$R = \frac{\rho \Delta r}{(r_0 + \Delta r/2)\frac{2\pi}{n}}, \qquad (1)$$

where r_0 is the distance from P to the resistance, Δr is the radial length of the resistance. For two arbitrary resistances i, j on the same radius, the distance from P to i, j are r_i, r_j, respectively, and V_{ij} is the voltage between i, j. Supposing that $r_j > r_i$, i.e., j locates outside of i, there are m resistances between i and j, and the length of every resistance is Δr, we have

$$V_{ij} = \frac{I}{n}(R_1 + R_2 + \cdots + R_m)$$

$$= \frac{I}{n}\sum_{k=1}^{m} \frac{\rho \Delta r}{[r_i + (2k-1)\Delta r/2]\frac{2\pi}{n}}. \qquad (2)$$

Let $\Delta r \to 0$, we have

$$V_{ij} = \frac{I\rho}{2\pi}\int_{r_i}^{r_j} \frac{dr}{r} = \frac{I\rho}{2\pi}\ln\frac{r_j}{r_i}. \qquad (3)$$

If the current input locations and voltage output locations are fixed, i.e. r_i and r_j are fixed, we have $V_{ij} = c \cdot I$, where $c = \frac{\rho}{2\pi}\ln\frac{r_j}{r_i}$ is a positive constant.

2.2 Extensions of the uEAC Unit

A single uEAC unit is very limited in nonlinear mapping because its input-output curve is linear, we may add some feedbacks to its input to enhance the computational capability to some degree, including minus-feedback and

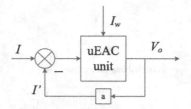

Fig. 2. uEAC with minus-feedback

multiplication-feedback. To describe explicitly, we divide the currents into input current I and field constructive current I_w, V_o is the output voltage.

1) Minus-feedback uEAC

A uEAC unit with minus-feedback is shown as Fig. 2,

$$V_o = \sum_{i=1}^{d} k_i(I_i - I') + \sum_{j=1}^{v} w_j I_{wj}. \tag{4}$$

where I is the input current, I_w is the field constructive current, V_0 is the output current, I' is the feedback current, a is the feedback coefficient. Supposing I_i is the i_{th} output voltage, I_{wj} is the j_{th} field constructive current, d and v are the number of input currents and field constructive currents, respectively. Substituting $I' = a \cdot V_o$, we have

$$V_o = \sum_{i=1}^{d} k_i(I_i - a \cdot V_o) + \sum_{j=1}^{v} w_j I_{wj}, \tag{5}$$

$$V_o = (\sum_{i=1}^{d} k_i I_i + \sum_{j=1}^{v} w_j I_{wj})/(1 + \sum_{i=1}^{d} a k_i), \tag{6}$$

where k_i and w_j are the coefficients of I_i and I_{wj}, respectively.

2) Multiplication-feedback uEAC

A uEAC unit with multiplication-feedback is shown as Fig. 3, the definitions of these variables are the same with those in the minus-feedback uEAC, we have

$$V_o = \sum_{i=1}^{d} k_i I_i + \sum_{j=1}^{v} w_j(I_{wj} \cdot I'). \tag{7}$$

Substituting $I' = a \cdot V_o$, we have

$$V_o = \sum_{i=1}^{d} k_i I_i + \sum_{j=1}^{v} w_j(I_{wj} \cdot a \cdot V_o), \tag{8}$$

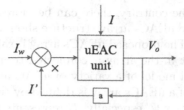

Fig. 3. uEAC with multiplication-feedback

$$V_o = \sum_{i=1}^{d} k_i I_i / (1 - a \sum_{j=1}^{v} w_j I_{wj}). \qquad (9)$$

We can rewrite Eq. 9 in vector form, as

$$g(I, I_w) = V_o = K^T I / (1 - a \cdot W^T I_w), \qquad (10)$$

where now K and W are vectors of size i and j, respectively. We shall note that, by adding feedbacks to the input of uEAC unit, its output formation is totally different and $g(I, I_w)$ now is a nonlinear function of the input I and I_w, but this enhancement of computational capability is far from satisfactory.

3 The Fully-Connected uEACs Array

In this section, we propose a fully-connected uEACs array that contains several uEAC units. Every uEAC units in the array is connected to all the other units by some weights. For a uEACs array with 6 uEAC units, its structure is as Fig. 4, where x is input of the array and y is the output.

Notice that the definition of $uEAC\_1$, $uEAC\_2$,...$uEAC\_6$ in the array is implicit because the terms $1, 2, \ldots, 6$ are just symbols of the uEAC units and do not convey any specific meaning. Moreover, it is not necessary to limit that there

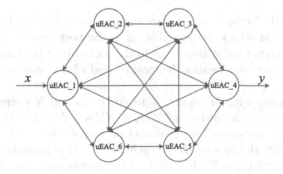

Fig. 4. General structure of the fully-connected uEACs array

is only one input x, on the contrary, there can be more inputs and outputs for the uEACs array, and the uEACs array structure should be designed according to the specific problems. The proposed uEACs array is said to be fully connected because every uEAC unit in the array is connected to all the other units, thus this structure is a general model of a variety of uEAC arrays structures.

This inner structure of a uEACs array is defined by two matrices, connection matrix Γ and weight matrix Ψ, especially, Γ represents the connections among uEAC units in the array while Ψ represents the strength of these connections. For a uEACs array with M inputs, N uEAC units and K outputs, the connection matrix Γ and weight matrix Ψ are $(K + N) \times (K + N + M)$-dimensional matrices. Connection matrix Γ is binary, and the element γ_{ij} in Γ represents the connection from the j_{th} uEAC unit to the i_{th} one. If the j_{th} uEAC unit is connected to the i_{th} uEAC unit, $\gamma_{ij} = 1$, or $\gamma_{ij} = 0$.

3.1 Optimization Based on PSO

To get an optimal uEACs array structure for a specific problem, we have to optimize the connection matrix Γ and weight matrix Ψ simultaneously because there is a one-to-one correspondence among the elements of these two matrices. From the mathematical point of view, all the heuristic optimization algorithms can be used to optimize the uEACs array, such as hill climbing algorithm, genetic algorithm and tabu search, but we must investigate their computational capability and complexity, time consumption, efficiency, etc.

In this study, we choose PSO as the optimizer because its update functions are simple and convergent very fast. We shall also note that for a specific problem, there may exist more than one "good enough" uEACs arrays, thus we shall analyze their structures form the statistical perspective but not limited to one or several running results. The updating rule of the i_th particle at step $k + 1$ is :

$$v_i^{k+1} = w \cdot v_i^k + c_1 \cdot r_1^k \cdot (p_l^k - x_i^k) + c_2 \cdot r_2^k \cdot (g_i^k - x_i^k) \tag{11}$$

$$x_i^{k+1} = x_i^k + v_i^k \tag{12}$$

where w is the inertia factor; x_i^k and v_i^k are the position vector and velocity vector of the i_th particle at step k, respectively; p_l^k is the best position vector *lbest* in the neighborhood structure at step k; g_i^k is the global best position vector *gbest* at step k; c_1 and c_2 are acceleration factors; r_1^k and r_2^k are uniformly distributed over $[0, 1]$.

For a uEACs array with M inputs, N uEAC units and K outputs, the particles position vector X is defined as a $2 \times (K + N)(K + N + M)$-dimensional real number vector that contains all the elements of connection matrix Γ and weight matrix Ψ. Especially, the former $(K + N)(K + N + M)$ elements correspond to matrix Γ and the rest $(K + N)(K + N + M)$ elements correspond to matrix Ψ. Moreover, the incorrect classification amount is chosen as the fitness function.

Algorithm 1. Optimize the uEACs array by PSO

1: Initializing particles number n, maximum iterations number m, target error e.
2: **for** $i - 1, 2, \cdots, n$ **do**
3:    initializing particle position vector X_i and velocity vector V_i.
4: **end for**
5: Arbitrarily selecting the best particle.
6: **for** $j = 1, 2, 3, \cdots, m$ **do**
7:    **for** $i = 1, 2, 3, \cdots, n$ **do**
8:        updating X_i and V_i;
9:        calculating particle fitness;
10:       updating the best particle.
11:    **end for**
12: **end for**
13: **while** ($error < e$)or($iterations > m$) **do**
14:    outputting the best position vector X_i;
15:    ending the algorithm.
16: **end while**

---

4   Simulations

We apply the proposed uEACs array and optimization strategy to the Iris database [23] to investigate its computational capability. There are totally 150 instances in the database, including 3 classes (50 instances each) that refer to 3 different iris plants, and every instance has 4 attributes. We design 6 independent simulation experiments. In the first experiment, 25 instances are picked as the training sample, and after training, we use the optimized uEACs array to classify 6 different testing samples, respectively. Especially, we execute the classification algorithm 20 times for every testing sample and take the average classification accuracy as the final accuracy. In the other 5 experiments, 50, 75, 100, 125, 150 instances are picked as the training sample, respectively, and the rest procedure is the same with the first experiment.

The following connection matrix Γ represents one of the optimal uEACs array structures in the experiment with 150 training samples and 150 testing samples, and the actual uEACs array structure is shown as Tab. 1 and Fig. 5. We shall note that all the elements on the diagonal are zero, it's not an optimization result but a prior setting, which means that all the uEAC unit in the array are basic unit without feedback. Moreover, the connections of uEAC units in the array are directed, i.e., elements γ_{ij} and γ_{ji} in the connection matrix Γ represent two different connections.

$$\Gamma = \begin{bmatrix} 0 & 1 & 1 & 1 & 1 & 0 & 1 & 0 & 0 & 0 & 0 \\ 0 & 0 & 1 & 1 & 1 & 1 & 0 & 0 & 0 & 1 & 1 \\ 1 & 1 & 0 & 1 & 0 & 1 & 0 & 1 & 0 & 0 & 1 \\ 0 & 1 & 1 & 0 & 0 & 0 & 0 & 0 & 1 & 1 & 0 \\ 1 & 0 & 1 & 1 & 0 & 1 & 1 & 1 & 0 & 0 & 1 \\ 1 & 0 & 1 & 1 & 1 & 0 & 1 & 1 & 1 & 1 & 1 \\ 1 & 0 & 0 & 1 & 1 & 1 & 0 & 1 & 0 & 1 & 1 \end{bmatrix}$$

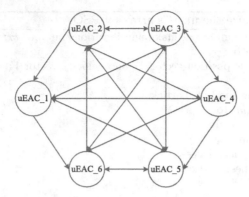

Fig. 5. The actual uEACs array structure

Table 1. uEACs array structure by Γ

	u_1 [1]	u_2	u_3	u_4	u_5	u_6
u_1	0	1	1	1	1	0
u_2	0	0	1	1	1	1
u_3	1	1	0	1	0	1
u_4	0	1	1	0	0	0
u_5	1	0	1	1	0	1
u_6	1	0	1	1	1	0

[1] u_1: $uEAC\_1$, the same below.

The average uEACs array classification accuracy is shown as Tab. 2, Fig. 6 and Fig. 7.

Table 2. uEACs array Iris database classification accuracy

ACA [3] TESS [4] TRSS [2]	50	75	100	125	150
25	80.7%	90%	92.1%	93.64%	95.73%
50	89.9%	93.53%	94.45%	95.76%	96.77%
75	97%	97.73%	98.45%	98.64%	98.87%
100	95.8%	98.67%	99.35%	99.16%	99.53%
125	99.1%	99.73%	99.3%	99.8%	99.73%
150	99.8%	100%	99.55%	99.92%	99.97%

[2]TRSS: Training Sample Size;
[3]ACA: Average Classification Accuracy;
[4]TESS: Testing Sample Size.

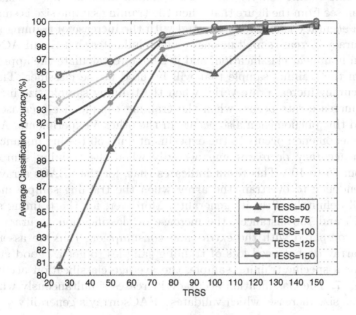

Fig. 6. Classification accuracy for the same testing sample

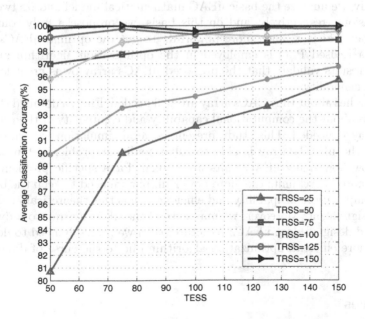

Fig. 7. Classification accuracy for the same training sample

For a specific testing sample, the average classification accuracy is shown as Fig. 6. We can see from the figure that when the training sample size is small, the classification accuracy is relatively low and with the increase of training sample size, the accuracy increase simultaneously, which indicates that the uEACs array is efficient on nonlinear classification, and it is able to calculate the appropriate features from the training sample that will be used in the classifier. The classification accuracy increase rate varies when the training sample size increases, the biggest increase rate occurs when the training sample size increases from 50 to 75, and this phenomenon also occurs in the other 5 experiments. Another interesting phenomenon occurs in the experiment with 50 testing instances, the classification accuracy decreases unexpectedly when the training sample size increases from 75 to 100. This result indicates that, for the uEACs array, there may be a tendency to overtrain the array when the training sample increase, and that will definitely effect its generality. Moreover, for the training sample with 100, 125 and 150 instances, the increase of classification accuracy seems insignificant compared with the extra time consumption, thus we assert that the uEACs array do not need lots of training samples to get a "good enough" structure. For a specific training sample, the average classification accuracy is shown as Fig. 7. The classification accuracy increases simultaneously when the testing sample size increase, which validates uEACs array's generality.

5   Conclusion

In this study, we analyze the basic uEAC mathematical model and its two feedback extensions, respectively, and on this basis, we propose a fully-connected uEACs array, which is much more powerful. To get the optimal uEACs array structure, a revised PSO is employed in the optimization algorithm and the simulation results validate that the proposed uEACs array owns great learning ability and generality.

More studies about the uEACs array are still needed. First, a detailed theoretical analysis about the computational capability and complexity of the proposed uEACs array is needed. This study presents several simulation experiments to prove that the uEACs array is able to solve nonlinear problems, but we still do not know how many uEAC units are enough for a specific problem. Definitely the more uEAC units the more powerful, but we need to keep the balance between computational capability and efficiency, thus this theoretical analysis is extremely significant. Moreover, we may extend the uEACs array to a dynamic network and design a deep uEACs structure, then we probably need to design a new, and more efficient, optimization algorithm due to the curse of dimensionality.

References

1. Small, J.S.: General-purpose electronic analog computing: 1945–1965. IEEE Annals of the History of Computing **15**(2), 8–18 (1993)

2. De Garis, H., Shuo, C., Goertzel, B., Ruiting, L.: A world survey of artificial brain projects, Part I: Large-scale brain simulations. Neurocomputing **74**(1), 3–29 (2010)
3. Goertzel, B., Lian, R., Arel, I., De Garis, H., Chen, S.: A world survey of artificial brain projects, Part II: Biologically inspired cognitive architectures. Neurocomputing **74**(1), 30–49 (2010)
4. Brooks, J.: Dreadnought Gunnery at the Battle of Jutland: The Question of Fire Control. Routledge (2004)
5. http://commons.wikimedia.org/wiki/File:Fermiac.jpg
6. Arel, I., Rose, D., Coop, R.: DeSTIN: a scalable deep learning architecture with application to high-dimensional robust pattern recognition. In: AAAI Fall Symposium: Biologically Inspired Cognitive Architectures (2009)
7. Arel, I., Rose, D., Karnowski, T.: A deep learning architecture comprising homogeneous cortical circuits for scalable spatiotemporal pattern inference. In: NIPS 2009 Workshop on Deep Learning for Speech Recognition and Related Applications, pp. 23–32 (2009)
8. Turing, A.M.: On computable numbers, with an application to the Entscheidungsproblem. J. of Math. **58**(345–363), 5 (1936)
9. Shannon, C.E.: Mathematical theory of the differential analyzer. J. Math. Phys. MIT **20**, 337–354 (1941)
10. Bush, V.: The differential analyzer. A new machine for solving differential equations. Journal of the Franklin Institute **212**(4), 447–488 (1931)
11. Rubel, L.A.: The extended analog computer. Advances in Applied Mathematics **14**(1), 39–50 (1993)
12. Mycka, J.: Analog computation beyond the Turing limit. Applied mathematics and computation **178**(1), 103–117 (2006)
13. Mills, J.W., Beavers, M.G., Daffinger, C.A.: Lukasiewicz logic arrays. In: Proceedings of the Twentieth International Symposium on Multiple-Valued Logic, 1990, pp. 4–10. IEEE (1990)
14. Mills, J.W., Daffinger, C.A.: CMOS VLSI Lukasiewicz logic arrays. In: Proceedings of the International Conference on Application Specific Array Processors, 1990, pp. 469–480. IEEE (1990)
15. Mills, J.W., Walker, T.O.N.Y., Himebaugh, B.: Lukasiewicz'Insect: Continuous-Valued Robotic Control After Ten Years. Journal of Multiple Valued Logic And Soft Computing **9**(2), 131–146 (2003)
16. Mills, J.W.: The nature of the extended analog computer. Physica D: Nonlinear Phenomena **237**(9), 1235–1256 (2008)
17. Himebaugh, B.: Design of EAC (2005). http://www.cs.indiana.edu/bhimebau/
18. Mills, J.W.: The continuous retina: Image processing with a single-sensor artificial neural field network. In: Proceedings IEEE Conference on Neural Networks (1995)
19. Parker, M., Zhang, C., Mills, J., Himebaugh, B.: Evolving letter recognition with an extended analog computer. In: IEEE Congress on Evolutionary Computation, CEC 2006, pp. 609–614. IEEE (2006)
20. Pan, F., Zhang, R., Long, T., Li, Z.: The research on the application of uEAC in XOR problems. In: 2011 International Conference on Transportation, Mechanical, and Electrical Engineering (TMEE), pp. 109–112. IEEE (2001)
21. Zhu, Y., Pan, F., Li, W., Gao, Q., Ren, X.: Optimization of Multi-Micro Extended Analog Computer Array and its Applications to Data Mining. International Journal of Unconventional Computing **10**(5–6), 455–471 (2014)

22. Mills, J.W., Himebaugh, B., Allred, A., Bulwinkle, D., Deckard, N., Gopalakrishnan, N., Zhang, C.: Extended analog computers: A unifying paradigm for VLSI, plastic and colloidal computing systems. In: Workshop on Unique Chips and Systems (UCAS-1). Held in conjunction with IEEE International Symposium on Performance Analysis of Systems and Software (ISPASS05), Austin, Texas (2005)
23. Bache, K., Lichman, M.: UCI machine learning repository, p. 901 (2013). http://archive.ics.uci.edu/ml

A Population-Based Clustering Technique Using Particle Swarm Optimization and K-Means

Ben Niu[1,2] (✉), Qiqi Duan[1], Lijing Tan[3], Chao Liu[4(✉)], and Ping Liang[5]

[1] College of Management, Shenzhen University, Shenzhen 518060, China
drniuben@gmail.com
[2] Department of Industrial and System Engineering,
The Hong Kong Polytechnic University, Hung Hom, Hong Kong
[3] Department of Business Management,
Shenzhen Institute of Information Technology, Shenzhen 518172, China
[4] School of Economic and Management, Bejing University of Technology, Bejing, China
liuchao@bjut.edu.cn
[5] Health Science Center, Shenzhen University, Shenzhen 518060, China

Abstract. Population-based clustering techniques, which attempt to integrate particle swarm optimizers (PSOs) with K-Means, have been proposed in the literature. However, the performance of these hybrid clustering methods have not been extensively analyzed and compared with other competitive clustering algorithms. In the paper, five existing PSOs, which have shown promising performance for continuous function optimization, are hybridized separately with K-Means, leading to five PSO-KM-based clustering methods. Numeric experiments on nine real-life datasets show that, in the context of numeric data clustering, there exist no significant performance differences among these PSOs, though they often show significantly different search abilities in the context of numeric function optimization. These PSO-KM-based clustering techniques obtain better and more stable solutions than some individual-based counterparts, but at the cost of higher time complexity. To alleviate the above issue, some potential improvements are empirically discussed.

Keywords: Population-based clustering · Optimization-based clustering · Particle swarm optimizer (PSO) · K-Means

1   Introduction

Typically, data clustering problems that partition N data points into K clusters can be modeled as continuous optimization problems where the centroids of K clusters and cluster indexes of N data points act as inputs and outputs [1]. However, most of the partitioning-based clustering methods aimed at solving K-Means-type optimization problems are sensitive to the initial centroids and easily get trapped into local optima [2]. More effective K-Means-type clustering algorithms are urgently required to uncover the hidden patterns for large-scale clustering.

Recently, population-based clustering techniques become increasing popularity. One advantage of the population-based clustering methods over the individual-based

© Springer International Publishing Switzerland 2015
Y. Tan et al. (Eds.): ICSI-CCI 2015, Part I, LNCS 9140, pp. 145–152, 2015.
DOI: 10.1007/978-3-319-20466-6_16

counterparts may lie in their potentially global exploration abilities. As powerful optimization tools, particle swarm optimizers (PSOs) have been widely studied and improved in the context of data clustering [3]. Note that real-life clustering problems may show fully different fitness landscapes, which cannot be represented adequately by existing benchmark functions. Although different PSOs could show significantly different performance on some benchmark functions, it is unknown whether the significant performance differences on benchmark functions still hold on diverse clustering optimization problems. To answer the above unsolved yet important problem for PSO-based clustering is the main objective of the paper.

In the paper, five existing PSOs, which have been widely cited in the optimization field and shown promising performances for continuous optimization, are hybridized separately with K-Means, giving rise to five PSO-KM-based clustering methods. Numeric comparative experiments on nine real-world datasets show some unexpected but valuable conclusions. Finally, some potential improvement directions for PSO-based clustering algorithms are empirically discussed.

2   Literature Overview

Many research efforts have been devoted to the application of PSOs on a variety of clustering problems. Among them, hybridization strategies are a hot study direction for optimization-based clustering. In this section, the hybridization methods including PSO and K-Means are mainly concentrated. Merwe and Engelhrecht [4] first proposed a PSO-based clustering algorithm, which combined PSO with K-Means. Specifically, in order to initialize the particle population of p individuals, K-Means needs to be run p times. In each run, the centroids of clusters obtained by K-Means are used as the initial seed of an individual. In terms of algorithmic design, the paper can be regarded as a repetition and extension to Merwe and Engelhrecht's work. However, the following differences between the original work [4] and the paper should be highlighted: 1) the original work only took into account the global PSO version (i.e., GPSO), whereas in the paper more advanced PSOs (e.g., FIPS and CLPSO) are combined with K-Means. 2) In the numeric experiments conducted by Merwe and Engelhrecht, only six small-size datasets ($N \leq 500$ and $D \leq 15$) were tested and analyzed. However, on medium-to-large size datasets, which frequently occurs in the data mining field, the effectiveness and efficiency of PSO-KM-based clustering algorithms should be further analyzed, which is one focus of the paper. Similar studies can be found in the literature (e.g., [5-9]). Chen and Ye [10] directly used PSO to solve the K-Means-type clustering problems, with lower clustering performance than most well-designed hybridization strategies.

Cohen and Castro [11] developed a particle swarm clustering (PSC) algorithm, where the entire population rather than an individual represents a clustering solution. The authors claimed to replace the fitness function with the performance index to measure the quality of each particle. Unfortunately, the formal definition of the performance index was not provided clearly in the original paper. Omran et al [12] presented a PSO-KM-based clustering algorithm called DCPSO, where a binary PSO was used to find the optimal number of clusters while K-Means was employed to update the centroids. The authors considered the maximization of inter-cluster distances as well as the minimization of intra-cluster distances. Although DCPSO showed promising clustering results in the context of image segmentation, the exponential increase in the search space gives rise to higher time complexity intractable for large-scale clustering. Kao et al [13] hybridized the Neld-Mead simplex search with PSO and K-Means, and proposed a population-based clustering approach named as K-NM-PSO. In practice, we can imagine that K-NM-PSO obtains better clustering accuracy but at the cost of time complexity. The efficiency of K-NM-PSO has not been validated on high-dimensional clustering applications. The above argument may be illustrated that the fact that, in the original paper, the authors only chose some small-scale datasets ($N \leq 1500$ and $D \leq 15$) to show its superiority. Alam et al [14] proposed a PSO-based hierarchical clustering method called HPSO. The individual representation way of HPSO is very similar with that of PSC. More recently, a state-of-the-art PSC version abbreviated as RCE was developed by Yuwono et al [15].

For more details about PSO-KM, please refer to the literature survey [3, 16-18]. To the best of our knowledge, there are few open source codes available for PSO-KM-based clustering, which increases the difficulties of conducting the repetition or comparative experiments. Another aim of the paper is to provide a comparative baseline for PSO-based clustering.

3   PSO-Based Clustering Algorithms

In this section, a population-based algorithmic framework, which combines PSO with K-Means, is proposed. In the framework, the following five PSOs are taken into account independently:

1) PSO with a *global* topology and a *random* inertia weight (GPSO-RW) [19]: the inertia weight W is randomly distributed in [0.5, 1].

2) PSO with a *global* neighborhood structure and the inertia weight W *decreasing linearly* from 0.9 to 0.4 (GPSO-WV) [20].

3) GPSO with a *constriction factor* (GPSO-CF) [21]: the differences between it and the above two PSO versions lie in the introduction of the constriction factor that prevents explosion of the particle system [22].

4) *Fully informed* PSO (FIPS) [23]: the weight-based *pbests* of all the neighbors are used to update each particle' position, which reduces the influence of the best-performing one (i.e., *gbest*). Note that only the *Ring* topology is employed for FIPS in the paper, as suggested in the literature [23].

5) *Comprehensive learning* PSO (CLPSO) [24]: for different dimensions, each individual learns towards different neighbors in a dynamic population topology.

To save space, assume that the readers have been familiar with the above PSOs. Otherwise, please refer to the corresponding papers for more details. Whether these PSOs could work well and show significantly different optimization performances on diverse clustering problems as on many continuous benchmark functions is an open problem worthy of further investigation. In the paper, all the clustering problems are converted to K-Means-type continuous optimization problems which may have fully different landscape characteristics as compared with benchmark functions.

3.1 Real-Coded Representation Strategy

The representation ways of a clustering solution can be simply classified into two groups. The first category adopts a direct (*integer/binary*-based) encoding strategy [25]. This representation way leads easily to two serious drawbacks, i.e., expensive computation cost (prohibitively for large–scale clustering) and redundancy [25]. The second one only takes into account the cluster centroids as K-Means. It encodes all the centers as an agent, which is stored in a real-valued array. Related studies have shown that the latter is better than the former with regard to computation complexity.

The real-coded representation strategy for particle P_t ($t = 1, ..., p$, where p is the swarm size) is presented in the following vector:

$$P_t = (C_1{}^t, ..., C_K{}^t), \text{ where } C_k{}^t = (M_{k1}^t, ..., M_{kd}^t). \tag{1}$$

Note that $C_k{}^t = (M_{k1}^t, ..., M_{kd}^t)$ is a real-coded vector corresponding to centroid k ($k = 1, ..., K$), and M_{kd}^t is the d-dimensional position of centroid $C_k{}^t$ for particle t.

3.2 Population Initialization

For optimization-based clustering techniques, two initialization ways are commonly used. One is to locate all the individuals at random in the entire search space, while the other is to choose k random samples from the entire dataset as the initial centroids. Some other advanced centroid initialization methods can be found in [26-28], which have investigated the effects of different initialization methods on the clustering performance. However, different conclusions were drawn from different research papers, which may be explained by the fact that different datasets were used by different papers while the performance of clustering methods depend heavily on the chosen datasets, the settings of system parameters, and so on. For refining the clustering accuracy and accumulating the convergence, the centroids obtained by K-Means can be organized as an initial particle before executing PSO.

3.3 Fitness Evaluation

In practice, different fitness evaluation functions (e.g., Davies–Bouldin index [29], Silhouettes index [30]) can be used for measuring the quality of a clustering solution [25]. For partitioning clustering algorithms, mean squared error (*MSE*), is widely used, according to the literature [31]. *MSE* is calculated by

$$MSE = \sum_{k=1}^{K} \sum_{X_i \in C_k} ||X_i - C_k||^2 \tag{2}$$

where $X_i = (X_{i1}, ..., X_{id})$ is data point i ($i = 1, ..., N$) in the real space $\Re^d$, and C_k is the centroid of cluster k, and $|| \cdot ||^2$ denotes the squared Euclidean distance between two points. The centroid of cluster k is updated by

$$C_k = \sum_{i \in C_k} X_i / |C_k| \tag{3}$$

where $|C_k|$ is the number of instances in cluster k. Note that the above fitness function is only suitable to clustering problems where the number of clusters is predefined (see [32] for detailed explanation). How to determine the optimal number of clusters is a hot research direction (refer to [25, 33, 34] for more information). In this paper, the number of clusters is supposed to be known *a prior*. In fact, multiple fitness functions can be taken into account during the optimization process, leading to multiobjective clustering [35], which are beyond the scope of the paper.

3.4   Particle Position Updating Rule

Each particle t with the velocity itself moves stochastically toward its personally historically best position (PB_t) and its neighbors' best positions (NB_t), until the maximum number of iterations is reached. The velocity V_t and position P_t adjustment rule for particle t are presented in the following:

$$V_t = cf * (W * P_t + c_1 * R_{t1} * (PB_t - X_t) + c_2 * R_{t2} * (NB_t - X_t)) \quad (4)$$

$$P_t = P_t + V_t \quad (5)$$

where cf is the constriction factor, W is the inertia weight, c_1 and c_2 are the cognitive and social learning coefficients, and R_{t1} and R_{t2} are two separately generated random number in $[0,1)$. The updating rule of PB_t^{g+1} at generation $(g + 1)$ is illustrated as following:

$$PB_t^{g+1} = \begin{cases} P_t^g, & if\ f(PB_t^g) \geq f(P_t^g); \\ PB_t^g, & otherwise. \end{cases} \quad (6)$$

Note that six different PSOs have many differences with regard to the definition of neighbors, implementation procedures, and parameter configures. For more details, the readers are encouraged to refer to the original papers. In the paper, only the maximum number of iterations is set as the stopping condition.

4   Experimental Studies

To evaluate the effectiveness and efficiency of PSO-KM, 9 benchmark datasets from the famous UCI machine learning repository [36] are chosen for comparative experiments. Almost all the datasets (except Coil2) need to be standardized (*z-score*) before clustering, for eliminating the discrepancy among different scales of different attributes. Furthermore, numeric experiments choose five commonly cited individual-based clustering algorithms and one recently proposed state-of-the-art counterpart as benchmark algorithms: namely, 1) the agglomerative hierarchical clustering method with average link (AHC-AL), 2) Lloyd's K-Means which adopts the batch update scheme (Lloyd-KM) [37], 3) MacQueen's K-Means which uses the online update scheme (Mac-KM) [38], 4) AS-136 [39], 5) BF-KM [26], and 6) PSO-RCE [15]. Owing to the limitation of pages, for more experimental details, the readers are suggested to scan the website freely available at [40] and download the related source code and data. The parameter settings of all the algorithms involved follow the suggestions of the original papers.

Table 1. Comparative Results of 11 Algorithms on UCI Datasets

	Iris	Wine	Coil2	Cancer	Credit	Opt	Musk	Magic
AHC-AL	196.86	2199.65	154.03	4987.70	22836	309332	1167550	---
Lloyd-KM	154.02	1305.54	155.00	2724.42	22400	230000	901000	137000
Mac-KM	147.57	1281.00	154.53	2724.16	22298	228754	899887	136919
AS-136	140.03	1649.44	153.88	2724.16	22171	235213	869655	136919
BF-KM	140.03	1649.44	154.03	2724.16	22354	224857	869655	136919
PSO-RCE	142.92	1318.84	155.85	2798.06	22967	288987	869655	136919
GPSO-RW	138.89	1270.75	156.47	2724.16	22171	288987	869655	136919
GPSO-WV	138.89	1270.75	156.47	2724.16	22171	289006	869655	136919
GPSO-CF	138.89	1270.75	156.47	2724.16	22171	289439	869655	136919
FIPS	138.89	1270.75	156.47	2724.16	22171	289851	869655	136919
CLPSO	138.89	1270.75	156.47	2724.16	22171	289851	869655	136919

Based on Table 1, in term of MSE, five PSO-KM-based clustering algorithms obtain better, more stable clustering performances in most cases except on the Opt dataset, followed by AS-136, BF-K-Means. For all the datasets, Mac-KM show slight better clustering performance than Lloyd-KM, owing to the fact that the online update scheme of Mac-KM is guaranteed to converge to local optimum. We can imagine that AHC-AL has the worst clustering performance and is not applicable for large-scale datasets (e.g., the Magic and Road Network dataset). Note that the state-of-the-art PSC-based algorithm, PSORCE, cannot show superiority clustering performance with regard to MSE. This might be due to the fact MSE is not its direct objective function. It is not guaranteed that, due to the stochastic nature, PSO-RCE must converge to the local optima in each run.

5   Conclusion

In the paper, five PSO-KM-based clustering algorithms have been compared with five widely cited individual-based clustering algorithms as well as one recently proposed state-of-the-art PSC-based counterpart. In the above comparative experiments, PSO-KM obtained more stable, better clustering performance in most cases in terms of MSE, but at the cost of higher time complexity. During optimization, each particle can often include *local* (rather than *global*) clustering structures. However, these five particle updating rules with individual interactions may disturb some locally valuable information for numeric data clustering, although they have shown success on continuous function optimization. More efficient particle updating rules (e.g., it might still utilize the concept of *pbest* and *nbest*, but it needs to adopt special definitions for them to uncover the hidden clustering structures) should be elaborated in the context of numeric data clustering. Furthermore, for PSO-KM, the process of fitness function evaluations involves excessive numeric calculations, which may be reduced by using some advanced data structures (e.g., kd-tree) or algorithms (e.g., k-NN). The above two potential improvement directions on PSO-KM are our ongoing research.

Acknowledgments: This work is partially supported by The National Natural Science Foundation of China (Grants nos. 71001072, 71271140, 71471158, 61273230, 61473266), Shenzhen Science and Technology Plan Project (Grant no. CXZZ20140418182638764), and Program for New Century Excellent Talents in University (Grant no. NCET-12-1027).

References

1. Jain, A.K.: Data Clustering: 50 Years Beyond K-Means. Pattern Recognition Letters. **31**(8), 651–666 (2010)
2. Tzortzis, G., Likas, A.: The MinMax K-Means Clustering Algorithm. Pattern Recognition. **47**(7), 2505–2516 (2014)
3. Alam, S., et al.: Research on Particle Swarm Optimization Based Clustering: A Systematic Review of Literature and Techniques. Swarm and Evolutionary Computation. **17**, 1–13 (2014)
4. Merwe, D.W., Engelbrecht, A.P.: Data clustering using particle swarm optimization. In: The 2003 Congress on Evolutionary Computation, CEC 2003, pp. 215–220. IEEE (2003)
5. Chen, J., Zhang, H.: Research on application of clustering algorithm based on PSO for the web usage pattern. In: The 2007 International Conference on Wireless Communications, Networking and Mobile Computing, WiCom 2007, pp. 3705–3708. IEEE (2007)
6. Lin Y., et al: K-means optimization clustering algorithm based on particle swarm optimization and multiclass merging. In: Advances in Computer Science and Information Engineering, pp. 569–578. Springer (2012)
7. Alam, S., et al: An evolutionary particle swarm optimization algorithm for data clustering. In: The Proceedings of IEEE Swarm Intelligence Symposium, pp. 1–6 (2008)
8. Abbas, A., Fakhri, K., Mohamed, S.K.: Flocking Based Approach for Data Clustering. Natural Computing **9**(3), 767–794 (2010)
9. Niknam, T., Amiri, B.: An Efficient Hybrid Approach Based on PSO, ACO and K-Means for Cluster Analysis. Applied Soft Computing **10**(1), 183–197 (2010)
10. Chen, C.Y., Fun, Y.: Particle swarm optimization algorithm and its application to clustering analysis. In: The Proceedings of 17th Conference on Electrical Power Distribution Networks (EPDC), pp. 789–794 (2012)
11. Cohen, S.C.M., Castro, L.N.: Data clustering with particle swarms. In: The Proceedings of the IEEE Congress on Evolutionary Computation, pp. 1792–1798. IEEE (2006)
12. Omran, M.G.H., Salman, A., Engelbrecht, A.P.: Dynamic Clustering Using Particle Swarm Optimization with Application in Image Segmentation. Pattern Analysis and Applications. **8**(4), 332–344 (2005)
13. Kao, Y.T., Zahara, E., Kao, I.W.: A Hybridized Approach to Data Clustering. Expert Systems with Applications. **34**(3), 1754–1762 (2008)
14. Alam, S., et al: Particle swarm optimization based hierarchical agglomerative clustering. In: The 2010 IEEE/WIC/ACM International Conference on Web Intelligence and Intelligent Agent Technology (WI-IAT), pp. 64–68. IEEE (2010)
15. Yuwono, M., et al.: Data Clustering Using Variants of Rapid Centroid Estimation. IEEE Transactions on Evolutionary Computation **18**(3), 366–377 (2014)
16. Rana, S., et al.: A Review on Particle Swarm Optimization Algorithms and Their Applications to Data Clustering. Artificial Intelligence Review **35**(3), 211–222 (2011)
17. Radha, T., et al.: Particle Swarm Optimization: Hybridization Perspectives and Experimental Illustrations. Applied Mathematics and Computation **217**(12), 5208–5226 (2011)
18. Mohamed, J.A.H., Sivakumar, R.: A Survey: Hybrid Evolutionary Algorithms for Clustering Analysis. Artificial Intelligence Review **36**(3), 179–204 (2011)
19. Ratnaweera, A., Halgamuge, S., Watson, H.C.: Self-Organizing Hierarchical Particle Swarm Optimizer with Time-Varying Acceleration Coefficients. IEEE Transactions on Evolutionary Computation **8**(3), 240–255 (2004)
20. Shi, Y., Eberhart, R.C.: A modified particle swarm optimizer. In: The Proceedings of IEEE Congress on Evolutionary Computation, pp. 69–73. IEEE (1998)

21. Eberhart, R.C., Shi, Y.: Comparing inertia weights and constriction factors in particle swarm optimization. In: The Proceedings of the Congress on Evolutionary Computation, pp. 84–88. IEEE (2000)
22. Clerc, M., Kennedy, J.: The Particle Swarm-Explosion, Stability, and Convergence in A Multidimensional Complex Space. IEEE Transactions on Evolutionary Computation 6(1), 58–73 (2002)
23. Mendes, B., Kennedy, J., Neves, J.: The Fully Informed Particle Swarm: Simpler, Maybe Better. IEEE Transactions on Evolutionary Computation 8(3), 204–210 (2004)
24. Liang, J.J., Qin, A.K., Suganthan, P.N., Baskar, S.: Comprehensive Learning Particle Swarm Optimizer for Global Optimization of Multimodal Functions. IEEE Transactions on Evolutionary Computation 10(3), 281–295 (2006)
25. Das, S., Abraham, A., Konar, A.: Automatic Clustering Using an Improved Differential Evolution Algorithm. IEEE Transactions on System, Man, and Cybernetics-Part A: Systems and Human 38(1), 218–237 (2008)
26. Bradley, P.S., Fayyad, U.M.: Refining Initial Points for K-Means Clustering. Microsoft Research (1998). http://research.microsoft.com/apps/pubs/default.aspx?id=68490 (MSR-TR-98-36)
27. Peñal, J.M., et al.: An Empirical Comparison of Four Initialization Methods for the K-Means Algorithm. Pattern Recognition Letters 20(10), 1027–1040 (1999)
28. Zhang, H., Yang, Z.R., Oja, E.: Improving Cluster Analysis by Co-initializations. Pattern Recognition Letters. 45(1), 71–77 (2014)
29. Davies, D., Bouldin, D.: A Cluster Separation Measure. IEEE Transactions on Pattern Analysis and Machine Intelligence. 1(2), 224–227 (1979)
30. Rousseeuw, P.J.: Silhouettes: A Graphical Aid to the Interpretation and Validation of Cluster Analysis. Journal of Computational and Applied Mathematics. 20, 53–65 (1987)
31. Tsai, C.W., et al: A Fast Particle Swarm Optimization for Clustering. Soft Computing, 1–18 (2014)
32. Chioua, Y.C., Lan, L.W.: Genetic Clustering Algorithms. European Journal of Operational Research 135(2), 413–427 (2001)
33. Pham, D.T., Dimov, S.S., Nguyen C.D.: Selection of K in K-means Clustering (2005). http://www.ee.columbia.edu/~dpwe/papers/PhamDN05-kmeans.pdf
34. Milligan, G.W., Cooper, M.C.: An Examination of Procedures for Determining the Number of Clusters in a Data Set. Psychometrika 50(2), 159–179 (1985)
35. Hruschka, E.R., et al.: A Survey of Evolutionary Algorithms for Clustering. IEEE Transactions on Systems, Man, and Cybernetics, Part C: Applications and Reviews. 39(2), 133–155 (2009)
36. UCI Machine Learning Repository. University of California, Irvine, CA. http://archive.ics.uci.edu/ml/ (lasted visited April 15, 2015)
37. Lloyd, S.P.: Least Squares Quantization in PCM. IEEE Transaction Information Theory. 28(2), 129–137 (1982)
38. MacQueen, J.: Some Methods for Classification and Analysis of Multivariate Observations. In: The Proceedings of the Fifth Berkeley Symposium on Mathematics Statistics and Probability, pp. 281–296 (1967)
39. Hartigan, J.A., Wong, M.A.: Algorithm AS 136: A K-Means Clustering Algorithm. Applied Statistics, 100–108 (1979)
40. Source Code of five PSO-KM-based clustering algorithms and Related Data. http://share.weiyun.com/31ed301f38f47c7de9e84d2ba88ec0f9 (last update April 15, 2015)

A Novel Boundary Based Multiobjective Particle Swarm Optimization

Fushan Li, Shufang Xie, and Qingjian Ni[✉]

School of Computer Science and Engineering,
Southeast University, Nanjing, China
nqj@seu.edu.cn

Abstract. A novel boundary based multiobjective particle swarm optimization is presented in this paper. The proposed multiobjective optimization algorithm searches the border of the objective space unlike other current proposals to look for the Pareto solution set to solve such problems. In addition, we apply the proposed method to other particle swarm optimization variants, which indicates the strategy is highly applicatory. The proposed approach is validated using several classic test functions, and the experiment results show efficiency in the convergence performance and the distribution of the Pareto optimal solutions.

Keywords: Multiobjective optimization · Multiobjective particle swarm optimization · Particle swarm optimization

1 Introduction

Multiobjective optimization problems (MOP), optimization with two or more objective functions, are crucial and commonly seen both in engineering applications and in academic research, as well as in real world. To illustrate, the quality of products and the cost of production constitute a multiobjective optimization problem during enterprise activities, therefore, how to balance the two parts becomes an important issue for business managers.

Until 2004, Coello Coello *et al.* [1] proposed multiobjective particle swarm optimization (MOPSO), believed as a milestone in the history. Two-level of non-dominated solutions approach to multiobjective particle swarm optimization was claimed by Abido *et al.* [2] in 2007, searching locally and globally at two levels on the current Pareto front surface, and Koduru *et al.* [3] put forward a multi-objective hybrid particle swarm optimization (PSO) using μ-fuzzy dominance. Abido [4] designed a new multiobjective particle swarm optimization technique by proposing redefinition of global best and local best individuals. A self-adaptive learning based particle swarm optimization was proposed by Yu Wang *et al.* [5] to simultaneously adopt four PSO based search strategies. Daneshyari, M. *et al.* [6] stated a cultural-based multiobjective particle swarm optimization, where a cultural framework was used to adapt the personalized flight parameters of the mutated particles.

© Springer International Publishing Switzerland 2015
Y. Tan et al. (Eds.): ICSI-CCI 2015, Part I, LNCS 9140, pp. 153–163, 2015.
DOI: 10.1007/978-3-319-20466-6_17

Multiobjective optimization problems still have strong vitality so far due to its numerous applications. In recent years, Pierluigi Siano *et al.* [7] implemented fuzzy logic controllers for DC-DC converters using multiobjective particle swarm optimization. A large-scale portfolio optimization using multiobjective dynamic mutli-swarm particle swarm optimizer was proposed by Liang, J.J. *et al.* [8].

In this paper we propose a novel multiobjective optimization algorithm, particle swarm optimization based on boundary stated in the following parts. Section 2 describes the basic concepts in multiobjective optimization problems, and the proposed algorithm based on PSO is presented in Section 3. Experiments are conducted and analyzed in Section 4, and the conclusion shows up in Section 5.

2   Basic Concepts

Multiobjective optimization problems are also called multicriteria problems. Generally, a MOP with n variables, k objective functions can be expressed by formula (1) as follows:

$$
\begin{cases}
\min y = F(x) = (f_1(x), f_2(x), ..., f_k(x))^T \\
s.t.g_i(x) \leq 0, i = 1, 2, ..q \\
h_j(x) = 0, j = 1, 2, ..., p
\end{cases}
\tag{1}
$$

As the following algorithm is based on Pareto optimality, we introduce the concepts related to it. In this paper, we mainly employ the definitions by Coello Coello *et al.* [1].

Definition 2.1 (Pareto Dominance): A decision vector x_1 dominates another decision vector x_2, defining as $x_1 \prec x_2$, if and only if x_1 is no less than x_2 in every objective, that is to say: $f_i(x_1) \leq f_i(x_2), i = 1, 2, ..., k$, and x_1 is strictly more than x_2 on at least one objective, in other words, $\exists i = 1, 2, ..., k, f_i(x_1) < f_i(x_2)$.

Definition 2.2 (Pareto Optimality): A decision vector $x^* \in F$ is *Pareto optimal*, if there is not a decision vector $x \neq x^* \in F$ dominating it.

Definition 2.3 (Pareto Optimal Set): All Pareto optimal decision vectors compose a Pareto optimal set P^*, that is $P^* = \{x^* \in F| \not\exists x \in F, x \prec x^*\}$.

Therefore, a Pareto optimal set is a set of all solutions, and the corresponding objective vectors are defined as *Pareto front*.

3   Proposed Boundary Based Method for Multiobjective Particle Swarm Optimization Problems

3.1   Basic Principle of PSO

Particle swarm optimization is an evolutionary algorithm that was inspired by the social behavior of bird folks, proposed by J. Kennedy and R.C. Eberhart. One classic PSO method [9] is:

$$
v_{ij}(t + 1) = wv_{ij}(t) + c_1r_{1j}(t)[y_{ij}(t) - x_{ij}(t)] + c_2r_{2j}(t)[\hat{y}_j(t) - x_{ij}(t)]
\tag{2}
$$

$$x_{ij}(t+1) = x_{ij}(t) + v_{ij}(t+1) \tag{3}$$

where w is the inertia weight, and c_1, c_2 are positive acceleration constants denoting the contribution of cognitive and social factors, respectively. $r_{1j}(t)$ and $r_{2j}(t)$ are random numbers between zero and one with uniform distribution, introducing uncertainty to the algorithm. $v_{ij}(t)$ represents the speed of particle i on dimension j at moment t, y_i is the best position particle i has reached so far and $\hat{y}(t)$ is the global optimal location at moment t. Ni et $al.$ [10] proposed some improvement on the PSO method, the update formula shows as follows:

$$CT_{ij}(t) = \frac{\sum_{p=1}^{K} y_{pj}(t)}{K} - x_{ij}(t) \tag{4}$$

$$OT_{ij}(t) = \frac{\sum_{p=1}^{K} y_{pj} - y_{ij}}{K} \tag{5}$$

$$x_i(t+1) = x_i(t) + \alpha \times (x_i(t) - x_i(t-1)) + \beta \times CT_i(t) + \gamma \times Gen() \times OT_i(t) \tag{6}$$

In formula (4)(5)(6), α, β, γ are all weights and α is a positive number generally. $Gen()$ is a random number generating function of specific distribution. When $Gen()$ is the Gaussian function the variant is GDPSO, while it is LDPSO when that is the Logist function. Such variants will be used to show applicatory of the method later.

3.2   Proposed Boundary Based MOPSO

Most scholars look for the Pareto solution set to tackle multiobjective optimization problems. The Pareto solution set in the objective space is actually the boundary of the space. Therefore, we can convert the problem of finding the Pareto solution set to how to find the boundary of the objective space. We will then take a two-dimension space as an example to illustrate.

First, we consider two functions

$$g(x) = |w_1 f_1(x) + w_2 f_2(x)| \tag{7}$$

$$h(x) = e^{\frac{f_1^2(x) + f_2^2(x)}{b}} \tag{8}$$

where w_1, w_2 and $b(b > 0)$ are all parameters and $f_1(x), f_2(x)$ are both objective functions. In the objective space, the first function can be seen as the absolute value of the weighted sum of $f_1(x)$ and $f_2(x)$. In other way it is the absolute value of the projection of the vector $(f_1(x), f_2(x))$ on the direction of vector (w_1, w_2). Obviously, when the vector (w_1, w_2) is vertical to the vector $(f_1(x), f_2(x))$, $g(x)$ has the minimum value zero. It is evident to know from the second function that when the distance between point $(f_1(x), f_2(x))$ and point $(0, 0)$ is the smallest, $h(x)$ gets the minimum. Thus, we can comprehensively combine $g(x)$ and $h(x)$ to acquire a new function

$$f'(x) = (g(x) + a)h(x) \tag{9}$$

Assume that all the points $(f_1(x), f_2(x))$ are located in the first quadrant, then with regard to (9), the minimum value must be obtained on the boundary. If not, suppose the minimum is $(f_1(x_0), f_2(x_0))$, noted as A in Fig. 1. Connect point A with the original point intersecting the boundary at point B $(f_1(x_1), f_2(x_2))$. We can easily find $g(x_1) = g(x_0)$ and $h(x_1) < h(x_0)$ and get the conclusion $f'(x_1) < f'(x_0)$ which is contradictory to our previous assumption. Therefore, we have proved the minimum value of function $f'(x)$ must be acquired on the boundary.

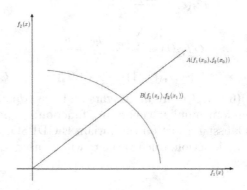

Fig. 1. The schematic plot of the objective space

As for the selection of parameter a, we can adopt the following method. Under the condition that w_1 and w_2 are certain, we wish to get the minimum value within θ to the vertical direction of the vector (w_1, w_2), which is shown in Fig. 2.

Make a straight line through the original point vertical to the vector (w_1, w_2) crossing the boundary at $(f_1(x_2), f_2(x_2))$ and choose a point $(f_1(x_3), f_2(x_3))$ randomly on the boundary meeting the requirement that the inclined angle with $(f_1(x_2), f_2(x_2))$ is larger than θ. Let $r = f_1^2(x_2) + f_2^2(x_2)$ and $r_1 = f_1^2(x_3) + f_2^2(x_3)$, r_{max} is the maximum distance between the points on the boundary and the original point, r_{min} is the minimum distance between them and θ_1 is the inclined angle between vector $(f_1(x_2), f_2(x_2))$ and vector $(f_1(x_3), f_2(x_3))$. If $f'(x_2) < f'(x_3)$, then

$$(g(x_2) + a)h(x_2) < (g(x_3) + a)h(x_3)$$

$$ae^{\frac{r^2}{b}} < (|r_1 \sin \theta_1| + a)e^{\frac{r_1^2}{b}} \tag{10}$$

$$a(e^{\frac{r^2}{b}} - e^{\frac{r_1^2}{b}}) < |r_1 \sin \theta_1|e^{\frac{r_1^2}{b}}$$

If $r < r_1$, our expectation can be reached.

If $r > r_1$, then it needs to meet the condition

$$a < \frac{|r_1 \sin \theta_1|}{e^{\frac{r^2 - r_1^2}{b}} - 1} \tag{11}$$

So when

$$a = \frac{r_{min} \sin \theta}{e^{\frac{r_{max}^2 - r_{min}^2}{b}} - 1} \tag{12}$$

we have got $f'(x_2) < f'(x_3)$, assuring that we obtain the minimum value within θ to the vertical direction of the vector (w_1, w_2).

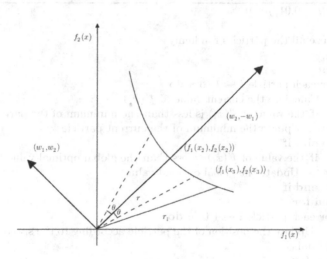

Fig. 2. The selection of parameter a

Due to the above-mentioned property, θ is set to control the distribution of the space. Pay attention to the previous hypothesis at the beginning that all the points $(f_1(x), f_2(x))$ are located in the first quadrant which is feasible. Since the minimum values of both $f_1(x)$ and $f_2(x)$ exist, we can add a constant large enough, making $(f_1(x), f_2(x))$ all in the assumed position.

To sum up, the algorithm can be easily described by enumerating (w_1, w_2) to acquire the minimum of $f'(x)$ with the current parameter. And we can transform the procedure to the enumeration of α. Let $w_1 = \cos \alpha, w_2 = \sin \alpha$ and as all the points $(f_1(x), f_2(x))$ are located in the first quadrant, $\frac{\pi}{2} < \alpha < \pi$. Enumerating averagely α might give rise to less Pareto solutions on the boundary, thus, we randomly select α by controlling the distance as follows:

1) Initialize $\alpha = \pi, delt = 0.01$;
2) Calculate the minimum under α and obtain the boundary point p_i;
3) Calculate the distance between p_i and p_{i-1}. If larger than argument t, then $delt = \frac{delt}{2}$, otherwise, $delt = delt * 1.5$ (when $i = 1$ and $delt = 0.01$);
4) $\alpha = \alpha - delt$;
5) if $\alpha > \frac{\pi}{2}$, return to step 2 , otherwise jump out of the program.

We can obtain the basic boundary according to the process, but the boundary is not the same as Pareto solutions. In the end some non-Pareto solutions need to be kicked out, and the Pareto solution set is gained. The concrete implementation can be referred to as the pseudocode in Algorithm 1.

Algorithm 1. Multiobjective Optimization Algorithm Based on Boundary

---

1: $\alpha = \pi, delt = 0.01, j = 0$
2: **repeat**
3:      Initialize all the particles randomly
4:      $j + +$;
5:      **repeat**
6:          **for** each particle $i \leftarrow 1$ to n **do**
7:              Calculate the current value of $f'(x_i)$
8:              **if** the value of $f'(x_i)$ is less than the minimum of the particle i **then**
9:                  Update the minimum of the current particle
10:             **end if**
11:             **if** the value of $f'(x_i)$ is less than the global optimal value **then**
12:                 Update the global optimal value
13:             **end if**
14:         **end for**
15:         **for** each particle $i \leftarrow 1$ to n **do**
16:             Update the position of the particle according to the speed and position update formulas
17:         **end for**
18:     **until** the terminal condition 1 is true
19:     **if** $j = 1$ **then**
20:         $delt = 0.01$
21:     **else**
22:         Calculate the distance dis between p_j and p_{j-1}
23:         **if** $dis > t$ **then**
24:             $delt = delt/2$
25:         **else**
26:             $delt = delt * 1.5$
27:         **end if**
28:     **end if**
29:     $\alpha = \alpha - delt$
30: **until** the terminal condition 2 is true

---

4 Experiment Results and Analysis

4.1 Test Functions

We use eight standard test functions listed in Table 1.

Table 1. Test problems used in this study

name	test function	limit
binh1	$F = (f_1(x,y), f_2(x,y))$ $f_1(x,y) = x^2 + y^2$ $f_2(x,y) = (x-5)^2 + (y-5)^2$	$-5 \leq x, y \leq 10$
fonseca1	$F = (f_1(x,y), f_2(x,y))$ $f_1(x,y) = 1 - e^{-(x-1)^2 - (y+1)^2}$ $f_2(x,y) = 1 - e^{-(x+1)^2 - (y-1)^2}$	none
fonseca2	$F = (f_1(x,y), f_2(x,y))$ $f_1(x,y) = 1 - e^{-(x-\frac{1}{\sqrt{2}})^2 - (y-\frac{1}{\sqrt{2}})^2}$ $f_2(x,y) = 1 - e^{-(x+\frac{1}{\sqrt{2}})^2 - (y+\frac{1}{\sqrt{2}})^2}$	$-4 \leq x, y \leq 4$
rendon1	$F = (f_1(x,y), f_2(x,y))$ $f_1(x,y) = \frac{1}{x^2 + y^2 + 1}$ $f_2(x,y) = x^2 + 3y^2 + 1$	$-3 \leq x, y \leq 3$
rendon2	$F = (f_1(x,y), f_2(x,y))$ $f_1(x,y) = x + y + 1$ $f_2(x,y) = x^2 + 2y - 1$	$-3 \leq x, y \leq 3$
ZDT1	$F = (f_1(x), f_2(x))$ $f_1(x) = x_1$ $f_2(x) = g(1 - \sqrt{\frac{f_1}{g}})$ $g(x) = 1 + 9 \sum_{i=2}^{m} \frac{x_i}{m-1}$ $m = 30$	$0 \leq x_i \leq 1$
ZDT2	$F = (f_1(x), f_2(x))$ $f_1(x) = x_1$ $f_2(x) = g(1 - \frac{f_1}{g})^2$ $g(x) = 1 + 9 \sum_{i=2}^{m} \frac{x_i}{m-1}$ $m = 30$	$0 \leq x_i \leq 1$
ZDT3	$F = (f_1(x), f_2(x))$ $f_1(x) = x_1$ $f_2(x) = g(1 - \sqrt{\frac{f_1}{g}} - \frac{f_1}{g} \sin 10\pi f_1)$ $g(x) = 1 + 9 \sum_{i=2}^{m} \frac{x_i}{m-1}$ $m = 30$	$0 \leq x_i \leq 1$

4.2 Experiment Parameter Settings

We put forward a novel algorithm for MOP in this paper, and adopt two variants of PSO, the classic PSO and GDPSO. Therefore, there are two implementations altogether, which includes the classic PSO based on boundary and the GDPSO based on boundary.

With regard to the two multiobjective optimization algorithm based on boundary, for each α the number of iteration is 200 and the number of particles is 300. Argument a and b have to be set flexibly according to different problems.

To obtain more precise results, we run the experiment 20 times for each test function.

4.3   Evaluation Metrics

We mainly employ two metrics to evaluate the performance of the algorithm: convergence metric [11] and spacing metric [12].

(1) Convergence metric

Let $P = (p_1, p_2, ..., p_{t_1})$ be the ideal Pareto set(generally, the ideal Pareto set cannot be reached, so we can replace it with some existing reference sets), and the Pareto solution set through calculation is $Q = (q_1, q_2, ..., q_{t_2})$. Calculate

$$d_i = \min_{j=1}^{t_1} \sqrt{\sum_{m=1}^{k} (\frac{f_m(p_j) - f_m(q_i)}{f_m^{max} - f_m^{min}})^2} \tag{13}$$

and f_m^{max}, f_m^{min} are the maximum and minimum value of the mth objective function in set P, then the convergence metric is shown in (14).

$$C = \frac{\sum_{i=1}^{t_2} d_i}{t_2} \tag{14}$$

The convergence metric reflects the performance of the convergence. Usually when it is nearer to the ideal Pareto front, C is smaller.

(2) Spacing metric

$$B = \sqrt{\frac{1}{t_2 - 1} \sum_{i=1}^{t_2} (u_i - \bar{u})^2} \tag{15}$$

where

$$u_i = \min_{j=1}^{t_2} \left\{ \sum_{m=1}^{k} |f_m(q_i) - f_m(q_j)| \right\}$$

$$\bar{u} = \frac{\sum_{i=1}^{t_2} u_i}{t_2}$$

It can be judged from the spacing metric how well the nondominated solutions in the objective space are distributed. And if $B = 0$, it means the distribution of the nondominated solutions we have got is uniform.

4.4 Experiment Data and Analysis

Boxplots are employed to display the experiment data. In the following figures, 1 represents the performance parameter of the classic PSO based on boundary, and 2 shows that of the GDPSO based on boundary.

(1) Convergence metric

Seen from Fig.3, the classic PSO algorithm based on boundary has the advantage of less fluctuation on the convergence metric: we get almost the same consequences during the twenty independent runs of every test function. However, the GDPSO based on boundary does not perform well due to the insufficient number of iterations.

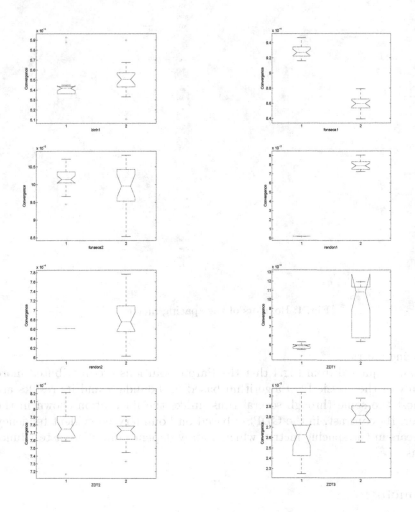

Fig. 3. Boxplots of the convergence metric

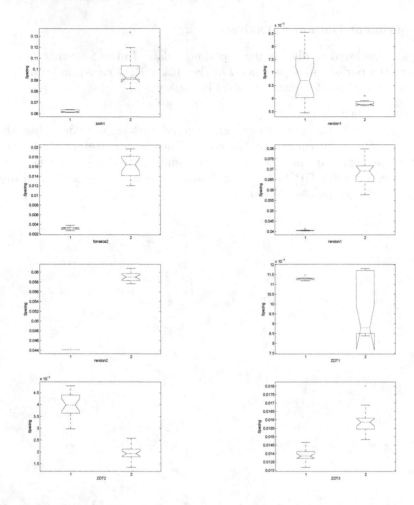

Fig. 4. Boxplots of the spacing metric

(2) Spacing metric

It can be judged from Fig.4 that the Pareto solutions are distributed more evenly in the classic PSO algorithm based on boundary and its results are almost the same through several runs unlike the fluctuation shown in the other. By contrast, in the GDPSO based on boundary no evident tendency appears in the spacing metric, which mainly depends on different test functions.

5   Conclusion

On the basis of the convergence and spacing metrics, the classic PSO algorithm based on boundary performs better. Although the novel boundary based multiobjective particle swarm optimization is of high performance and has fast convergence speed, it has deficiencies more or less. As for the classic PSO based on boundary we can improve on the choice of parameter a and b, but it will be hard to have large development on the convergence metric as we have to decrease the number of iteration considering runtime. The extension of this work would be a further discussion of the parameter settings and to apply the proposed method to high-dimensional objective space.

References

1. Coello, C.A.C., Pulido, G.T., Lechuga, M.S.: Handling multiple objectives with particle swarm optimization. IEEE Transactions on Evolutionary Computation 8(3), 256–279 (2004)
2. Abido, M.A.: Two-level of nondominated solutions approach to multiobjective particle swarm optimization. In: Proceedings of the 9th Annual Conference on Genetic and Evolutionary Computation, pp. 726–733. ACM (2007)
3. Koduru, P., Das, S., Welch, S.M.: Multi-objective hybrid pso using μ-fuzzy dominance. In: Proceedings of the 9th Annual Conference on Genetic and Evolutionary Computation, pp. 853–860. ACM (2007)
4. Abido, M.A.: Multiobjective particle swarm optimization for environmental/economic dispatch problem. Electric Power Systems Research 79(7), 1105–1113 (2009)
5. Wang, Y., Li, B., Weise, T., Wang, J., Yuan, B., Tian, Q.: Self-adaptive learning based particle swarm optimization. Information Sciences 181(20), 4515–4538 (2011)
6. Daneshyari, M., Yen, G.G.: Cultural-based multiobjective particle swarm optimization. IEEE Transactions on Systems, Man, and Cybernetics, Part B: Cybernetics 41(2), 553–567 (2011)
7. Siano, P., Citro, C.: Designing fuzzy logic controllers for dc-dc converters using multi-objective particle swarm optimization. Electric Power Systems Research 112, 74–83 (2014)
8. Liang, J.J., Qu, B.Y.: Large-scale portfolio optimization using multiobjective dynamic mutli-swarm particle swarm optimizer. In: IEEE Symposium on Swarm Intelligence (SIS), pp. 1–6. IEEE (2013)
9. Shi, Y., Eberhart, R.: A modified particle swarm optimizer. In: Proceedings of IEEE World Congress on Computational Intelligence, pp. 69–73. IEEE (1998)
10. Ni, Q., Deng, J.: Two improvement strategies for logistic dynamic particle swarm optimization. In: Dobnikar, A., Lotrič, U., Šter, B. (eds.) ICANNGA 2011, Part I. LNCS, vol. 6593, pp. 320–329. Springer, Heidelberg (2011)
11. Schott, J.R.: Fault tolerant design using single and multicriteria genetic algorithm optimization. Technical report, DTIC Document (1995)
12. Deb, K., Jain, S.: Running performance metrics for evolutionary multi-objective optimizations. In: Proceedings of the Asia-Pacific Conference on Simulated Evolution and Learning (SEAL), pp. 13–20 (2002)

A Novel Particle Swarm Optimization for Portfolio Optimization Based on Random Population Topology Strategies

Xushan Yin, Qingjian Ni[✉], and Yuqing Zhai

College of Software Engineering, Southeast University, Nanjing, China
nqj@seu.edu.cn

Abstract. The problem of portfolio selection in the field of financial engineering has received more attention in recent years and many portfolio selection models has been proposed in succession. To solve a generalized Markowitz mean-variance portfolio selection model, this paper proposed four improved particle swarm optimization algorithms (RTWPSO-AD, RTWPSO-D, DRWTPSO-AD, DRWTPSO-D) based on the strategies of Random Population Topology. We abstract the topology of particle swarm optimization (PSO) into an undirected connected graph which can be generated randomly according to a predetermined degree. The topology changes during the evolution when Dynamic Population Topology strategy is adopted. By setting the degree, we can control the communication mechanisms in the evolutionary period, enhancing the solving performance of PSO algorithms. The generalized portfolio selection model is classified as a quadratic mixed-integer programming model for which no computational efficient algorithms have been proposed. We employ the proposed four algorithms to solve the model and compare the performance of them with the classic PSO variant. The computational results demonstrate that the population topologies of PSO have direct impacts on the information sharing among particles, thus improve the performance of PSO obviously. In particular, the proposed DRTWPSO-D shows an extraordinary performance in most set of test data, providing an effective solution for the portfolio optimization problem.

Keywords: Portfolio optimization · PSO · Random population topology · Cardinality constrained mean-variance (CCMV) model

1 Introduction

The rapid development of economy has brought to investors not only opportunities, but also serious challenges and risks. The instability of the financial markets leads to more researches in the field of financial investment, among which portfolio selection has become a significant problem.

Markowitz [1] established a quantitative framework for portfolio selection, proposing the Mean-Variance model (MV model). However, the portfolio

© Springer International Publishing Switzerland 2015
Y. Tan et al. (Eds.): ICSI-CCI 2015, Part I, LNCS 9140, pp. 164–175, 2015.
DOI: 10.1007/978-3-319-20466-6_18

selection problem proved to be a typical NP-hard problem. In addition, when cardinality constraint is included, the model becomes a mixed quadratic and integer programming for which no computational efficient algorithms have been proposed.

In recent years, computational intelligence (CI) has developed at lightening speed and has already been introduced into the field of portfolio optimization [3] [4] [5]. Particle swarm optimization (PSO) is a collaborative population-based meta-heuristic algorithm introduced by Kennedy and Eberhart in 1995 and has proven to be effective in many empirical studies. There has been some researches introducing PSO into portfolio optimization problems, but few of them lay an emphasis on the improvement of the performance for PSO when applied to this specific problem.

In this paper, we propose four improved PSO algorithms based on the strategy of Random Population Topology for an generalized MV model which considers cardinality constraints and bounding constrains.

This paper is divided into five sections. A brief introduction of the study has been presented in Section 1. In section 2, we introduce the extended cardinality constrained mean-variance (CCMV) model. We give a detailed description of the proposed strategies and algorithms in Section 3. The computational experiments are described in Section 4. Finally, Section 5 is devoted to the conclusions.

2   Portfolio Selection Model

The portfolio selection model CCMV we adopted in this paper generalizes the standard MV model to include cardinality and bounding constraints. The model can be described as:

$$\min \lambda \sum_{i=1}^{N} \sum_{j=1}^{N} z_i x_i z_j x_j \sigma_{ij} - (1 - \lambda) \sum_{i=1}^{N} z_i x_i \mu_i \qquad (1)$$

subject   to:

$$\sum_{i=1}^{N} x_i = 1, \qquad (2)$$

$$\sum_{i=1}^{N} z_i = K, \qquad (3)$$

$$\varepsilon_i z_i \leq x_i \leq \delta_i z_i \qquad (4)$$

$$z_i \in \{0, 1\}, i = 1, 2, \cdots, N. \qquad (5)$$

Where N is the number of different assets, x_i is the proportion of asset i in the portfolio, σ_{ij} is the covariance between returns of asset i and asset j, μ_i is the mean return of asset i and R is the desired mean return of the portfolio, $\lambda \in [0, 1]$ is the risk aversion parameter, K is the desired number of different assets included in the portfolio, ε_i and δ_i are the lower and upper bounds for the proportion of the assets in the portfolio, respectively. $z_i \in \{0,1\}$ is defined as the decision variable, asset i will be included in the portfolio if $z_i = 1$, otherwise, it will not be.

With different values of λ, we can get different objective function values which are composed of mean return and variance. The objective function values of all these Pareto optimal solutions form what is called the efficient frontier in the Markowitz theory.

3 The Proposed PSO Algorithms Based on Random Population Topology

3.1 Random Population Topology

For convenience of analysis, a population topology in PSO could be usually abstracted into an undirected connected graph. The degree of a particle in the population means the number of neighboring particles it maintains. The average degree of a population topology means the average number of neighboring particles, which stands for the socializing degree of the swarm.

Researchers have conducted the research on Random Population Topology and confirmed that the structures of population topology of PSO have direct influence on its performance [6] [7] [8] [9] [10].

3.2 Strategies of Random Population Topology

In this paper, four strategies are designed for the population topology of the PSO algorithm. These strategies can be expressed as follows.

Algorithm 1. Random Population Topology based on the average degree (RT-AD)

1: For a population with the size of S, set up a matrix L of $S \times S$, and let $L(i; i) = 1$.
2: Determine the value of AD.
3: Arrange the index of all particles in PSO algorithm randomly, generate a connected topology by setting $L(i; j) = 1$ if i and j are two consecutive indexes.
4: Set $CurrentAD = 2$.
5: **repeat**
6:     for every row i in the L matrix, generate a random number $m(m \neq i)$.
7:     Let $L(m; i) = L(i; m) = 1$.
8:     increase $CurrentAD$.
9: **until** $CurrentAD = AD$

---

(1) Random Population Topology based on the average degree
    Random Population Topology based on the average degree (RT-AD), means generating a population topology with a given average degree AD randomly for a PSO algorithm and keep the topology during the evolution. The detailed strategy of RT-AD is shown in Algorithm 1.

Algorithm 2. Random Population Topology based on the degree (RT-D)

1: For a population with the size of S, set up a matrix L of $S \times S$, and let $L(i; i) = 1$.
2: Determine the value of D.
3: Arrange the index of all particles in PSO algorithm randomly, generate a connected topology by setting $L(i; j) = 1$ if i and j are two consecutive indexes.
4: Set $CurrentD = 2$ for each particle.
5: **repeat**
6:    For every position of row i, column j in the L matrix, search for an unconnected particle m whose degree is less than D for particle i.
7:    Let $L(m; i) = L(i; m) = 1$.
8:    increase $CurrentD$ of particle i.
9: **until** $CurrentD$ of each particle i is equal to D

(2) Random Population Topology based on the degree
Unlike RT-AD, Random Population Topology based on the degree (RT-D) generates a population topology with a certain degree D for each particle, which means all the particles have the same number of neighbours. The detailed strategy of RT-D is shown in Algorithm 2.

Algorithm 3. Dynamic Random Population Topology based on the average degree (DRT-AD)

1: For a population with the size of S, set up a matrix L of $S \times S$, and let $L(i; i) = 1$.
2: Determine the value of AD.
3: Arrange the index of all particles in PSO algorithm randomly, generate a connected topology by setting $L(i; j) = 1$ if i and j are two consecutive indexes.
4: Set $CurrentAD = 2$.
5: Set the evolution generation counter to 0.
6: **repeat**
7:    Evolve the particles in the population.
8:    Generation counter increases 1.
9:    **if** the generation counter is evenly divisible by M **then**
10:       Increase the value of AD linearly.
11:       Generate a new Random Population Topology with average degree=AD.
12:    **end if**
13: **until** The termination condition is satisfied

(3) Dynamic Random Population Topology based on the average degree
Dynamic Random Population Topology based on the average degree (DRT-AD), means generating a new population topology randomly by every certain number of generations which can be represented by a variable M. To avoid premature convergence and increase the possibility of exploring the whole solution space at the early stage of evolution, the average degree AD can be set to 2 (which generates a ring topology). With the evolution generations growing, particles tend to converge to an optimum and AD should

---

Algorithm 4. Dynamic Random Population Topology based on the degree (DRT-D)

1: For a population with the size of S, set up a matrix L of $S \times S$, and let $L(i;i) = 1$.
2: Determine the value of D.
3: Arrange the index of all particles in PSO algorithm randomly, generate a connected topology by setting $L(i;j) = 1$ if i and j are two consecutive indexes.
4: Set $CurrentD = 2$ for each particle.
5: Set the evolution generation counter to 0.
6: **repeat**
7:     Evolve the particles in the population.
8:     Generation counter increases 1.
9:     **if** the generation counter is evenly divisible by M **then**
10:         Increase the value of D linearly.
11:         Generate a new Random Population Topology with the degree=D.
12:     **end if**
13: **until** The termination condition is satisfied

---

be increased gradually to produce a new random population topology. The detailed strategy of DRT-AD is as Algorithm 3.
(4) Dynamic Random Population Topology based on the degree
     Similar to DRT-AD, the Random Population Topology based on the degree (DRT-D) generates new topologies with a given degree D for all the particles in the population. The detailed strategy of DRT-D is as Algorithm 4.

3.3 The Proposed Approaches: RTWPSO-AD, RTWPSO-D, DRTWPSO-AD, DRTWPSO-D

In our research, the strategies of Random Population Topology are implemented in a classic variant of PSO: PSO with inertia weight (WPSO) [10]. We propose four improved PSO algorithms which are: RTWPSO-AD, RTWPSO-D, DRTWPSO-AD, DRTWPSO-D based on the four strategies: RT-AD, RT-D, DRT-AD, DRT-D respectively and employ them to solve the CCMV model.

4 Computational Experiments

To prove the effectiveness of the Random Population Topology strategy, we adopt the proposed four algorithms: RTWPSO-AD, RTWPSO-D, DRTWPSO-AD, DRTWPSO-D and WPSO to solve the optimization problem and compare the performance of them.

4.1 Settings of Experiments

All the algorithms search for the efficient frontiers for the portfolio optimization problem based on CCMV model formulated in Eqs.(1)-(5).

The datasets correspond to weekly prices from March, 1992 to September, 1997 including the following indices: Hang Seng in Hong Kong, DAX 100 in Germany, FTSE 100 in UK, S&P 100 in USA and Nikkei 225 in Japan. The number of assets considered for each index was 31, 85, 89, 98 and 225, respectively.

To get the Pareto optimal set under different values of risk aversion parameter λ, we set $\Delta\lambda = 0.02$ thus the number of Pareto optimal solutions, denoted by ξ, is 51. The cardinality constraints are set as follows: $K = 10$, $\varepsilon_i = 0.01$, $\delta_i = 1$. The swarm size is set to 100 and the number of iteration is set to 1000 for all experiments.

We trace out the efficient frontiers of CCMV model. For comparison between the efficient frontiers we get and the standard efficient frontier, we use four criteria which are mean Euclidian distance, variance of return error, mean return error and execution time. The mean Euclidian distance, variance of return error and mean return error are defined in [11].

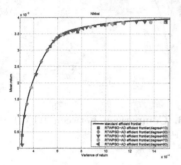

Fig. 1. RTWPSO-AD with different values of AD

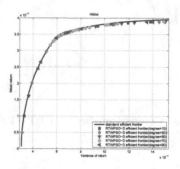

Fig. 2. RTWPSO-D with different values of D

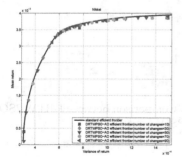

Fig. 3. DRTWPSO-AD with different change numbers

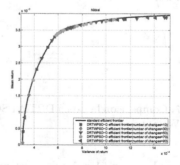

Fig. 4. DRTWPSO-D with different change numbers

4.2   Experiment 1: Parameter Adjustment of RTWPSO-AD, RTWPSO-D, DRTWPSO-AD, DRTWPSO-D

In this section, parameter adjustment of the four algorithms is carried out to obtain the best performance of each algorithm, proving the effectiveness of them as well as providing the basis for subsequent experiments.

For RTWPSO-AD and RTWPSO-D, five set of experiments with different number of degree as 10, 30, 50, 70, 90 for the topology are carried out based on Nikkei index. For DRTWPSO-AD and DRTWPSO-D, five set of experiments in which the number of topology changes is set to be 10, 30, 50, 70, 90 respectively.

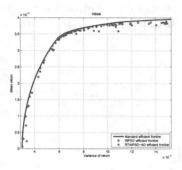

Fig. 5. Comparison between RTWPSO-AD&WPSO

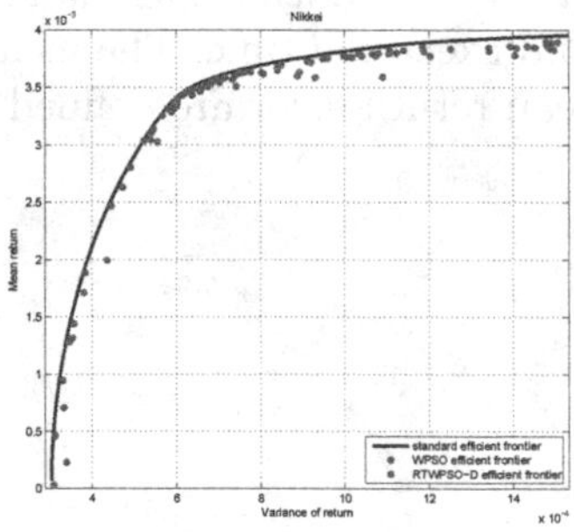

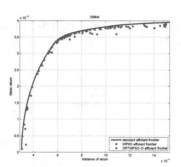

Fig. 6. Comparison between RTWPSO-D&WPSO

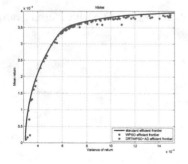

Fig. 7. Comparison between DRTWPSO-AD&WPSO

Fig. 8. Comparison between DRTWPSO-D&WPSO

Based on the 51 points we get under different values of risk aversion parameter λ, we trace out the efficient frontiers for each experiment. To get a clear impression of the results we got, we merged the five efficient frontiers into a single one for each of the four algorithms. The corresponding results are exhibited in Figs.(1)-(4). We find out that RTWPSO-AD and RTWPSO-D performs well when the degree is around 70, DRTWPSO-AD performs well when the number of topology changes is around 50 while DRTWPSO-D performs better around 70. We compare the best performance of each algorithm with WPSO and trace out the efficient frontiers which are shown in Figs.(5)-(8).

From Figs.(5)-(8), we can see that all the four algorithms perform better than WPSO and get a more smoother efficient frontier which is more closer to the standard efficient frontier.

4.3 Experiment 2: Comparison Between the Proposed Four Algorithms and WPSO

In this section, we compare the performance of the proposed four algorithms and WPSO by solving the five problems in different stock markets.

Since the efficient frontiers we got are very close to each other, to achieve a better comparision result, we merged the four efficient frontiers of RTWPSO-AD, RTWPSO-D, DRTWPSO-AD and DRTWPSO-D into one for different markets. The results are presented in Figs.(9)-(18) as well as the box-plots of Euclidian distance for each stock market.

Table 1. The contribution percentage of the four proposed algorithms

Index	Assets	Contribution percentage(%)			
		RTWPSO-AD	RTWPSO-D	DRTWPSO-AD	DRTWPSO-D
Hang Seng	31	0.33	0.16	0.14	**0.37**
DAX 100	85	0.27	0.24	0.16	**0.33**
FTSE 100	89	0.20	0.27	0.20	**0.33**
S&P 100	98	0.24	0.20	0.22	**0.35**
Nikkei	225	0.22	0.16	0.20	**0.43**

From the mixed efficient frontiers and the contribution percentage of each algorithm shown in Table 1, we can conclude that DRTWPSO-D got most optimal solutions in all stock markets and the other three algorithms show similar performances in most cases. Besides, for the Hang Seng index in Fig.(9), FTSE 100 index in Fig.(13), Nikkei index in Fig.(17), the mixed efficient frontiers are almost coincide with the standard efficient frontier despite the fact that we are faced with a much larger search space in the Nikkei market when the dimension increases to 225. For the DAX 100 index in Fig.(11) and S&P index in Fig.(15), the mixed efficient frontiers are quite close to the standard efficient frontier when the value of parameter λ is high. But when the value of parameter λ is low, the mixed efficient frontiers tend to deviate from the standard efficient frontier, in which the situation is worse especially for the DAX 100 index.

To take a look at the results from a numerical point, we exhibit the corresponding results in Table 2. We can see clearly that the proposed DRTWPSO-D obtained the best result in most problems using the three criteria and the

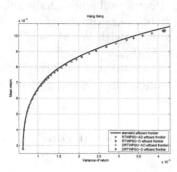

Fig. 9. The mixed efficient frontier of Hang Seng

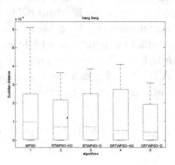

Fig. 10. The box-plot of Hang Seng

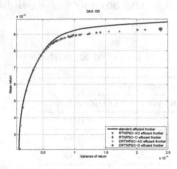

Fig. 11. The mixed efficient frontier of DAX 100

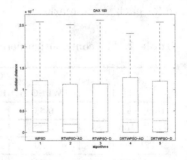

Fig. 12. The box-plot of DAX 100

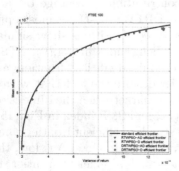

Fig. 13. The mixed efficient frontier of FTSE 100

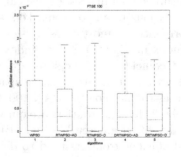

Fig. 14. The box-plot of FTSE 100

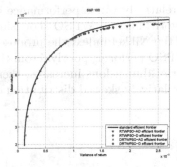

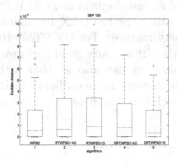

Fig. 15. The mixed efficient frontier of S&P 100

Fig. 16. The box-plot of S&P 100

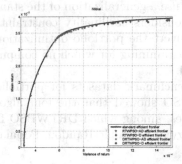

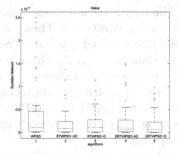

Fig. 17. The mixed efficient frontier of Nikkei 225

Fig. 18. The box-plot of Nikkei 225

Table 2. The experimental results of the four proposed algorithms and WPSO

Index		WPSO	RTWPSO-AD	RTWPSO-D	DRTWPSO-AD	DRTWPSO-D
Hang Seng	Mean Euclidian distance	9.51E-05	8.80E-05	9.83E-05	9.57E-05	**7.90E − 05**
	Variance of return error(%)	2.01188	1.90392	2.00295	2.0429	**1.85689**
	Mean return error(%)	0.748491	0.685853	0.775047	0.748999	**0.63821**
	Time(s)	119	119.964	120.864	123.815	120.3
DAX 100	Mean Euclidian distance	**1.75E − 04**	1.77E-04	1.84E-04	1.86E-04	1.86E-04
	Variance of return error(%)	8.53576	**8.10246**	8.55663	8.76838	8.46739
	Mean return error(%)	**1.51591**	1.52205	1.62269	1.57935	1.86482
	Time(s)	459	435.939	413.405	435.175	422.044
FTSE 100	Mean Euclidian distance	5.91E-05	5.87E-05	5.78E-05	5.72E-05	**5.43E − 05**
	Variance of return error(%)	4.04585	4.05032	3.94084	3.92245	**3.64471**
	Mean return error(%)	0.493448	0.489157	0.479852	0.475298	**0.455364**
	Time(s)	475	492.114	457.887	449.514	463.111
S&P 100	Mean Euclidian distance	9.56E-05	1.01E-04	1.04E-04	9.88E-05	**8.68E − 05**
	Variance of return error(%)	3.84698	3.97382	4.00802	3.97534	**3.83954**
	Mean return error(%)	0.920094	1.08019	1.17502	1.08721	**0.830604**
	Time(s)	568	560.264	531.764	537.864	559.701
Nikkei	Mean Euclidian distance	5.18E-05	3.64E-05	3.58E-05	3.51E-05	**3.07E − 05**
	Variance of return error(%)	3.20306	1.99452	2.10767	1.98833	**1.83554**
	Mean return error(%)	1.11824	1.30823	3.6018	0.761275	**0.669599**
	Time(s)	2242	2042.63	2190.45	2070.53	2223.36

box-plots just confirm it while WPSO shows a relatively worse performance in most cases. For the DAX 100 index, the four algorithms and WPSO show similar performances. For the FTSE 100 index and the Nikkei index, the proposed four algorithms show good performance while WPSO performs poorly. From the box-plots, we can also see that with the number of dimensions increases, there begins to occur more outliers which indicates a relatively skewed distribution of the Euclidian distances.

5 Conclusion

In this paper, we proposed four improved particle swarm optimization algorithms (RTWPSO-AD, RTWPSO-D, DRTWPSO-AD, DRTWPSO-D) based on the strategy of Random Population Topology to solve the portfolio optimization problem and traced out its efficient frontier. We use a generalization of the standard Markowitz mean-variance model which considers cardinality constraints and bounding constrains. These constraints convert the portfolio optimization problem into a mixed quadratic and integer programming for which no computational efficient algorithms have been proposed.

The proposed algorithms were tested on benchmark datasets for portfolio optimization problems. Comparisons with WPSO showed that the four algorithms perform better than WPSO in most cases, among which DRTWPSO-D shows an excellent performance with the strategy of Dynamic Random Population Topology.

All the experimental results presented in this paper lead us to conclude that the strategy of Random Population Topology is an important method for the improvement of the performance for PSO algorithm and the proposed algorithms did provide an effective solution for the portfolio optimization problem.

References

1. Markowitz, H.: Portfolio selection. The Journal of Finance **7**(1), 77–91 (1952)
2. Chang, T.-J., Meade, N., Beasley, J.E., Sharaiha, Y.M.: Heuristics for cardinality constrained portfolio optimisation. Computers & Operations Research **27**(13), 1271–1302 (2000)
3. Armananzas, R., Lozano, J.A.: A multiobjective approach to the portfolio optimization problem. In: IEEE Congress on Evolutionary Computation (CEC), vol. 2, pp. 1388–1395. IEEE Press, Edinburgh, UK (2005)
4. Crama, Y., Schyns, M.: Simulated annealing for complex portfolio selection problems. European Journal of Operational Research **150**(3), 546–571 (2003)
5. Kennedy, J.: Small worlds and mega-minds: effects of neighborhood topology on particle swarm performance. In: IEEE Congress on Evolutionary Computation (CEC), vol. 3, pp. 1931–1938. IEEE Press, Washington, D. C (1999)
6. Kaminakis, N.T., Stavroulakis, G.E.: Topology optimization for compliant mechanisms, using evolutionary-hybrid algorithms and application to the design of auxetic materials. Composites Part B-Engineering **43**(6), 2655–2668 (2012)

7. Tsujimoto, T., Shindo, T., Kimura, T., Jin'no, K.: A relationship between network topology and search performance of pso. In: IEEE Congress on Evolutionary Computation (CEC), pp. 1–6. IEEE Press, Brisbane, Australia (2012)
8. Ni, Q., Deng, J.: A new logistic dynamic particle swarm optimization algorithm based on random topology. The Scientific World Journal **2013** (2013)
9. Jianan, L., Chen, Y.: Particle swarm optimization (pso) based topology optimization of part design with fuzzy parameter tuning. Computer-Aided Design & Applications. **11**(1), 62–68 (2014)
10. Shi, Y., Eberhart, R.C.: Empirical study of particle swarm optimization. In: IEEE Congress on Evolutionary Computation (CEC), vol. 3, pp. 1945–1950. IEEE Press, Washington, D. C(1999)
11. Cura, T.: Particle swarm optimization approach to portfolio optimization. Nonlinear Analysis: Real World Applications. **10**(4), 2396–2406 (2009)

Directional Analysis of Slope Stability Using a Real Example

Zhe-Ping Shen[1,2] and Walter W. Chen[2(✉)]

[1] Department of Construction and Spatial Design, Tungnan University, New Taipei, Taiwan
fishfishfishgoo@gmail.com
[2] Department of Civil Engineering, National Taipei University of Technology, Taipei, Taiwan
waltchen@ntut.edu.tw

Abstract. PSO is a powerful but rarely used tool in slope stability analysis until very recently. Despite its simplicity, PSO can be integrated with existing program effortlessly and improves the performance and accuracy of the resulting analysis. In this study, a real landslide site was selected as an example. The problem slope was represented as a digital elevation model using laser scanning, and the model was cut in parallel lines 45 degrees to the North by a custom program. The resulting 19 profiles were inputted to the STABL program for stability analysis using a PSO scripting program. The results showed that the computed factor of safety varied from profile to profile, but PSO improved the results consistently for all profiles. A comparison was made with the previous study in which the slope was cut in the South-North direction. Both studies showed that the directional analysis of slope stability is an important topic for future research.

Keywords: Slope stability · Slope profile · PSO · STABL

1 Background

Computers have been used in slope stability analysis ever since the mainframe era and soon after the introduction of the FORTRAN language. In fact, the most noteworthy slope stability analysis program to date, STABL, was written in FORTRAN IV at Purdue University in 1975 [1] "for the general solution of slope stability problems by a two-dimensional limiting equilibrium method." The program, which was originally released in card deck form, remains the de facto standard today (after several major revisions) and is licensed by many engineering consulting companies and used in innumerable construction projects. To augment present capabilities of STABL, the authors devised a technique to add PSO (Particle Swarm Optimization) functionality [2, 3] to the analysis without modifying the STABL program itself [4]. The technique was tested on a pseudo slope [4] and then verified on real landslide slope [5]. This study continues on the work previous presented and expands to cover the analysis in a different perspective.

© Springer International Publishing Switzerland 2015
Y. Tan et al. (Eds.): ICSI-CCI 2015, Part I, LNCS 9140, pp. 176–182, 2015.
DOI: 10.1007/978-3-319-20466-6_19

2   Method

There are a number of different versions of STABL. Some come with a graphical front end, while the others can only be run in a terminal mode. Following the methodology described in [4, 5], this study used a scripting program to drive the terminal mode STABL to perform the required analysis. First, the 3-dimensional topography of the slope under analysis was produced by means of laser scanning. Then, the problem slope was cut in parallel strips programmatically using a 1-m distance between slices. This way many different profiles were created and an input file could be generated for each of the cross-sectional cut as shown in Figure 1. Finally, the input files were sent to STABL for analysis, and output files containing the analysis results were sent back by STABL. Afterwards, all output files were gathered and analyzed by the driving scripting program. Through the PSO logic, a new batch of input files were created for the next iteration of STABL analysis. The process continued until a convergence of the Factor of Safety (FS, the objective function) had been reached and a solution had been found.

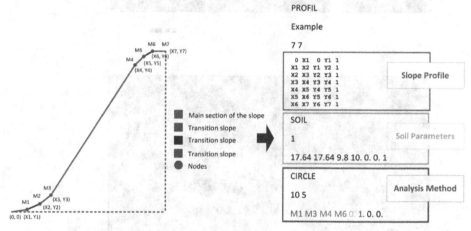

Fig. 1. Automated generation of STABL input files from cross-sectional cuts of the slope topography

3   Site of Study

The site of this study is shown in Figure 2 (enclosed in red square). It was a landslide area near the Houshanyue hiking trail in the vicinity of Taipei [6]. Terrestrial laser scanning was performed in this area to obtain the 3D representation of the landscape and the Digital Elevation Model (DEM) was created. Subsequent analysis was based on this DEM with soil parameters obtained from nearby boreholes.

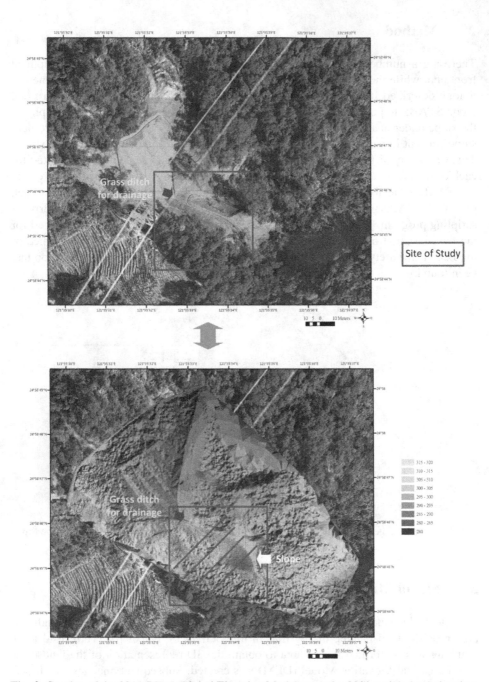

Fig. 2. Overlay of the 2011 DEM (Digital Elevation Model) over the 2009 aerial photo showing the site of study and the lines of 45-degree cut

4   Results

One of the advantages of using a 3D model of the problem slope in the analysis is that it is much more accurate than relying on only a few surveying points on the surface of the hill. This improvement would not have been attainable until the appearance of low-cost and high-speed terrestrial laser scanners very recently. Using a laser scanner, hundreds of thousands of points of measurement covering the entire slope could be made in a very short time interval. Therefore, profiles such as that shown in Figure 1 could be generated and used in the analysis. For the problem slope of Houshanyue, the lines of cut could be made in any direction. For this study, a set of parallel lines 45 degrees to the North were chosen as shown in the bottom figure of Figure 2. Since the final output of STABL (after PSO iterations) only contained textual results, a Matlab program was written to plot the corresponding slope profiles and the most critical sliding surfaces determined by PSO. An example is shown in Figure 3.

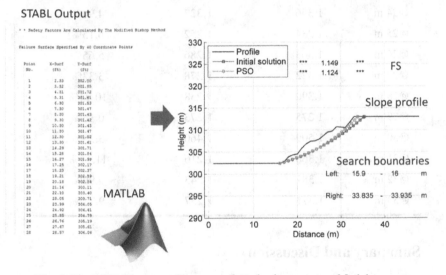

Fig. 3. Final STABL output files were plotted using a custom Matlab program

Table 1 summarizes the results of the analysis. When the problem slope in this study was cut in parallel lines 45 degrees to the North, 19 profiles were created as tabulated in Table 1. The profiles were named (for example, @15 m) in relative to an arbitrarily chosen boundary at the site. Note that some of the profiles generated by the automatic program were not continuous due to the undulating nature of the slope surface, therefore they were not included in the consequent analysis. Only profiles that were complete and well-defined were used in the computation. It can be seen from the results of Table 1 that the use of PSO improved the computation results and reduced the FS for every single profile. The improvement was substantial, ranging from 0.16% to 11.11%. The lowest FS was found at the profile 19 m from the boundary, and the FS was equal to 1.130.

Table 1. The % improvement of FS's using PSO on parallel profiles cut 45 degrees to the North

Slice	Initial FS (Fig. 1)	PSO FS	% improvement
@15 m	1.337	1.240	7.26%
@16 m	1.250	1.164	6.88%
@17 m	1.359	1.267	6.77%
@18 m	1.408	1.343	4.62%
@19 m	1.244	1.130	9.16%
@20 m	1.323	1.192	9.90%
@21 m	1.492	1.427	4.36%
@22 m	1.340	1.217	9.18%
@23 m	1.354	1.263	6.72%
@24 m	1.396	1.327	4.94%
@25 m	1.334	1.252	6.15%
@26 m	1.363	1.286	5.65%
@27 m	1.322	1.278	3.33%
@28 m	1.390	1.238	10.94%
@29 m	1.275	1.273	0.16%
@30 m	1.509	1.460	3.25%
@31 m	1.368	1.216	11.11%
@32 m	1.381	1.243	9.99%
@33 m	1.371	1.237	9.77%

5   Summary and Discussion

This is a continuing study of the use of PSO in the slope stability analysis. A real landslide was chosen as the example and the site of study. The surface topography was re-created faithfully using a laser scanner. The resulting model was cut in parallel lines 45 degrees to the North to generate 19 profiles for STABL analysis. The final results were plotted together in Figure 4, and they showed clear differences between profiles. Therefore, it can be concluded that the FS positively depends on the location of the profile. However, in contrast to the substantial variation of FS of the earlier study where the slope was cut in the South-North direction [7], the FS in this study showed relatively little variation with respect to the profiles (Figure 4). The percentage difference between the largest and the smallest FS's in this study was (1.460-1.130) / 1.130 = 29.2%. On the other hand, the percentage difference in [7] was (2.324-0.924) / 0.924 = 151.5%. Three points are worth making about these results. First, both studies showed that multiple parallel profiles were indeed useful in the

3-dimensional analysis of the problem slope. Second, the due North profiles yielded a lower FS (0.924) than the 45-degree profiles (1.130). Finally, since the 45-degree and the 0-degree cut profiles returned different minimum FS's, it indicated that the FS is directionally dependent property of the slope. Therefore, a more thorough directional analysis of slope stability (covering other scenarios) will be needed in the next stage of the research.

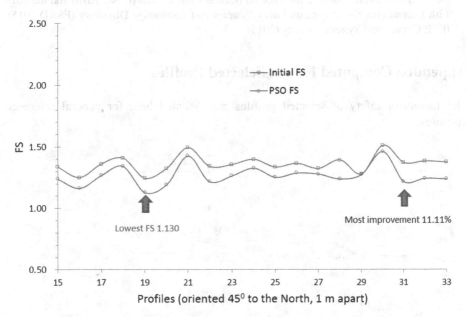

Fig. 4. A plot combining the results of 19 profiles shown in Table 1. The lines of the variation of FS's before and after the use of PSO were shown in blue and red, respectively.

Acknowledgments. This study was partially supported by grant numbers NSC 102-2221-E-027-077 and MOST 103-2221-E-027-052 from the National Science Council and the Ministry of Science and Technology of Taiwan (ROC). The authors acknowledge and highly appreciate the financial support.

References

1. Siegel, R.A.: STABL User Manual. Purdue University, West Lafayette (1975)
2. Kennedy, J., Eberhart, R.: Particle swarm optimization. In: Proc. IEEE International Conf. on Neural Networks. IEEE Service Center, pp. 1942–1948 (1995)
3. Eberhart, R., Kennedy, J.: A new optimizer using particle swarm theory. In: Proc. Sixth Symposium on Micro Machine and Human Science. IEEE Service Center, pp. 39–43 (1995)
4. Chen, W.W., Shen, Z.-P., Wang, J.-A., Tsai, F.: Scripting STABL with PSO for Analysis of Slope Stability. Neurocomputing, 167–174 (2015)

5. Shen, Z.-P., Hsu, C.-K., Chen, W.W.: Slope stability analysis of houshanyue landslide using STABL and PSO. In: Proc. the 34th Asian Conference on Remote Sensing, Asian Association on Remote Sensing (2013)
6. Hsu C.-K. Chen, W.W.: Topography analysis of houshanyue landslide. In: Proc. the 34th Asian Conference on Remote Sensing, Asian Association on Remote Sensing (2013)
7. Shen, Z.-P., Chen, W.W.: Slope stability analysis using multiple parallel profiles. In: Proc. the 2015 11th International Conference on Natural Computation (ICNC 2015) and the 2015 12th International Conference on Fuzzy Systems and Knowledge Discovery (FSKD 2015). IEEE Circuits and Systems Society (2015)

Appendix: Computed FS's of Selected Profiles

The factors of safety of selected profiles are included here for general reference purposes.

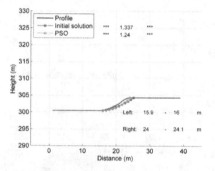
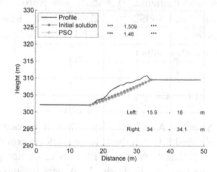

Fig. 5. Analysis results of the profile 15 m from the boundary

Fig. 6. Analysis results of the profile 20 m from the boundary

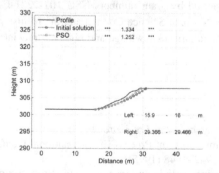
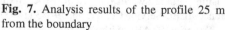

Fig. 7. Analysis results of the profile 25 m from the boundary

Fig. 8. Analysis results of the profile 30 m from the boundary

Obtaining Classification Rules Using lvqPSO

Laura Lanzarini, Augusto Villa Monte$^{(\boxtimes)}$, Germán Aquino,
and Armando De Giusti

Institute of Research in Computer Science LIDI (III-LIDI),
Faculty of Computer Science, National University of La Plata (UNLP),
La Plata, Buenos Aires, Argentina
{laural,avillamonte,gaquino,degiusti}@lidi.info.unlp.edu.ar

Abstract. Technological advances nowadays have made it possible for
processes to handle large volumes of historic information whose manual processing would be a complex task. Data mining, one of the most
significant stages in the knowledge discovery and data mining (KDD)
process, has a set of techniques capable of modeling and summarizing
these historical data, making it easier to understand them and helping
the decision making process in future situations. This article presents a
new data mining adaptive technique called lvqPSO that can build, from
the available information, a reduced set of simple classification rules from
which the most significant relations between the features recorded can
be derived. These rules operate both on numeric and nominal attributes,
and they are built by combining a variation of a population metaheuristic
and a competitive neural network. The method proposed was compared
with several methods proposed by other authors and measured over 15
databases, and satisfactory results were obtained.

Keywords: Classification rules · Data mining · Adaptive strategies ·
Particle swarm optimization · Learning vector quantization

1 Introduction

Data mining is a research field that in recent years has gained attention from various sectors. Government employees, business people and academics alike, for very
different reasons, have contributed to the development of various techniques that
can summarize the information that is available. This is one of the most important stages in the knowledge discovery and data mining process, and it is characterized for producing useful and novel information without any prior hypotheses.
It encompasses a set of techniques capable of modeling available information and,
even though there are different types of models, decision makers usually choose
those that are self-explanatory. For this reason, rules, i.e., statements of the *IF
condition1 THEN condition2* type, are preferred when characterizing that huge

---

A. Villa Monte—Post-Graduate Fellow, UNLP
G. Aquino—Post-Graduate Fellow, CONICET

© Springer International Publishing Switzerland 2015
Y. Tan et al. (Eds.): ICSI-CCI 2015, Part I, LNCS 9140, pp. 183–193, 2015.
DOI: 10.1007/978-3-319-20466-6_20

volume of historical data that were automatically saved. In particular, we were interested in obtaining classification rules, i.e., rules whose consequence is formed by a single condition with the same attribute being involved: the class. However, most of the existing methods produce sets of rules that are so large and complex that, despite having the *IF-THEN* structure, rules become almost unreadable. For this reason, a new method to obtain classification rules is proposed in this article, with two essential features: the cardinality of the set of rules obtained is low, and the antecedent of the rules that are generated is reduced. To this end, the method proposed combines a competitive neural network with an optimization technique. The former is responsible for the supervised grouping of the examples with the purpose of identifying the most relevant attributes for building the rule. Then, by means of an optimization technique, the search process is guided towards the appropriate set of rules.

This paper is organized as follows: Section 2 lists some related articles, Sections 3 and 4 briefly describe the neural network and metaheuristic used, respectively, Section 5 details the method proposed, Section 6 presents the results obtained, and Section 7 presents a summary of the conclusions.

2   Related Work

The literature describes several methods for building classification rules that can operate with numerical and nominal attributes. The most popular one is the method known as C4.5, defined by Quinlan in [16], which can be used to generate a pruned classification tree whose branches allow obtaining the desired set of rules.

It should be noted that the lvqPSO method proposed in this article, unlike the C4.5 method, generates a list of classification rules. That is, when using the rules to classify new examples, they are not analyzed separately but rather in their order of appearance. Therefore, the rules are inspected one by one until the one that is applicable to the case at hand is found. This is related to how the set of rules was built, and has been used in other methods, such as PART, defined Witten in [2]; PSO/ACO, defined by Holden in [3]; and cAnt-MinerPB, proposed by Medland in [14].

In the case of PART, a pruned partial tree is built to determine each rule, following the ideas proposed by Quinlan in C4.5. The difference lies in that the tree is not built in full, but rather an error quota is applied to stop tree generation and select the best brach obtained so far. On the other hand, the PSO/ACO and cAnt-MinerPB method share with lvqPSO the idea of using a population-based metaheuristic to search for the best rules.

In particular, lvqPSO presents an approach that is based on Particle Swarm Optimization (PSO). This technique has already been used in previous works [1,5,8,19]. Unquestionably, one of the main issues when operating with nominal attributes is the impossibility of adequately covering all areas of the search space with the examples that are available. This results in a poor start for the population and a premature convergence to a local optimum. As a way to solve this problem, and at the same time reducing rule generation time, the initial state is obtained from an LVQ (Learning Vector Quantization) competitive neural network.

The literature describes methods that optimize a competitive neural network with PSO and significantly reduce the calculation time for the training phase [15], or methods that use PSO to determine the optimal number of competitive neurons to be used in the network, such as [4]. Unlike these papers, our proposal is using PSO to obtain the set of rules, and an LVQ network to avoid the premature convergence of the population.

3 Learning Vector Quantization (LVQ)

Learning Vector Quantization (LVQ) is a supervised classification algorithm that is based on centroids or prototypes [9]. It can be interpreted as a competitive neural network formed by three layers. The first layer is just an input layer. The second layer is where competence takes place. The output layer is responsible for the classification process. Each neuron in the competitive layer is associated to a number vector whose dimension is the same as that of the input examples, and a label that indicates the class that it is going to represent. Once the adaptive process finishes, these vectors will contain the information related to the classification centroids or prototypes. There are several versions of the training algorithm. The one used in this article is described below.

When the algorithm is started, the number K of centroids to be used must be indicated. This allows defining the architecture for the network, since the number of input entries and output results are given by the problem.

Centroids are initialized taking K random examples. Examples are then entered one by one, and centroid position is then adapted. To do so, the centroid that is closest to the example being analyzed is selected using a preset distance measurement. Since this is a supervised process, it is possible to determine if the example and the centroid belong or not to the same class. If the centroid and the example do belong to the same class, the centroid is "moved closer" to the example in order to strengthen representation. If, on the contrary, they belong to different classes, the centroid is "moved away". These movements are done by means of a factor or adaptation speed that allows weighing the distance for the move.

This process is repeated until modifications are below a preset threshold, or until the examples are identified with the centroids themselves in two consecutive iterations, whichever happens first.

For the implementation used in this article, the second nearest centroid is also analyzed and, should it belong to a different class than that of the example, and should it be at a distance that is less than 1.2 times the distance to the first centroid, the "moving away" step is applied.

Several variations of LVQ are described in [10].

4 Obtaining Classification Rules with PSO

Particle Swarm optimization or PSO is a population-based metaheuristic proposed by Kennedy and Eberhart [6] where each individual in the population, called particle, represents a possible solution to the problem and adapts by following three factors: its knowledge of the environment (its fitness value), its historical knowledge or previous experiences (its memory), and the historical knowledge or previous experiences of the individuals in its neighborhood (its social knowledge).

PSO was originally defined to work on continuous spaces, so a few considerations should be taken into account when working on discrete spaces. For this reason, Kennedy and Eberhart defined in [7] a new binary version of the PSO method. On of the key problems of this last method is its difficulty to change from 0 to 1 and from 1 to 0 once it has stabilized. This has resulted in different versions of binary PSO that seek to improve its exploratory capacity. In particular, the variation defined by Lanzarini et al. [13] will be used in this article.

Using PSO to generate classification rules that can operate on nominal and numerical attributes requires a combination of the methods mentioned above, since the attributes that will be part of the antecedent (discrete) have to be selected and the value or range of values they can take (continuous) has to be determined.

Since this is a population-based technique, the required information has to be analyzed for each individual in the population. A decision has to be made between representing a single rule or the entire set for each individual, and the representation scheme has to be selected for each rule. Given the objectives proposed for this work, the Iterative Rule Learning (IRL) [18] approach was followed, where each individual represents a single rule and the solution to the problem is built from the best individuals obtained after a sequence of runs. Using this approach implies that the population-based technique will be applied iteratively until achieving the desired coverage and obtaining a single rule in each iteration: the best individual in the population. Additionally, a fixed-length representation was chosen, where only the antecedent of the rule will be coded and, given the approach adopted, an iterative process will be carried out to associate all individuals in the population to a preset class, which does not require consequent codification. The code used for each particle is described in detail in [11].

The efficacy of population-based optimization techniques is closely related to the size of the population. For this reason, the method proposed here uses the variable population strategy defined in [12]. Thus, a minimum-size population can be used to initiate the process and then adjust the number of particles during the adaptive process.

The fitness value for each particle is calculated as follows:

$$Fitness = \alpha * balance * support * confidence - \beta * lengthAntecedent \quad (1)$$

where

- *support*: it is the support value for the rule. That is, the quotient between the number of examples that fulfill the rule and the total number of examples being analyzed.
- *confidence*: it is the confidence value for the rule. That is, the quotient of the number of examples that satisfy the rule and the number of those that satisfy the antecedent.
- *lengthAntecedent*: it is the quotient between the number of conditions used in the antecedent and the total number of attributes. It should be noted that each attribute can only be used once within the rule antecedent.
- α, β: these are two constants that represent the significance assigned to each term.
- *balance*: it takes values between (0,1], and it is used to compensate the effect of the imbalance between classes when calculating the support value. It is applied only when working with classes that have a number of examples that is above the mean. Let $C_1, C_2, ..., C_i, ..., C_N$ be the classes into which the examples are divided. N is the total number of classes. Let E_i be the number of examples in the nth class. Let T be the total number of examples being used. That is,

$$T = \sum_{i=1}^{N} E_i \tag{2}$$

Let j be the class to which the rule corresponding to the phenotype of the particle belongs. Let S_i be the number of examples in class C_i covered by the rule. Note that S_j corresponds to the support of the rule and

$$\sum_{i=1,i\neq j}^{N} S_i \tag{3}$$

is the total number of examples incorrectly covered by that rule. Then, the value of this factor is calculated as follows

$$balance = \sum_{i=1,i\neq j}^{N} \frac{E_i - S_i}{T - E_j} \tag{4}$$

That is, *balance* will have a value of 1 if the rule is perfect, i.e., its confidence is 1. On the other hand, Balance will be 0 if the rule covers all of the examples being used regardless of their class.

5   Proposed Method for Obtaining Rules: lvqPSO

Rules are obtained through an iterative process that analyzes non-covered examples in each class, starting with the largest classes. Each time a rule is obtained, the examples that are correctly covered by the rule are removed from the input data set. The process continues until all examples are covered or until the number of non-covered examples in each class is below the corresponding established minimum support or until a maximum number of tries has been done to obtain a rule, whichever happens first. It should be noted that, since the examples are removed from the input data set as they are covered by the rules, the rules operate as a classification list. That is, in order to classify a new example, the rules must be applied in the order in which they were obtained, and the example will be classified with the class that corresponds to the consequent of the first rule whose antecedent is verified for the example at hand.

Before starting the iterative process for obtaining the rules, the method starts with the supervised training of an LVQ neural network using the entire set of examples. The purpose of this training is identifying the most promising areas of the search space.

Since neural networks only operate with numerical data, nominal attributes are represented by means of dummy code that uses both binary digits and the different options that may be present in such nominal attribute. Also, before starting the training process, each dimension that corresponds to a numerical attribute is linearly scaled in [0,1]. The similarity measure used is the Euclidean distance. Once training is finished, each centroid will contain approximately the average of the examples it represents.

To obtain each of the rules, the class to which the consequent belongs is first determined. Seeking high-support rules, the method proposed will start by analyzing those classes with higher numbers of non-covered examples. The minimum support that any given rule has to meet is proportional to the number of non-covered examples in the class upon rule generation. That is, the minimum required support for each class decreases as iterations are run, as the examples in the corresponding class are gradually covered. Thus, it is to be expected that the first rules will have a greater support than the final ones.

After selecting the class, the consequent for the rule is determined. To obtain the antecedent, a swarm population will be optimized using the process described in Section 4. This swarm will be initialized with the information from the centroids. In Algorithm 1, the pseudo-code of the method proposed is shown.

6   Results Obtained

In this section, the performance obtained with the method proposed is compared against that of the cAnt-MinerPB, J48 (implementation of C4.5) and PART methods mentioned in Section 2 for generating classification rules for a known set of 15 databases of the UCI repository [17].

Thirty separate runs of ten-fold cross-validation were performed for each method, and an LVQ network with 9 neurons was used. In the case of the PART and C4.5 methods, a confidence factor of 0.3 and 0.25, respectively, was used for pruning the tree. The cAnt-MinerPB method was used with a colony of 9 ants. For the rest of the parameters, their default values were used.

Tables 1, 2 and 3 summarize the results obtained with each method for each of the databases, calculating mean and deviation. In each case, not only the

Algorithm 1. Pseudocode of the proposed method

Train LVQ network using all training examples.
Calculate the minimum support for each class.
while (termination criterion is not reached) **do**
    Choose the class with the highest number of non-covered examples.
    Build a reduced population of individuals from centroids.
    Evolve the population using variable population PSO.
    Obtain the best rule for the population.
    if (the rule meets support and confidence requirements) **then**
        Add the rule to the set of rules.
        Consider the examples classified by this rule as correctly covered.
        Recalculate the minimum support for this class.
    end if
end while

coverage accuracy of the set of rules was considered (Table 1), but also the clarity of the model obtained, which is reflected in the average number of rules obtained (Table 2) and the average number of terms used to form the antecedent (Table 3). In each case, a two-tailed mean difference test with a significance level of 0.05 was carried out, where the null hypothesis establishes that the means are equal. Based on the results obtained, when the difference is signficant according to the specified level, the best option was shaded and highlighted in bold in the table and when the difference is not significant equivalent solutions were only highlighted in bold.

As shown in Table 2, in most of the cases, the number of rules used by the method proposed is lower than with the other methods. This is due to the emphasis placed on the simplification of the model. The purpose is not only solving a classification problem, but also building a specific help tool for decision making. The expectation is generating a model that can identify the most relevant attributes and exposes how they are related among themselves when classifying available information.

In Table 1, in 5 of the cases, the accuracy achieved with lvqPSO is equivalent to or better than that obtained with other methods. In the case of the "Breast cancer", "Credit-a", "Heart disease" and "Zoo" databases, accuracy is better or approximately the same, and the number of rules is still lower. In the case of the "Iris" database, the number of rules is similar to that used by the deterministic methods C4.5 and PART, and lower than with cAnt-MinerPB.

However, when the problem to solve requires a large number of rules, the method proposed is less accurate. This happens in other databases that were tested. If these cases are analyzed in average, it can be stated that for a 2% improvement, a five-fold increase in rule set cardinality was required. For example, for the "Diabetes" database, the best accuracy is obtained with the cAnt-MinerPB method, with 26 rules, followed by C4.5, with 20 rules. There is a 2% difference in accuracy when compared to lvqPSO, but the latter uses less than 4 rules to solve the problem, i.e., only one sixth of the number of rules used by cAnt-MinerPB. The same happens with the "Breast-w" database, with lvqPSO

Table 1. Accuracy of the rule set obtained when applying the lvqPSO, PART, cAnt-MinerPB and J48 methods

Database	lvqPSO	cAnt-MinerPB	C4.5 (J48)	PART
Balance scale	$0,7471 \pm 0,0231$	$0,7696 \pm 0,0105$	$0,7732 \pm 0,0073$	$\mathbf{0,8219 \pm 0,0129}$
Breast cancer	$\mathbf{0,7203 \pm 0,0121}$	$0,7088 \pm 0,0208$	$0,5641 \pm 0,0466$	$0,6631 \pm 0,0207$
Breast-w	$0,9490 \pm 0,0079$	$0,9485 \pm 0,0049$	$\mathbf{0,9549 \pm 0,0050}$	$\mathbf{0,9566 \pm 0,0065}$
Credit-a	$\mathbf{0,8569 \pm 0,0055}$	$0,8493 \pm 0,0087$	$0,8515 \pm 0,0047$	$0,7454 \pm 0,0444$
Credit-g	$0,7060 \pm 0,0128$	$\mathbf{0,7374 \pm 0,0091}$	$0,7095 \pm 0,0072$	$0,6998 \pm 0,0111$
Diabetes	$0,7242 \pm 0,0152$	$\mathbf{0,7409 \pm 0,0050}$	$\mathbf{0,7438 \pm 0,0116}$	$\mathbf{0,7377 \pm 0,0156}$
Heart disease	$\mathbf{0,7650 \pm 0,0151}$	$\mathbf{0,7598 \pm 0,0164}$	$0,7433 \pm 0,0169$	$\mathbf{0,7647 \pm 0,0264}$
Heart statlog	$0,7626 \pm 0,0180$	$0,7648 \pm 0,0141$	$\mathbf{0,7811 \pm 0,0083}$	$0,7667 \pm 0,0150$
Iris	$\mathbf{0,9427 \pm 0,0218}$	$0,9380 \pm 0,0122$	$\mathbf{0,9453 \pm 0,0103}$	$\mathbf{0,9440 \pm 0,0110}$
Kr-vs-kp	$0,9365 \pm 0,0037$	$0,9812 \pm 0,0010$	$\mathbf{0,9929 \pm 0,0007}$	$0,9917 \pm 0,0013$
Mushroom	$0,9676 \pm 0,0046$	$\mathbf{0,9969 \pm 0,0001}$	$0,9843 \pm 0,0252$	$\mathbf{0,9962 \pm 0,0106}$
Promoters	$0,6977 \pm 0,0261$	$\mathbf{0,7917 \pm 0,0250}$	$0,7082 \pm 0,0389$	$0,6782 \pm 0,0600$
Soybean	$0,8735 \pm 0,0156$	$\mathbf{0,9066 \pm 0,0056}$	$\mathbf{0,9082 \pm 0,0053}$	$0,8936 \pm 0,0090$
Wine	$0,8727 \pm 0,0146$	$\mathbf{0,9192 \pm 0,0124}$	$0,8800 \pm 0,0144$	$0,8867 \pm 0,0091$
Zoo	$\mathbf{0,9417 \pm 0,0204}$	$\mathbf{0,9408 \pm 0,0153}$	$0,3290 \pm 0,0277$	$0,3190 \pm 0,0307$

Table 2. Number of rules obtained when applying the lvqPSO, PART, cAnt-MinerPB and J48 methods

Database	lvqPSO	cAnt-MinerPB	C4.5 (J48)	PART
Balance scale	$\mathbf{9,6700 \pm 0,5945}$	$14,6900 \pm 0,4483$	$41,3300 \pm 1,3300$	$38,5100 \pm 1,2215$
Breast cancer	$\mathbf{5,8600 \pm 0,4742}$	$25,4100 \pm 1,0744$	$11,4500 \pm 1,1128$	$18,8900 \pm 1,4098$
Breast-w	$\mathbf{2,8900 \pm 0,3281}$	$12,9000 \pm 0,5477$	$10,8500 \pm 0,7091$	$10,3500 \pm 0,4950$
Credit-a	$\mathbf{3,4100 \pm 0,1595}$	$25,0000 \pm 0,8654$	$18,0400 \pm 1,8404$	$33,3200 \pm 1,3028$
Credit-g	$\mathbf{8,3600 \pm 0,8113}$	$41,4100 \pm 2,0776$	$85,4500 \pm 3,4574$	$70,5700 \pm 1,5868$
Diabetes	$\mathbf{3,7900 \pm 0,3178}$	$26,6400 \pm 0,8996$	$22,1900 \pm 2,7517$	$7,5200 \pm 0,4264$
Heart disease	$\mathbf{4,9200 \pm 0,2821}$	$16,0800 \pm 0,7406$	$23,9000 \pm 0,9043$	$19,6400 \pm 0,4248$
Heart statlog	$\mathbf{4,6400 \pm 0,3777}$	$15,1200 \pm 0,6663$	$18,1900 \pm 1,2556$	$17,8700 \pm 0,4191$
Iris	$\mathbf{3,0600 \pm 0,0843}$	$8,3500 \pm 0,3808$	$4,6600 \pm 0,0966$	$3,7800 \pm 0,2658$
Kr-vs-kp	$\mathbf{3,6500 \pm 0,3274}$	$18,0900 \pm 0,9049$	$29,0700 \pm 0,6717$	$22,1100 \pm 0,6385$
Mushroom	$\mathbf{3,2300 \pm 0,1059}$	$22,6900 \pm 2,0328$	$18,6100 \pm 0,2378$	$11,2400 \pm 0,2591$
Promoters	$7,4250 \pm 0,3775$	$11,5300 \pm 0,6430$	$16,7500 \pm 0,5874$	$\mathbf{7,2800 \pm 0,3765}$
Soybean	$\mathbf{24,6857 \pm 0,7151}$	$62,9700 \pm 2,1334$	$43,5400 \pm 0,2366$	$31,7300 \pm 0,4322$
Wine	$\mathbf{3,7818 \pm 0,0874}$	$6,6900 \pm 0,3900$	$7,6600 \pm 0,4766$	$5,6200 \pm 0,1317$
Zoo	$\mathbf{6,9500 \pm 0,0548}$	$10,9700 \pm 0,2214$	$8,3500 \pm 0,0707$	$7,6300 \pm 0,0483$

Table 3. Antecedent average length for each rule obtained when applying the lvqPSO, PART, cAnt-MinerPB and J48 methods

Database	lvqPSO	cAnt-MinerPB	C4.5 (J48)	PART
Balance scale	$1,8035 \pm 0,0910$	$1,5900 \pm 0,0527$	$6,3494 \pm 0,0668$	$3,0807 \pm 0,0694$
Breast cancer	$1,7373 \pm 0,0497$	$1,8464 \pm 0,0578$	$2,1210 \pm 0,0939$	$2,0048 \pm 0,0570$
Breast-w	$3,3173 \pm 0,3546$	$1,1622 \pm 0,0329$	$3,8899 \pm 0,1581$	$2,1249 \pm 0,0800$
Credit-a	$1,4049 \pm 0,1186$	$1,2247 \pm 0,0504$	$4,8299 \pm 0,2014$	$2,4844 \pm 0,0746$
Credit-g	$2,0349 \pm 0,1276$	$1,9258 \pm 0,1227$	$5,6443 \pm 0,1209$	$2,9695 \pm 0,0890$
Diabetes	$2,4647 \pm 0,2027$	$1,1627 \pm 0,0238$	$5,7461 \pm 0,3253$	$1,9255 \pm 0,1043$
Heart disease	$1,8766 \pm 0,1039$	$1,7455 \pm 0,0682$	$3,9391 \pm 0,0952$	$2,5415 \pm 0,0692$
Heart statlog	$1,8491 \pm 0,0715$	$1,2927 \pm 0,0346$	$4,6700 \pm 0,1629$	$2,8783 \pm 0,1099$
Iris	$1,2083 \pm 0,0518$	$1,1793 \pm 0,0329$	$2,6118 \pm 0,0540$	$0,9949 \pm 0,0135$
Kr-vs-kp	$2,4750 \pm 0,1236$	$1,4986 \pm 0,1125$	$7,7738 \pm 0,0318$	$3,1275 \pm 0,0689$
Mushroom	$1,6400 \pm 0,0733$	$1,0899 \pm 0,0294$	$2,6302 \pm 0,0384$	$1,2609 \pm 0,0153$
Promoters	$1,1131 \pm 0,0302$	$1,0409 \pm 0,0527$	$2,2847 \pm 0,0364$	$0,9976 \pm 0,0416$
Soybean	$3,1299 \pm 0,2543$	$3,3858 \pm 0,0537$	$6,0213 \pm 0,0349$	$2,7127 \pm 0,0709$
Wine	$2,7811 \pm 0,2248$	$1,0626 \pm 0,0340$	$3,1027 \pm 0,1232$	$1,5529 \pm 0,0759$
Zoo	$1,6675 \pm 0,0462$	$1,6888 \pm 0,0601$	$4,0057 \pm 0,0242$	$1,4693 \pm 0,0145$

being less accurate than C4.5 and PART by approximately 1% but using only one third of the number of rules with less queries in each antecedent.

7 Conclusions

A novel method for obtaining classification rules has been presented. This method is based on PSO and can operate with numerical and nominal attributes.

An LVQ neural network was used to adequately initialize the population of rules. The centroids obtained when grouping available data allow identifying the relevance of each attribute for the examples. In any case, this metric is not enough to select the attributes that will form the rule, and it is at this point where PSO takes control to carry out the final selection.

A representation for the rules was used, combining a binary representation that allows selecting the attributes that are used in the rule with a continuous representation used only to determine the boundaries of the numerical attributes that are part of the antecedent. A variation of binary PSO was used whose population is adequately initialized with the information from the centroids in the previously trained LVQ network and which has the ability of adjusting population size.

The results obtained when applying the method proposed on a set of test databases show that the lvqPSO method obtains a simpler model. In average, it uses approximately 40% of the number of rules generated by the other methods, with antecedents formed by just a few conditions and an acceptable accuracy given the simplicity of the model obtained.

Although not included in this article, the measurements performed using the method proposed but using fixed-size population PSO resulted in a less accurate set of rules. This is because the architecture of the LVQ network must be indicated beforehand, which affects grouping quality.

References

1. Chen, M., Ludwig, S.: Discrete particle swarm optimization with local search strategy for rule classification. In: 2012 Fourth World Congress on Nature and Biologically Inspired Computing (NaBIC), pp. 162–167 (2012)
2. Frank, E., Witten, I.H.: Generating accurate rule sets without global optimization. In: Proceedings of the Fifteenth International Conference on Machine Learning, ICML 1998, pp. 144–151. Morgan Kaufmann Publishers Inc., San Francisco, CA, USA (1998)
3. Holden, N., Freitas, A.A.: A hybrid pso/aco algorithm for discovering classification rules in data mining. Journal of Artificial Evolution and Applications **2008**, 2:1–2:11 (2008)
4. Hung, C., Huang, L.: Extracting rules from optimal clusters of self-organizing maps. In: Second International Conference on Computer Modeling and Simulation, ICCMS 2010, vol. 1, pp. 382–386 (2010)
5. Jiang, Y., Wang, L., Chen, L.: A hybrid dynamical evolutionary algorithm for classification rule discovery. In: Second International Symposium on Intelligent Information Technology Application, vol. 3, pp. 76–79 (2008)
6. Kennedy, J., Eberhart, R.C.: Particle swarm optimization. In: Proceedings of the IEEE International Conference on Neural Networks, pp. 1942–1948 (1995)
7. Kennedy, J., Eberhart, R.C.: A discrete binary version of the particle swarm algorithm. In: Proceedings of the IEEE International Conference on Systems, Man, and Cybernetics, vol. 5, pp. 4104–4108. IEEE Computer Society, Washington, DC, USA (1997)
8. Khan, N., Iqbal, M., Baig, A.: Data mining by discrete pso using natural encoding. In: 2010 5th International Conference on Future Information Technology (FutureTech), pp. 1–6 (2010)
9. Kohonen, T.: The self-organizing map. Proceedings of the IEEE **78**(9), 1464–1480 (1990)
10. Kohonen, T., Schroeder, M.R., Huang, T.S. (eds.): Self-Organizing Maps, 3rd edn. Springer, New York (2001)
11. Lanzarini, L., Monte, A.V., Ronchetti, F.: Som+pso. a novel method to obtain classification rules. Journal of Computer Science & Technology **15**(1), 15–22 (2015)
12. Lanzarini, L., Leza, V., De Giusti, A.: Particle swarm optimization with variable population size. In: Rutkowski, L., Tadeusiewicz, R., Zadeh, L.A., Zurada, J.M. (eds.) ICAISC 2008. LNCS (LNAI), vol. 5097, pp. 438–449. Springer, Heidelberg (2008)
13. Lanzarini, L., López, J., Maulini, J.A., De Giusti, A.: A new binary PSO with velocity control. In: Tan, Y., Shi, Y., Chai, Y., Wang, G. (eds.) ICSI 2011, Part I. LNCS, vol. 6728, pp. 111–119. Springer, Heidelberg (2011)

14. Medland, M., Otero, F.E.B., Freitas, A.A.: Improving the cAnt-Miner$_{PB}$ classification algorithm. In: Dorigo, M., Birattari, M., Blum, C., Christensen, A.L., Engelbrecht, A.P., Groß, R., Stützle, T. (eds.) ANTS 2012. LNCS, vol. 7461, pp. 73–84. Springer, Heidelberg (2012)
15. Özçift, A., Kaya, M., Gülten, A., Karabulut, M.: Swarm optimized organizing map (swom): A swarm intelligence based optimization of self-organizing map. Expert Systems with Applications **36**(7), 10640–10648 (2009)
16. Quinlan, J.R.: C4.5: programs for machine learning. Morgan Kaufmann Publishers Inc., San Francisco (1993)
17. UCI: Machine learning repository. http://archive.ics.uci.edu/ml
18. Venturini, G.: Sia: a supervised inductive algorithm with genetic search for learning attributes based concepts. In: Brazdil, P.B. (ed.) ECML-93. LNCS, vol. 667, pp. 280–296. Springer, Berlin Heidelberg (1993)
19. Wang, H., Zhang, Y.: Improvement of discrete particle swarm classification system. In: 2011 Eighth International Conference on Fuzzy Systems and Knowledge Discovery (FSKD), vol. 2, pp. 1027–1031 (2011)

Ant Colony Optimization

A Modified Ant Colony Optimization
for the Multi-objective Operating Room Scheduling

Chonglei Gu, Qingzhao Liu, and Wei Xiang[✉]

Faculty of Mechanical Engineering and Mechanics, Ningbo University,
Ningbo 315211, People's Republic of China
xiangwei@nbu.edu.cn

Abstract. Operating room (OR) scheduling plays a decisive role in providing timely treatment for the patients, reducing the operation cost and increasing the hospital resources' utilization. It can be regarded as a multi-objective combinatorial optimization problem. The objectives involved in the problem are defined from different perspectives and often conflicting. A modified ant colony optimization (ACO) algorithm with Pareto sets construction and two types of pheromone setting is proposed to solve the multi-objective OR scheduling problem. The scheduling results by three different approaches, i.e. the simulation, the ACO with single objective of the makespan (ACO-SO), and the proposed ACO with multi-objectives (ACO-MO) are compared. The computational results show that the ACO-MO achieved good results in shortening makespan, reducing nurses' overtime and balancing resources' utilization in general.

Keywords: Operating room scheduling · Multi-objective optimization · Pareto set · Ant colony optimization

1 Introduction

It is important to ensure hospital providing a satisfied healthcare service. The satisfaction should be reflected in three perspectives, i.e. patients, medical staffs, and hospital management. Those three parties may focus on different performance measurements which sometime conflict with each other. For example, the demands for surgical service have been constantly increased because of aging population. Patients want to have the quality surgeries as soon as possible, which causes the excessive overtime on medical staffs. In turn, such stressful work in operating room (OR) leads to the loss of medical staffs. It turns to be a vicious cycle in demands increasing and resources shortage. Under this situation, patients care for the timely surgery service and the time is their criteria. Medical staffs care for the workload and less overtime is their preference. OR management care for the resources' utilization and operation cost. But the time, the workload, and the operation cost are mutually incompatible. It is a challenge for OR management to solve such multi-objective optimization (MO) problem.

OR planning and scheduling have been a hot research topic recently [1-3]. Most OR scheduling researches in operation research community described the problem as an optimization model with single objective, i.e. the patients' waiting time[4][5],

Y. Tan et al. (Eds.): ICSI-CCI 2015, Part I, LNCS 9140, pp. 197–204, 2015.
DOI: 10.1007/978-3-319-20466-6_21

the utilization of OR [6][7], and the Makespan [8][9]. There also exist some OR scheduling problems considered multiple objectives [10-12]. Although not necessarily, the performances often conflict with each other, meaning that building an optimal schedule with respect to one objective goes at the cost of the other objectives. In addition, the combinatorial nature and the nonlinearity in constraints make it extremely difficult to optimize. In this paper, a modified ACO is proposed to solve such multi-objective OR scheduling problem.

2   Multi-objective OR Scheduling Problem

Since OR scheduling problem is an optimization problem involved multiple surgery stages and multiple resources, several measurements should be taken into account when sequencing the individual surgery and allocating the required resources. We assume that there is a set of surgeries I to be performed in an operating system with different types of resources C. The OR scheduling problem is to minimize three objectives (described in function f_1, f_2 and f_3) involving the makespan, the overtime, and the equilibrium degree.

$$\min \vec{F} = (f_1, f_2, f_3) \tag{1}$$

The three objective functions are:

1. Objective function 1: is to minimize the end time to finish all surgeries, so-called makespan. T_{i3} is the end time of last stage of a surgery i.

$$f_1 = \min \max_{i \in I} T_{i3} \tag{2}$$

2. Objective function 2: is to minimize the variation of resources working time. It is used to evaluate the balance of resource utilization and defined as the ratio of the standard deviation to mean as shown in equation (4-5). x_{ij}^{cm} is a 0-1 variable indicating whether resource m of resource type c is assigned to stage j of surgery i.

$$f_2 = \min \max_{c \in C} (\mu_c / \sigma_c) \tag{3}$$

$$\mu_c = \frac{1}{|M_c|} \sum_{m \in M_c} \sum_{j \in J_i} \sum_{i \in I} T_{ij}^{cm} x_{ij}^{cm} \tag{4}$$

$$\sigma_c = \sqrt{\frac{1}{|M_c|} \sum_{m \in M_c} \left(\sum_{j \in J_i} \sum_{i \in I} T_{ij}^{cm} x_{ij}^{cm} - \mu_c \right)^2} \tag{5}$$

3. Objective function 3: is to minimize the total overtime of all resources. R_{cm} is the regular working time for a specific resource m in resource type c.

$$f_3 = \min \left\{ \sum_{c \in C} \sum_{m \in M_c} \left(\max_{i \in I, j \in J_i} \left(T_{ij} \cdot x_{ij}^{cm} \right) - R_{cm} \right) \right\} \tag{6}$$

3   The Modified ACO Algorithm for Multi-objective OR Scheduling

The combinatorial nature of the OR scheduling optimization make it challenging to obtain a global optimal solution. Instead, we aim at a meta-heuristic approach for sub-optimal solution. Further to consider the conflicted objectives in OR scheduling, a modified ACO algorithm is integrated with the Pareto set to efficiently solve the multi-objective optimization problem.

ACO has been successfully applied to solve several combinatorial optimization problems, like the traveling salesman problem (TSP) and the job-shop scheduling problem (JSP). The traditional ant graph structure in those problems is represented as a graph with all nodes to be traversed. And the ant foraging path is the sequence of cities/jobs for TSP/JSP. However, the OR scheduling determines not only the surgeries sequence but also the resource allocation for each of the surgery stages. A two-level ant graph has to be designed. In our previous research, an ACO with such two-level ant graph was proposed for solving surgery scheduling optimization with single objective of makespan [9]. The outer level ant graph is the same as traditional ACO ant graph for TSP/JSP and defined as a surgery graph with all surgery nodes. The ant foraging path is the scheduling sequence of the surgeries. The inner level graph is composed of all available resources along the three-stage surgery procedure. Ant foraging path in inner level graph determines the resources selection for each specific stage during a surgery.

3.1   Construct Pareto Set

Individual ant traverses the two-level ant graph to build a feasible schedule solution. Due to the multiple objectives and often conflicting objectives in OR scheduling problem, there does not exist a single solution that simultaneously minimizes each individual objective. The Pareto Set is introduced. OR scheduling problem is described as a MO problem with three objectives: $\min \vec{F} = (f_1, f_2, f_3)$. Assume S is the feasible solution set. A feasible solution $s_1 \in S$ is said to dominate another solution $s_2 \in S$, represented as ($s_1 \succ s_2$), if 1) $f_i(s_1) \leq f_i(s_2), i \in \{1,2,3\}$ and 2) $f_j(s_1) < f_j(s_2), \exists j \in \{1,2,3\}$. Solution s_1 is called Pareto optimal, or Pareto optimal, if none of the objective functions can be improved in value without degrading some of the other objective values. All Pareto optimal solutions are considered equally good and form a Pareto set.

3.2   Pheromone Setting and Update Strategy

For ACO solving for the single objective optimization (ACO-SO), the pheromone is released on the specific path belonging an optimal solution [9]. However, in the multi-objective OR scheduling, there no longer exists such single optimal solution, but a number of Pareto optimal solutions. Therefore, a special pheromone releasing should be introduced to take into account the impact of several Pareto optimal solutions due to multiple objectives. Two kinds of pheromone setting, single-path-single-pheromone (SPSP) and single-path-multi-pheromone (SPMP) are introduced here.

Single-Path-Single-Pheromone (SPSP)
Like ACO-SO, SPSP allows only one pheromone value releasing on a path. Within one cycle, ants traverse the ant graph and return an iteration Pareto set. In order to let following ants cluster to paths with Pareto optimal solutions, an iteration-Pareto-optimal (IPO) update strategy is introduced in this work, i.e. only those paths with best objectives (like makespan, overtime, and balance utilization) are reinforced with an incremental comprehensive pheromone which is determined by weighted objective value. We set Q_r as the pheromone strength vector associated with three objectives, define weights for each individual objectives as ω_r, r=1,2,3, and group solutions with individual best objective as a set PS_{best}.

The pheromone (τ_{ij}) indicates the strength on a path from node i to node j in the ant graph. Only pheromone on paths of ant solution in PS_{best} are enhanced and updated according to equation (7). ρ denotes the pheromone evaporation rate. The incremental comprehensive pheromone ($\Delta\tau_{ij}(s)$) on the edge (i,j) of a solution s in PS_{best} is determined by a weighted objective shown in equation (8). L_s^r is the rth objective value of a solution s. Decision maker can regulate the emphasis on pheromone according to weights setting for different objectives. Such pheromone IPO update strategy enhances the pheromone on the paths with best individual objective.

$$\tau_{ij} = (1-\rho) * \tau_{ij} + \sum_{s \in PS_{best}} \Delta\tau_{ij}(s) \tag{7}$$

$$\Delta\tau_{ij}(s) = \begin{cases} \sum_{r=1}^{3}\left(\omega_r \dfrac{Q_r}{L_s^r}\right), & \text{if ant s goes through } (i, j) \text{ in this iteration} \\ 0, & \text{otherwise} \end{cases} \tag{8}$$

Single-Path-Multi-Pheromone (SPMP)
SPMP allows multiple pheromone value to be laid on a single ant path. since three objectives are proposed for OR scheduling problem, there are three pheromones associated (r=1,2,3) on a single path. Therefore, a pheromone vector which includes three pheromone values is defined for all pheromones in a two-level ant graph.

Equation (9) defines the pheromone from node i to node j in the ant graph. Its value is determined by the individual objective strength τ_{ij}^r and the associated weights ω_r.

$$\tau_{ij} = \sum_{r=1}^{3} \omega_r \tau_{ij}^r \tag{9}$$

$$\tau_{ij}^r = (1-\rho) * \tau_{ij}^r + \Delta\tau_{ij}^r(s_r) \tag{10}$$

$$\Delta\tau_{ij}^r(s_r) = \begin{cases} \dfrac{Q_r}{L_{s_r}^r}, & \text{if ant } s_r \text{ goes through } (i, j) \text{ in this iteration} \\ 0, & \text{otherwise} \end{cases} \tag{11}$$

IPO update strategy (Equation (10) and (11)) is adopted to cluster ants to paths with Pareto optimal solutions. Only the pheromone value associated to the individual best objective is reinforced, the other two pheromone value in vector keep unchanged. However, evaporation will be happened for all pheromones in pheromone vector.

4   Computational Study

The proposed ACO-MO algorithm is implemented with MATLAB and is run on a PC running Windows XP with Intel Core5 @2.79GHz and 3GB of memory. The same test cases from our previous research [9] are adopted here. The surgeries are classified into five types: small (S), medium (M), large (L), extra-large (EL), and special (SE), according to surgery duration. The durations of these five surgery types, the pre-surgery stage and the post-surgery stage are represented as normal distribution (μ, σ) and are listed in Table 1. Table 2 shows those three test cases with detail problem size and resources required. Three performance measurements are used this comparison and they are the makespan, the variation coefficient of resources working time (VCWT), the total overtime of nurses.

Table 1. The duration of pre/post-surgery stage and duration of a surgery in different types

	Pre-surgery	Surgery case					Post-surgery
		S	M	L	EL	SE	
Duration	(8, 2)	(33,15)	(86,17)	(153,17)	(213,17)	(316,62)	(28,17)

Table 2. Three test cases

Case	surgeries	PHU beds	Nurses	Surgeons	ORs	PACU beds	Anesthetist	Surgery type (S:M:L:E:SE)
1	10	2	8	6	4	4	6	2:6:1:1:0
2	20	3	15	10	5	4	8	4:12:3:1:0
3	30	4	20	10	6	5	10	7:18:3:1:1

4.1   ACO-MO Parameters Setting

Since there are two pheromone settings (SPSP and SPMP) proposed in ACO-MO for OR scheduling problem, experiments has to be done to identify the more appropriate pheromone setting way. In sum, the results in Table 3 indicate the superior performance of the SPMP in all three performance measurements. It demonstrates that to build an objective-specific pheromone for each individual objective function has advantage in clustering and guiding ants to the preferred direction of a decision-maker. Therefore, all following computational experiments on ACO-MO are adopting SPMP.

Table 3. Comparison between pheromone settings SPSP and SPMP for test case #3

Types	Makespan	Over time	VCWT		
			OR	Nurse	Anesthetist
ACO-MO-SPSP	812	62	0.16	0.48	0.38
ACO-MO-SPMP	789	55	0.11	0.44	0.35

The basic ACO-MO parameters include the number of ants (m), pheromone factor (α), heuristic factor (β), evaporate rate (ρ), pheromone intensity (Q), decremented pheromone value (q0), and weights of multiple objectives (ω_r). Those parameters have impact on the algorithm's convergence and solution quality. The final optimal ACO-MO parameters for three test cases are summarized in Table 4.

Table 4. The optimal ACO-MOB parameters

Test case#	m	α	β	ρ	Q	q_0	NC_max	w1:w2:w3
1	30	1	5	0.5	[40,1,50]	0.1	200	0.2:0.2:0.6
2	60	1	5	0.5	[40,1,50]	0.2	200	0.2:0.2:0.6
3	50	1	5	0.5	[40,1,50]	0.2	300	0.2:0.2:0.6

4.2   Computational Result Discussion

A comparison experiment is built to evaluate the performance of three different scheduling approaches. These three approaches are receptively, the scheduling result in simulation model (named as 'Simulation'), the ACO approach with single objective of makespan (named as 'ACO-SO'), and the proposed ACO-MO in this work. The computational results are achieved by the same three test cases. The data of the first two approaches are obtained from our pilot research work [9].The comparison results by three different scheduling approaches are shown in Table 5.

Table 5. Scheduling comparison between original, ACO-SO and ACO-MO

		Makespan	Over time	VCWT		
				OR	Nurse	Anesthetist
	Simulation	478	0	0.276	0.551	0.237
Test1	ACO-SO	341	0	0.08	0.55	0.13
	ACO-MO-SPMP	341	0	0.08	0.55	0.13
	Simulation	602	122	0.12	0.24	0.50
Test2	ACO-SO	489	9	0.09	0.46	0.18
	ACO-MO-SPMP	489	9	0.06	0.26	0.15
	Simulation	1002	142	0.21	0.47	0.47
Test3	ACO-SO	789	42	0.09	0.44	0.31
	ACO-MO-SPMP	789	42	0.09	0.38	0.27

For three test cases, both ACO-MO-SPMP and ACO-SO show an outstanding performance than simulation result in every measurements, which indicates the efficiency of using ACO in solving such combinatorial optimization problem. As to the comparison on ACO-MO-SPMP and ACO-SO, for test case #1, because of the small scale of the problem, there exists no difference in two approaches. For test case #2 and #3, both report the same Makespan and total overtime. As to the resources utilization, the VCWT of all resources including ORs and nurses and anesthetists all show varying improvements. For example, the VCWT of ORs in ACO-SO and ACO-MO-SPMP is 0.09 and 0.06 in test case #2, and the VCWT of nurses in ACO-SO and ACO-MO-SPMP is 0.46(0.44) and 0.26(0.38) in test case #2(#3). Such improvement indicates that ACO-MO can achieve the optimal makespan and the overtime, while at the same time let the schedule have more balanced resources allocation.

5   Conclusion and Ongoing Work

Due to the combinatorial nature of the OR scheduling problem, and the conflicting objectives considered, a modified ACO algorithm integrating Pareto set by aiming at achieving sub-optimal solutions is proposed in this paper. The multiple objectives include minimizing makespan, the total overtime and the equilibrium degree of resources utilization. Two kinds of pheromone setting, SPSP and SPMP for MO are presented to make traditional ACO suit for such MO problem. The scheduling result of the proposed ACO-MO algorithm is compared with the simulation scheduling result, and ACO with single objective of makespan. Three measurements, i.e. the makespan, the total overtime of nurses and the variation coefficient of working time of resource, are evaluated. Comparison results indicate a better performance for the proposed ACO-MO in general. Future research will be in the direction of extending our ACO algorithm to solve the OR scheduling problems with more realistic constraints arise in actual OR management in hospital, like nurse rostering constraints and surgeons/nurses preference constraints in medical team etc.

Acknowledgment. The Project is supported by Zhejiang Provincial Natural Science Foundation of China (LY12G01007), Ningbo Natural Science Foundation (2013A610109) and K.C.Wong Magna Fund in Ningbo University.

References

1. Cardoen, B., Demeulemeester, E., Beliën, J.: Operating room planning and scheduling: a literature review. Eur J Oper Res. **201**(3), 921–932 (2010)
2. Guerriero, F., Guido, R.: Operational research in the management of the operating theatre: a survey. Health Care Manag Sci. **14**, 89–114 (2011)
3. May, J.H., Spangler, W.E., Strum, D.P., Vargas, L.G.: The Surgical Scheduling Problem: Current Research and Future Opportunities. Prod Oper Manag. **20**(3), 392–405 (2011)
4. Jebali, A., Alouane, A.B.H., Ladet, P.: Operating rooms scheduling. International Journal of Production Economics. **99**, 52–62 (2006)
5. Krempels, K.H., Panchenko A.: An approach for automated surgery scheduling. In: Proceedings of the Sixth International Conference on the Practice and Theory of Automated Timetabling (2006)
6. Dexter, F., Epstein, R.H.: Operating room efficiency and scheduling. Current Opinion in Anesthesiology. **18**, 195–198 (2005)
7. Denton, B., Viapiano, J., Vogl, A.: Optimization of surgery sequencing and scheduling decisions under uncertainty. Health Care Management Science. **10**, 13–24 (2007)
8. Marcon, E., Dexter, F.: Impact of surgical sequencing on post anesthesia care unit staffing. Health Care Management Science. **9**, 87–98 (2006)
9. Xiang, W., Yin, J., Lim, G.: Modified ant colony algorithm for surgery scheduling under multi-resource constraints. Advances in Information Sciences and Service Sciences. **5**(9), 810–818 (2013)
10. Ogulata, S., Erol, R.: A hierarchical multiple criteria mathematical programming approach for scheduling general surgery operations in large hospitals. Journal of Medical Systems **27**(3), 259–270 (2003)
11. Cardoen, B., Demeulemeester, E., Belien, J.: Optimizing a multiple objective surgical case sequencing problem. Int. J. Production Economics **119**, 354–366 (2009)
12. Beliën, J., Demeulemeester, E., Cardoen, B.: A decision support system for cyclic master surgery scheduling with multiple objectives. J Sched **12**, 147–161 (2009)

Multi-Colony Ant Algorithm Using a Sociometry-Based Network and Its Application

Sheng Liu[✉] and Xiaoming You

School of Management, Shanghai University of Engineering Science, Shanghai 201620, China
{ls6601,yxm6301}@163.com

Abstract. In this paper, the social fabric approach is weaved into multi-behavior based multi-colony ant colony system (MBMC-ACS) to construct pheromone diffusion model. According to the propagation characteristics of knowledge in the social fabric, the Cobb-Dauglas production function is introduced to describe the increase of pheromone caused by pheromone diffusion. The pheromone diffused inter-colonies based on sociometry-based networks can simulate the knowledge evolution mechanism in organizational learning network, which allows the algorithm to avoid premature convergence and stagnation problems. The experimental results for TSP show the validity of this algorithm.

Keywords: Social fabric · Pheromone diffusion · Cobb-dauglas production function · MBMC-ACS (Multi-Behavior based Multi-Colony ant colony System)

1 Introduction

Swarm intelligence is a field, which studies "the emergent collective intelligence of groups of simple agents" [1]. In groups of insects, which live in colonies, such as ants and bees, an individual can only do simple task on its own, while the colony's cooperative work is the main reason determining the intelligent behavior it shows.

ACO algorithm is inspired by social behavior of ant colonies and was represented in the early 1990's by M. Dorigo and colleagues [2]. In recent years, many modified ant algorithms have been developed based on the ACO technique and applied to numerous difficult problems [3],[4],[5],[6]. In the previous version of the ACO, while successful, but individuals had no social presence, little research has been done on the improvement of optimization ability from pheromone diffusion based on network structure.

In this new approach, the social fabric theory is introduced into a multi-behavior based multi-colony ant colony system (MBMC-ACS) [6] to construct pheromone diffusion model where the pheromone and the density of the pheromone along the trail is presented as the knowledge that the ACS share among its individual ants. Through the notion of a social fabric, pheromone diffusion will expand the ability of a knowledge source to influence a population. Thus, the innovative potential of optimization will greatly be promoted.

The organization of the paper is as follows: Section 2 describes how the ant colony is divided into several sub-colonies according to different behavior options. Section 3

© Springer International Publishing Switzerland 2015
Y. Tan et al. (Eds.): ICSI-CCI 2015, Part I, LNCS 9140, pp. 205–212, 2015.
DOI: 10.1007/978-3-319-20466-6_22

introduces the social fabric basics, and how it is used as a computational tool to influence the diffusion of pheromone in Section 4. Finally, experimental study and conclusions are given in Section5 and 6, respectively.

2   Our Previous Work

The basic idea of multi-behavior based multiple colonies is that all individuals will be decomposed into some independent sub-colonies in parallelization schemes by using different behavioral characteristics and inter-colonies migration strategies [6].

The four kinds of behavior have been briefly identified: exploitation, moderate, exploration, exploitation and exploration.

Exploitation: an ant selects the next city in a stochastic selection manner every time, which is used to provide a new hyper plane, to overcome the premature convergence.

Moderation: an ant selects the next city in a hybrid behavior every time, which is used to maintain stability in the evolution of colony. The probability of an ant move from the city i to the city j was determined by the following two values:

1) The attractiveness η_{ij} of the move, as computed by some heuristic indicating the a priori desirability of that move;

2) The pheromone trail level τ_{ij} of the move, indicating how useful it has been in the past to make that particular move; it represents, therefore, an a posteriori indication of the desirability of that move.

Given the attractiveness and the pheromone trail level, the probability of an ant move from the city i to the city j was determined by the following pre-specified random distribution:

$$j = \begin{cases} \arg\max_{l \in U}[(\tau_{il})^{\alpha}(\eta_{il})^{\beta}], & q \leq q_0 (\text{exp}loitation) \\ V, & otherwise(\text{exp}loration) \end{cases} \tag{1}$$

where q is a randomly number uniformly distributed in [0..1], q_0 is a parameter ($0 \leq q_0 \leq 1$), and V is the transition probability rule defined as follow.

$$P_{ij} = \begin{cases} \dfrac{[\tau_{ij}]^{\alpha} \cdot [\eta_{ij}]^{\beta}}{\sum_{l \subset U}[\tau_{il}]^{\alpha} \cdot [\eta_{il}]^{\beta}}, & if \ j \in U \\ 0, & otherwise \end{cases} \tag{2}$$

where U is the set of cities which can be visited starting from city i. The parameters α and β weigh the influence of trails and attractiveness.

Exploration: an ant selects the next city in a greedy selection technique every time, which is used to look for and store outstanding individual in the local area.

$$P_{ij} = \begin{cases} \dfrac{\eta_{ij}}{\sum\limits_{l \in U} \eta_{il}}, & if \ j \in U \\ 0, & otherwise \end{cases} \tag{3}$$

Exploitation and Exploration: an ant selects the next city in a "follow the crowd" manner every time that underlies ant's behavior can be guided by the perceived behavior of other individuals, which is used to make a delicate balance between progress and stability.

$$P_{ij} = \begin{cases} \dfrac{\tau_{ij}}{\sum\limits_{l \subset U} \tau_{il}}, & if \ j \in U \\ 0, & otherwise \end{cases} \tag{4}$$

According to four kinds of behavior defined here, the ant colony is divided into four sub-colonies, each sub-colony of ants have different behavioral characteristics, different sub-colonies evolved independently. The parallel evolution example of four sub-colonies is shown in Figure1.

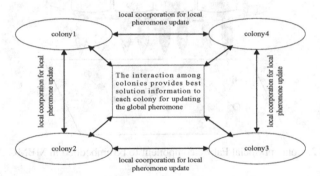

Fig. 1. A framework of multi-colony parallel evolution

Ants in each colony perform the same task, that is, to find the solution in the search space. The exploration of the search space in each colony may be guided by different patterns of behavior for different purposes. So, the differences in the direction of evolution can be maintained and the adequacy of the search is ensured.

Meanwhile, the sub-colonies have their own population evolved independently and in parallel according to four different behavior options, and update their local pheromone and global pheromone level respectively according to designed rules in the execution process.

3   Weaving the Social Fabric Influence Into MBMC-ACS

The social fabric is a living skin created out of engineered emergence of agents illustrating the tension between the individual and the community in a context of interaction between them. Here the pheromone interaction of ants will be abstracted into the "social fabric" (a sociometry-based network), which is viewed as a computational tool that influences pheromone diffusion between colonies [7].

The experimental framework for the social fabric component which is to be added to the MBMC-ACS is illustrated in Figure 2. The figure shows how the edges which individuals have visited are represented as different community's nodes in the functional landscape produced by different behavioral characteristics and the pheromone interaction between them will be as "network connection". We use the probability q_1 to produce the links between the communities. In that sense, the social fabric description can be a very useful way of making hypotheses between visible and less visible networks through their links. The size of the communities may be different, reflecting the heterogeneous topological structure of community networks. In detail, the community network can be constructed as follows:

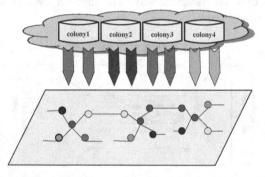

Fig. 2. Concept Social Fabric component to be embedded in MBMC-ACS

   1) Consider a MBMC-ACS with N_0 ants and divide them into m communities with random n_i ($i = 1$, . k . , m) ants in each community and $\sum_{i=1}^{m} n_i = N_0$. The exploration of the search space in each colony may be guided by different patterns of behavior for different purposes.

   2) In community k, the locally best solution r_{best}^k found in every s generations since the start of algorithm is computed and stored (i.e., a feasible solution to the TSP).

   3) Between community k and p, we use the probability q_1 to add a link between every two same edges (i, j), which belong to r_{best}^k and r_{best}^p respectively, r_{best}^k and r_{best}^p are similarity.

4) The edge (i, j) in community k is represented as a $Node_{ij}^k$ in the landscape, the node talks to its neighbors determined by the chosen topology, over which it can transmit its information to its neighbors will correspond to its influence.

5) The nodes are connected through a topology that determines connectivity type between nodes. Topology used in our approach is Lbest ring topology.

4 Pheromone Diffusion Model

The model we propose explores how the pheromone interaction and network structure affect the pheromone amount of the same edges in solutions which belong to different communities.

Formally, let $\tau_{ij}^k(t)$ denote the amount of pheromone on $Node_{ij}^k$ at time t. Though we employ the word pheromone, $\tau_{ij}^k(t)$ should rather be seen as a form of node's capital or competence which is characterized by knowledge endowments, whose accumulation results from nodes performing learning and innovative activities.

$Node_{ij}^k$ and $Node_{ij}^{k'}$ interact if and only if there is a direct connection between them and pheromone level in which $\tau_{ij}^k(t)$ strictly dominates $\tau_{ij}^{k'}(t)$. It is assumed that the nodes involved in pheromone diffusion are randomly chosen, with uniform probability q_1 after the exchange, the increase of pheromone caused by pheromone diffusion is described by the Cobb-Dauglas production function.

$$\Delta\tau_{ij}^{k'}(t+1) = \begin{cases} 0, \tau_{ij}^k(t) \le \tau_{ij}^{k'}(t) \\ A(\tau_{ij}^{k'}(t))^\alpha(\tau_{ij}^k(t)-\tau_{ij}^{k'}(t))^\beta, \tau_{ij}^k(t) > \tau_{ij}^{k'}(t) \end{cases} \tag{5}$$

$$\tau_{ij}^{k'}(t+1) = \tau_{ij}^{k'}(t) + \Delta\tau_{ij}^{k'}(t+1) \tag{6}$$

where $0 < A$, $\alpha, \beta < 1$, and $\alpha + \beta \le 1$. Here, we assume $\alpha < \beta$ to ensure that for the improvement of pheromone, the exogenous pheromone diffusion is greater than intrinsic level of pheromone, which is on the premise of pheromone diffusion lead to increase of pheromone. Similarly, assuming $0 < A < 1$ to adjust the increment of pheromone, in order to avoid the increment is too high, which is not coincide with common sense.

This diffusion process continues until all possible diffusions have been made, which holds when for all possible pairs, one node weakly dominates the other. Since with real pheromone levels never become identical, to implement this weak domination, we consider them identical if they differ by less than one percent: $0.99 < \tau_{ij}^k(t) / \tau_{ij}^{k'}(t) < 1.01$, which describes the steady state of the process.

5   Experimental Study

In this section, we performed tests with different instances of TSP to test the ACS performs using the social fabric. We used berlin52, bier127, and Chc144 to verify the difference between proposed method and the conventional method.

The parameters used by our approach are the following: the number of ants is 120, the ant colony is divided into four sub-colonies, the ratio of ants in each sub-colony is $r_1 : r_2 : r_3 : r_4$, where $r_1 : r_2 : r_3 : r_4$ =0.05:0.1:0.4:0.45, β =5, q_0 =0.90, q_1 =0.6, $\rho = \partial$ =0.1, $\tau_0 = (N*L_{NN})^{-1}$, where L_{NN} is the tour length produced by the nearest neighbor heuristic and N is the number of cities. The results of comparison are shown in Table 1. G.B. (Global Best) represents the minimum length obtained by different methods, BRKSF means best result known so far, N.E. (Normalized Error) is calculated as the following:

$$Normalized\ Error=1-(Best\ Result\ so\ Far/Average)$$

where Average are over 15 trials, from Table 1 it can be noticed that improved method offers the best performances in nearly every case using real-valued distance as compared with other heuristics [8],[9]. For example, on the berlin52 problem the improved method reached a better result in global best, with a much lower average normalized error of 0.000314.

Figure 3 shows the convergence diagrams of Chc144 using the proposed algorithm and MBMC-ACS. They illustrate that the speed of convergence of the proposed algorithm is faster than that of MBMC-ACS.

Figures 4a–b show the trajectories of 52 and 127 nodes of the improved method, Simulation results show that the proposed approach is working appropriately, and also providing better results.

Table 1. Comparison of improved method with other heuristics on instances of the symmetric TSP

Problem	BRKSF	Improved method		MMAS		EAS	
		GB	%N.E	GB	% N.E	GB	%N.E
berlin52	7542	**7544.37**	**0.000314**	7549.29	0.050707	7544.37	0.011321
bier127	118282	**118293.52**	**2.89*10⁻⁴**	123903.00	0.045	123903.00	0.045
Chc144	30354.3	**30355.78**	**7.9*10⁻⁵**	30356.75	1.84*10⁻⁴	30356.7	0.0074

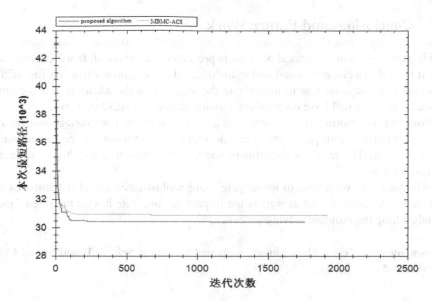

Fig. 3. Convergent diagram of Chc144 using the proposed algorithm and MBMC-ACS

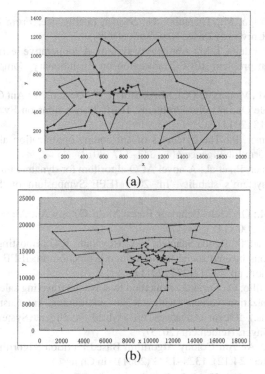

Fig. 4. The shortest trajectory of the improved method. (a) 52 nodes and (b) 127 nodes.

6   Conclusions and Future Work

In this paper, a sociometry-based network is presented via the social fabric concept and how it is used as a computational tool to influence the pheromone diffusion for ACS is demonstrated. Our goal was to investigate the impact that the addition of a sociometry-based network will have on problem solving ability of a ACO system.

From above exploring, it is obvious that a sociometry-based network is an effective facility for optimization problems. It can accelerate the solution process when configured properly. The result of experiment has shown proposed algorithm is a precise method for TSP.

In future work, we intend to investigate more sophisticated social structures at the pheromone diffusion model as well as the impact the structure has on the role of individuals during the problem solving process.

Acknowledgment. This work is partially supported by Natural Science Foundation of China (No.61075115).

References

1. Bonabeau, E., Dorigo, M., Theraulaz, G.: Swarm Intelligence: From Natural to Artificial Systems. Oxford University Press, New York (1999)
2. Dorigo, M., Gambardella, L.M.: Ant colony system: A cooperative learning approach to the traveling salesman problem. IEEE Transaction on Evolutionary Computation 1(1), 53–66 (1997)
3. Liao, K., Socha, M., Montes de Oca, M.A., Stützle, T., Dorigo, M.: Ant Colony Optimization for Mixed-Variable Optimization Problems. IEEE Transactions on Evolutionary Computation 18(4), 503–518 (2014)
4. Iacopino, C., Palmer, P.: The dynamics of ant colony optimization algorithms applied to binary chains. Swarm Intelligence 6(4), 343–377 (2012)
5. Iacopino, C., Palmer, P., et al.: A novel ACO algorithm for dynamic binary chains based on changes in the system's stability. In: 2013 IEEE Symposium on Swarm Intelligence, pp. 56–63 (2013)
6. Liu, S., You, X.M.: On Multi-Behavior Based Multi-Colony Ant Algorithm for TSP. IEEE Computer Society, 2009.11
7. Mostafa, Z.A., Ayad, S., Randa Snanieh, T.A., Reynolds, R.G.: Boosting cultural algorithms with an incongruous layered social fabric influence function. In: IEEE Congress on Evolutionary Computation, pp. 1225–1232 (2011)
8. Lizárraga, E., Castillo, O., Soria, J.: A method to solve the traveling salesman problem using ant colony optimization variants with ant set partitioning. In: Castillo, O., Melin, P., Kacprzyk, J. (eds.) Recent Advances on Hybrid Intelligent Systems. SCI, vol. 451, pp. 237–246. Springer, Heidelberg (2013)
9. Wu, B., Shi, Z.Z.: An Ant Colony Algorithm Based Partition Algorithm for TSP. Chinese journal of computers 24(12), 1328–1333 (2001). (in Chinese)

A Hybrid Multi-Cell Tracking Approach with Level Set Evolution and Ant Colony Optimization

Dongmei Jiang[1], Benlian Xu[1(✉)], and Long Ge[2]

[1] School of Electrical and Automatic Engineering, Changshu Institute of Technology, Changshu 215500, China
xu_benlian@cslg.cn
[2] Changshu National University Science and Technology Park, Changshu 215500, China
ge_long@163.com

Abstract. In this paper, we propose a hybrid multi-cell tracking approach to accurately and jointly estimate the state and its contour of each cell. Our approach consists of level set evolution and ant colony optimization, representing, respectively, the deterministic and stochastic methods for cell tracking. Firstly, birth ants are directly distributed into the regions depicted by raw curves achieved by the traditional level set evolution. Then, the ants move towards potential regions based on the pheromone deposited by ants and the gradient information of current image. Finally, the resulting pheromone field is embedded in the variational level set to drive the evolution of cell curve to yield an accurate one and correspond cell position estimate. The experiment results show that our method could automatically track multi-cell and achieve an accurate contour estimation of each cell.

Keywords: Level set · Ant colony · Cell tracking · Contour estimate

1 Introduction

The analysis of cellular behavior is meaningful for biomedical research, such as oncological studies, drug discovery, proteomics, proteomics and tissue engineering. During the past few years, various algorithms [1-4] have been developed for automated or half-automated cell tracking for different cell images. In general they can be roughly classified into two categories: deterministic methods and stochastic methods. Existing literatures show an increasing interest in the latter category. Deterministic methods usually divide into two kinds, one is model –based evolution approaches and another is detection-based association approaches. This method could produce good result for high image quality and low cell density, but tracking may fail under problematic imaging conditions such as large cell density, cell division events, or segmentation errors. Stochastic methods which track based on Bayesian probabilistic framework methods has also been proposed in recent years and are more robust to low resolution and signal-to-noise(SNR) scenarios than other tracking methods.

© Springer International Publishing Switzerland 2015
Y. Tan et al. (Eds.): ICSI-CCI 2015, Part I, LNCS 9140, pp. 213–221, 2015.
DOI: 10.1007/978-3-319-20466-6_23

The level set [5-7] is another paradigm of the deterministic methods, which is based on the representation and evolution of curves or surfaces and it provides the advantage of identifying multiple cells in a single frame. A desirable feature is that level set methods can represent contours of complex topology and are able to handle topological changes (such as splitting, merging or moving into regions of other cells) in a natural and efficient way.

Cell contour is represented as the boundary of a cell in an image, and it makes us understand the contents of multiple cells in an easy way. Due to technical limitations or artifacts, intensity inhomogeneity usually happens for most of cells in an image, thus the traditional level set is usually hard to evolve to the real boundary of cell. Inspired by ant stochastic searching behavior [8-10], we propose a hybrid multi-cell tracking approach, which applies the pheromone depositing by ant colony into a variational level set formulation without reinitialization proposed by Li et al. [11, 12] to automatically track multiple cells and achieve an accurate contour estimation of each cell.

2   Methods

This section gives the principle of the proposed hybrid multi-cell tracking approach with level set evolution and ant colony optimization in details. Birth ants are directly distributed into the regions represented by raw curves achieved by the level set evolution without reinitialization. Then, the ants move towards potential regions according to the proposed decision strategy and deposit pheromone accordingly. The resulting pheromone field is used to drive the evolution of cell curve to yield an accurate one and corresponding cell position estimate. Fig.1 illustrates the overview of our proposed algorithm.

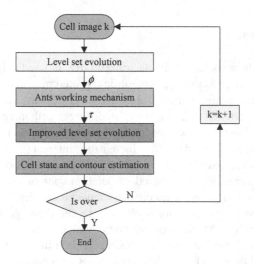

Fig. 1. The overview of our proposed algorithm

2.1   Ants Colony Initial Distribution

Our approach is based on the variation level set formulation of curve evolution that completely eliminates the need of the re-initialization. The variational formulation is as followed

$$\varepsilon(\phi) = \mu p(\phi) + \lambda L_g(\phi) + v A_g(\phi) \tag{1}$$

where $p(\phi) = \int_\Omega \frac{1}{2}(|\nabla\phi|-1)^2 dxdy$ is to penalize the deviation of the level set function ϕ from a signed distance function; $L_g(\phi) = \int_\Omega g\delta(\phi)|\nabla\phi|dxdy$ computes the length of the zero level curve; $A_g(\phi) = \int_\Omega gH(-\phi)dxdy$ is introduced to speed up curve evolution; μ, λ and v are the constants. We can minimize the function ϕ by satisfying the Euler-Lagrange equation $\frac{\partial\varepsilon}{\partial\phi} = -\frac{\partial\phi}{\partial t} = 0$. Then, the steepest descent process for minimization of the function ε is the gradient flow by following:

$$\frac{\partial\phi}{\partial t} = \mu\left[\Delta\phi - div\left(\frac{\nabla\phi}{|\nabla\phi|}\right)\right] + \lambda\delta(\phi)div\left(g\frac{\nabla\phi}{|\nabla\phi|}\right) + vg\delta(\phi) \tag{2}$$

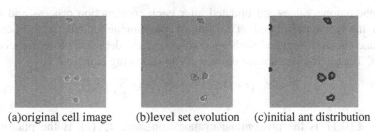

(a)original cell image   (b)level set evolution   (c)initial ant distribution

Fig. 2. Ant colony initial distribution

The variation level set formulation has been applied to a variety of synthetic and real images in different modalities. Fig.2 b) shows a result of multi-cell evolution of the contour on a 201×201 pixel microscope image of multi-cell according to Fig.2 a). It can be observed that the yielding cell contours can only give the approximate estimates, and we, therefore, resort to ant colony optimization to guide the curve evolution to the real boundary of each cell. Specifically, a given number of ants are assigned on each point (or pixel) of the level set curve illustrated by fig.3 b) (denoted by blue points), and we will use the proposed ant searching information to help the variational level set obtain an accurate contour estimation of each cell.

2.2   Ant Decision Strategy

Each ant selects the following pixel to be visited according to a stochastic mechanism by the pheromone mechanism, i.e. ant can move directly towards one of its neighbors at each time, and any pixel can be visited simultaneously by several ants. The ant working environment in the current cell image is the pheromone filed deposited by ants that can be read and modified by ants. For any pixel index, 4-neighboring configuration is considered. Therefore, supposed that an ant is now located a pixel (i, j), in the $t-$th iteration, the movement from its location to one of its neighboring pixels may happen with the following probability

$$p_{i,j}(t) = \begin{cases} 0 & otherwise \\ \dfrac{\left[\tau_{i,j}(t)\right]^{\alpha}\left[\theta_{i,j}(t)\right]^{\beta}}{\sum_{j\in N(i,j)}\left[\tau_{i,j}(t)\right]^{\alpha}\left[\theta_{i,j}(t)\right]^{\beta}} & if\ j\in N(i,j) \end{cases} \tag{3}$$

where α and β are adjustment parameters related to pheromone intensity and importance of heuristic, respectively; $N(i,j)$ is the set of neighbors for pixel (i,j); $\tau_{i,j}(t)$ is the total sum of pheromone amount left by all ants passed pixel (i,j); $\theta_{i,j}(t)$ is the heuristic information of pixel (i,j) to be defined later.

The pheromone values are updated after each construction process and the ants' decision has been performed. The aim of pheromone update is to increase the pheromone values associated with good solution, and decrease those associated with bad ones. The pheromone update defines as the following equation:

$$\tau_{i,j}(t+1) = (1-\rho)\tau_{i,j}(t) + \sum_{\Omega_{i,j}} r_{i,j}(t) \tag{4}$$

where $\rho \in (0,1]$ is the pheromone evaporation rate; $\tau_{i,j}(t)$ is the pheromone on pixel (i,j) at the $t-$th iteration; $\sum_{\Omega_{i,j}} r_{i,j}(t)$ describes the amount of pheromone external input to pixel j at the $t-$th iteration.

From the ant transition probability, we know that the heuristic function $\eta_{(i,j)}(t)$ is an important parameter for ant decision. That is, the closer to the cell edge the greater value of the probability. Naturally, the grayscale distribution of current image is utilized. We define pixel (i,j) with intensity $G(i,j)$ illustrated in Fig.3, and its neighboring region grayscale variance is computed by the following equation:

$$\theta(i,j) = \frac{1}{|N(i,j)|} \sum_{(i',j')\in N(i,j)} \left(G_{i',j'} - \overline{G}(N(i,j))\right)^2 \tag{5}$$

where $\overline{G}(N(i,j))$ is the average gray intensity of $N(i,j)$; $|N(i,j)|$ is the number of neighboring pixels of pixel (i,j).

According to the equation, the grayscale variance has a smaller value in the area of background and interior of cell, whereas a larger value is probably taken between two sides of edge, as well as in the vicinity of each edge pixel.

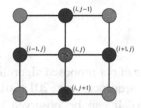

Fig. 3. 4-neighboring pixel

2.3 The Improved Level Set Evolution

As for the intensity inhomogeneity in cell images, the variation level set formulation is hard to obtain an accurate cell boundary. The pheromone field formed by ants can extract of the accurate local image information. We will focus on applying the pheromone field in Eq. (1) to active contour for image segmentation, so that the zero level curve can evolve to the desired features.

Note that parameter τ has the local property that takes a lager value at the point close to the cell edge, whereas a smaller value at the point is taken for those points far away from the cell edge. Therefore, we define the following local energy

$$P_\tau(\phi) = \int_\Omega \tau(\phi(t) - \phi(t-1))^2 H(\phi) dxdy \tag{6}$$

Obviously, the local energy can be minimized when the contour is exactly on the object boundary and the entire energy functional of the variational level set evolution is redefined as:

$$\varepsilon(\phi) = \mu p(\phi) + \lambda L_g(\phi) + v A_g(\phi) + \xi P_\tau(\phi) \tag{7}$$

Then we use the standard gradient descent method to minimize the energy functional Eq. (7) to find the object boundary. The gradient descent flow is following

$$\frac{\partial \phi}{\partial t} = \mu \left[\Delta\phi - div\left(\frac{\nabla\phi}{|\nabla\phi|}\right)\right] + \lambda\delta(\phi)div\left(g\frac{\nabla\phi}{|\nabla\phi|}\right) + vg\delta(\phi) - \xi\tau e\delta(\phi) \tag{8}$$

where g is the edge indicator function; $\delta(\cdot)$ is the Dirac function used to limit the evolutional value around the zero level set function; e is the function as below

$$e(x) = \int_\Omega \tau(\phi(t) - \phi(t-1))^2 dy \tag{9}$$

The above Eq. (8) is the proposed implicit active contour model, and the energy drives the zero level set toward the object boundaries. Once the level set evolves to the object contours, an accurate contour estimation of each cell is achieved. By calculating the center of each cell contour, the location of each cell is then available. Then, the detect cells are associated between two or more consecutive frames using the nearest neighboring method, and the motion trajectory of each cell is finally obtained.

3   Results

In this section, the performance of our proposed algorithm is tested on a real low-SNR image sequences. The image sequences are 201×201 pixels RGB image with 20 frames as shown in Fig. 4(a). It can be observed that the images are intensity inhomogeneity and the upper cell moves slowly, while the lower cells move with nearly round shape. Some related parameters are set to be $\mu = 0.18/time\,step$, $\lambda = -3.88, \upsilon = -0.98, \xi = 1.2$.

Fig.4 presents an example of successfully tracking results using our proposed algorithm. It can be observed that our method can get the accurate contour estimates of cells in the cell image sequence and automatically track cells with different dynamics and shapes, i.e., cell 5 entering in frame 2, cell 4 in frame 13 leaving. Each cell trajectories are drawn in the image sequence.

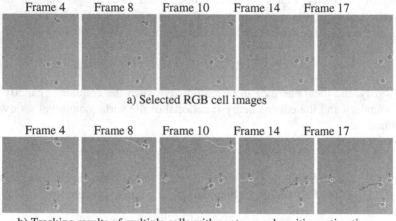

a) Selected RGB cell images

b) Tracking results of multiple cells with contour and position estimations

Fig. 4. Tracking results of a given cell image sequence (201pix×201pix)

As we all known some cells in an image sequence move in a small region, but some cells move faster and even out of field of view. So the cell trajectory is another measure to analyze cell migration behavior. Cell trajectory can oversee the entire tracking history, and it can detect potential problems among all the tracking process. Fig.5 gives the cell trajectories of the cell moving. Fig.6 plots the corresponding position estimate of each cell in x and y directions per frame, respectively. The position estimate indicates that the method is sensitive some cells which are partially entering or leaving the image, such as cell 3 and 5.

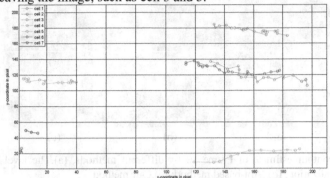

Fig. 5. The estimated results of cell trajectories

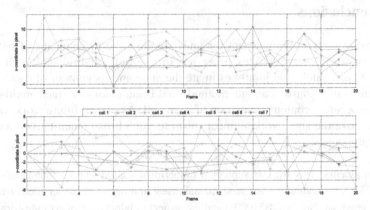

Fig. 6. The estimated velocity of each cell in x and y directions

Figs.7 and 8 give the comparisons of multi-cell contour estimates in frames 6 and 10 by the level set evolution without reinitialization [12], the ant-based method [8] and our proposed approach, respectively. It can be observed that our method outperforms the other two methods for most of cells in the image.

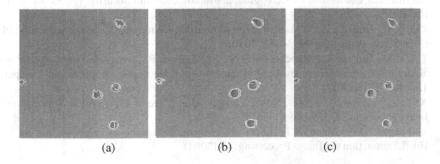

(a)   (b)   (c)

Fig. 7. Cell contour estimate of Frame 6 by different methods, (a).The level set evolution without reinitialization; (b). Ant-based method; (c).Our method

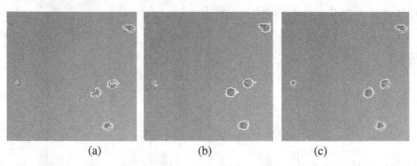

(a)   (b)   (c)

Fig. 8. Cell contour estimate of Frame 10 by different methods, (a).The level set evolution without reinitialization; (b). Ant-based method; (c).Our method

4   Conclusions

Cell behavior analysis is important for understanding the mechanisms of biomedical research or drug discovery. This paper has introduced a hybrid multi-cell tracking approach to accurately and jointly estimate the state and its contour of each cell. The experiment results show that our method could automatically track multi-cell and achieve an accurate contour estimation of each cell. As part of future work, we would like to improve the ants' working mechanism to get a broad quantitative view of cell cycle progression.

Acknowledgments. This work is supported by national natural science foundation of China (No.61273312), the natural science fundamental research program of higher education colleges in Jiangsu province (No. 14KJB510001) and the project of talent peak of six industries (DZXX-013).

References

1. Zhou, X., Lu, Y.: Efficient mean shift particle filter for sperm cells tracking. In: International Conference on Computational Intelligence and Security, pp. 335–339. IEEE Press, Beijing (2009)
2. Zhang, H., Jing, Z., Hu, S.: Localization of multiple emitters based on the sequential PHD filter. Signal Processing **90**, 34–43 (2010)
3. Thida, M., Eng, H.-L., Monekosso, D.N., Remagnino, P.: A particle swarm optimisation algorithm with interactive swarms for tracking multiple targets. Applied Soft Computing **13**, 3106–3117 (2013)
4. Xu, C., Prince, J.L.: Snakes, Shapes, and Gradient Vector Flow. IEEE Transaction On Image Processing **7**, 359–369 (1998)
5. Mukherjee, D.P., Acton, S.T.: Level set analysis for leukocyte detection and tracking. IEEE Transaction On Image Processing, **13** (2004)

6. Dzyubachyk, O., Meijering, E.: Advanced Level-Set-Based Cell Tracking in Time-Lapse Fluorescence Microscopy. IEEE Trans. Med. Imag. **29**, 852–867 (2010)
7. Shariat, F.: Object Segmentation Using Active Contours: A Level Set Approach (2009)
8. Xu, B., Lu, M., Zhu, P., Shi, J.: An accurate multi-cell parameter estimate algorithm with heuristically restrictive ant system. Signal Processing **101**, 104–120 (2014)
9. van Kaick, O., Hamarneh, G., Zhang, H., Wighton, P.: Contour correspondence via ant colony optimization. In: Pacific Conference on Computer Graphics and Applications, IEEE Computer Society, Hawaii (2007)
10. Tian, J., Yu, W., Xie, S.: An ant colony optimization algorithm for image edge detection. In: IEEE World Congress on Computational Intelligence, pp. 751–756 (2008)
11. Li, C., et al: Implicit active contours driven by local binary fitting energy. In: IEEE Computer Society Conference on Computer Vision and Pattern Recognition (2007)
12. Li, C., et al: Level set evolution without re-initialization: a new variational formulation. In: Computer Vision and Pattern Recognition (2005)

Ant Algorithm Modification
for Multi-version Software Building

Igor Kovalev, Pavel Zelenkov, Margarita Karaseva[✉], and Dmitry Kovalev

Siberian State Aerospace University, Krasnoyarsk, Russia
kovalev.fsu@mail.ru, {zelenkow,karaseva-margarita}@rambler.ru,
grimm7jow@gmail.com

Abstract. The multi-version software and problem of its building as an optimization problem are considered. The ant algorithm as a way to solve the problem of multi-version software building is presented. The results of the standard and modified ant algorithms are given and compared.

Keywords: Optimization · Ant algorithms · Multi-version software

1 Introduction

Nowadays "natural algorithms" which are the optimization algorithms based on the natural ways of decision making are actively investigated by a lot of scientists. One of those algorithms is the ant colony optimization algorithm (ACO)[1-3]. That algorithm is a result of the combined work of scientists who study the behavior of social insects and the IT specialists. The base of that algorithm is the ant behavior and their ability to find the smallest way to a food source.

The ant colony is a distributed system. Despite the simplicity of its separate parts, that system can solve very sophisticated problems. Every single member of the colony tries to find the smallest way to a source of food. While doing that, it does not have an access to the knowledge of other members; therefore there should be a way that can help them to combine their knowledge. The ant ability to mark a route with pheromones is a way to combine their knowledge. If an ant finds a source of food, it marks its route using pheromones on the way back to the colony. The other ants will use that signal while searching for food. The more pheromones are used to mark the way the higher probability that an ant will choose that way in his search for food is [1].

That mechanism of self-organization became a base for the ant colony algorithm. The main idea of the algorithm is that the agents having the behavior that models the behavior of ants are united into a set to solve the optimization problem. The agents coordinate their work with the help of stigmergy which is a mechanism of the indirect collaboration using the alterations in the common environment. In case of ACO that mechanism is pheromones. The agents mark the traversed path with the help of pheromones increasing the probability of choosing that way among other variants. There is a mechanism which is known as the evaporation of the pheromones and it is used to prevent a situation when the algorithm goes irreversibly to the area of the local extremum. That mechanism is used to make the paths which were chosen as a solution by a mistake, less attractive by evaporating the pheromones on them. At the same time, the

© Springer International Publishing Switzerland 2015
Y. Tan et al. (Eds.): ICSI-CCI 2015, Part I, LNCS 9140, pp. 222–228, 2015.
DOI: 10.1007/978-3-319-20466-6_24

paths which were chosen by agents during the decision making process will increase their attractiveness and that should lead to a situation where all the agents will choose the general solution.

2   Generic Ant Colony Optimization Algorithm

Consider the generic ant colony optimization algorithm [2]:

1. Create some ants. The start point where an ant is placed depends on the restrictions imposed by a task. That is due to the way the ant placement is the most important factor for every task. They can be all placed at the same spot or different spots with or without repetitions. At the same stage of the algorithm, the start level of the pheromone is determined. It is initialized with a small positive value in order to achieve non-zero probabilities of jumping to the next node to the start of procedure.

2. Find solutions. The probability of jumping from a node i to a node j is defined by the following formulae:

$$p_{ij}(t) = \frac{[\tau_{ij}(t)]^\alpha \left[\frac{1}{d_{ij}}\right]^\beta}{\sum_{j \in allowedNodes} [\tau_{ij}(t)]^\alpha \left[\frac{1}{d_{ij}}\right]^\beta} \tag{1}$$

where $\tau_{ij}(t)$–the level of the pheromones, d_{ij} - a heuristic distance, a $\alpha u \beta$ -the constant parameters. If $\alpha = 0$, then the selection of the nearest node is the most probable, and that means that the algorithms becomes greedy. If $\beta = 0$, the choice occurs only when the pheromones are used for the selection process, and that leads to suboptimal solutions.

3. Refresh the levels of the pheromones. The level of the pheromones is refreshed according to the following formulae:

$$\tau_{ij}(t+1) = (1-\rho)\tau_{ij}(t) + \sum_{\substack{k \in antthat \\ usededge\,(i,j)}} \frac{Q}{L_k} \tag{2}$$

where ρ– the intensity of the evaporation, L_k– the cost of the current solution for k-th ant, Q–parameter describes the order of magnitude of the cost of the optimal solution, therefore $\frac{Q}{L_k}$ is a pheromone which is used by k-th ant to mark edge (i, j).

4. Additional actions. Usually the local search algorithm is used here.

5. Checking for the end of the search. In case of all restrictions were complied, the search process stops, otherwise return to step 1.

3   Ant Algorithm Modification for N-version Software

The N-version software structure forming is a task that can be solved using ACO. The N-version programming methodology is one of the most promising and already effectively used the methodology to build the fault-tolerant software [4-7]. This methodology is based on adding the redundant programming and it allows us to increase software reliability significantly. The N-version of the executed program modules assumes the generation of a set of functionally equal versions for each module, according to the specifications. The tools for concurrent execution are provided for all versions of a program module. The input data are identical for each version of a single module. The results of execution can vary due to a lot of reasons. The selection of the correct answer is calculated in the node of estimation and decision making.

Therefore, there is a task of choosing an optimal set of the program components using a set of criterions [8,9]. This task is a set covering problem (SCP) [10,11]. A $m \times n$ matrix $A = [a_{ij}]$ with all elements equal to 0 or 1 is given. Additionally, every column has the positive cost b_j. If $a_{ij} = 1$ that means that column j covers row i. The goal of SCP is to choose a set of columns with the minimal cost and to cover every row at the same time. Define j as a set of columns and y_j as a binary variable. If $j \in J y_j = 1$ else $y_j = 0$. Here is a formal definition of SCP:

$$min \quad f(y) = \sum_{j=1}^{n} b_j y_j \tag{3}$$

$$\sum_{j=1}^{n} a_{ij} y_j \geq 1, i = 1, \ldots, m \tag{4}$$

$$y_j \in \{0,1\}, j = 1, \ldots, n \tag{5}$$

The MAX-MIN ant system algorithm will be used for experiments because it is one of the most investigated and effective algorithms of the ant algorithm family. The main features of this algorithm are the existence of the high and low bounds for the pheromone level value as well as the existence of the pheromone level value renewing method. Only the best solution is counted and renewed.

4   Example of the Test Task

A schematic of a test N-version program is shown in fig. 1.

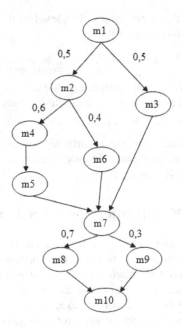

Fig. 1. A schematic of a test N-version program

The parameters of the versions for each module are presented in Table 1.

Table 1. Cost and reliability parameters of the versions for each module

Module	v1	...	v5	...	v10
Module 1	C=54 R=0,6	...	C=70 R=0,75	...	C=90 R=0,8
Module j	...	...	...	...	...
Module 10	C=20 R=0,5	...	C=40 R=0,57	...	C=65 R=0,68

Here is the algorithm:

1. Create ants. The amount of ants is equal the amount of the modules (N=10).
2. Set the minimal pheromone value for each version.
3. Create a minimal solution which contains every module. The probability of choosing the i-version is calculated according to the following formula:

$$\frac{\tau_i\left[\frac{P_i}{C_i}\right]^2}{\sum_{\substack{each\,version \\ in\,module}} \tau_i\left[\frac{P_i}{C_i}\right]^2} \tag{6}$$

4. Calculate the parameters of resulting solutions (P - reliability, C - cost). Check conditions. If one of the conditions is exceeded, then an agent is marked as bad one and it will not be used later.
5. Try to add a version of a module. If conditions are exceeded then agent completes its search.
6. Check for agents that have not completed their search. If they exist go to 5.
7. Compare agents that have finished their search and comply with the conditions with the global best solution. If a new one is better, replace the global best with a new one.
8. Renew the pheromone levels on the versions. Clear the agents' parameters.
9. Check for exceeding the maximal amount of iterations. If not, go to 3 [2].

Table 2. Results of the experiment

Iteration	Parameters
1	Cost: 1609Reliability: 0,952533008881286 Amount of versions: 33
198	Cost: 1353Reliability: 0,954864981124059 Amount of versions: 25

A set of experiments was conducted using the following parameters: $P_{min} = 0.95, C \rightarrow min$.The best and fastest result is presented in Table 2. Later, another set of experiments was conducted using different parameters $C_{max} = 1500, P \rightarrow max$.The best and fastest result is presented in Table 3.

Table 3. Results of the experiment

Iteration	Parameters
1	Cost: 1499Reliability: 0,9380492330715
	Amount of versions: 25
181	Cost: 1486Reliability: 0,961068949324989
	Amount of versions: 29

The standard formula 6 does not take into consideration the probability the of module usage and the amount of already chosen versions of that module. So formula 6 was changed:

$$p_{ij} = \frac{K_i \tau_{ij} \left[\frac{P_{ij}}{C_{ij}}\right]^2}{\sum_{\substack{notused \\ version}} \tau_i \left[\frac{P_{ij}}{C_{ij}}\right]^2} \tag{7}$$

$$K_i = P_i \frac{1}{n_i}, \tag{8}$$

where P_i is the probability that the I-th module will be used, n_i is the amount of already chosen versions of the I-th module.

A set of experiments was conducted using the following parameters: $P_{min} = 0.95, C \rightarrow min$.The best and fastest result is presented in Table 4.

Table 4. Results of the experiment

Iteration	Parameters
2	Cost: 1830Reliability: 0,95155703596872
	Amount of versions: 36
43	Cost: 1344Reliability: 0,950684456755364
	Amount of versions: 27
80	Cost: 1340Reliability: 0,953686684105021
	Amount of versions: 26
536	Cost: 1302Reliability: 0,952320253504339
	Amount of versions: 25

As it can be seen, the better result than the result provided by the standard algorithm was achieved on the *80-th* iteration. The best achieved result is shown in Table 5.

Table 5. Results of the experiment

Step	Parameters
1	Cost: 1776Reliability: 0,954688224332445
	Amount of versions: 34
8	Cost: 99Reliability: 0,951325746678937
	Amount of versions: 28
1884	Cost: 1268Reliability: 0,950307343800687
	Amount of versions: 25

It is worth to mention that even the worst result obtained using the modified algorithm is better than the best result obtained using the standard algorithm. Later, another set of experiments was conducted using different parameters $C_{max} = 1500$, $P \rightarrow max$.

The best and fastest result is presented in Table 6.

Table 6. Results of the experiment

Iteration	Parameters
1	Cost: 1465Reliability: 0,946156596647648
	Amount of versions: 26
32	Cost: 1490Reliability: 0,948035498484535
	Amount of versions: 27
112	Cost: 1480Reliability: 0,957518364561218
	Amount of versions: 27
1142	Cost: 1469Reliability: 0,963395416341579
	Amount of versions: 30
1651	Cost: 1488Reliability: 0,964324367894845
	Amount of versions: 27

As it can be seen, the better result than the result provided by the standard algorithm was achieved on the 118-th iteration. The best achieved result is shown in Table 7.

Table 7. Results of the experiment

Step	Parameters
1	Cos: 1435Reliability: 0,944523875543017
	Amount of versions: 27
406	Cost: 1497Reliability: 0,963332687235713
	Amount of versions: 29
1278	Cost: 1500Reliability: 0,96460060723887
	Amount of versions: 28

5   Conclusion

In this paper we propose a modification of the generic ant colony optimization algorithm. It allows solving the problem of multi-version software development. The modified algorithm shows the best results of the test task. Although those modifications slow the calculation process, the better solution finding compensates that disadvantage.

Moreover, the research is supported by the Ministry of Education and Science of the Russian Federation in accordance with the agreement № 14.574.21.0041

References

1. Dorigo, M., Stutzle, T.: Ant Colony Optimization. Massachusetts Institute of Technology (2004)
2. Dorigo, M., Birattari, M., Stitzle, T.: Ant Colony Optimization: Artificial Ants as a Computational Intelligence Technique. IEEE computational intelligence magazine (November 2006)
3. Dorigo, M., Di Caro, G., Gambardella, L.M.: Ant algorithm for discrete optimization. Artificial Life 5(2), 137–172 (1999)
4. Avizienis, A.: The N-Version approach to fault-tolerant software. IEEE Trans. on Software Engineering SE11(12), 1491–1501 (1985)
5. Tai, A., Meyer, J., Avizienis, A.: Performability Enhancement of Fault-Tolerant Software. IEEE Trans. on Reliability 42(2), 227–237 (1993)
6. Lyu, M.R.: Handbook of Software Reliability Engineering. In: Lyu, M.R. (ed.) IEEE Computer Society Press and McGraw-Hill Book Company (1996)
7. Lyu, M.R.: Software Fault Tolerance. In: Lyu, M.R. (ed.) John Wiley & Sons Ltd (1996)
8. Kovalev, I.V., Karaseva, M.V., Slobodin, M.J.: The mathematical system model for the problem of multi-version software design. In: Proceedings of AMSE International Conference on Modelling and Simulation, MS 2004. AMSE, French Research Council, CNRS, Rhone-Alpes Region, Hospitals of Lyon. Lyon-Villeurbanne (2004)
9. Kovalev, I.V.: Fault-tolerant software architecture creation model based on reliability evaluation / Kovalev, I.V., Younoussov, R.V.: Advanced in Modeling & Analysis, 48(3-4). Journal of AMSE Periodicals, 31–43 (2002)
10. Lessing, L., Dumitrescu, I., Stützle, T.: A comparison between ACO algorithms for the set covering problem. In: Dorigo, M., Birattari, M., Blum, C., Gambardella, L.M., Mondada, F., Stützle, T. (eds.) ANTS 2004. LNCS, vol. 3172, pp. 1–12. Springer, Heidelberg (2004)
11. Ren, Z., Feng, Z., Ke, L., Chang, H.: A fast and efficient ant colony optimization approach for the set covering problem. In: Proceedings of the IEEE Congress on Evolutionary Computation, CEC 2008 (IEEE World Congress on Computational Intelligence), pp. 1839–1844. IEEE (2008)

Artificial Bee Colony Algorithms

An Improved Artificial Bee Colony with Hybrid Strategies

Hong Yang$^{(\boxtimes)}$

Department of Computer, Hunan Normal University,
Changsha 410081, China
yanghonghnu@126.com

Abstract. In this paper, we propose an improved artificial bee colony (ABC) algorithm for function optimization. The new approach is called IABC, which employs two strategies to further enhance the performance of the original ABC algorithm. The first strategy utilizes the search information of the global best solution to guide the search of other bees, and the second one introduces a new solution updating model to generate candidate solutions. To test the performance of our algorithm, experiments are conducted a set of well-known functions. Computational results show that IABC achieves better performance than the original ABC and *gbest*-guided ABC (GABC) in terms of solution accuracy and convergence rate.

Keywords: Artificial bee colony (ABC) · Swarm intelligence · Hybrid strategies · Function optimization · Global optimization

1 Introduction

Optimization problems exists in different kinds of areas, including engineering design, project scheduling, economic dispatch, portfolio selection, and so on. With the development of society, optimization problems become more and more complex. Effective optimization algorithms are always required to suitable for solving the increasingly complex problems. In the past decades, several different nature-inspired algorithms have been proposed to solve various kinds of optimization problems, such as evolutionary algorithm (EA) [1], particle swarm optimization (PSO) [2], ant colony optimization (ACO) [3], artificial bee colony (ABC) [4], cat swarm optimization (CSO) [5], etc.

The ABC is a swam intelligence based algorithm, which simulates the foraging behavior of honey bees. Due to its simple concept, easy implementation yet effectiveness, it has bee successfully applied to solve many real-world and benchmark optimization problems [6–9]. Zhu and Kwong [8] proposed a *gbest*-guided ABC called GABC, in which the information of the global best solution (*gbest*) is utilized to guide the current search. Simulation results show that GABC performs better than the original ABC. Differs from the original ABC, Akay and Karaboga [9] introduced a parameter, called *MR* (0<*MR*<1), to control the frequency of perturbation. Experimental results show that this modification can effectively improve the

© Springer International Publishing Switzerland 2015
Y. Tan et al. (Eds.): ICSI-CCI 2015, Part I, LNCS 9140, pp. 231–238, 2015.
DOI: 10.1007/978-3-319-20466-6_25

performance of the original ABC. Inspired by the mutation strategy of differential evolution (DE) [10], Gao and Liu [11] presented a new ABC algorithm called ABC/best/1. Computational results show that the new approach outperforms the original ABC and GABC. To accelerate the convergence of ABC, ABC with an external archive is proposed [12]. In this approach, some best solutions are stored in an archive during the search process. When generating new solutions, some good solutions are selected from the archive to guide the search. Experiments demonstrate the effectiveness of this new algorithm. Wang et al. [13] presented a multi-strategy ensemble artificial bee colony (MEABC) algorithm. In MEABC, a pool of three distinct solution search equations are used to generate new solutions. Experiments are conducted a large set of benchmark functions. Results show that MEABC achieves better solutions than the original ABC, GABC, ABC/best/1, and several other PSO and DE algorithms. Kiran et al. [14] proposed another multi-strategy ABC algorithm, which integrates five search strategies to update solutions. A counter is used to determine the selection of strategies for the bees. Experiments on 28 numerical benchmark functions show that the new algorithm obtains better performance than several other recently proposed algorithms. Karaboga and Gorkemli [15] described a quick ABC (qABC) algorithm by defining a new behavior of onlooker bees of ABC. Comparisons between qABC and some popular algorithms show the effectiveness of the new approach.

In this paper, an improved ABC algorithm, called IABC, is proposed to enhance the performance of ABC. The IABC employs two strategies. The first strategy utilizes the search information of the global best solution to guide the search of other bees. The second strategy introduces a new solution updating model to generate candidate solutions.

The rest paper is organized as follows. Section 2 gives a brief introduction of the original ABC. Section 3 describes the proposed approach. Section 4 presents the experimental results and discussions. Finally, the work is summarized in Section 5.

2   Artificial Bee Colony (ABC)

In ABC, there are three kinds of bees: employed bees, onlooker bees, and scout bees. The number of employed bees is equal to half of the swarm size. The onlooker bees have the same size as the employed bees. The employed bees search new food in the neighborhood of food sources, and they share the search information of these food sources with the onlooker bees. The onlooker bees tend to select good food sources from those found by the employed bees, and then re-search the neighborhood of the selected food source. The scout bees can abandon their food sources and search new ones [8,11].

Each food source represents a candidate solution in the search space. Let $X_i=\{x_{i,1},x_{i,2},\ldots,x_{i,D}\}$ be the ith food source (solution) in the swarm, where D is the dimensional size. For each food source X_i, an employed searches its neighborhood to generate a new candidate solution V_i as follows [4]:

$$v_{i,j} = x_{i,j} + \phi_{i,j} \cdot \left(x_{i,j} - x_{k,j} \right). \tag{1}$$

where X_k is a randomly selected solution, which is different from X_i. The dimension index j is randomly generated with the range of $[1, D]$. The parameter $\phi_{i,j}$ is a random number with the range of $[-1,1]$. If the new solution V_i is better than its parent solution X_i, then update X_i with the new V_i.

After all employed bees complete the search, each onlooker bee begins to select a food source to conduct further search. The selection is based a probabilistic mode as described below [4]:

$$P_i = \frac{fit_i}{\sum_{j=1}^{SN} fit_j}. \tag{2}$$

where fit_i is the fitness value of the ith solution, and SN is the number of solutions in the swarm.

If a food source (solution) can not be improved under a predefined number of generations, the food source is abandoned. The predefined number is called *limit*, which is an important control parameter in ABC. For a abandoned food source X_i, the scout bee randomly generates a new solutions to replace it [4].

$$x_{i,j} = x_{min,j} + rand(0,1) \cdot \left(x_{max,j} - x_{min,j} \right). \tag{3}$$

where $rand(0,1)$ is a random number with the range of $[0,1]$, and $[X_{min}, X_{max}]$ is the search interval for a given problem.

3   Proposed Approach

It has been pointed out in [8] that new solutions in the original ABC are generated by updating only one dimension of their corresponding parent solutions. Then, there are many similarities between the parent solutions and offspring. As a result, the original ABC show slow convergence rate during the search process. To accelerate the search, Zhu and Kwong [8] defined a new solution search equation by utilizing the information of *gbest* to guide the search. The new solution search equation is described as follows [8]:

$$v_{i,j} = x_{i,j} + \phi_{i,j} \cdot \left(x_{i,j} - x_{k,j} \right) + \varphi_{i,j} \cdot \left(gbest - x_{i,j} \right). \tag{4}$$

where $\varphi_{i,j}$ is a random number in the range $[0,C]$, and C is a constant number. Based on the analysis of [8], the parameter $C=1.5$ is a good setting.

Updating only one dimension may not be suitable for the search of ABC. So, we propose a new method to decide how many dimensions to be updated in the solution search equation. First, we introduce a new parameter M, which determines the number of dimensions to be changed. In this paper, M is set to $0.1 \cdot D$, where D is the dimensional size. For each solution X_i, a new solution V_i is generated as described in Algorithm 1.

Algorithm 1. New solution search mechanism

Begin
  for j=1 to M **do**
    Update the jth dimension of X_i according to equation (4);
  end for
end

Table 1. Ten benchmark functions used in the experiments

Functions	Search range	Global optimum
$f_1 = \sum_{i=1}^{D} x_i^2$	[−100, 100]	0
$f_2 = \sum_{i=1}^{D} \|x_i\| + \prod_{i=1}^{D} x_i$	[−10, 10]	0
$f_3 = \sum_{i=1}^{D} \left(\sum_{j=1}^{i} x_j \right)^2$	[−100, 100]	0
$f_4 = \max_i \left(\|x_i\|, 1 \le i \le D \right)$	[−100, 100]	0
$f_5 = \sum_{i=1}^{D-1} \left[100(x_{i+1} - x_i^2)^2 + (x_i - 1)^2 \right]$	[−30, 30]	0
$f_6 = \sum_{i=1}^{D} \left(\lfloor x_i + 0.5 \rfloor \right)^2$	[−100, 100]	0
$f_7 = \sum_{i=1}^{D} i x_i^4 + rand[0,1)$	[−1.28, 1.28]	0
$f_8 = \sum_{i=1}^{D} -x_i \sin\left(\sqrt{\| x_i \|} \right)$	[−500, 500]	−418.98·D
$f_9 = \sum_{i=1}^{D} [x_i^2 - 10\cos(2\pi x_i) + 10]$	[−5.12, 5.12]	0
$f_{10} = -20\exp\left(-0.2\sqrt{\frac{1}{D}\sum_{i=1}^{D} x_i^2} \right) - \exp\left(\frac{1}{D}\sum_{i=1}^{D}\cos\left(2\pi x_i\right) \right) + 20 + e$	[−32,32]	0

The main steps of our new approach IABC algorithm are described as follows.

Step 1. Randomly initialize the swarm.

Step 2. Find the current *gbest*.

Step 3. For each employed bee, generate a new V_i according to Algorithm 1. Compute the fitness value of V_i. The better one between X_i and V_i is selected as the new X_i.

Step 4. Calculate p_i for each onlooker bee according to equation (2).

Step 5. Generate a new solution V_i according to Algorithm 1 based on p_i. Compute the fitness value of V_i. The better one between X_i and V_i is selected as the new X_i.

Step 6. The scout bee determines the abandoned X_i, if exists, update it by equation (3).

Step 7. Update the *gbest* found so far, and *cycle=cycle+1*.

Step 8. If the number of cycles reaches to the maximum value *MCN*, then stop the algorithm and output the results; otherwise go to Step 3.

Compared to the original ABC algorithm, our approach IABC does not add extra loop operations. Therefore, both IABC and the original ABC have the same computational complexity.

4   Experimental Study

In this section, we present an experimental study on the proposed IABC. There are ten well-known benchmark functions used in the following experiments [16–17]. The specific descriptions of the involved functions are listed in Table 1.

In the experiments, the IABC is compared with the original ABC and GABC for bout $D=30$ and 50. To have a fair competition, the same parameter settings are employed for common parameters. For the above three ABC variants, the swarm size *SN* and the parameter *limit* are set to 100, and 100, respectively. For GABC and IABC, the parameter C is set to 1.5 by the suggestions of [8]. For IABC, the parameter M is set to 0.1·D based on empirical studies. The maximum number of cycles (*MSN*) is set to 1000 for both $D=30$ and 50. All results reported in this paper are averaged on 30 independent runs.

Tables 2 and 3 present the comparison results of the three ABC algorithms for $D=30$ and $D=50$, respectively, where "Mean" indicates the mean best function values. As shown in Table 2, IABC outperforms ABC on all test functions except for f_5 and f_6. For f_5, ABC performs better than IABC. The three algorithms can find the global optimum for f_6. GABC achieves better results than IABC on f_5 and f_8. For the rest of 7 functions, IABC can search better solutions.

Table 2. Results achieved by ABC, GABC and IABC for $D=30$

Functions	ABC	GABC	IABC
	Mean	Mean	Mean
f_1	1.58E-09	7.54E-16	**2.87E-29**
f_2	2.59E-06	1.35E-11	**3.66E-17**
f_3	1.07E+04	7.71E+03	**4.65E+03**
f_4	4.57E+01	2.52E+01	**6.98E-01**
f_5	5.32E+00	**1.82E-01**	1.30E+01
f_6	**0.00E+00**	**0.00E+00**	**0.00E+00**
f_7	2.05E-01	8.07E-02	**3.25E-02**
f_8	-12209.8	**-12569.5**	-12451.3
f_9	6.62E-09	5.32E-15	**3.55E-15**
f_{10}	1.37E-05	9.45E-11	**2.54E-14**

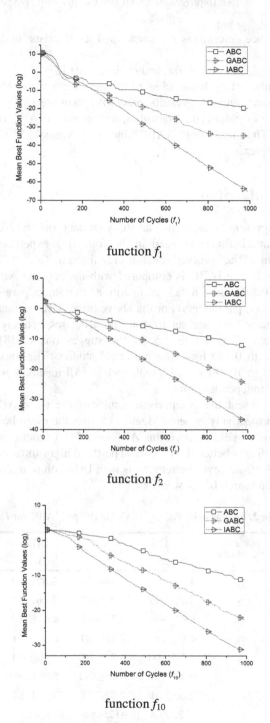

function f_1

function f_2

function f_{10}

Fig. 1. The convergence curves of ABC, GABC and IABC on three functions

When the dimension increases to 50, we use the same stopping condition for all algorithms. This helps to investigate the effects of dimension on the performance of these ABC algorithms. It can be seen that IABC still achieves better results than ABC and GABC on the majority of test function even if the dimension is 50.

Fig. 1 presents the convergence processes of ABC, GABC, and IABC on some representative functions. It can be seen that IABC shows faster convergence speed than ABC and GABC.

Table 3. Results achieved by ABC, GABC and IABC for D=50

Functions	ABC Mean	GABC Mean	IABC Mean
f_1	9.35E-06	6.71E-10	**1.09E-14**
f_2	1.42E-03	6.23E-06	**4.73E-09**
f_3	4.07E+04	3.77E+04	**3.20E+04**
f_4	7.42E+01	6.69E+01	**2.26E+01**
f_5	1.13E+01	**4.32E+00**	4.14E+01
f_6	**0.00E+00**	**0.00E+00**	**0.00E+00**
f_7	6.64E-01	3.06E-01	**1.29E-01**
f_8	-19389.2	-20197.2	**-20829.8**
f_9	5.49E+00	6.42E-03	**3.21E-03**
f_{10}	9.87E-03	2.76E-05	**3.78E-08**

5   Conclusions

In this paper, we present an improved ABC algorithm called IABC for numerical optimization. The new approach employs two strategies. The first strategy employs the new solution equation of GABC to guide the search of swarm. The second strategy determines how many dimensions to be updated when generating new candidate solutions. In order to verify the performance of IABC, ten well-known benchmark functions are utilized in the experiments. Computational results show that IABC performs better than the original ABC and GABC.

For the parameter M, we did not investigate its effects on the performance of IABC. How to adjust this parameter will be studied in the future work.

References

1. Bäck, T.: Evolutionary algorithms in theory and practice: evolution strategies, evolutionary programming, genetic algorithms. Oxford University Publisher, New York (1996)
2. Kennedy, J., Eberhart, R.C.: Particle swarm optimization. In: Proceedings of International Conference on Neural Networks, vol. IV, pp. 1942–1948. IEEE Press, Piscataway (1995)

3. Dorigo, M., Maniezzo, V., Colorni, A.: The Ant System: Optimization by a Colony of Cooperating Agents. IEEE Transactions on Systems, Man and Cybernetics-Part B: Cybernetics **26**(1), 29–41 (1996)
4. Karaboga, D.: An Idea Based on Honey Bee Swarm for Numerical Optimization. Technical Report-TR06, Erciyes University, Engineering Faculty, Computer engineering Department (2005)
5. Chu, S.C., Tsai, P.W.: Computational Intelligence Based on the Behavior of Cats. International Journal of Innovative Computing, Information and Control **3**, 163–173 (2007)
6. Yeh, W.C., Hsieh, T.J.: Solving reliability redundancy allocation problems using an artificial bee colony algorithm. Computers & Operations Research **38**, 1465–1473 (2011)
7. Szeto, W.Y., Wu, Y., Ho, S.C.: An artificial bee colony algorithm for the capacitated vehicle routing problem. European Journal of Operational Research. **215**(1), 126–135 (2011)
8. Zhu, W., Kwong, S.: Gbest-guided artificial bee colony algorithm for numerical function optimization. Applied Mathematics and Computation **217**, 3166–3173 (2010)
9. Akay, B., Karaboga, D.: A modified Artificial Bee Colony algorithm for real-parameter optimization. Information Sciences **192**, 120–142 (2012)
10. Storn, R., Price, K.: Differential evolution—a simple and efficient heuristic for global optimization over continuous spaces. Journal of Global Optimization **11**, 341–359 (1997)
11. Gao, W., Liu, S.: A modified artificial bee colony algorithm. Computers & Operations Research **39**, 687–697 (2012)
12. Wang, H., Wu, Z.J., Zhou, X.Y., Rahnamayan, S.: Accelerating artificial bee colony algorithm by using an external archive. In: Proceedings of the IEEE Congress on Evolutionary Computation, IEEE Press, Cancún, pp. 517–521 (2013)
13. Wang, H., Wu, Z.J., Rahnamayan, S., Sun, H., Liu, Y., Pan, J.Y.: Multi-strategy ensemble artificial bee colony algorithm. Information Sciences **279**, 587–603 (2014)
14. Karaboga, D.: Beyza Gorkemli, B.: A quick artificial bee colony (qABC) algorithm and its performance on optimization problems. Applied Soft Computing **23**, 227–238 (2014)
15. Kiran, M.S., Hakli, H., Gunduz, M., Uguz, H.: Artificial bee colony algorithm with variable search strategy for continuous optimization. Information Sciences **300**, 140–157 (2015)
16. Wang, H., Rahnamayan, S., Sun, H., Omran, M.G.H.: Gaussian bare-bones differential evolution. IEEE Transactions on Cybernetics **43**(2), 634–647 (2013)
17. Wang, H., Sun, H., Li, C.H., Rahnamayan, S., Pan, J.S.: Diversity enhanced particle swarm optimization with neighborhood search. Information Sciences **223**, 119–135 (2013)

An Artificial Bee Colony Algorithm
with History-Driven Scout Bees Phase

Xin Zhang[1] and Zhou Wu[2(✉)]

[1] College of Electronic and Communication Engineering,
Tianjin Normal University, Tianjin, China
xinzhang9-c@my.cityu.edu.hk
[2] Department of Electrical Electronic and Computer Engineering,
University of Pretoria, Hatfield, Pretoria, South Africa
wuzhsky@gmail.com

Abstract. The scout bees phase of artificial bee colony (ABC) algorithm emulates a random restart and cannot make sure the quality of the solution generated. Thus, we propose to use the entire search history to improve the quality of regenerated solutions, called history-driven scout bee ABC (HdABC). The proposed algorithm has been tested on a set of 28 test functions. Experimental results show that ABC cannot significantly outperforms HdABC on all functions; while HdABC significantly outperforms ABC in most test cases. Moreover, when the number of restarts increases, the performance of HdABC improves.

Keywords: Artificial bee colony · Search history · Binary space partitioning tree

1 Introduction

Artificial bee colony (ABC) is a simple and powerful metaheuristic for solving global optimization problems [1]. It is based on the intelligent behavior of honey bees. Many researchers have tried to improve its performance and make it better. For example, researchers propose to use the global best solution found so far to generate candidate solutions [2], [3]. Inspired by particle swarm optimization, Zhu et al. propose Gbest-guided ABC (GABC) algorithm [4]. Kang et al. propose Rosenbrock ABC which combines Rosenbrock's rotational direction method with the ABC algorithm [5]. Zhang et al. propose one-position inheritance and opposite directional search methods respectively for the employed bees phase and onlooker bees phase [6]. Karaboga et al. create a quick ABC algorithm which imitates the behavior of onlooker bees in a better way than standard ABC [7]. Kiran et al. modify ABC with a directed method [8]. Applications of discrete variants of ABC include [9], [10], [11], [12].

None of the above mentioned algorithms focus on the scout bees phase. Actually, the scout bees phase can be seen as a random search; the quality of the regenerated solution is unpredictable and low quality solution causes a waste of resources. Thus, we will concentrate on modifying the scout bees phase to improve the performance of the ABC algorithm. In this paper, we propose a novel ABC algorithm which uses the entire search history to improve the quality of solutions, called History-driven scout

© Springer International Publishing Switzerland 2015
Y. Tan et al. (Eds.): ICSI-CCI 2015, Part I, LNCS 9140, pp. 239–246, 2015.
DOI: 10.1007/978-3-319-20466-6_26

bee Artificial Bee Colony (HdABC). It has been noticed that history is a good reference to help the search. There are already some works on applying the entire search
history in the EA field [13], [14], [15]. In our proposed algorithm, we apply BSP tree
to improve the performance of the ABC algorithm.

The paper is organized as follows. Section 2 describes standard ABC algorithm.
Section 3 explains the proposed HdABC algorithm in detail. Experimental results are
shown in Section 4 and Section 5 gives the conclusion.

2   Artificial Bee Colony Algorithm

Standard ABC algorithm can be divided into four phases as follows:

1. Initialization Phase
   In the initialization phase, a population NP of solutions (food sources), i.e.,
   $\mathbf{x}_i = \{x_{i,1}, x_{i,2}, x_{i,3}, \ldots, x_{i,D}\}$, is initialized randomly in the search space.
2. Employed Bees phase
   Each employed bee randomly communicates with another employed bee to search
   a new location, i.e., $\mathbf{v}_i = \{v_{i,1}, v_{i,2}, v_{i,3}, \ldots, v_{i,D}\}$. The equation to generate new location is shown in (1).

$$v_{i,j} = x_{i,j} + \phi_{i,j}\{x_{i,j} - x_{k,j}\} . \tag{1}$$

where the indices $j \in \{1,2,\ldots,D\}$ and $k \in \{1,2,\ldots,NP\}, k \neq i$ are randomly generated. A coefficient $\phi_{i,j}$ is a random number between [-1, 1].

The employed bees evaluate the new food source $\mathbf{v}_i$, and compared with their current food source $\mathbf{x}_i$ by the fitness of solutions. The equation to calculate the fitness
is shown in (2), where $f(\mathbf{x}_i)$ represents the objective value of the solution $\mathbf{x}_i$.

$$fit(\mathbf{x}_i) = \begin{cases} \frac{1}{1+f(\mathbf{x}_i)}, & if\ f(\mathbf{x}_i) \geq 0 \\ 1 + abs(f(\mathbf{x}_i)), & if\ f(\mathbf{x}_i) < 0 \end{cases} \tag{2}$$

3. Onlooker Bees phase
   Onlooker bees receive the information from the employed bees, and make decision
   on selecting some food sources for further search. By using the equation shown in
   (3), the probability p_i is calculated by the fitness of the food sources. The onlooker bees go to the better food sources with higher probability.

$$p_i = \frac{fit_i}{\sum_{n=1}^{NP} fit_n} . \tag{3}$$

4. Scout bees phase
   During the search, some food sources will be abandoned. A user defined parameter
   limit is introduced to control when to abandon a food source. The employed bee of
   the abandoned food sources will become scout bee. A scout bee searches a new
   food source randomly to replace the abandoned food source.

3   The Proposed Algorithm

3.1   Idea of the Proposed Algorithm

In most of optimization problems, except that with illness condition, the landscape of function is generally smooth. Thus starting at any point, if we move towards the nearest optimal point, we may get a better solution. In this paper, all the evaluated solutions and their fitness are memorized. By these search history, we estimate the fitness landscape of the function. By that, the estimated local optimal point of any solution can be found. Then we can find the better restarted solutions by moving them towards the local optimal point. Therefore the efficiency of ABC algorithm can be improved.

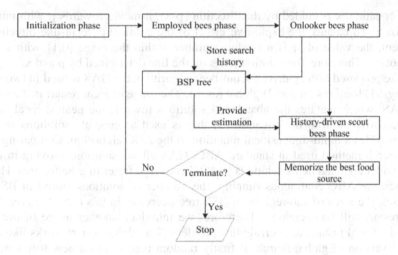

Fig. 1. Block diagram of HdABC

3.1   Structure of the Proposed Algorithm

The proposed algorithm has two main parts: 1) memory, and 2) history-driven scout bees algorithm. The function of the memory is to store the entire search history. The entire search history is used to build the estimated fitness landscape of the objective function (surrogate model), which can be used to estimate the fitness of solutions. Fig. 1 shows the block diagram of our proposed algorithm. In the beginning, the location of food sources (solutions) are randomly generated and evaluated by the objective function. Once a solution is evaluated, it will be stored with its fitness value in BSP tree. Then, starts the cycle of employed bees phase, onlooker bees phase and our proposed history-driven scout bees phase. The differences between standard ABC algorithm and our HdABC are that: HdABC stores all the evaluated solutions in BSP tree and substitutes history-driven scout bees phase for the scout bees phase.

3.2 History-Driven Scout Bees Algorithm

Binary space partitioning (BSP) tree is built as a surrogate model for providing fitness estimation. In [14], Chow and Yuen apply the idea of the nearest-neighbor search into BSP tree. They stored both the evaluated solution and their fitness into BSP tree. In this paper, we use the same tree as that in [14], [15].

To improve the quality of restarted solutions, the guided anisotropic search (GAS) module [14] is used in this paper. The GAS module is a novel parameter-less adaptive mutation operator, which applies a randomized gradient descent-like method to mutate solutions towards the better solutions. Equation (4) shows the equation used to generate the new solution $\mathbf{v}$ by using solution $\mathbf{x}$, and the corresponding local optimal point $\mathbf{p}$.

$$\mathbf{v} = \mathbf{x} + \alpha(\mathbf{p} - \mathbf{x}) . \tag{4}$$

In the equation, $\mathbf{x}$ is guided by the direction $(\mathbf{p}\text{-}\mathbf{x})$, it moves towards $\mathbf{p}$ with mutation step size α. To balance the exploitive effect of the gradient descent-like direction assignment, the value of α is a random number within the range [0,1] with uniform distribution. Therefore, the solution $\mathbf{v}$ lies on the line determined by $\mathbf{p}$ and $\mathbf{x}$.

In the proposed history-driven scout bees algorithm, the GAS is used in two different ways: 1) local restart and 2) global restart. The usage of local restart is the same as the GAS, which mutates the abandoned solutions towards the nearest local optimal region. Different from the equation (3) that is used to generate solutions in ABC, GAS provides a multi-dimensional mutation to the ABC algorithm. Comparing to the local search method used in standard ABC, GAS allows solutions moving in a relatively wider area and helps solutions in ABC converge faster to a better area. However, when the ABC continuous running, the number of solutions stored in BSP tree increases, the sizes of sub-regions in BSP tree decrease. In this case, the area for the local restart will be decreased. Therefore, we introduce another usage of the GAS, which is global restart, to override this problem. The global restart works like an enhanced version of global search. It firstly random regenerates a new solution in the whole search space, and then applies the GAS to mutate it towards to a better area, which helps the search to escape from the local optimal and ensures the new solution has a relatively better quality. Nevertheless, the function of global restart is mainly focus on preventing the premature convergence. It is expected that the improvement is not significant when there is only global restart applied into the ABC algorithm. Therefore, in the proposed history-driven scout bees algorithm, we introduce a parameter r to control the ratio of using local restart and global restart, where r is increasing with the number of generations:

$$r = \frac{number\ of\ generation}{Total\ number\ of\ generation} . \tag{5}$$

In each time a food source is abandoned, r will be calculated by (5). The history-driven restarted algorithm generates a random number $d \in [0, 1]$ to make decision on performing local restart or global restart, i.e.,

$$operation = \begin{cases} local\ restart, & if\ d \geq r \\ global\ restart, & if\ d < r \end{cases} . \tag{6}$$

In the beginning of the search, the value is relatively smaller. The history-driven scout bees algorithm concentrates more in local restart, which helps the search converge faster. Then, at the middle of the search, the ratio of performing local restart and global restart is 1:1, which balances the exploration and exportation in the search. At the later stage of the search, the search is converged. Therefore the history-driven scout bees algorithm focuses on the global restart to prevent the premature convergence.

4   Experiment

In this paper, 28 real valued benchmark functions f_1-f_{28} are used to test the performance of test algorithms as shown in Table 1. All the functions are tested in $D = 30$. Each test algorithm tries to solve each of the test function 30 times. For the maximum number of evaluation in each trial, we run experiment with two different values: one is $MFEs$=5000D, and the other is $MFEs$=10000D.

To verify the performance of the proposed HdABC algorithm, it is compared with standard ABC. To compare the performance of test functions, the Mann-Whitney Utest (U-test) is used. The detailed setup of the test algorithms is NP=25 and $limit$=100.The values of NP and $limit$ follow [2], [3], [6].

Table 1. Test function set

f_1	Sphereical	f_{15}	Pathological
f_2	Schwefel's Problem 2.22	f_{16}	InvertedCosineWave
f_3	Schwefel's Problem 1.02	f_{17}	InvertedCosineMixture
f_4	Schwefel's Problem 2.21	f_{18}	EpistaticMichalewicz
f_5	Rosenbrock	f_{19}	LevyMontalvo2
f_6	Quartic	f_{20}	Neumaier3
f_7	Generalized Rastrigin	f_{21}	OddSquare
f_8	Generalized Griewank	f_{22}	Paviani
f_9	Schwefel's Problem 2.26	f_{23}	Periodic
f_{10}	Ackley	f_{24}	Salomon
f_{11}	High Conditioned Elliptic	f_{25}	Shubert
f_{12}	Levy	f_{26}	Sinusoidal
f_{13}	Zakharov	f_{27}	Michalewicz
f_{14}	Alpine	f_{28}	Whitely

Tables 2 shows the average values of the best fitness, standard deviation and p-values of the test algorithms found in 30 independent runs. To increase the readability, the best results are typed in **boldface**. The performance of standard ABC is compared with HdABC by using U-test to determine the significance. The p-value shows the result of the U-test. By definition, the result is said to be significant if p<0.05. A marker "*" is added to the p-values which shows the result is significant. Table 8 and Table 9 illustrates the mean values of number of scout bees done by the test algorithms in 30 independent runs with $MFEs$ =150000 and $MFEs$ = 300000, respectively.

From Table 2, compared with ABC, the performance of HdABC is clearly better than ABC. HdABC obtains the best results in total 23 out of 28 functions. Among these 23 functions, the result is significant in 14 functions. Besides, HdABC performs worse than original ABC in only 5 functions, but all of them are not significant. It shows that HdABC outperforms standard ABC in most cases.

In fact, the difference between standard ABC and HdABC is that HdABC replaces the original scout bees phase to the proposed algorithm. Therefore, the results shown in Table 2 are directly reflecting the contribution made by the proposed algorithm. The proposed algorithm gives a significant improvement to original ABC, it makes the HdABC greatly outperforms standard ABC. To further prove that the proposed algorithm improves the original ABC, we have run another experiment with *MFEs* = 300000. In general, with the larger *MFEs*, the maximum number of generations will be larger. Thus more employed bees will be abandoned during the search, and then the contribution of our proposed algorithm should be shown clearly.

Table 2. The mean standard deviation (Std. D) and *p*-value (*p*) of the best function values found by ABC and HdABC in f_1–f_{28}, *MFES* = 5000*D*

	f_1	f_2	f_3	f_4	f_5	f_6
ABC	6.99E-16	3.55E-16	5.28E+03	3.89E+01	3.52E-01	1.20E+01
Std.D	*1.08E-16*	*8.46E-16*	*1.38E+03*	*5.86E+00*	*6.06E-01*	*5.17E-01*
HdABC	**2.38E-16**	**2.96E-16**	**5.09E+03**	**7.02E+00**	**1.99E-01**	**9.74E+00**
Std.D	*8.40E-17*	*6.62E-16*	*1.47E+03*	*1.50E+00*	*2.71E-01*	*6.17E-01*
p	**3.69E-11*	*9.64E-01*	*5.59E-01*	**3.02E-11*	*7.39E-01*	**3.69E-11*
	f_7	f_8	f_9	f_{10}	f_{11}	f_{12}
ABC	1.07E-14	**5.59E-14**	-1.38E+04	5.81E-14	7.51E-16	6.71E-16
Std.D	*2.29E-14*	*1.96E-13*	*1.63E-02*	*7.15E-15*	*1.61E-16*	*1.25E-16*
HdABC	**2.61E-15**	3.87E-13	-1.38E+04	**5.22E-14**	**7.02E-16**	**2.41E-16**
Std.D	*8.02E-15*	*1.31E-12*	*2.37E-02*	*6.40E-15*	*1.61E-16*	*1.05E-16*
p	**1.60E-03*	*7.73E-01*	*6.20E-01*	**2.00E-03*	*1.41E-01*	**3.69E-11*
	f_{13}	f_{14}	f_{15}	f_{16}	f_{17}	f_{18}
ABC	2.19E+02	**1.09E-08**	4.45E+00	-2.54E+01	-8.73E-16	-2.41E+01
Std.D	*3.02E+01*	*2.75E-08*	*3.41E-01*	*8.63E-01*	*2.69E-16*	*6.63E-01*
HdABC	**1.54E+02**	1.74E-08	**3.81E+00**	**-2.63E+01**	**-9.92E-16**	**-2.57E+01**
Std.D	*3.49E+01*	*5.68E-08*	*4.63E-01*	*1.04E+00*	*2.96E-16*	*6.03E-01*
p	**8.35E-08*	*8.53E-01*	**1.16E-07*	**5.56E-04*	*1.10E-01*	**4.20E-10*
	f_{19}	f_{20}	f_{21}	f_{22}	f_{23}	f_{24}
ABC	**4.20E-03**	-2.63E+03	-2.80E-03	-9.98E+05	1.00E+00	1.73E+00
Std.D	*2.17E-02*	*8.55E+02*	*1.17E-02*	*5.18E-10*	*1.26E-05*	*2.41E-01*
HdABC	3.35E-02	**-3.78E+03**	**-5.90E-03**	**-9.98E+05**	**1.00E+00**	**1.40E+00**
Std.D	*1.75E-01*	*5.42E+02*	*1.30E-02*	*5.37E-10*	*6.04E-06*	*1.49E-01*
p	*2.28E-01*	**7.04E-07*	*2.06E-01*	*2.49E-01*	*2.06E-01*	**7.60E-07*
	f_{25}	f_{26}	f_{27}	f_{28}		
ABC	**-2.32E+34**	-1.74E+00	-2.96E+01	1.33E+02		
Std.D	*5.05E+33*	*8.23E-01*	*2.17E-02*	*1.04E+02*		
HdABC	-2.29E+34	**-3.50E+00**	**-2.96E+01**	**1.30E+02**		
Std.D	*4.01E+33*	*1.70E-03*	*2.72E-02*	*9.89E+01*		
p	*8.42E-01*	**6.52E-09*	**4.51E-02*	*9.59E-01*		

The performance of ABC and HdABC in the case of *MFEs* = 300000 are not shown in this paper for the saving of space. The results show that ABC gives the same performance as HdABC in 3 functions. It illustrates that both of them obtain the global optimal solution. As it is no differences between ABC and HdABC in these 3 functions, we simply discard these results. Compared to ABC in the rest of 25 functions, HdABC obtains the best results in 22 out of 25 functions. Among these 22 functions, HdABC performs significantly better than ABC in 17 functions. In only 3 test functions, ABC outperforms HdABC, but the results of all of them are not significant. Compared to the results in Table 2, it is clearly shown that the number of the test functions that HdABC outperforms ABC is significantly increased, and the number of the test functions that HdABC performs worse than ABC is decreased. The experimental results prove that the proposed algorithm brings a positive effect to the original ABC.

Table 3. The mean of the number of scout bees done by ABC and HdABC in f_1–f_{28}, $MFES =$ 5000D and 10000D, respectively

5000D	f_1	f_2	f_3	f_4	f_5	f_6	f_7	f_8	f_9	f_{10}
ABC	50	25	2.87	559	4.87	312	25.1	30.3	2.43	25.1
HdABC	81.4	28	3.17	572	4.73	518	35.8	46.4	2	32.3
	f_{11}	f_{12}	f_{13}	f_{14}	f_{15}	f_{16}	f_{17}	f_{18}	f_{19}	f_{20}
ABC	25.1	50	15.8	0.2	97.1	37.7	50	102	20.1	15.3
HdABC	36.1	80.7	14.2	0.17	103	40.7	75.1	103	24.3	18.7
	f_{21}	f_{22}	f_{23}	f_{24}	f_{25}	f_{26}	f_{27}	f_{28}		
ABC	76.6	50	26.7	85.3	32.7	26.9	6.17	57.8		
HdABC	95.4	63.7	21.7	137	26	34.6	4.83	67.8		
10000D	f_1	f_2	f_3	f_4	f_5	f_6	f_7	f_8	f_9	f_{10}
ABC	96	48	8.7	982	21.6	591	68.1	72.6	26.2	48.3
HdABC	184	56.3	8.2	1030	24.9	1003	102	132	38.9	75.2
	f_{11}	f_{12}	f_{13}	f_{14}	f_{15}	f_{16}	f_{17}	f_{18}	f_{19}	f_{20}
ABC	72.1	96	31.1	24.2	193	78.5	96.1	198	48.7	34
HdABC	105	193	27.6	32.6	202	88.7	181	200	80.4	53.4
	f_{21}	f_{22}	f_{23}	f_{24}	f_{25}	f_{26}	f_{27}	f_{28}		
ABC	141	96	57.5	169	63.3	55.4	31	114		
HdABC	187	148	63.1	290	59.9	101	50.7	133		

Table 3 shows the mean of numbers of scout bees done by ABC and HdABC in the 28 test functions. Through comparison, it shows that with the larger value of *MFEs*, the number of scout bees is significantly increased.

5   Conclusion

History-driven scout bees Artificial Bee Colony (HdABC) algorithm is proposed. It uses a memory archive called Binary Space Partitioning tree to store the entire search history, and applies the Guided Anisotropic Search (GAS) module to find a better solution. The proposed algorithm contains two parts. One is for global search and the other is for local search. Experimental results show that the use of entire search history and the GAS module bring positive effects to ABC algorithm.

For the future direction, the GAS module can be applied to employed bees phase or onlooker bees phase to further improve the performance of ABC algorithm. Hybrid gravitational evolution [16], [17] or neighborhood field optimization [18] methods with ABC are also interesting.

References

1. Karaboga, D., Basturk, B.: A powerful and efficient algorithm for numerical function optimization: artificial bee colony (ABC) algorithms. Journal of Global Optimization. **39**, 459–471 (2007)
2. Diwold, K., Aderhold, A., Scheidler, A., Middendorf, M.: Performance evaluation of artificial bee colony optimization and new selection schemes. Memetic Computing. **3**, 149–162 (2011)
3. Zhang, X., Zhang, X., Ho, S.L., Fu, W.N.: A modification of artificial bee colony algorithm applied to loudspeaker design problem. IEEE Transactions on Magnetics. **50**, 737–740 (2014)
4. Zhu, G., Kwong, S.: Gbest-guided artificial bee colony algorithm for numerical function optimization. Applied Mathematics and Computation. **217**, 3166–3173 (2010)
5. Kang, F., Li, J., Ma, Z.: Rosenbrock artificial bee colony algorithm for accurate global optimization of numerical function. Information Sciences. **181**, 3509–3531 (2011)
6. Zhang, X., Zhang, X., Yuen, S.Y., Ho, S.L., Fu, W.N.: An improved artificial bee colony algorithm for optimal design of electromagnetic devices. IEEE Transactions on Magnetics. **49**, 4811–4816 (2013)
7. Karaboga, D., Gorkemli, B.: A quick artificial bee colony (qABC) algorithm and its performance on optimization problems. Applied Soft Computing **23**, 227–238 (2014)
8. Kiran, M.S., Findik, O.: A directed artificial bee colony algorithm. Applied Soft Computing **26**, 454–462 (2015)
9. Ozturk, C., Hancer, E., Karaboga, D.: Dynamic clustering with improved binary artificial bee colony algorithm. Applied Soft Computing **28**, 69–80 (2015)
10. Ozturk, C., Hancer, E., Karaboga, D.: Improved clustering criterion for image clustering with artificial bee colony algorithm. Pattern Analysis and Applications, 1–13 (2014)
11. Pan, Q.K., Wang, L., Li, J.Q., et al.: A novel discrete artificial bee colony algorithm for the hybrid flowshop scheduling problem with makespan inimisation. Omega **45**, 42–56 (2014)
12. Cui, Z., Gu, X.: An improved discrete artificial bee colony algorithm to minimize the makespan on hybrid flow shop problems. Neurocomputing **148**, 248–259 (2015)
13. Yuen, S.Y., Chow, C.K.: A genetic algorithm that adaptively mutates and never revisits. IEEE Transactions on Evolutionary Computation. **13**, 454–472 (2009)
14. Chow, C.K., Yuen, S.Y.: An Evolutionary Algorithm that Makes Decision Based on the Entire Previous Search History. IEEE Transactions on Evolutionary Computation **15**, 741–769 (2011)
15. Leung, S.W., Yuen, S.Y., Chow, C.K.: Parameter control system of evolutionary algorithm that is aided by the entire search history. Applied Soft Computing. **12**, 3063–3078 (2012)
16. Lou, Y., Li, J., Shi, Y., Jin, L.: Gravitational co-evolution and opposition-based optimization algorithm. International Journal of Computational Intelligence Systems **6**, 849–861 (2013)
17. Lou, Y., Li, J., Jin, L., Li, G.: A coEvolutionary algorithm based on elitism and gravitational evolution strategies. Journal of Computational Information Systems **8**, 2741–2750 (2012)
18. Wu, Z., Chow, T.W.S.: Neighborhood field for cooperative optimization. Soft Computing **17**, 819–834 (2013)

Multiobjective RFID Network Planning by Artificial Bee Colony Algorithm with Genetic Operators

Milan Tuba[1]([✉]), Nebojsa Bacanin[1], and Marko Beko[2]

[1] Faculty of Computer Science, Megatrend University,
Bulevar umetnosti 29, 11070 Belgrade, Serbia
tuba@ieee.org, nbacanin@megatrend.edu.rs
[2] Computer Science Engineering Department,
Universidade Lusòfona de Humanidades e Tecnologias, Lisbon, Portugal
marko@isr.ist.utl.pt

Abstract. This paper introduces genetically inspired artificial bee colony algorithm adapted for solving multiobjective radio frequency identification (RFID) network planning problem, which is a well-known hard optimization problem. Artificial bee colony swarm intelligence metaheuristic was successfully applied to a wide range of similar problems. In our proposed implementation, we incorporated genetic operators into the basic artificial bee colony algorithm to enhance the intensification process in the late iterations. Such improved version was previously tested and proved to be better than the basic variant of the artificial bee colony algorithm. In the practical experiments, we tested our proposed approach on six benchmark instances used in the literature, with clustered and random tag sets. In comparative analysis with other state-of-the-art approaches our proposed algorithm exhibited superior performance and potential for further improvements.

Keywords: RFID network planning · Artificial bee colony · Swarm intelligence · Metaheuristics · Multi-objective optimization

1 Introduction

Radio frequency identification (RFID) technology belongs to the group of relatively new technological achievements and its use has significantly increased during last ten years along with technical, economical and commercial development. The purpose of the RFID network is to enable transmission of data by a portable device called tag, which is read by RFID reader and processed according to the demands of a particular application [4].

Due to the limited range of a single reader, the problem of deployment of optimal RFID network emerges. Important questions arise, some of them interconnected and dependent, like: 1) how many readers are needed; 2) where should

Milan Tuba–This research is supported by Ministry of Education, Science and Technological Development of Republic of Srbia, Grant No. III-44006

© Springer International Publishing Switzerland 2015
Y. Tan et al. (Eds.): ICSI-CCI 2015, Part I, LNCS 9140, pp. 247–254, 2015.
DOI: 10.1007/978-3-319-20466-6_27

the readers be placed 3) what is the efficient parameter settings for each reader; 4) how to avoid readers collision etc. Thus, the deployment of RFID system generates RFID network planning problem (RNP) and since it belongs to the group of hard optimization problems, classical, non-deterministic methods could not obtain satisfactory results in a reasonable amount of time. This justifies metaheuristic approach for tackling the RNP problem. Swarm intelligence metaheuristics have been proven as successful optimizers in this area [12], [3], [11], [13], [8], [6], [5], [16].

In this paper, we adjusted artificial bee colony algorithm improved with genetic operators (GI-ABC) for tackling the RNP. GI-ABC proved to be better approach than the basic artificial bee colony (ABC) algorithm according to the conducted tests [1]. In this implementation, the intensification process is improved in the later stages of the algorithm by adapting uniform crossover and mutation operators from genetic algorithms (GA).

We tested the proposed GI-ABC on standard RNP benchmark instances and performed a comparative analysis using [7] as reference.

After Introduction, a mathematical model for multi-objective RFID network planning problem is presented in Section 2. Section 3 provides details about GI-ABC algorithm. Results and parameter settings, along with the comparative analysis, are given in Section 4. Finally, remarks and conclusion are presented in Section 5.

2   RFID Network Planning Problem Mathematical Model

The assignment of the RNP problem is to deploy RFID readers in the working domain while reaching the following goals: maximum tag coverage, minimum number of readers, minimum interference and minimum sum of transmitted (or radiated) power [7].

Obtaining maximum level of *tag coverage* is the main goal of the RNP problem. The power received by the tag is defined as:

$$P_t[dBm] = P_1[dBm] + G_r[dBi] + G_t[dBi] - L[dB], \tag{1}$$

where P_1 is reader's transmitted power, G_r and G_t are reader's and tag's antenna gains respectively, and L denotes attenuation factor calculated by Friis transformation [14]:

$$L[dB] = 10log[(4\pi/\lambda)^2 d^n] + \delta[dB], \tag{2}$$

where d is a physical distance between the reader and the tag, n is environmental factor that varies from 1.5 to 4 due to changes in physical conditions, while δ represents losses in wireless communication.

The power received by the reader can be calculated as:

$$P_r[dBm] = P_b[dBm] + G_r[dBi] + G_t[dBi] - 20log(4\pi d/\lambda), \tag{3}$$

where P_b represents backscatter power sent by the tag. P_b depends on the tag reflection coefficient Γ and on the tag received power P_t (in watts):

$$P_b = (\Gamma_{tag})^2 P_t \tag{4}$$

Finally, the coverage rate of a network is defined as:

$$COV = \sum_{t \in TS}^{max} Cv(t)/N_t \cdot 100\%, \tag{5}$$

where

$$CV_t = \begin{cases} 1 & if \; \ni r_1, r_2 \in RS, \; PT_{r_1,t} \geq T_t \wedge PR_{t,r_2} \geq T_r \\ 0 & otherwise, \end{cases} \tag{6}$$

where $N_t = |TS|$ represents the number of tags distributed in the working domain.

If the network coverage goal is achieved, the minimization of the *number of readers* becomes priority due to the fact that the network cost strongly depends on the number of deployed readers.

In situations when several readers interrogate the tag at the same time, *interference* could occur. As a consequence, misreading and lower level of quality of service happens. Therefore, the avoidance of interference represents an important goal in the RNP. Total amount of interference in the RFD network is calculated as the sum of interference levels of all deployed tags [7]:

$$INT = \sum_{t \in TS} \gamma(t), \tag{7}$$

where

$$\gamma(t) = \sum PT_{r,t} - max\{PT_{r,t}\}, \; r \in RS \wedge PT_{r,t} \geq T_t \tag{8}$$

The sum of the *transmitted power* of all readers should be reduced in the context of power saving. But, according to Eq. (3) and Eq. (1), low transmitted power could directly jeopardize the goal of obtaining tag coverage. Thus, this objective takes the lowest priority in the RNP. The sum of the transmitted power of all readers is defined as [7]:

$$SPOW = \sum_{r \in RS} PS_r, \tag{9}$$

where PS_r denotes the transmitted power of the reader r.

All of the mentioned objectives have to be satisfied simultaneously so the RFID network planning is a multi-objective optimization problem (MORNP).

3   Artificial Bee Colony Algorithm with Genetic Operators for the MORNP

The main source of inspiration for the ABC algorithm, which was devised by Karaboga [9], was the foraging behavior of honey bee swarms. ABC was implemented for many problems, such as global [15] and constrained [10] optimization, industrial problems [2], and others.

The intensification and diversification of the search space are guided by three types of artificial agents (bees): employed, onlookers and scouts. ABC creates initial population which consists of randomly distributed solutions. In every iteration, each employed bee in the population discovers a food source in its neighborhood. This process is modeled with the following equation [10]:

$$v_{i,j} = \begin{cases} x_{i,j} + \phi * (x_{i,j} - x_{k,j}), & R_j < MR \\ x_{i,j}, & otherwise \end{cases} \tag{10}$$

where $x_{i,j}$ is j-th parameter of the old solution i, $x_{k,j}$ is j-th parameter of a neighbor solution k, ϕ is a random number between 0 and 1, and MR is modification rate. MR is ABC control parameter that prevents algorithm to converge to suboptimal region of the search space, which is particularly important in early iterations.

When a neighborhood solution is found, its fitness is evaluated and if it is higher than the old one, new solution is retained in the population.

Upon completion of the intensification process employed bees share information about the quality of food sources with the onlookers. Onlookers select a food source i with a probability that is proportional to its fitness. Onlooker selection process is modeled as [10]:

$$p_i = \frac{fit_i}{\sum_{j=1}^{m} fit_j}, \tag{11}$$

where p_i represents the probability that the food source i will be selected, m is the total number of food sources and fit is the value of fitness.

According to Eq. (11), better food sources will attract more onlookers than the bad ones. When all onlookers select a food source for exploitation, they search around its neighborhood in the same way as employed bees Eq. (10).

As mentioned before, when an employed bee can not improve particular food source, it abandons it and becomes a scout. Abandoned food source is replaced with a random one. ABC control parameter $limit$ determines which solution will be abandoned.

By analyzing the original ABC algorithm a deficiency in the late phase of execution has been noticed [1]. After significant number of iterations, when optimal solution is almost found, scout bees, which perform the exploration process, are not useful any more and function evaluations are being wasted. This problem can be treated by better adjustment of exploration and exploitation balance.

To improve the exploitation process in the late iterations, uniform crossover and mutation operator from GA have been adopted during the replacement process of the exhausted food sources. Empirical point in algorithm's execution has been found where some of the scout bees are being replaced by a new class of agents named guided onlookers that perform strong exploitation around the current best solution [1].

To implement new mechanism, few additional control parameters were adopted. *Breakpoint* (abbr. *bp*) parameter defines the number of iterations after which the guided onlookers are triggered. *Replacement rate* (abbr. *rr*) defines the probability that scouts will be replaced by the guided onlookers. If replacement occurs, the best fit and one random individual are chosen from the population as parents for uniform crossover process. After crossover, mutation operator takes place. Each solution parameter is being mutated with certain probability *mutation probability rate* (abbr. *mpr*) according to [1]:

$$offsp[i] = offsp[i] + \phi_1 \cdot (rndsol[i] - offspr[i]), \qquad (12)$$

where $offsp$ is a child solution, $rndsol$ is random solution, and ϕ_1 is a random number in the range $[-0.1, 0.1]$. In this implementation we did not use *second break point* from [1].

Each bee in the population is represented as a real number vector with the dimension of $3M$, where M is the number of used RFID readers. Each reader is described by two coordinates of the readers' $2D$ position, and the third component represents transmitted power of that reader. All readers used in the benchmarks are mobile, and tags are static. The number of readers is also a parameter of optimization, however it was not included in the representation of bees. Multiobjective problems can be solved using different techniques. Reduction to single-objective problem by introducing weighted sum of all objectives is common but also rather criticized approach. For the MORNP we have a specific situation where objectives are clearly prioritized. Tag coverage is the ultimate goal; only when it is achieved reduction in the number of deployed readers becomes significant (without disturbing the tag coverage). Power reduction is possible only at the end of this optimization procedure, resulting at the same time in interference reduction. In such situation, weighted sum heavily favoring the tag coverage proved to be adequate. Reader number reduction was separately done considering the very limited search space for that parameter and the fact that change in the number of readers destroys intensification and effectively starts a new search.

4   Experimental Results

In the empirical testing we used six RNP instances: $C30$, $C50$, $C100$, $R30$, $R50$ and $R100$, the same ones that were used in [7]. All tests were performed in the working domain which was a square of the size $50\ m$ by $50\ m$, where $C30$ and $R30$ contain 30 tags, $C50$ and $R50$ contain 50 tags, and $C100$ and $R100$ have 100 tags. C instances are clustered distributed tags, while R instances have tags that are distributed uniformly, and as such are harder to solve. All benchmark instances are taken from the public URL: http://www.ai.sysu.edu.cn/GYJ/RFID/TII/.

For the sake of objective comparative analysis with [7], we used readers whose power is adjustable in the range of $[20, 33]$ dBm (0.1 to 2 watts). The power used in calculations was transmitted power, with antenna gain included, while some other papers use transmitted power. In the backscatter communication from tag to reader, wave length λ was set to 0.328 m (915 MHz). The receiver sensitivity thresholds for tags and readers are $T_t = -14\,dBm$ and $T_r = -80\,dBm$, with corresponding antenna gains of $G_t = 3.7\,dBi$ and $G_r = 6.7\,dBi$. We set δ to 2, n to 2, and Γ_{tag} to 0.3.

For the GI-ABC, the size of population N was set to 20, with 20,000 iterations (MIN), yielding total of 400,000 function evaluations. The same number of

Table 1. Experimental results

Algorithm	Mean				Best			
	Coverage	ReaderN	Interfer.	Power	Coverage	ReaderN	Interfer.	Power
Results for benchmark C30								
GPSO	100.00 %	6	0.000	35.074	100.00 %	6	0.000	31.865
VNPSO	100.00 %	6	0.000	34.762	100.00 %	6	0.000	31.951
GPSO-RNP	100.00 %	3.18	0.000	35.511	100.00 %	3	0.000	33.948
VNPSO-RNP	100.00 %	3.04	0.000	35.034	100.00 %	3	0.000	33.535
ABC	100.00 %	2	0.000	16.246	100.00 %	2	0.000	15.297
GI-ABC	100.00 %	2	0.000	**15.739**	100.00 %	2	0.000	**15.050**
Results for benchmark C50								
GPSO	95.60 %	6	0.000	35.170	100.00 %	6	0.000	31.852
VNPSO	99.20 %	6	0.000	35.023	100.00 %	6	0.000	31.742
GPSO-RNP	100.00 %	5.04	0.000	36.244	100.00 %	5	0.000	33.418
VNPSO-RNP	100.00 %	5.06	0.000	36.565	100.00 %	5	0.000	34.522
ABC	100.00 %	4	0.000	26.673	100.00 %	4	0.000	23.800
GI-ABC	100.00 %	4	0.000	**26.271**	100.00 %	4	0.000	**23.320**
Results for benchmark C100								
GPSO	98.34 %	6	0.002	38.652	100.00 %	6	0.000	37.374
VNPSO	99.72 %	6	0.000	38.167	100.00 %	6	0.000	36.803
GPSO-RNP	100.00 %	5.16	0.000	38.800	100.00 %	5	0.000	37.513
VNPSO-RNP	100.00 %	5.04	0.000	38.513	100.00 %	5	0.000	37.449
ABC	100.00 %	4	0.000	33.638	100.00 %	4	0.000	30.066
GI-ABC	100.00 %	4	0.000	**33.113**	100.00 %	4	0.000	**29.851**
Results for benchmark R30								
GPSO	92.13 %	6	0.000	38.849	100.00 %	6	0.000	38.842
VNPSO	94.53 %	6	0.000	38.849	100.00 %	6	0.000	38.655
GPSO-RNP	99.87 %	7.46	0.002	39.821	100.00 %	6	0.000	39.265
VNPSO-RNP	100.00 %	6.86	0.003	40.143	100.00 %	6	0.000	39.574
ABC	100.00 %	5	0.000	37.475	100.00 %	5	0.000	32.747
GI-ABC	100.00 %	5	0.000	**36.961**	100.00 %	5	0.000	**32.551**
Results for benchmark R50								
GPSO	92.52 %	6	0.000	39.692	98.00 %	6	0.000	40.520
VNPSO	93.96 %	6	0.000	39.690	98.00 %	6	0.000	39.595
GPSO-RNP	99.84 %	8.26	0.012	40.652	100.00 %	7	0.000	40.315
VNPSO-RNP	100.00 %	7.66	0.030	40.667	100.00 %	7	0.000	40.080
ABC	100.00 %	5	0.006	41.273	100.00 %	5	0.000	39.017
GI-ABC	100.00 %	5	**0.005**	**41.132**	100.00 %	5	0.000	**38.900**
Results for benchmark R100								
GPSO	91.18 %	6	0.014	40.074	95.00 %	6	0.000	40.098
VNPSO	94.14 %	6	0.012	40.333	97.00 %	6	0.043	40.657
GPSO-RNP	99.74 %	9.24	0.118	41.505	100.00 %	8	0.000	40.925
VNPSO-RNP	100.00 %	8.44	0.242	41.462	100.00 %	8	0.000	41.031
ABC	100.00 %	5	0.015	44.721	100.00 %	5	0.006	40.011
GI-ABC	100.00 %	5	0.015	**44.188**	100.00 %	5	0.006	**39.961**

evaluations was used in [7]. MR was set to 0.8, *limit* to 1000, *bp* to 3000, *rr* to 0.9 and *mpr* to 0.01. More details about parameter adjustment for the GI-ABC are available in [1].

A comparative analysis was performed with the pure ABC algorithm, GPSO (traditional PSO with the global topology), VNPSO (traditional PSO with the von Neumann topology), and GPSO-RNP and VNPSO-RNP as corresponding algorithms with incorporated tentative reader elimination (TRE) and mutation [7]. In this research we wanted to examine basic algorithm behavior so we did not include any elaborate TRE mechanism into the GI-ABC. However, simple mechanism described in Section 3 facilitated excellent results.

All experiments were conducted using 50 independent runs with different random number seeds. We show the best and mean values for all objectives. Treatment of objectives described in Section 3, completely satisfies real world requirements and leads to expected high quality results.

Table 1 shows experimental results. For easier comparison, best results from each category are marked bold. From Table 1 we conclude that GI-ABC obtains 100% tag coverage for all benchmark instances, hence the most important objective was perfectly satisfied, which other algorithms were not able to achieve. The number of readers is much better for our proposed algorithm in all cases. For the third objective, interference, GI-ABC obtains optimal results in all cases, except $R50$ and $R100$, but even in these cases interference is insignificant and better than the interference obtained by other compared algorithms. Finally, for the objective with the lowest priority, transmitted power, GI-ABC also exhibits best performance in all cases. In some cases significantly fewer readers were deployed and that facilitated huge improvements in transmitted power reduction. In other cases, transmitted power for our proposed algorithm was not significantly better since fewer readers had to work with higher power to cover the whole working area, but nevertheless the total power was lower.

5   Conclusion

In this paper, we presented adjustment of genetically inspired artificial bee colony (GI-ABC) for the RFID network planning problem. We used simple technique to control and reduce the number of readers while optimizing three other objectives. Proposed algorithm was tested on 6 standard RNP benchmark instances and comparative analysis with other state-of-the-art metaheuristics, which were tested on the same benchmarks, show that GI-ABC exhibited uniformly better performance. Future implementation of the more elaborated number of readers reduction mechanism will certainly make it even more superior algorithm across all objectives.

References

1. Bacanin, N., Tuba, M.: Artificial bee colony (ABC) algorithm for constrained optimization improved with genetic operators. Studies in Informatics and Control **21**(2), 137–146 (2012)
2. Brajevic, I., Tuba, M.: An upgraded artificial bee colony algorithm (ABC) for constrained optimization problems. Journal of Intelligent Manufacturing **24**(4), 729–740 (2013)
3. Chen, H., Zhu, Y., Hu, K.: Multi-colony bacteria foraging optimization with cell-to-cell communication for RFID network planning. Applied Soft Computing **10**, 539–547 (2010)
4. Chen, H., Zhu, Y., Hu, K., Ku, T.: RFID network planning using a multi-swarm optimizer. Journal of Network and Computer Applications **34**(3), 888–901 (2011)
5. Di Giampaolo, E., Forni, F., Marrocco, G.: RFID-network planning by particle swarm optimization. Applied Computational Electromagnetics Society Journal **25**(3), 263–272 (2010)
6. Gao, X., Gao, Y.: TDMA grouping based RFID network planning using hybrid differential evolution algorithm. In: Wang, F.L., Deng, H., Gao, Y., Lei, J. (eds.) AICI 2010, Part II. LNCS, vol. 6320, pp. 106–113. Springer, Heidelberg (2010)
7. Gong, Y.J., Shen, M., Zhang, J., Kaynak, O., Chen, W.N., Zhan, Z.H.: Optimizing RFID network planning by using a particle swarm optimization algorithm with redundant reader elimination. IEEE Transactions on Industrial Informatics **8**(4), 900–912 (2012)
8. Gu, Q., Yin, K., Niu, B., Chen, H.: RFID networks planning using BF-PSO. In: Huang, D.-S., Ma, J., Jo, K.-H., Gromiha, M.M. (eds.) ICIC 2012. LNCS, vol. 7390, pp. 181–188. Springer, Heidelberg (2012)
9. Karaboga, D.: An idea based on honey bee swarm for numerical optimization. Technical Report - TR06, pp. 1–10 (2005)
10. Karaboga, D., Akay, B.: A modified artificial bee colony (ABC) algorithm for constrained optimization problems. Applied Soft Computing **11**(3), 3021–3031 (2011)
11. Lu, S., Yu, S.: A fuzzy k-coverage approach for RFID network planning using plant growth simulation algorithm. Journal of Network and Computer Applications **39**, 280–291 (2014)
12. Ma, L., Chen, H., Hu, K., Zhu, Y.: Hierarchical artificial bee colony algorithm for RFID network planning optimization. The Scientific World Journal **2014**(Article ID 941532), 21 (2014)
13. Ma, L., Hu, K., Zhu, Y., Chen, H.: Cooperative artificial bee colony algorithm for multi-objective RFID network planning. Journal of Network and Computer Applications **42**, 143–162 (2014)
14. Rao, K.V.S., Nikitin, P.V., Lam, S.F.: Antenna design for UHF RFID tags: a review and a practical application. IEEE Transactions on Antennas and Propagation **53**(12), 3870–3876 (2005)
15. Subotic, M., Tuba, M.: Parallelized multiple swarm artificial bee colony algorithm (MS-ABC) for global optimization. Studies in Informatics and Control **23**(1), 117–126 (2014)
16. Yang, Y., Wu, Y., Xia, M., Qin, Z.: A RFID network planning method based on genetic algorithm. In: Proceedings of the International Conference on Networks Security, Wireless Communications and Trusted Computing, vol. 1, pp. 534–537 (2009)

Evolutionary and Genetic Algorithms

Genetic Algorithm Based Robust Layout Design
By Considering Various Demand Variations

Srisatja Vitayasak and Pupong Pongcharoen[✉]

Centre of Operations Research and Industrial Applications,
Department of Industrial Engineering, Faculty of Engineering, Naresuan University,
Phitsanulok 65000, Thailand
{srisatjav,pupongp}@nu.ac.th

Abstract. Placement of machines in a limited manufacturing area plays an important role to optimise manufacturing efficiency. Machine layout design (MLD) involves the arrangement of machines into shop floor area to optimise performance measures. The MLD problem is classified as Non-deterministic Polynomial-time hard (NP-hard) problem, in which, the amount of computation required to solve the NP-hard problem increases exponentially with problem size. In the manufacturing context, customers' demands are periodically varied and therefore have an influence on changing production flow between machines for each time-period. With high variation between periods, the volume of material flow changes significantly. Machine layout can be robustly designed under demand uncertainty over time period so that no machine movement is needed. The objective of this paper was to investigate the effect of five degrees of demand variation on Genetic Algorithm based robust layout design that minimises total material handling distance. The experimental results showed that the degrees of demand variation had significantly affected average material handling distance with 95% confident interval except the largest-size problem. Considering standard deviation, increasing in variability of material handling distance had resulted from the higher degrees of variation especially in the small-size problems. This suggested that designing the robust machine layout should recognise the variation of customer demand.

Keywords: Genetic algorithm · Robust layout · Stochastic demand · Variance demand

1 Introduction

Uncertainties in manufacturing environment, such as the change in product design, shorter product life cycles, elimination of existing products, and the introduction of new products, have effected customer's demand variation. Material flow intensities between machines on the manufacturing shop floor can be changed. The more customer demand is fluctuated, the higher material handling distance is varied. Tompkin et al. (2010) mentioned that about 20-50% of the total operating expenses within manufacturing are attributed to material handling. Material flow between machines relate to transportation distance and time on the shop floor area, productivity,

© Springer International Publishing Switzerland 2015
Y. Tan et al. (Eds.): ICSI-CCI 2015, Part I, LNCS 9140, pp. 257–265, 2015.
DOI: 10.1007/978-3-319-20466-6_28

production cost, and a competitive edge in the market. Material handling cost can be reduced by an effective facility layout design [1]. Machine layout design is placement of machines in a limited manufacturing area plays an important role to optimise manufacturing efficiency.

According to demand changes over time periods, machine layout design is classified as robust layout and re-layout. Re-layout process is to rearrange machines to minimise handling distances for each period but it is time consuming for machine repositioning and produce rearrangement costs [2, 3]. The layout can be robustly designed to overcome the rearrangement costs. Designing a robust layout is aimed at minimising the total material handling distance based on the predicted demands through a multi-period planning horizon. Profiles of product demand can be in the form of scenarios with different probabilities [4], forecast [5] or statistical distribution functions, such as uniform distribution [6], normal distribution [7], and exponential distribution [8]. Fuzzy number has also been used to introduce the stochastic flow between facilities, and the fuzzy cost has been represented by the triangular membership function [9]. However, there has been no report on the investigation of influence of demand variation on robust machine layout. The objective of this paper was to investigate the effect of degrees of demand variation on robust machine layout design that minimises total material handling distance based on multiple time-periods with uncertainty of demand on a planning horizon.

The remaining sections of this paper are organised as follows: Section 2 describes the Genetic Algorithm for solving the machine layout design (MLD) problem and its pseudo-code followed by the robust MLD under demand uncertainty in section 3; Section 4 presents the experimental design and analysis on computational results; and finally, discussions and conclusions are drawn in section 5.

2   Genetic Algorithm Based Robust Machine Layout Design Tool

MLD problems are categorised as NP-hard problems [10]. Solving these problems using full numerical methods especially for large size require the longer computational times. Amount of computational time increases exponentially with problem size [11]. In the contexts of operations research and computational intelligence, a computational method optimises a problem by iteratively trying to improve a candidate solution considering a given measure of quality. The approximation algorithms applied to solve the MLD problem, such as Genetic Algorithm [12, 13], Simulated Annealing [14], Ant Colony Optimisation [15], Bat Algorithm [16] and Backtracking Search Algorithm have been successfully applied to solve the MLD problem, but they do not guarantee the optimum solutions [17].

Genetic Algorithm (GA) [18, 19] is a biological based stochastic search algorithm for approximating the optimal solution in search space. Exploitation and exploration processes are carried out simultaneously via crossover and mutation operations, respectively. These features play an important role in terms of getting trap or escape from local optimal. The GA has been extensively applied to solve production and operations management problems [20]. The pseudo-code of the proposed GA for robust MLD shown in Fig. 1 [16] can be described as follows: i) encode the problem to produce a list of genes using a numeric string [21]. Each chromosome contains a number of genes, each representing a machine number so that the length of the chromosome is equal to the total number of machines needed to be arranged; ii) prepare input data: number of machines (M), dimension of machines (width: M_W x length: M_L), number of products (N), and their machine sequences (M_S), and identify parameters: population size (Pop), number of generations (Gen), probability of crossover (P_c), probability of mutation (P_m), floor length (F_L), floor width (F_W), gap between machines (G), and number of periods (P); iii) create the demand levels of each product in each period (D_{gk}). iv) randomly generate an initial population based on the defined Pop; iv) apply crossover and mutation operators to generate new offspring respecting P_c and P_m respectively; v) arrange machines row by row based on F_L and F_W; vi) evaluate the fitness function value; vii) select the best chromosome having the shortest material handling distance using the elitist selection; viii) choose chromosomes for the next generation by using roulette wheel selection; and ix) stop the GA process according to the number of generations. When the GA process is terminated, the best-so-far solution is reported.

It has been mentioned that the GA's parameters always play an important role on its performance [11]. The appropriate setting of P_c and P_m be set at 0.9 and 0.5, respectively [22]. The number of chromosomes and generations were 50. Genetic operators adopted in this work were the Two-point Centre Crossover (2PCX) and Two Operation Random Swap (2ORS) [23].

```
Input problem dataset (M, Mw, ML, Ms, N)
    Parameter setting (Pop, Gen, Pc, Pm, FL, Fw, G, P)
    Create demand level (Dgk) for each product associated with demand distribution
    Randomly create initial population (Pop)
    Set a  = 1 (first generation)
    While a ≤ Gen do
            For b =  1 to cross do (cross = round ((Pc x Pop)/2)))
                    Crossover operation
            End loop for b
            For c =  1 to mute do (mute = round(Pm x Pop))
                    Mutation operation
            End loop for c
            Arrange machines row by row based on FL , Fw and G
            For k = 1 to Number of periods (P) do
                    Calculate material handling distance based on demand level
(Dgk)
                    k = k+1
            End loop for k
            Selection of the best solution using elitist selection
            Chromosome selection using roulette wheel method
            a = a + 1
    End loop while
Output the best solution
```

Fig. 1. Pseudo code of GA for MLD with stochastic demand modified from Dapa et al. (2013) [16]

3   Robust Machine Layout Design Under Demand Uncertainty

Machines are usually designed in rectangular shape, different sizes and different models. The multi-row layout configuration is often found in the literature [24-28]. Arranging non-identical rectangular machines in multiple-row environment is where machines are placed row by row within a restricted area such as that shown in Fig. 2. Machines are placed in parallel row by row based on F_L and gap between machines. Flow path means the movement of material handling equipment, e.g. automated guided vehicles, which can move to left or right side of the row and then move up or down to the destination row. The distance of material flow was evaluated for the shortest distance such as transportation of materials from M4 to M11. There are two choices: route A or B. Because route A is shorter than B, thus they are transported with route A. The operated point of each machine is centroid.

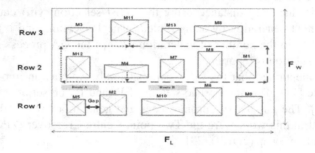

Fig. 2. Example of multiple-row machine layout design modified from Leechai et al. (2009) [26]

In this study, the following assumptions were made in order to formulate the problem: i) the material handling distance between machines was determined from the machine's centroid, ii) machines were arranged in multiple rows, iii) there was enough space in the shop floor area for machine arrangement, iv) the movement of material flow was a straight line, v) the gap between machines was predefined and similar, and vi) the processing time and moving time were not taken into consideration.

The evaluation function for the efficiency of robust layout design can be used to minimise total of the material handling distance (MHD) for all periods as shown in Eq. 1[13, 29].

$$\text{Total MHD of robust layout} = \sum_{i=1}^{M}\sum_{j=1}^{M}\sum_{g=1}^{N}\sum_{k=1}^{P} d_{ij} f_{ijgk} D_{gk} \tag{1}$$

M is a number of machines, i and j are machine indexes (i and j = 1, 2, 3,..., M). N is the number of product types, g is the product index (g = 1, 2, 3, ..., N). P is the number of time periods, k is the time period index (k = 1, 2, 3, ..., P), d_{ij} is the distance from machines i to j ($i \neq j$), f_{ijgk} is the frequency of material flow of product g from machines i to j in period k, and D_{gk} is the customer demand of product g in period k.

In order to investigate the degree of demand variation, demand profiles on each product type were proposed with normal distribution which is the most widely used model for the distribution of a random variable. The normal N (μ, σ^2) is used to denote a normal distribution with mean μ and variance σ^2. A normal distribution with μ = 0 and σ^2 = 1 is called a standard normal random variable. Demand profiles for experimental study were generated using five forms of normal distribution: $\mu \pm 0.5\sigma$, $\mu \pm 1\sigma$, $\mu \pm 1.5\sigma$, $\mu \pm 2\sigma$, $\mu \pm 2.5\sigma$ and $\mu \pm 3\sigma$. With a higher variance, the probability of product demand moves further from μ is increased as shown in Fig. 3.

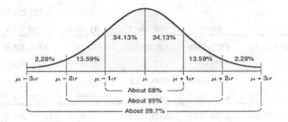

Fig. 3. Areas under a Normal distribution curve [30]

4   Experimental Design and Analysis on Computational Results

In this work, the computational experiments were conducted using eight testing datasets which have different numbers of non-identical machines and number of product types as shown in Table 1. Each dataset was tested with five degrees of demand variation: 0.5, 1, 1.5, 2, 2.5 and 3. The number of time periods was set to twelve periods.

Table 1. Testing datasets

Datasets	Number of machines (M)	Number of products (N)
10M5N	10	5
20M10N	20	10
20M20N	20	20
20M40N	20	40
30M15N	30	15
40M20N	40	20
40M40N	40	40
50M25N	50	25

With eight datasets, each of which took thirty replications and five degrees of variation, 1,200 computational runs in total were carried out. The machine layout designing program was developed and coded in modular style using the Tool Command Language and Tool Kit (Tcl/Tk) programming. An experiment was designed and conducted on a personal computer with Intel Core i5 2.8 GHz and 4 GB DDR3 RAM. The experiment was aimed at minimising the material handling distance. The computational results were analysed in terms of the minimum, maximum, mean and standard deviation (SD) as shown in Table 2.

Table 2. Values of total material handling distance (unit: metres) in each dataset

Data set	Value	Degree of demand variation						P-value
		0.5	1.0	1.5	2.0	2.5	3.0	
10 5N	Mean	798,455.6	778,054.3	752,610.2	741,292.1	716,177.0	715,278.2	0.000
	SD	12,661.0	13,814.9	8,504.2	12,157.3	14,097.9	13,807.1	
	Min	790,680.9	769,917.0	749,245.0	734,615.2	708,259.9	706,364.2	
	Max	826,500.9	805,887.3	783,688.3	773,198.6	749,327.3	745,071.4	
20M 10N	Mean	6,090,233.3	6,029,309.6	5,960,297.3	5,934,004.7	5,941,205.8	5,895,145.7	0.000
	SD	136,503.5	175,405.7	177,763.3	146,028.0	207,073.0	172,407.5	
	Min	5,852,922.4	5,640,605.0	5,548,461.6	5,640,703.8	5,458,338.5	5,563,535.2	
	Max	6,405,797.5	6,396,783.7	6,520,753.3	6,277,827.2	6,520,753.3	6,397,712.9	
20 20N	Mean	14,120,216.9	13,958,173.0	13,721,142.9	13,642,369.0	13,534,394.7	13,473,219.4	0.000
	SD	190,095.1	244,759.9	169,846.0	149,338.3	242,901.1	206,879.1	
	Min	13,770,010.8	13,567,935.1	13,357,533.5	13,154,515.8	13,022,012.2	12,967,093.1	
	Max	14,489,679.9	14,641,982.9	14,125,857.1	13,919,343.6	14,003,164.5	13,914,647.7	
20 40N	Mean	26,779,950.4	26,331,906.1	26,089,612.3	25,863,853.4	25,821,388.3	25,976,173.0	0.000
	SD	334,451.5	290,473.0	309,042.6	337,976.8	407,368.1	351,672.8	
	Min	26,091,179.4	25,744,858.1	25,497,690.8	25,182,286.5	25,091,339.0	25,395,546.0	
	Max	27,454,041.7	27,125,796.0	26,905,331.1	26,412,402.0	26,547,708.5	26,706,696.5	
30 15N	Mean	12,861,905.4	12,633,409.0	12,497,233.1	12,297,043.7	12,313,545.8	12,287,517.0	0.000
	SD	293,809.5	226,673.0	255,494.2	311,474.8	210,624.6	272,412.7	
	Min	12,262,049.3	12,264,522.9	12,017,526.3	11,710,277.6	11,987,376.8	11,769,353.3	
	Max	13,526,366.2	13,099,550.3	12,979,324.3	12,856,160.6	12,929,108.4	12,842,082.1	
40 20N	Mean	19,400,243.7	19,190,781.5	18,750,884.1	18,603,557.1	18,587,821.7	18,706,237.9	0.000
	SD	584,399.2	542,183.4	601,789.9	585,018.7	708,056.3	583,785.3	
	Min	18,336,836.3	18,102,922.0	17,590,348.5	17,416,714.9	16,880,685.0	17,366,065.2	
	Max	20,325,492.2	20,307,448.7	19,932,879.2	19,907,592.9	20,345,596.6	19,704,154.5	
40 40N	Mean	38,121,671.8	37,951,791.0	37,413,438.3	36,698,521.3	36,868,117.9	37,312,198.7	0.000
	SD	1,127,208.1	1,168,882.8	1,040,185.6	879,492.4	1,230,505.5	1,004,793.5	
	Min	35,797,412.2	35,503,204.0	34,948,160.6	35,530,237.0	35,119,602.3	35,341,525.0	
	Max	40,319,647.6	41,219,926.3	40,116,558.9	39,218,020.8	40,102,416.1	39,231,231.1	
50 25N	Mean	31,815,766.2	31,794,316.0	31,639,579.2	32,033,331.0	32,037,873.0	31,951,724.7	0.289
	SD	836,672.3	832,774.6	743,358.5	758,783.8	751,141.0	676,924.9	
	Min	30,202,549.6	29,823,536.0	29,616,562.0	30,692,224.0	30,357,873.7	31,042,342.4	
	Max	33,752,084.3	34,243,146.2	33,708,784.9	33,681,905.5	33,536,816.8	33,715,837.3	

From Table 2, the mean of material handling distance is shown as increasing when the problem size (number of machines and products) is larger. In each degree of variation, the problem dataset 40M40M had the highest values of mean and SD because of the number of machines and the type of products. When the number of machines is increased, the feasible solutions are increased. A variety of solutions had an effect on the standard variation. Also, the average computational time required to solve each dataset depends on the problem size. In case of a 1.0 degree of variation, the first three ranks of datasets according to the execution time were 40M40N, 50M25N and 20M40N which was about 32.5, 25.6, and 22.5 minutes, respectively, while 10M5N took only 1.7 minutes.

For each dataset, the distance decreased respecting to increasing in the degrees of variation because of demand values and machine position. The results were examined using the analysis of variance (ANOVA). The P value of ANOVA equaled 0.000 in all datasets except 50M25N, in which the P-values was 0.289. Degrees of variation significantly affected the material handling distance with 95% confident interval since the P values are less than 0.05 in almost all datasets. For 50M25N dataset, degree of variation had no statistically significant effect on the distance. Applying the student's t-test to compare the differences in mean of the distance within degrees of variations, the results showed that there were statistically significant differences (P values < 0.05) between 1.5 and 2.0, 1.5 and 2.5, and 2.0 and 2.5 degrees of variation. Influence of changes in demand values between periods relates to problem size and degree of variation. In datasets with 10 and 20 machines, when a 0.5 degree increased to 3.0, the SD value was higher. Wider variability in material handling distance had resulted from the higher degrees of demand variation especially in the small-size problems.

5   Discussion and Conclusions

This paper presents the investigation of the effect of five levels of stochastic demand on the designing a robust machine layout with shortest material handling distance. Demand profiles with five degrees of demand variation were based on normal distributions. The experimental results indicated that degrees of demand variation had statistically significant effect in material handling distance in almost all datasets. The wider fluctuation in customer demand between periods affected the existing layout. Robust layout for the largest-size problem (50M25N) can withstand demand variation. For other datasets, the layout may be redesigned in uncertain demand environment to maintain the shorter material handling distance. Future research can also investigate this kind of the effect on machine re-layout design, in which some machines can be repositioned at the end of time periods.

Acknowledgement. This work was part of the research project supported by the Naresuan University Research Fund under the grant number R2558C129.

References

1. Tompkins, J.A., White, J.A., Bozer, Y.A., Tanchoco, J.M.A.: Facilities Planning, 4th edn. John Wiley & Sons, Inc., New York (2010)
2. Balakrishnan, J.: Dynamic layout algorithms: a state-of-the-art survey. Omega-Int. J. Manage. S. **26**, 507–521 (1998)
3. McKendall, Jr., A.R., Shang, J., Kuppusamy, S.: Simulated Annealing heuristics for the dynamic facility layout problem. Comput. Oper. Res. **33**, 2431–2444 (2006)
4. McKendall, Jr., A.R., Hakobyan, A.: Heuristics for the dynamic facility layout problem with unequal-area departments. Eur. J. Oper. Res. **201**, 171–182 (2010)
5. Ertay, T., Ruan, D., Tuzkaya, U.R.: Integrating data envelopment analysis and analytic hierarchy for the facility layout design in manufacturing systems. Inf. Sci. **176**, 237–262 (2006)
6. Krishnan, K.K., Jithavech, I., Liao, H.: Mitigation of risk in facility layout design for single and multi-period problems. Int. J. Prod. Res. **47**, 5911–5940 (2009)
7. Tavakkoli-Moghaddam, R.S., Javadian, N., Javadi, B., Safaei, N.: Design of a facility layout problem in cellular manufacturing systems with stochastic demands. Appl. Math. Comput. **184**, 721–728 (2007)
8. Chan, W., Malmborg, C.J.: A Monte Carlo simulation based heuristic procedure for solving dynamic line layout problems for facilities using conventional material handling devices. Int. J. Prod. Res. **48**, 2937–2956 (2010)
9. Enea, M., Galante, G.M., Panascia, E.: The facility layout problem approached using a fuzzy model and a genetic search. J. Intell. Manuf. **16**, 303–316 (2005)
10. Loiola, E.M., de Abreu, N.M., Boaventura-Netto, P.O., Hahn, P., Querido, T.: A survey for the quadratic assignment problem. Eur. J. Oper. Res. **176**, 657–658 (2007)
11. Pongcharoen, P., Warattapop, C., Thapatsuwan, P.: Exploration of genetic parameters and operators through travelling salesman problem. Science Asia **33**, 215–222 (2007)
12. Jithavech, I., Krishnan, K.K.: A simulation-based approach for risk assessment of facility layout designs under stochastic product demands. Int. J. Adv. Manuf. Tech. **49**, 27–40 (2010)
13. Vitayasak, S., Pongcharoen, P.: Identifying Optimum Parameter Setting for Layout Design Via Experimental Design and Analysis. Adv. Mater. Res. **931-932**, 1626–1630 (2014)
14. Balakrishnan, J.: Soluition for the constrainted dynamic facility layout problem. Eur. J. Oper. Res. **57**, 280–286 (1992)
15. Corry, P., Kozan, E.: Ant Colony Optimisation for machine layout problems. Comput. Optim. Appl. **28**, 287–310 (2004)
16. Dapa, K., Loreungthup, P., Vitayasak, S., Pongcharoen, P.: Bat algorithm, genetic algorithm and shuffled frog leaping algorithm for designing machine layout. In: Ramanna, S., Lingras, P., Sombattheera, C., Krishna, A. (eds.) MIWAI 2013. LNCS, vol. 8271, pp. 59–68. Springer, Heidelberg (2013)
17. Pongcharoen, P., Hicks, C., Braiden, P.M., Stewardson, D.J.: Determining optimum Genetic Algorithm parameters for scheduling the manufacturing and assembly of complex products. Int. J. Prod. Econ. **78**, 311–322 (2002)
18. Gen, M., Cheng, R., Lin, L.: Network models and optimization: Multiobjective Genetic Algorithm approach (Decision engineering), 2008th edn. Spinger, London (2008)
19. Goldberg, D.: The design of innovation (Genetic Algorithms and evolutionary computation), 1st edn. Springer, London (2002)
20. Aytug, H., Knouja, M.J., Vergara, E.F.: Use of Genetic Algorithms to solve production and operations management problems: A review. Int. J. Prod. Res. **41**, 3955–4009 (2003)

21. Vitayasak, S., Pongcharoen, P.: Machine selection rules for designing multi-row rotatable machine layout considering rectangular-to-square ratio. J. Appl. Oper. Res. **5**, 48–55 (2013)
22. Vitayasak, S.: Multiple-row rotatable machine layout using Genetic Algorithm Research report (in Thai). Naresuan Univeristy, Phitsanulok (2011)
23. Vitayasak, S., Pongcharoen, P.: Interaction of crossover and mutation operations for designing non-rotatable machine layout. In: Operations Research Network Conference (2011)
24. Chiang, W.-C., Kouvelis, P., Urban, T.L.: Single- and multi-objective facility layout with workflow interference considerations. Eur. J. Oper. Res. **174**, 1414–1426 (2006)
25. Drira, A., Pierreval, H., Hajri-Gabouj, S.: Facility layout problems: A survey. Annu. Rev. Control **31**, 255–267 (2007)
26. Leechai, N., Iamtan, T., Pongcharoen, P.: Comparison on Rank-based Ant System and Shuffled Frog Leaping for design multiple row machine layout. SWU Engineering Journal **4**, 102–115 (2009)
27. Singh, S.P., Singh, V.K.: An improved heuristic approach for multi-objective facility layout problem. Int. J. Prod. Res. **48**, 1171–1194 (2010)
28. Sirinaovakul, B., Limudomsuk, T.: Maximum weight matching and Genetic Algorithm for fixed-shape facility layout problem. Int. J. Prod. Res. **45**, 2655–2672 (2007)
29. Vitayasak, S., Pongcharoen, P.: Backtracking Search Algorithm for designing a robust machine layout. WIT Trans. Eng. Sci. **95**, 411–420 (2014)
30. Bluman, A.G.: Elementary statistics: A brief version. 4th edn. McGraw-Hill (2008)

Design Index-Based Hedging: Bundled Loss Property and Hybrid Genetic Algorithm

Frank Xuyan Wang[✉]

Validus Research Inc., Waterloo, ON N2J1R1, Canada
frank.wang@validusresearch.com

Abstract. For index-based hedging design, the scatter plot of the hedging contract losses versus the to-be-hedged losses is generally used to visualize and quantify basis risk. While studying this scatter plot, which does not cluster along the diagonal as desired, a "bundled loss" phenomenon is found. In a setting where both the hedging and the hedged contracts have 100,000 years of simulated losses, this shows that if we need to hedge one loss in a year for the hedged contract, we may need to pay for other losses in other years in the hedging contract, which are unnecessary and unwanted. The reason is that the index used in the hedging may have identical loss values in different years while the hedged contract may not. This finding is a guiding principle for forming the risk measures and solution frameworks. To solve the problem so formed, a hybrid multi-parent and orthogonal crossover genetic algorithm, GA-MPC-OX, is used and pertinent adjustments are studied. For a problem with hundreds of dimensions, using eleven parents seems best, while a problem with tens of dimensions would prefer nine parents. Depending on the dimensions, relevant best strategies of the orthogonal crossover are also suggested by experimental results. To combat the stagnation of the algorithm, the perturbation by Lévy stable distribution is studied. This reveals possible effective parameters and forms. Numerical comparison with other algorithms is also conducted that confirms its competence for the hedging problem.

Keywords: Hedging problem · Genetic algorithm · Multi-parent crossover · Orthogonal crossover · Lévy stable distribution

1 Introduction

In the reinsurance industry, we frequently need to mimic a client company's losses by an insurance industry loss index, which are functions of the collective losses from all insurance companies across all geography, peril, and line of business. The latter is used to construct index-based hedging contracts for the client.

If the index loss is an accurate approximation of the client's actual loss, we should naturally expect their scatter plot closely clustering along the diagonal, their expected losses around the same, and their empirical CDF and PDF plots not far apart. More specifically, we would want the risk, as given by quantiles of the loss differences distribution for various probabilities, or by probabilities of these differences above

© Springer International Publishing Switzerland 2015
Y. Tan et al. (Eds.): ICSI-CCI 2015, Part I, LNCS 9140, pp. 266–275, 2015.
DOI: 10.1007/978-3-319-20466-6_29

given losses on condition that the client's losses greater than a list of thresholds, to be within some expected limit. Since these differences, especially where the losses from the client contract are above the losses from the index-based contract, are the residual risk of un-hedged losses. The expected loss of the index-based contract is a key determinant of the cost of the hedging. So we first attempt to quantify the effectiveness of the hedge. Next, we introduce methods to optimally balance the effectiveness and the cost of hedging.

The accompanying mathematical problem is finding the forms of the function used to construct the hedging, and the objective value function we should use to optimize the index, as well as which algorithm or what problem-related adjustment to the algorithm we should adopt for solving the hedging problem.

In a previous study of the insurance-linked securities portfolio optimization [1], a domain-specific property, that many of the candidate contracts are either the best or the worst and their contribution should be kept constant, is found. In it, a hybrid multi-parent crossover, orthogonal crossover genetic algorithm and catfish algorithm, GA-MPC-OX, which can utilize said property, is proposed. Its superiority is established through numerical comparison studies. Similarly, for the hedging problem, we found a "bundled loss" property, which worked as a guiding principle in forming our solution framework, as well as in selecting and evolving algorithms to solve it. We will then check its efficiency by comparing with results from using such algorithms as the Firefly, Bat, Cuckoo, Flower Algorithm [6], and the Wind Algorithm [5].

2   Hedging Problem Solution

2.1   Bundled Loss

In our experiment, the scatter plot of index loss vs. client loss never clusters along the diagonal within a narrow band. We always see the points spread out horizontally, such as in Fig.1.

A close examination of the points reveals a "bundled loss" phenomenon in the hedging problem, and this discovery gives us empirical rules on how to address these problems.

The "bundled loss" principle can be explained most clearly in the top end of the to-be-hedged client contract loss (called V), which occupies more than 30% of the non-zero loss years: all values are the same number, 10^9. To hedge the 10^9 loss of one of these years of V with a to-be-constructed industry loss portfolio, called the index or the hedging contract, the portfolio needs to have a loss of 10^9 for that year. At the same time, the portfolio (called Y, we will not differentiate between the portfolio and the index formed from it) will have many other years that have the same loss or almost the same loss as that year, possibly in the years where V has zero losses, since V has zero losses in 90% of the years. The additional losses of Y are the "bundled losses", for hedging the loss of the needed year of V, and will be the additional cost in the expected payoff of Y. Because of these bundled losses, we will see a horizontal scatter in the V-Y scatter plot: same Y but different V (the special case mentioned of V=0 is in the y-axis condensation shown in our results).

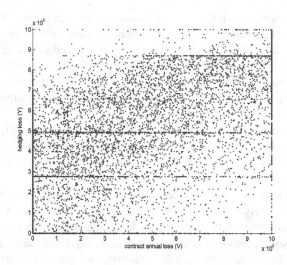

Fig. 1. Why does the scatter plot not cluster along the diagonal? The x-axis is the contract annual loss, and the y-axis is the hedging annual loss, in the same year, for a simulated set of 100,000 years

To overcome these bundled losses, we need to improve the industry portfolio discrimination power: the most ideal case is Y will have different losses for different years (and we can use an unlimited number of functions of Y to cut it into slim slices, which have value 0 outside of the narrow domains). To achieve this we can use two approaches:

- Using more industry loss contracts, such as by Country and by Cresta contracts; when that is not available, we can split the industry losses by risk group, and use them as our universe.
- Using multiple portfolios rather than just one to form Y, so that each portfolio may address different portions of the V loss range (In our study, we used 10 portfolios so our variable dimension is 480 plus the 10 weights of the portfolios); this strategy shifts the difficulty of the problem to that of the capability of the algorithm for finding high dimensional optimal solutions.

Since the scatter plot will always have some spill-out due to the bundled loss, we should not rely on it alone for deciding whether we get a good solution; we should use additional means for comparing or checking the results. Other than the CDF plot of V-Y on condition that V greater than some given threshold losses and the PDF plot of V and Y, we can also look at various numerical criteria.

When it is hard to discern from the scatter plot which solution is better, we can check the risk represented by 0.96, 0.99, 0.996 quantiles of abs(V-Y) (we used abs since the positive part means un-hedged loss and the negative part means over-hedged loss that is the additional cost); the count of years with abs(V-Y) falls into ten different sub-intervals of the range 0 to 10^9 with each band a length of 10^8; and the cost represented by E(Y), and vice versa. For example, for two solutions with similar risk (and scatter plot), the solution with smaller cost $E(Y)$ would be better. If two solutions have similar risk and cost, then the one with higher conditional CDF plot is better.

If we want the count of years when abs(V-Y) near the higher end of 10^9 to be small, the count of   years when abs(V-Y) near the lower end of 0 will be large due to the bundled loss. To get an overall number for this trade-off, we need a weighting scheme to sum up these counts. We can use this number as our objective, which tested to be more robust than the p-norm or any other forms, considering that a few years' losses may be outliers out of the simulation data generating process.

The 1-norm, 2-norm, or other higher p-norm appears to be more sensitive to outliers in the data, and will produce solutions with bouncing Y PDF while the original V PDF is very smooth and slowly-changing. Various weighting schemes are tested with some exponential form weighting function the best.

Using the count of years abs(V-Y) falling to different intervals in the objective value function will increase the stability, but may not differentiate solutions accurately. Two solutions with objective values so constructed, and differing by a few percent, may not have the property that the better solution will have the better objective value, especially when the function forms used are different. We then need to consider and compare all the different measures discussed above.

As for the forms or the formulas that define Y, we get these empirical principles: the best form mimics the V payoff function, i.e., using event limit, annual limit and big annual aggregate deductible (better than any piecewise linear, piecewise constant, power, or other highly nonlinear functions, such as combined min and max functions); when designing the multiple portfolios, each portfolio is better to cover a different loss range and has tens of times less non-zero loss years than V.

There is also another empirical finding of giving different weighting to V>Y and V<Y losses: if we consider the un-hedged V>Y portion as worse than the over-hedged portion where V<Y, and want to give it a relatively large factor, we can then find that we will reduce the un-hedged conditional probability such as $1-F(V-Y<=4e8|V>7.5e8)$. At the same time, it will increase the cost as given by $E(Y)$. The relationship between the factor and the probability is almost linear. For example, when using a factor of 1.1, we get a probability of 1-0.81 and when we use a factor of 1.35, we get a probability of 1-0.85.

When the objective values of solutions only differ by a small percentage, their risk measures and costs, as well as the scatter plots, CDF and PDF plots, will be similar. It seems that a decrease in some portions of the risk curve will be offset by increases in other portions. However, to reveal some emerging pattern in the solution form, the tiny difference matters. Only the best solution has a low enough noise to show the true figure.

The search for good Y form and a good algorithm is a reciprocal process. We do not want the noise in the algorithm to affect the decision about which form is better, and we want the adjustment of the algorithm to be pertinent to the objective function form. So we first fix several forms of the Y and test the algorithm; the settings that are constantly better are adopted, and with the new algorithm, we test more Y forms using the criteria of various plots and abs(V-Y) counts. This process is then repeated.

2.2   GA-MPC-OX Adjusted

In [1], the GA-MPC-OX algorithm, which performed better than any of the other algorithms tested for the portfolio optimization problem, such as PSwarm, MOEAD, ENSMOEAD, DyHF, CMODE, ICDE, PSO-DE, DSA, DECC-G, CoDE, ETLBO, OXDE, MBA, IRM-MEDA, TLBO, MMEA, RM-MEDA, ABC, IABC, NBIPOPaCMA, SHADE_CEC2013, DRMA-LSCh-CMA, and iCMAES-ILS, [10]-[30], is proposed. This prompts us to adapt it to the hedging problem.

Numerical experiments showed that the GA-MPC-OX can easily become stagnant, so we adjusted the number of parents used in its crossover operator. We found that the strategy of using nine parents generating nine children is best in a problem with 59 free-to-change variables, followed by eleven parents, and then by six, seven, or fourteen parents. These numbers seem related to the dimension, for example, for a problem with 490 dimensions, using eleven parents is best.

The Catfish Algorithm from [3] as used in [1] is akin to the dominance property of the portfolio optimization problem, and may not work in our hedging problem. Replacing it with the original normal perturbation operator from [4] produced better results.

The original interpolation method for orthogonal crossover in [2] tested better than [1]'s shortcut of table lookup method, so we followed the original method. However, instead of using Catfish Algorithm's method of taking candidates from the lower half, we took them from the upper half of the candidate pool. For the levels used, we tested methods of using increasing, decreasing, or equal probabilities of selecting a level from a number of levels. The three best strategies are using three levels with equal probabilities, always taking three levels, and using six levels with equal probabilities. But for higher dimensions, the last strategy seems best, followed by the first and then the second; adding randomness in selecting levels appears more effective.

With the four combinations of parent and levels numbers, when the best two for each were selected, for 58 dimensional problems, the precedence is (9,3),(9,6),(11,3), and (11,6). For 590 dimensional problems, the order is reversed.

Five other algorithms are used to solve the same problem as ours: the WDO [5], Firefly, Bat, Cuckoo Search, and Flower Pollination Algorithm [6]. In one form of the objective function, which is a weighted sum of the counts of the differences in loss belonging to different intervals, so that the smaller the objective value, the better the hedging should be, our algorithm finds the objective value after 100,000 function evaluations of 100,394, and after 1.4 million function evaluations of 64,169. The WDO gets the objective value after 100,000 function evaluations of 262,557, 2.61 times our number, and is stagnant after 40,000 function evaluations.

For another form of the objective function, our algorithm finds the objective value after 150,000 function evaluations of 109,551, after 200,000 function evaluations of 101,110, after 500,000 function evaluations of 85,781, and after one million function evaluations of 83,648. The Firefly Algorithm finds the objective value after 200,000 function evaluations of 372,697, stagnant after 150,000 function evaluations. The Bat Algorithm gets the objective value after 160,000 function evaluations of 327,017. The Cuckoo Search Algorithm gets the objective value after one million function evaluations of 130,405, 55.9% larger than that of 83,648. The Flower Pollination Algorithm gets the objective value after 500,000 function evaluations of 126,181, 47% larger than that of 85,781.

These comparisons may not absolutely show the superiority of our algorithm due to the example implementation of the other algorithms, but they do show the importance of the fine adjusting of the strategies and parameters used. Dr. Yang [6] emphasized the benefit of Lévy stable distribution, so we will try applying it to our algorithm.

2.3 Gauss or Lévy

The normal perturbation in the original GA-MPC algorithm is of the form $0.5U+0.25U*N$, where U is the uniform distribution in $(0,1)$ and N is the standard normal distribution. The Lévy flight perturbation Dr. Yang used is of the form $0.01N*S(1.5,0)*(x\text{-best})$, where $S(1.5,0)$ is the Lévy alpha-stable distribution (http://en.wikipedia.org/wiki/Stable_ distribution) with stability parameter 1.5, skewness parameter 0, scale parameter 1, and location parameter 0.

We performed three runs and saw one run using Lévy flight perturbation obtained better results than when using normal perturbation, while the other two runs were worse. It seems the Lévy flight has effect but it is not trivial to harness its power, or it is purely caused by chance and more due to the randomly selected initial population. We tested on the following additional forms of the perturbation: $U-0.5+0.25U*N$, $U-0.5+0.25U*\tan(\pi*(U-0.5))$, $U-0.5+0.25U*S(0.5,0)$, $0.5U+0.25U*\tan(\pi*(U-0.5))$, $0.5U+0.25U*S(0.5,0)$, $0.5U+0.25U*S(\alpha,0)$, $\alpha*N*S(0.5,1)$, $\alpha*(U-0.5)*S(0.5,1)$, and $\alpha*N*S(0.5,0)$, using the stable distribution code from [7], since it is faster than the other two implementations [8] and [9]. The test results are collected in Table 1.

Table 1. Effects Of Perturbation Forms

Perturbation Form	Objective Value
$0.5U+0.25U*N$	58369.3129228756[a]
$0.01N*S(1.5,0)*(x\text{-best})$	58588.2094140444
$U-0.5+0.25U*N$	58626.7825481966
$U-0.5+0.25U*\tan(\pi*(U-0.5))$	59253.6826758084
$U-0.5+0.25U*S(0.5,0)$, 1st run	58417.5946349593
$U-0.5+0.25U*S(0.5,0)$, 2nd run	58836.9978332785
$U-0.5+0.25U*S(0.5,0)$, 3rd run	58903.4202994694
$0.5U+0.25U*\tan(\pi*(U-0.5))$	58454.9400481151
$0.5U+0.25U*S(0.01,0)$	58533.4400598268
$0.5U+0.25U*S(0.05,0)$	58365.6438802964
$0.5U+0.25U*S(0.1,0)$	58487.5490576022

Table 1. (*Continued*)

Perturbation Form	Objective Value
0.5U+0.25U*S(0.3,0)	58846.7701952340
0.5U+0.25U*S(0.5,0)	58437.9835642726
0.5U+0.25U*S(0.7,0)	58921.7782284960
0.5U+0.25U*S(0.9,0)	58595.5569928767
0.5U+0.25U*S(1.01,0)	58721.6837996683
0.5U+0.25U*S(1.05,0)	58356.6284969962
0.5U+0.25U*S(1.1,0)	58396.0031511394
0.5U+0.25U*S(1.3,0)	58446.7420494971
0.5U+0.25U*S(1.5,0)	59143.4785428433
0.5U+0.25U*S(1.7,0)	58577.2452540111
0.5U+0.25U*S(1.9,0)	58334.9363537724
0.5U+0.25U*S(1.95,0)	58390.6540507207
0.5U+0.25U*S(1.99,0)	58798.0866233763
0.01*N*S(0.5,1)	58623.7008885335
0.01*(U-0.5)*S(0.5,1)	59379.2371618059
0.01*N*S(0.5,0)	58489.2841485367
0.05*N*S(0.5,0)	58782.6471115213
0.075*N*S(0.5,0)	58468.6686062425
0.1*N*S(0.5,0)	58412.1454721504
0.15*N*S(0.5,0)	58472.1776203138
0.2*N*S(0.5,0)	58399.2953524066
0.25*N*S(0.5,0)	58465.769182205
0.3*N*S(0.5,0)	58874.2030665987
0.35*N*S(0.5,0)	58653.8876928435
0.4*N*S(0.5,0)	59082.81743882
0.45*N*S(0.5,0)	58556.6515104805
0.5*N*S(0.5,0)	59047.7318800064

Table 1. (*Continued*)

Perturbation Form	Objective Value
1*N*S(0.5,0)	59020.9476755368
0.1*N*S(0.05,0)	58442.6398141161
0.1*N*S(1.05,0)	58436.1462185919
0.1*N*S(1.5,0)	59154.7577594512
0.1*N*S(1.9,0)	59028.6114845251

[a.] 58 dimensional problem, used 9-parent MPC and 3-levels orthogonal crossover operators.

Out of all the tested cases, the Gauss or normal distribution used by the original GA-MPC is at the higher quantile end, outperformed only by three cases that used general Lévy alpha-stable distribution for which the stability parameter α is near the Gauss end 2, or the Cauchy end 1, or near 0: 1.9, 1.05, and 0.05. The middle point of (1,2) 1.5 was the worst for that range, but 0.5 was the second best for the interval (0,1). Adding the symmetry perturbation term U-0.5 was not as good as adding the shifted-up term 0.5U, for the hedging problem: using more weights would match more losses with added costs. It may also be possible that our cases are mainly stochastic noises, and more experiments are needed for a definite conclusion.

3   Conclusion

For the hedging problem, a bundled loss property is found, which explains why the scatter plot is always blurred and cannot be used for the fine selection of the solution, except when the algorithms used are too inefficient or solutions found deviate too much from each other. When we cannot distinguish two solutions by their scatter plot, we can still differentiate between them by other means, such as using weighted counts of their differences for objective value, conditional probability plots, and etc. This property also guided us in adjusting the hybrid multi-parent, orthogonal crossover genetic algorithm GA-MPC-OX for the hedging problem, which tested far better than several example algorithms that may have not been fine-tuned or problem-tuned for performance. The normal perturbation used in the GA-MPC-OX generally performed well, but can be surpassed by some parameter Lévy alpha-stable distribution in some tests. Studies suggest some parameter values are effective. Out of many numerical tests, the following four parameters or combinations work better in more of the test cases, if not always, than all other tested combinations: 0.2*U*S(1.9,0),

0.2*U*S(0.5,0) or 0.25*U*S(1.9,0) or 0.2*U*S(1.9,0)*(x-b) in equal probability,
0.25*U*S(1.9,0) or 0.2*U*S(1.9,0) or 0.2*U*S(0.5,0)*(x-b) in equal probability,
0.25*U*S(1.9,0) or 0.2*U*S(1.9,0) or 0.2*U*S(0.5,0)*(x-b) or 0.5*U+0.2*S(1.9,0) or 0.5*U+0.2*U*S(0.5,0) in equal probability. However, a perturbation scheme that always performs better still requires more research.

Acknowledgments. F.X.Wang thanks Dr. Lixin Zeng for forming the hedging problem framework and guiding the research direction through questions, comments, suggestions, and expectations. All this work was done to fill in the details of that framework, answer the aforementioned questions, and explain the observed discrepancy.

References

1. Wang, F.X.: Relay Optimization Method (May 2014). http://www.optimization-online.org/DB_FILE/2014/05/4345.pdf
2. Wang, Y., Cai, Z., Zhang, Q.: Enhancing the search ability of differential evolution through orthogonal crossover. Information Sciences **185**(1), 153–177 (2012)
3. Chuang, L.Y., Tsai, S.W., Yang, C.H.: Chaotic catfish particle swarm optimization for solving global numerical optimization problems. Applied Mathematics and Computation **217**, 6900–6916 (2011)
4. Elsayed, S.M., Sarker, R.A., Essam, D.L.: GA with a new multi-parent crossover for solving IEEE-CEC 2011 competition problems. In: Proc. IEEE Congr. Evol. Comput. (CEC), pp. 1034–1040 (2011)
5. Bayraktar, Z., Komurcu, M., Bossard, J.A., Werner, D.H.: The Wind Driven Optimization Technique and its Application in Electromagnetics. IEEE Transactions on Antennas and Propagation 61(5), 2745–2757 (2013). http://wdo.cloudturkiye.com/wdo_matlab_03.m
6. Yang, X.S.: Nature-Inspired Metaheuristic Algorithms, 2nd edn. Luniver Press (2010). http://www.mathworks.com/matlabcentral/fileexchange/authors/119376
7. Weron, R.: STABLERND: MATLAB function to generate random numbers from the stable distribution (April 26, 2010). http://ideas.repec.org/c/boc/bocode/m429003.html
8. McCulloch, J.H.: Stable Random Number Generator (December 18, 1996). http://www.econ.ohio-state.edu/jhm/programs/STABRND.M
9. Veillette, M.: STBL: Alpha stable distributions for MATLAB (July 16, 2012). http://www.mathworks.com/matlabcentral/fileexchange/37514-stbl–alpha-stable-distributions-for-matlab
10. Vaz, A.I.F., Vicente, L.N.: A particle swarm pattern search method for bound constrained global optimization. Journal of Global Optimization **39,** 197–219 (2007). http://www.norg.uminho.pt/aivaz/pswarm/
11. Zhang, Q., Li, H.: MOEA/D: A Multi-objective Evolutionary Algorithm Based on Decomposition. IEEE Trans. on Evolutionary Computation **11**(6), 712–731(2007)
12. Zhang, Q., Liu, W., Li, H.: The Performance of a New Version of MOEA/D on CEC 2009 Unconstrained MOP Test Instances. Working Report CES-491, School of CS & EE. University of Essex (February 2009)
13. Zhao, S.Z., Suganthan, P.N., Zhang, Q.: MOEA/D with an Ensemble of Neighbourhood Sizes. IEEE Trans. on Evolutionary Computation (TEC) 16(3), 442–446 (2012)
14. Wang, Y., Cai, Z.: A dynamic hybrid framework for constrained evolutionary optimization. IEEE Transactions on Systems, Man, and Cybernetics, Part B: Cybernetics **42**(1), 203–217 (2012)
15. Wang, Y., Cai, Z.: Combining multiobjective optimization with differential evolution to solve constrained optimization problems. IEEE Transactions on Evolutionary Computation **16**(1), 117–134 (2012)
16. Jia, G., Wang, Y., Cai, Z., Jin, Y.: An improved $(\mu+\lambda)$-constrained differential evolution for constrained optimization. Information Sciences **222**, 302–322 (2013)

17. Liu, H., Cai, Z., Wang, Y.: Hybridizing particle swarm optimization with differential evolution for constrained numerical and engineering optimization. Applied Soft Computing **10**(2), 629–640 (2010)
18. Civicioglu, P.: Transforming Geocentric Cartesian Coordinates to Geodetic Coordinates by Using Differential Search Algorithm. Computers and Geosciences 46, 229–247 (2012). http://www.pinarcivicioglu.com/ds.html
19. Yang, Z., Tang, K., Yao, X.: Large Scale Evolutionary Optimization Using Cooperative Coevolution. Information Sciences **178**(15), 2985–2999 (2008)
20. Yang, Z., Tang, K., Yao, X.: Self-adaptive differential evolution with neighborhood search. In: Proceedings of the 2008 IEEE Congress on Evolutionary Computation (CEC 2008), Hongkong, China (2008)
21. Wang, Y., Cai, Z., Zhang, Q.: Differential evolution with composite trial vector generation strategies and control parameters. IEEE Transactions on Evolutionary Computation **15**(1), 55–66 (2011)
22. Rao, R.V., Patel, V.: An elitist teaching-learning-based optimization algorithm for solving complex constrained optimization problems. International Journal of Industrial Engineering Computations 3(4), 535–560(2012). https://sites.google.com/site/tlborao/
23. Wang, Y., Cai, Z., Zhang, Q.: Enhancing the search ability of differential evolution through orthogonal crossover. Information Sciences **185**(1), 153–177 (2012)
24. Sadollah, A., Bahreininejad, A., Eskandar, H., Hamdi, M.: Mine blast algorithm: A new population based algorithm for solving constrained engineering optimization problems. Applied Soft Computing **13**(5), 2592–2612 (2013)
25. Wang, Y., Xiang, J., Cai, Z.: A regularity model-based multiobjective estimation of distribution algorithm with reducing redundant cluster operator. Applied Soft Computing **12**(11), 3526–3538 (2012)
26. Zhou, A., Zhang, Q., Jin, Y.: Approximating the Set of Pareto Optimal Solutions in Both the Decision and Objective Spaces by an Estimation of Distribution Algorithm. IEEE Trans. on Evolutionary Computation **13**(5), 1167–1189 (2009)
27. Zhang, Q., Zhou, A., Jin, Y.: RM-MEDA: A Regularity Model Based Multiobjective Estimation of Distribution Algorithm. IEEE Trans. on Evolutionary Computation **12**(1), 41–63 (2008)
28. Karaboga, D., Basturk, B.: A powerful and Efficient Algorithm for Numerical Function Optimization: Artificial Bee Colony (ABC) Algorithm. Journal of Global Optimization 39(3), 459–171 (2007). http://www.mathworks.com/matlabcentral/fileexchange/27125-solution-to-economic-dispatch-by-artificial-bee-colony-algorithm/content/ABC-eld/runABC.m
29. Tsai, P.W., Pan, J.S., Liao, B.Y., Chu, S.C.: Enhanced artificial bee colony optimization. International Journal of Innovative Computing, Information and Control **5**(12), 5081–5092 (2009)
30. Special Session & Competition on Real-Parameter Single Objective Optimization at CEC 2013, June 21-23, Cancun, Mexico (2013). http://www.ntu.edu.sg/home/EPNSugan/index_files/CEC2013/CEC2013.htm

Economic Load Dispatch of Power System Using Genetic Algorithm with Valve Point Effect

Awodiji Olurotimi Olakunle$^{(\boxtimes)}$ and Komla A. Folly

Department of Electrical Engineering, University of Cape Town, Cape Town, South Africa
awdolu002@myuct.ac.za, komla.folly@uct.ac.za

Abstract. This paper presents the solution of economic load dispatch problem using quadratic cost functions with valve point effect by means of Genetic Algorithm (GA). GA technique is particularly useful for optimization problems with non-convex, discontinuous and non-differentiable solution. In this paper, three methods of GA are used: namely the Micro Genetic Algorithm (MGA), Classical Genetic Algorithm (GA) and Multipopulation (MPGA). The three methods were tested and validated on the Nigerian Grid system made of four thermal power plants and three hydro power stations. The simulation results with and without losses considered are compared. It is shown that the MPGA gives better results in term of minimized production cost than both MGA and GA. However, the MGA is faster in finding a quick feasible solution as a result of its small population size. The results demonstrate the applicability of the three techniques for solving economic load dispatch problem in power system operations.

Keywords: Genetic algorithm · Economic load dispatch · Micro genetic algorithm · Quadratic cost function · Valve point effect

1 Introduction

Economic load dispatch (ELD) is an important task in power systems operations. The aim of ELD is to allocate power generation to match load demand at minimal possible cost while satisfying all the power units and system operation constraints of the different generation resources [1]. Therefore, the ELD problem is a large scale constrained non-linear optimization problem.

For the purpose of economic dispatch studies, online generators are represented by functions that relate their production cost to their power output [2]. For simplicity, the generator cost function is mostly approximated by a single quadratic function. However, because the cost curve of a fossil fired plant is highly non-linear, containing discontinuities owing to valve point loading [3], the fuel cost function is more realistically denoted as a recurring rectified sinusoidal function [4] rather than a single quadratic cost function. The ELD problem is traditionally solved using conventional mathematical techniques such as lambda iteration and gradient schemes. These approaches require that fuel cost curves be increased monotonically to obtain the global optimal solution. Conversely, the units have naturally highly non-linear input-output properties due to valve point effect

© Springer International Publishing Switzerland 2015
Y. Tan et al. (Eds.): ICSI-CCI 2015, Part I, LNCS 9140, pp. 276–284, 2015.
DOI: 10.1007/978-3-319-20466-6_30

[5]. Also, different approaches such as linear programming and nonlinear programming have been applied to economic load dispatch problem. The main drawback of linear programming methods is that they are associated with the piecewise linear cost approximation although they are fast and reliable and the nonlinear programming approaches are complex [6].

However, with the advent of evolutionary algorithms such as Genetic Algorithm (GA), Simulated Annealing (SA), Particle Swarm Optimization (PSO), Differential Evolution (DE), Artificial Bee Colony (ABC) etc. which are stochastic based optimization techniques that search for the solution of problems using a simplified model of the evolutionary process found in nature, ELD problems can be solved easily. The success of Evolutionary Algorithms (EAs) is partly due to their inherent capability of processing a population of potential solutions simultaneously, which allows them to perform an extensive exploration of the search space [7]. Although the heuristic methods provide fast and reasonable solutions, they do not always guarantee globally optimal solutions in finite time [6].

Genetic Algorithm methods have been employed successfully to solve complex optimization problems, In this paper, three methods of GA are applied to solve economic load dispatch problem with valve point effect; namely, the Micro Genetic Algorithm (MGA), Classical Genetic Algorithm (GA) and Multipopulation Genetic Algorithm. The simulation results with and without transmission losses are considered in this paper. The performances of these methods are validated using the Nigerian grid system.

2   Problem Formulation

Consider an interconnected power system consisting of n thermal power stations as shown in Fig.1, the ELD problem seeks to find the optimal combination of thermal power plants that minimizes the total cost while satisfying the total demand and system constraints [8].

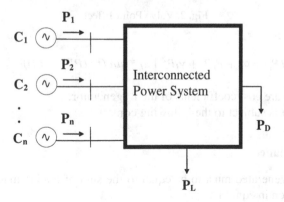

Fig. 1. Interconnected power system

The ELD problem is formulated as follows:

$$\text{Min} \quad C_T = \sum_{i=1}^{n} C_i(P_i) \tag{1}$$

where C_T is the total generation cost .and

$$C_i(P_i) = \alpha_i + \beta_i P + \gamma_i P^2 \tag{2}$$

is a quadratic cost function of the ith unit, α_i, β_i, and γ_i are cost coefficient of the ith generator, which are found from the input-output curves of the generators and are dependent on the particular type of fuel used. P_i is the power output of ith unit of thermal plants.

Note that, when the thermal generating unit changes its output, there is a nonlinear cost variation due to valve point effect. Typically, the valve point effect arises because of the ripple like effect of the valve point as each steam begins to open. This is illustrated in Fig. 2 [9]. The fuel cost of a thermal generation unit considering nonlinear effect of valve will be a nonlinear function as given in (3):

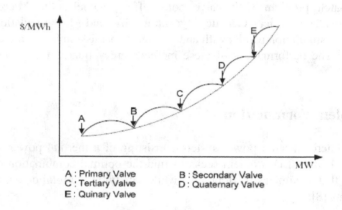

A : Primary Valve   B : Secondary Valve
C : Tertiary Valve   D : Quaternary Valve
E : Quinary Valve

Fig. 2. Valve Point Effect

$$C_i(P_i) = \alpha_i + \beta_i P_i + \gamma_i P_i^2 \mid e_1 * sin(f_1 *(P_i^{min} - Pi)) \mid \tag{3}$$

where e_i and f_i are cost coefficients of the ith generator.
The minimization is subject to the following constraints:

2.1   Power Balance

The total power generated must to be equal to the sum of load demand and transmission losses as given in eqn. (4):

$$P_D + P_L - \sum_{i=1}^{n} P_i = 0 \tag{4}$$

where:

P_D = the power demand

P_L = the transmission loss

The transmission losses can be represented by the B-coefficient method as shown in eqn. (5):

$$P_L = \sum_{i=1}^{n} \sum_{j=1}^{n} P_i B_{ij} P_j \tag{5}$$

where B_{ij} is the transmission loss coefficient.

2.2   Maximum and Minimum Power Limits

The power generated by each generator has some limits and can be expressed as:

$$P_i^{\min} \leq P_i \leq P_i^{\max}$$

$P_i^{\min}$ = the minimum power output

$P_i^{\max}$ = the maximum power output

3   Overview of Genetic Algorithm

Genetic Algorithm searches a solution space for optimal solutions to a problem. The key characteristic of GA is how the searching is done. The algorithm creates a "population" of possible solutions to the problem and lets them "evolve" over multiple generations to find better and better solutions. The following steps were used to solve a problem using GA:

1. Create a population of random candidate solution named *pop*.
2. Until the algorithm termination conditions are met, do the following (each iteration is called a generation):
   a. Create an empty population named *new-pop*.
   b. While *new-pop* is not full, do the following:
       i.   Select two individuals at random from *pop* so that individuals who are more fit are more likely to be selected.
       ii. Cross-over the two individuals to produce two new individuals.
   c. Let each individual in *new-pop* have a random chance to mutate.
   d. Replace *pop* with *new-pop*.
3. Select the individual from *pop* with the highest fitness as the solution to the problem.

The population is a collection of candidate solutions that are considered during the course of the algorithm. Over the generations of the algorithm, new and often "good" members are "born" into the population, while older and "bad" members "die" out of the population. A single solution in the population is referred to as an individual. The fitness of an individual is a measure of how "good" the solution represented by the individuals. The selection process is analogous to the survival of the fittest in the natural world. Individuals are selected for cross-over based upon their fitness value. The fitter the individual the more likely the individual will be able to reproduce and survive to the next generation. The cross-over occur by mingling the solutions together to produce two new individuals. During each generation there is a small chance for each individual to mutate, which will change the individual in some small ways.

3.1 Micro Genetic Algorithm

When dealing with high dimensionality problem, it may be difficult or too time consuming for all the model parameters to converge within a given margin of error. In particular, as the number of model parameters increases, so does the required population size. Also, large population sizes imply large numbers of cost-function evaluations. An alternative is the use of micro-genetic algorithms, which evolves very small populations that are very efficient in locating promising areas of the search space. But, the small populations are unable to maintain diversity for many generations. Therefore, whenever diversity is lost the algorithm is restarted while keeping only the very best fit individuals (usually we keep the best one that is elitism of one individual). Restarting the population several times during the run of the genetic algorithms has the added benefit of preventing further exploration of the search space and so may make the program converge to a local minimum. Also, since we are not evolving large populations, convergence can be achieved more quickly and less memory is required to store the population.

In principle, micro genetic algorithms are similar to the classical genetic algorithm in the sense that they share the same evolution parameters and similar features except that new genetic material is introduced in to the population every time the algorithm restarted. This is an important distinction: since without the restarting, the algorithm will lose its exploitation capability. It was also found out that the algorithm is much less sensitive to the choice of evolution parameters.

3.2 Multipopulation Genetic Algorithm

In GA, individuals are selected according to their fitness for the production of offspring. Parents are recombined to produce offspring. All offspring will be mutated with certain probability. The fitness of the offspring is then computed. The offspring are inserted into the population replacing the parents, producing a new generation. This circle is performed until the optimization criteria are reached. Such a single population genetic algorithm is powerful and performs well on a broad class of problems. However, better results can be obtained by introducing many populations called subpopulation. Every subpopulation evolves for a few generations isolated (like the single population genetic algorithm) before one or more individuals are exchanged between the subpopulations. The multipopulation genetic algorithm models the evolution of a species in a way more similar to nature than the single population genetic algorithm

4   Nigerian Grid System

The Nigerian national grid belongs to rapidly growing power systems faced with complex operational challenges at different operating regimes. Indeed, it suffers from inadequate reactive power compensation leading to wide spread voltage fluctuations coupled with high technical losses and component overloads during heavy system loading mode. The standardized 1999 model of the Nigerian network comprises 7 generators, out of which 3 are hydro whilst the remaining generators are thermal, 28 bulk load buses and 33 extra high voltage (EHV) lines. The typical power demand is 2,830.1MW and with technical power network loss of about 39.85MW. The single line diagram of the 330kV Nigerian grid system is shown in Fig.3.

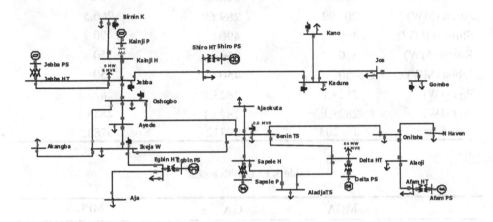

Fig. 3. Single line diagram of Nigerian 330kV 31-bus grid systems

The Nigerian thermal power plants characteristics are shown in Table 1 with each plants cost coefficients and their corresponding minimum and maximum power outputs.

Table 1. Nigerian thermal power plants characteristics

Units	α	β	γ	e	f	P_i^{min}	P_i^{max}
Sapele	6929	7.84	0.13	600	0.052	137.5	550
Delta	525.74	6.13	1.2	260	0.028	75	300
Afam	1998	56	0.092	450	0.048	135	540
Egbin	12787	13.1	0.031	850	0.094	275	1100

5   Discussion of Results

In this work, the ELD is applied to the four thermal plants of Egbin, Sapele, Delta and Afam. The results of the ELD with and without losses considered are shown in Table 2 and Table 3, respectively. With a total power generating capacity of 2823.1MW, the contribution of each thermal power plant to the total power generated and the cost of generation using the GA methods are shown in Table 2 and Table 3, respectively.

Table 2. ELD without losses

	MGA	GA	MPGA
Egbin (MW)	1070.71	937.5	942.8
Sapele (MW)	179.81	227.88	233.6
Delta (MW)	80.59	81.03	90.2
Afam (MW)	201.99	289.68	266.5
Shiroro (MW)	490	490	490
Kainji (MW)	350	350	350
Jebba (MW)	450	450	450
P_G (MW)	2823.1	2823.1	2823.1
P_D (MW)	2823.1	2823.1	2823.1
Cost $/hr	101203	99712	97832

Table 3. ELD with losses

	MGA	GA	MPGA
Egbin (MW)	838.39	814.56	817.47
Sapele (MW)	345.46	457.79	451.13
Delta (MW)	88.48	89.51	89.82
Afam (MW)	300.46	212.82	215.62
Shiroro (MW)	490	490	490
Kainji (MW)	350	350	350
Jebba (MW)	450	450	450
P_G (MW)	2862.79	2862.68	2864.04
P_D (MW)	2823.1	2823.1	2823.1
P_L (MW)	39.69	41.58	40.94
Cost $/hr	113410	112736	110324

The ELD was implemented on MATLAB 2012 platform. The contributions of the hydro power plants to the load demand were fixed, while the least cost schedule of the thermal power plants were determined. The minimum cost of production without losses given by MPGA is 97832$/hr compared to 101203$/hr for MGA and 99712$/hr for GA as shown in Table 2. With losses considered in Table 3, MPGA gives the least cost of production of $110324/hr and MGA gives lowest transmission losses of 39.69MW compared to 41.58MW of GA and 40.94MW of MPGA. The generators schedule reflects the best possible contribution of the individual generators based on the demand. From these results, it can be seen that the MPGA gives the lowest total cost of production compare to the other two algorithms.

6   Conclusion

In this paper, MGA, GA and MPGA methods have been applied to schedule generators of the Nigerian thermal power plants. The results show that these methods are capable of being applied successfully to the economic dispatch problem of larger thermal power plants. However, MPGA is shown to give the best cost minimization as a result of many populations called subpopulation used in the optimization process. Therefore, these methods can be applied to solve economic load dispatch of larger power system in future for economic operations and planning purposes.

References

1. Fang, Y., Ke, M., Zhao, X., Zhao, Y.D., Lu, H., Zhau, J.H., Kit, P.W.: Differential evolution algorithm for multi-objective economic load dispatch considering minimum emission costs. In: IEEE Power and Energy Society General Meeting, pp. 1–5. IEEE Press, New York (2011)
2. Perez-Guerrero, R.E., Cedeno-Maldonado, J.R.: Economic power dispatch with non-smooth cost function using differential evolution. In: 37th IEEE Annual North American Power Symposium, pp. 183–190. IEEE Press, New York (2011)
3. Sinha, N., Chakraarti, R., Chattopadhyay P.K..: Evolutionary Programming Techniques for Economic Load Dispatch. IEEE Transactions on Evolutionary Computation, 83–94 (2003)
4. Pereira-Neto, A., Unsihuay, C., Saavedra, O.R.: Efficient evolutionary strategy optimization procedure to solve the nonconvex economic dispatch problem with generator constraints. In: IEEE Proceeding on Generation, Transmission and Distribution, pp. 653–666. IEEE Press, New York (2005)
5. Chang, C.L.: Genetic Based Algorithm for Power Economic Dispatch. IET Generation, Transmission Distribution, 261–269 (2007)
6. Shaik, A., Sushil C.: A new intelligence solution for power system economic load dispatch. In:10th IEEE International Conference on Environmental and Electrical Engineering, pp. 1–5. IEEE Press, New York (2011)
7. Thitithamrongchai., C., Eua-Arpon. B.: Economic load dispatch for piecewise quadratic cost function using hybrid self- adaptive differential evolution with augmented lagrange multiplier method. In: IEEE International Conference on Power System Technology, pp. 1–8. IEEE Press, New York (2006)

8. Wood, A.J., Woolenberg, B.F.: Power Generation Operation and Control, 2nd edn., ch. 2, pp. 29–88. John Wiley and Sons, New York (1996)
9. Dozein, M.G., Ansari, J., Kalantar, M.: Economic dispatch incorporating wind power plant using particle swarm optimization. In: 2nd Iranian Conference on Renewable Energy and Distributed Generation, pp. 178–182. IEEE Press, New York (2012)
10. Wudil, T.S.G.: Development of Loss Formula for NEPA Transmission Network A.T.B. University Bauchi, Nigeria (2004)
11. University of Stanford. http://www.stanford.edu/public/docs/sep12
12. Hartmut Pohlheim System Technology Research Daimler Benz AG. www.pohlheim.com/papers/mpga/

Memetic Self-Configuring Genetic Programming for Fuzzy Classifier Ensemble Design

Maria Semenkina[✉] and Eugene Semenkin

Siberian State Aerospace University, Krasnoyarsk, Russia
semenkina88@mail.ru, eugenesemenkin@yandex.ru

Abstract. For a fuzzy classifier automated design a hybrid self-configuring evolutionary algorithm is implemented. For the tuning of linguistic variables a self-configuring genetic algorithm is used. Ensemble members and the ensembling method are generated automatically with the self-configuring genetic programming algorithm that does not need preliminary adjusting. A hybridization of self-configuring genetic programming algorithms with a local search in the space of trees is fulfilled to improve their performance for fuzzy rule bases and ensembles automated design. The local search is implemented with two neighbourhood systems, three strategies of tree scanning ("full", "incomplete" and "truncated") and two ways of movement between adjacent trees (transition by the first improvement and the steepest descent). The performance of all developed memetic algorithms is estimated on a representative set of test problems of the function approximation as well as on real-world classification problems. The numerical experiment results show the competitiveness of the approach proposed.

Keywords: Self-configuring evolutionary algorithms · Local search on discrete structures · Fuzzy classifier · Ensembles · Automated design · Performance estimation

1 Introduction

Classification is a well-known application of natural computing algorithms. Within the machine learning domain, problems in which the aim is to assign each input vector to one of a finite number of discrete categories are called classification problems [1]. Classification problem solving is usually described in terms of an optimization procedure that maximizes the number of correctly classified instances and minimizes the number of misclassified ones. This makes classification problems an appropriate area for the application of nature-inspired intellectual information processing technologies (IIT) like neural networks, fuzzy systems, evolutionary computations and many others.

The process of intelligent information technology (IIT) design and adjustment is rather complex even for experts in this area. For the automated implementation of IIT in classification it is necessary to consider its design as an optimization problem.

© Springer International Publishing Switzerland 2015
Y. Tan et al. (Eds.): ICSI-CCI 2015, Part I, LNCS 9140, pp. 285–293, 2015.
DOI: 10.1007/978-3-319-20466-6_31

This problem is very complicated for standard optimization tools which make evolutionary algorithms rather popular in this field [2, 3]. Genetic programming (GP) can be used for fuzzy logic system (FLS) design [4] and for IIT ensemble design.

Rapidly increasing computing power and technology made possible the use of more complex intelligent architectures, taking advantage of more than one intelligent system in a collaborative way. This is an effective combination of intelligent techniques that outperforms or competes to single standard intelligent techniques.

One of the hybridization forms, the ensemble technique, has been applied in many real world problems. Johansson et al. [5] used genetic programming (GP) [6] for building an ensemble from a predefined number of ANNs where the functional set consisted of the averaging and multiplying and the terminal set included the models (i.e., ANNs) and constants. In [7], a similar approach was proposed where first a specified number of neural networks are generated and then a GP algorithm is applied to build an ensemble making up the symbolic regression function from partial decisions of the specific members.

GP usually requires much effort to be adopted to a problem in hand. That is why before suggesting GP usage to end users for application in the development of classification tools, one must take care to avoid those main issues which are problems even for evolutionary computation experts. That is why we use in this study the self-configuring GP (SelfCGP) from [7] to avoid issues in the adjustment of the algorithm.

We suggest using a special local search for trees representing a fuzzy logic system or IIT ensemble in order to improve the SelfCGP convergence. Such hybridization of GP and local search is often referred to as memetic search [8]. However, although the local search is often used for real valued and discrete optimization problems, it is not commonplace to use it for such a data structure as a tree. Nonetheless some authors work in this direction. Some of them try to introduce a new heuristic inside the GP operator [9], others present a memetic GP for decision tree growing using domain specific heuristics [10], and a third group, the most widespread, use a local search only for numerical nodes from the terminal set. In this paper, we consider the local search on trees representing fuzzy logic systems and IIT ensembles. We implement the local search with different neighbourhood systems and different strategies of tree scanning. The use of the local search for only the best individual is also unsuitable as it can give a low algorithm performance [8].

The rest of the paper is organized as follows. Section 2 describes the method for the GP self-configuring for a fuzzy logic system and IIT ensemble automated design, with its testing results confirming the usefulness of the method. Section 3 describes the proposed local search techniques. In Section 4 we describe the testing results for the proposed memetic self-configuring algorithm. In Section 5 we apply the developed approach to hard real world problem solving. In the Conclusion we discuss the results and directions for further research.

2   Self-Configuring Evolutionary Algorithm for Automated Design of Fuzzy Logic Systems and IIT Ensembles

The self-configuring genetic algorithm (SelfCGA) and self-configuring genetic programming (SelfCGP) do not require any efforts of the end user for their adjustment.

Instead of tuning real parameters, variants of settings are used, namely types of selection (fitness proportional, rank-based, and tournament-based with three tournament sizes), crossover (one-point, two-point, as well as equiprobable, fitness proportional, rank-based, and tournament-based uniform crossovers [11]), population control and level of mutation (medium, low, high for all mutation types). Each of these has its own initial probability distribution, which is changed as the algorithm executes [11].

As was reported in [11] and [7] SelfCGA and SelfCGP demonstrate better reliability than the average reliability of the corresponding single best algorithm. Additionally, uniform crossovers in SelfCGP prevent a bloat of trees. Both algorithms can be used instead of conventional EA in complex problem solving.

We have to describe our way to model and optimize a rule base for a fuzzy logic system with GP [4] and linguistic variables adjusting with GA. The terminal set of our GP includes the terms of the output variable, i.e. class markers. The functional set includes a specific operation for dividing an input variable vector into subvectors or, in other words, for the splitting of the example set into parts through input variable values. It might be that our GP algorithm will ignore some input variables and will not include them in the resulting tree, i.e., a high performance rule base that does not use all problem inputs can be designed. This feature of our approach allows the use of our GP for the selection of the most informative combination of problem inputs.

The efficiency of the proposed approach was tested on a representative set of known test problems. The test results showed that the fuzzy logic systems designed with the suggested approach have a small number of rules in comparison with the full rule base [4]. These fuzzy systems have a small enough classification error. This is why we can recommend the developed approach for solving real world problems.

Having the developed appropriate tool for IIT automated design that does not require effort for its adjustment, we applied our self-configuring genetic programming technique to construct a formula that shows how to compute an ensemble decision using the component IIT decisions. The algorithm involves different operations and mathematical functions and uses models of different kinds providing diversity among the ensemble members. In our numerical experiments, we use fuzzy logic systems (FLS), automatically designed with the SelfCGP algorithm, as the ensemble members. The algorithm automatically chooses the component IIT which are important for obtaining an efficient solution and does not use the others.

3   Memetic Self-Configuring Genetic Programming

We can implement the following neighbourhood system for trees: trees with modified leaves (terminal set elements) will be called 1-level neighbours and trees with a modified functional element will be called 2-level neighbours. Neighbourhood systems for trees that represent fuzzy logic systems and IIT ensembles contain terminal set (T) on the 1-level (Output variable, i.e. class marker and ensemble members correspondently) and functional set on the 2-level.

The search in such neighbourhoods for the locally best-found solution should improve the efficiency of the problem solving without a significant increase in computational efforts. However, the effectiveness of the local search depends not only on the choice of neighbourhood but also on the method of its scanning. There are several ways of movement between adjacent trees: transition by the first improvement (FI) and steepest descent (SD) that mean an exhaustive search of neighbouring trees and the transit into the best found neighbour solution.

The local search procedure on the tree structure can be described as in Table 1.

Table 1. Local search algorithm for trees

Algorithm pseudo code

Enter:

Tree – initial tree that contains n nodes of different types, where n is number of nodes in *Tree*; *Tree(i)* – i-th node; *TY(i)* – type of i-th node (l – leaves, b – binary function, u – unary function); *F* – Functional set containing binary functions set (F_b) and unary function set (F_u); *T* – Terminal set;

Fitness_evol(Tree)– function of fitness evaluation of *Tree*; *Fit* – fitness of *Tree*;

Tree_s, *Tree_best* – modified trees; *Fit_s*, *Fit_best* – fitness of modified trees;

Transition – selected type of transition between trees; N – number of nodes that were scanning; F – number of fitness function evaluation;

Rand_node (TY) – function for new node value random generation.

Start: *Tree, Fit, Tree_s:=Tree, Tree_best:=Tree, Fit_s:=Fit, Fit_best:=Fit, N:=0, i:=1, F:=0.*

1. *Tree_s(i):=Rand_node(TY(i));* $N:=i$;
   Fit_s:=Fitness_evol(Tree_s); $i:=i+1$;
   F++;   **Go to 1.**
2. If $i>n$ **go to 4.**   3.4 *Tree_s:=Tree;*
   2.1 If *Fit_s<Fit* then **go to 3.4.**   *Fit_s:=Fit;*
3. If *Transition≠FI* then **go to 3.2.** $i:=i+1$;
   3.1 *Tree:=Tree_s;*   **Go to 1.**
   Fit:=Fit_s;   4. If *Transition=FI* then **go to End.**
   $i:=i+1$;   4.1 *Tree := Tree_best;*
   Go to 1.   *Fit :=Fit_best;*
   3.2 If *Fit_s<Fit_best* then **go to 3.4.**   4.2 If $N>n$ then **go to End.**
   3.3 *Tree_best:=Tree_s;*   4.3 $i:=N+1$;
   Fit_best:=Fit_s;   **Go to 1.**
   Tree_s:=Tree;   **End:** *Tree, Fit,* F.
   Fit_s:=Fit;

In this study, we will use both ways of movement and both systems of neighbourhood. In the first case, the 2-level neighbourhood will be used at the beginning of the algorithm execution and the 1-level neighbourhood will be used on the later stages. We call this method of search a "full" local search (FL). In the second case, only the 1-level neighbourhood will be used, this variant is named as an "incomplete" local search (IL). In the third case, a "truncated" local search (TL) will be considered that means scanning only n randomly chosen nodes in the tree. Changes in tree nodes that are closer to the top of the tree have a more significant impact on the result obtained. Therefore, when we use the 2-level neighbourhood, nodes which are closer to the top will be changed before others. This means that the tree will be considered in the top-down way.

During implementation and testing of the considered local search procedures the number of additional fitness function estimations must be taken into account. This number significantly depends on the way of the transition and the strategy for scanning the neighbourhood. In addition, the speed of the memetic algorithm depends on the selection of individuals to be improved by the local search (only the best individual or $p\%$ best in each generation, or once every t generations, etc.). We use the hybridization variant with 10% best individuals on each generation and name it as memetic self-configuring genetic programming (MSCGP). An individual improved by the local search is returned back to its population in the found form (Lamarckian approach). Each tree has an equal amount of computational resource for its adjustment. It means that the tree improved by the local search has less SelfCGA generations for its numerical parameter tuning.

4   Memetic Algorithm Testing Results

For the test of the proposed memetic algorithm, the test function set from [12] was used as for the self-configuring genetic programming algorithm for a symbolic regression problem [7]. Since local search algorithms precisely localize the optimum position, the comparison of the efficiency should be done with the reliability criterion. The reliability of the algorithm is the ratio of the number of successful algorithm runs to the total number of algorithm runs. The algorithm run is considered as successful if the desired accuracy is achieved. Each algorithm received the same computational resources to find a solution and was launched 100 times for each test problem. The statistical significance was estimated with ANOVA. The performance evaluation was fulfilled for the MSCGP with three types of local search ("full", "incomplete" and "truncated") and two strategies for movement ("first improvement" and "steepest descent") that can be designated as "MSCGP-FL-FI", "MSCGP-IL-FI", "MSCGP-TL-F", "MSCGP-FL-SD", "MSCGP-IL-SD" and "MSCGP-TL-SD" respectively, "+FLS" and "+E" mean a SelfCGP for automated design of fuzzy logic systems and ensembles, correspondingly. Partial illustration of the test experiment results is given in Table 2.

The following criteria for evaluating the algorithms were used:

- Reliabilities that were averaged over all test problems and the spread of their values in brackets ("Reliability").
- Information on the number of resources required to find the first suitable solutions in terms of accuracy that were averaged over all tasks and in brackets the spread on all tasks («Average number of evaluated trees»).

It is easy to see that the local search variant with a greater neighbourhood size and more detailed scanning (the most "greedy" variant) has the best reliability and the worst number of fitness function evaluations. If we change the way of transition to the first improvement then the corresponding algorithm demonstrates slightly less reliability but works faster.

Table 2. Algorithm reliability on test problems

Algorithm	Reliability	Average number of evaluated trees
SelfCGP+FLS	0.64 / [0.33, 0.96]	[4600, 21100]
MSCGP+FLS -FL-FI	0.68 / [0.37, 0.97]	[4340, 20500]
MSCGP+FLS -IL-FI	0.64 / [0.34, 0.96]	[4150, 19800]
MSCGP+FLS -TL-FI	0.65 / [0.35, 0.97]	[4210, 20050]
MSCGP+FLS-FL-SD	0.72 / [0.43, 0.99]	[4540, 21000]
MSCGP+FLS -IL-SD	0.66 / [0.38, 0.96]	[4380, 20650]
MSCGP+FLS-TL-SD	0.68 / [0.39, 0.96]	[4500, 20800]
SelfCGP+E	0.79 / [0.51, 1.00]	-
MSCGP+E-FL-FI	0.84 / [0.61, 1.00]	-
MSCGP+E-IL-FI	0.79 / [0.56, 1.00]	-
MSCGP+E-TL-FI	0.83 / [0.59, 1.00]	-
MSCGP+E-FL-SD	0.90 / [0.66, 1.00]	-
MSCGP+E-IL-SD	0.85 / [0.61, 1.00]	-
MSCGP+E-TL-SD	0.87 / [0.63, 1.00]	-

The proposed local search algorithms increased the efficiency of the previously existing self-configuring genetic programming algorithm. With the joint application of the self-configuring genetic programming and local search algorithms the performance is greater than one of the conventional genetic programming algorithms with the best setting with comparable computational resources.

5   Numerical Experiments with Real World Problems

The developed approach was applied to two credit scoring problems from the UCI repository [13] often used to compare the accuracy with various classification algorithms: Credit (Australia-1) (14 attributes, 690 examples) and Credit (Germany) (20 attributes, 1000 records). Both problems have two classes.

These classification problems were solved with fuzzy classifiers and IIT ensembles designed by MSCGP (MSCGP+FLS, MSCGP+E) with a different variant of the local search (LS). This technique was trained on 70% of the instances from the data base and validated on the remaining 30% of examples. The results of the validations, namely the portion of correctly classified instances from the test set (for scoring problems) are averaged over 40 independent runs. The statistical significance of all our experiments was estimated with ANOVA. The results for initial SelfCGP+FLS were taking from [4].

The proposed algorithms are compared with alternative classification techniques. The results for the alternative approaches have been taken from scientific literature. In [14] the performance evaluation results for two data sets of credit scoring are given for the two-stage genetic programming algorithm (2SGP) specially designed for bank scoring as well as for the following approaches taken from other papers: conventional genetic programming (GP), classification and regression tree (CART), C4.5 decision trees and k nearest neighbours (k-NN). We have taken additional material for comparison from [15] which includes evaluation data for the automatically designed fuzzy rule based classifier (Fuzzy). The results obtained with the best variants of MSCGPs in comparison with alternatives are given in Table 3.

As can be seen from Table 3 both proposed algorithms demonstrate the best performance. It is necessary to stress that fuzzy classifiers designed by MSCGP give additionally human interpreted linguistic rules which is not the case for the majority of other algorithms in Table 3. The designed rule bases usually contain 9-13 rules which do not include all given inputs.

Besides the performance evaluation we can derive some additional information. Analysis of the data sets shows that input variables can be divided into some groups so that the inputs of one group are highly correlated to each other but the correlation between inputs of different groups is weak. There are also inputs weakly correlated with the output. A fuzzy classifier and IIT ensemble designed with the suggested memetic SelfCGP do not usually include inputs of the last kind. Moreover, they usually include members of every group of inputs but only one input from each, i.e. they do not include highly correlated inputs into designed classifiers. This allows the algorithm to create relatively small rule bases with rather simple rules. The use of the local search gives an additional feature to our algorithms. They can find the best combination of the most informative inputs. This property can be used for the feature selection in the solved problems.

Table 3. The comparison of classification algorithms

Classifier	Australian credit	German credit
MSCGP+E	0.9068	0.8086
MSCGP+FLS with majority voting	0.9046	0.8075
MSCGP+FLS with weighted averaging	0.9046	0.8050
MSCGP+FLS	0.9041	0.8021
2SGP	0.9027	0.8015
SelfCGP+FL	0.9022	0.7974
SelfCGP+ANN	0.9022	0.7940
SelfCGP	0.8930	0.7850
Fuzzy	0.8910	0.7940
C4.5	0.8986	0.7773
CART	0.8986	0.7618
k-NN	0.8744	0.7565

6   Conclusions

A self-configuring genetic programming algorithm and a self-configuring genetic algorithm were hybridized to design fuzzy classifiers. Neither algorithm requires human efforts to be adapted to the problem in hand, which allows the automated design of classifiers. A special way of representing the solution gives the opportunity to create relatively small rule bases with rather simple rules. This makes possible the interpretation of the obtained rules by human experts. The self-configuring genetic programming algorithm and six local searches were hybridized to design fuzzy classifiers with high efficiency.

The quality of classification is high as well, which was demonstrated through the solving of two real world classification problems from the area of bank scoring. The results obtained allow us to conclude that the developed approach is workable and useful and should be further investigated and expanded.

With the approach developed an end user has no necessity to be an expert in the computational intelligence area but can implement a reliable and effective classification tool. It makes the approach very useful for different area experts freeing them from extra efforts on the intellectual information technology algorithmic core implementation and allowing them to concentrate their attention on the area of their expertise, e.g. medicine, finance, engineering, etc.

Acknowledgement. Research is fulfilled within governmental assignment of the Ministry of Education and Science of the Russian Federation for the Siberian State Aerospace University, project 140/14.

References

1. Bishop, C.: Pattern recognition and machine learning. Springer (2006)
2. Cordón, O., Herrera, F., Hoffmann, F., Magdalena, L.: Genetic Fuzzy Systems: Evolutionary Tuning and Learning of Fuzzy Knowledge Bases. World Scientific, Singapore (2001)
3. Herrera, F.: Genetic Fuzzy Systems: Taxonomy, Current Research Trends and Prospects. Evol. Intel. 1(1), 27–46 (2008)
4. Semenkina, M., Semenkin, E.: Hybrid self-configuring evolutionary algorithm for automated design of fuzzy classifier. In: Tan, Y., Shi, Y., Coello, C.A. (eds.) ICSI 2014, Part I. LNCS, vol. 8794, pp. 310–317. Springer, Heidelberg (2014)
5. Johansson, U., Lofstrom, T., Konig, R., Niklasson, L.: Building neural network ensembles using genetic programming. In: International Joint Conference on Neural Networks (2006)
6. Poli R., Langdon W.B., McPhee N.F.: A Field Guide to Genetic Programming. Published via http://lulu.com and freely available at http://www.gp-field-guide.org.uk (2008)
7. Semenkina, M., Semenkin, E.: Classifier ensembles integration with self-configuring genetic programming algorithm. In: Tomassini, M., Antonioni, A., Daolio, F., Buesser, P. (eds.) ICANNGA 2013. LNCS, vol. 7824, pp. 60–69. Springer, Heidelberg (2013)
8. Z-Flores, E., Trujillo, L., Schütze, O., Legrand, P.: Evaluating the Effects of Local Search in Genetic Programming. In: Tantar, A.-A., et al (eds.) EVOLVE - A Bridge between Probability, Set Oriented Numerics, and Evolutionary Computation V. AISC, vol. 288, pp. 213–228. Springer, Heidelberg (2014)
9. Eskridge, B., Hougen, D.: Imitating success: A memetic crossover operator for genetic programming. In: Proceedings of the 2004 IEEE Congress on Evolutionary Computation (CEC 2004), pp. 809–815. IEEE Press (2004)
10. Wang, P., Tang, K., Tsang, E. P. K., Yao, X.: A memetic genetic programming with decision tree-based local search for classification problems. In: Proceedings of the 2011 IEEE Congress on Evolutionary Computation, pp. 917–924 (2011)

11. Semenkin, E., Semenkina, M.: Self-configuring genetic algorithm with modified uniform crossover operator. In: Tan, Y., Shi, Y., Ji, Z. (eds.) ICSI 2012, Part I. LNCS, vol. 7331, pp. 414–421. Springer, Heidelberg (2012)

12. Finck, S., et al.: Real-Parameter Black-Box Optimization Benchmarking. Presentation of the noiseless functions. Technical Report Research Center PPE (2009)

13. Frank, A., Asuncion, A.: UCI Machine Learning Repository. University of California, School of Information and Computer Science, Irvine (2010). http://archive.ics.uci.edu/ml

14. Huang, J.-J., Tzeng, G.-H., Ong, C.-S.: Two-Stage Genetic Programming (2SGP) for the Credit Scoring Model. Applied Mathematics and Computation **174**, 1039–1053 (2006)

15. Sergienko, R., Semenkin, E.: Michigan and pittsburgh methods combination for fuzzy classifier design with coevolutionary algorithm. In: IEEE Congress on Evolutionary Computation (CEC 2013), pp. 3252–3259 (2013)

Reference Point Based Constraint Handling Method for Evolutionary Algorithm

Jinlong Li$^{(\boxtimes)}$, Aili Shen, and Guanzhou Lu

School of Computer Science and Technology,
University of Science and Technology of China (USTC), Hefei, China
jlli@ustc.edu.cn

Abstract. Many evolutionary algorithms have been proposed to deal with Constrained Optimization Problems (COPs). Penalty functions are widely used in the community of evolutionary optimization when coming to constraint handling. To avoid setting up penalty term, we introduce a new constraint handling method, in which a reference point selection mechanism and a population ranking process based on the distances to the selected reference point are proposed. The performance of our method is evaluated on 24 benchmark instances. Experimental results show that our method is competitive when compared with the state-of-the-art approaches and has improved the solution and the optima value of instance g22.

Keywords: Constrained optimization problem · Constraint handling techniques · Evolutionary strategy

1 Introduction

In many science and engineering fields, it is common to face many types of constrained optimization problems. Without loss of generality, the minimization is considered in this paper, and the constrained optimization problems can be formulated as follows:

$$\min \ f(\boldsymbol{x}) \quad \boldsymbol{x} = \{x_1, \ldots, x_n\} \in S \ . \tag{1}$$

S is the decision space bounded by lower and upper constraints. The objective function $f(\boldsymbol{x})$ is subject to inequality constraints and equality constraints: subject to:

$$g_j(\boldsymbol{x}) \leq 0, j = 1, \ldots, q \ . \tag{2}$$

$$h_j(\boldsymbol{x}) = 0, j = q + 1, \ldots, m \ . \tag{3}$$

The Evolutionary Algorithms (EAs) are widely used to solve the optimization problems [3,12]. However, EAs are mainly for unconstrained optimization problems, so an explicit constraint handling mechanism is needed to incorporate into evolutionary algorithms. When combing constraint handling mechanisms

© Springer International Publishing Switzerland 2015
Y. Tan et al. (Eds.): ICSI-CCI 2015, Part I, LNCS 9140, pp. 294–301, 2015.
DOI: 10.1007/978-3-319-20466-6_32

with EAs, whether obtaining a feasible solution takes precedence of optimizing the objective value, researchers have different views [4,9,12,15]. Deb [4] proposed three comparison criteria for ranking feasible solutions over infeasible solutions when pair-wise individuals are compared. There are also some other constraint handling techniques used in the EAs. Mezura-Montes and Coello [10] gave a review of present mostly used constraint handling methods: (1) Penalty function [1,3,5,11]; (2) Decoders [6]; (3) Separation of objective function and constraints [4]; (4) Feasibility rules [9]; (5) Stochastic ranking [12,13]; (6) ε-constrained method [14]; (7) Multiobjective concepts [2,15]; (8) Ensemble of constraint-handling technique [8]. Pareto dominance concept is adopted in algorithms that treat constrained optimization problems in a multiobjective way. The definition of Pareto dominance is: a vector $\mu = (\mu_1, ..., \mu_k)$ is Pareto dominate another vector $v = (v_1, ..., v_k)$, which is denoted as $\mu \prec v$, if $\forall i \in 1, ..., k, \mu_i \leq v_i$ and $\exists j \in 1, ..., k, \mu_i < v_i$.

In this paper, we propose a Reference Point based Constraint Handling method, named RPCH. In order to guide the population toward the global optima without trapping into local optima, a reference point is selected at every generation using a straightforward mechanism in which the solution with smaller objective value, regardless of the degree of violation, are favored at early stage while solutions with near optimal objective value are more inclined to be selected in the later stage. Experimental results on 24 widely used instances show that our algorithm can obtain competitive results on most of the instances. What's more, RPCH has improved the objective value in one instance.

Section 2 briefly reviews the related work of solving constrained optimization problems via evolutionary algorithms. Section 3 presents our proposed algorithm in detail. Experimental results on 24 benchmark instances and comparisons with state-of-the-art approaches are presented in section 4. In section 5 we give a brief summary of this paper and a few remarks.

2   Related Work

To overcome the weakness of penalty functions, Runarsson and Yao [12] pointed out that the proper balance between objective and penalty functions can be obtained by stochastically ranking the individuals, which is called SR. In their work, a probability parameter p_f was introduced when two adjacent individuals are compared as follows: (1) assuming both of the solutions are feasible, then the one which has a smaller objective value is preferred, otherwise (2) assuming a uniformly generated random value is less than p_f, then the one having smaller objective value ranks higher, otherwise, the one having a smaller constraint violation value ranks higher. Combined with the evolutionary strategy, the stochastic method acquires a competitive result on 13 well-known benchmark functions. However, its performance is affected by the probability parameter p_f. Later, the same author Runarsson and Yao published an improved version of SR [13].

There are some approaches [2,15] solving constrained optimization problems using multiobjective optimization techniques. A multiobjective optimization-based evolutionary algorithm for constrained optimization was proposed by Cai and Wang [2]. In [2], the author introduced three models of population-based algorithm-generator with different individual choosing mechanism and replacement mechanism. An Infeasible Solution Archiving and Replacement Mechanism (ISARM) is introduced. However, some parameters such as the expanding factor in simplex crossover are problem dependent, which restricts its use in real-world engineering fields [15]. Later, combining multiobjective optimization with differential evolution CMODE, was proposed by the same author Cai and Wang [15].

3   The Proposed Approach

In this section, the overall framework of the proposed approach is presented. Then we describe the details of some important steps.

Algorithm 1. Framework of the Proposed Approach

Input: λ: the size of population, μ: the number of parent individuals.
Output: $\boldsymbol{x}^*$:the best solution found in the evolution process and its objective
value $f(\boldsymbol{x}^*)$

1 $g = 0$;
2 initialization;
3 **while** $g < max\_G$ **do**
4   function evaluation, standard normalization of $g_j(\boldsymbol{x})$ and computing
    constraint violation;
5   finding reference point;
6   selection;
7   differential mutation and traditional mutation;
8   $g = g + 1$;
9 **end**

3.1   Framework of the Proposed Approach

The framework of RPCH is presented in Algorithm 1. g is the current number of generation and max_G is the maximum number of generation allowed in the evolution process. In the initialization step (line 2), the initial population of $\boldsymbol{x}$ is generated according to a uniform and random probability distribution function in order to cover the whole search space S. The initial step size is set to $\sigma_k = (U_k - L_k)/\sqrt{n}$. In every iteration, function evaluation, normalization of $g_k(\boldsymbol{x})$ and computing constraint violation are performed for every individual. Then the reference point is selected to guide the selection of best μ out of λ individuals as parent to reproduce. To exploit the best individual found at current generation, differential mutation is used in order to generate high quality individuals. At the same time, traditional mutation is used in order to explore the search space. Step 5-7 will be explained in the following section one by one.

3.2   Finding Reference Point

In order to explain the selection of reference point, three phases that most evolution process will experience are introduced first. In phase one, individuals in
the population are all feasible solutions; there are feasible and infeasible individuals in phase two; in phase three individuals in the population are all infeasible
solutions. To make it simple, we firstly give some definitions used in finding the
reference point of current generation. We denote A as the smallest objective value
of feasible individual, B as the objective value of infeasible individual with the
smallest constraint violation, C as the smallest objective value of infeasible individual regardless of the degree of constraint violation, O as the objective value
of reference point found at current generation. The approximation of optimum
(reference point) will be explained in detail as follows:

In phase one, A is selected as the reference point O.

In phase two,

$$O = \begin{cases} A & A < B \text{ and } rand(0,1) < P , \\ \frac{A+B}{2} & A > B \text{ and } rand(0,1) < P , \\ C & rand(0,1) > P . \end{cases} \tag{4}$$

In phase three,

$$O = \begin{cases} B & rand(0,1) < P , \\ C & rand(0,1) > P . \end{cases} \tag{5}$$

The probability P is dynamically adjusted during the evolution and its initial
value is set to 0.2. The implementation of probability P is as follows:

$$P(G) = 0.2(1 + 5(\frac{G}{max\_GEN})^{0.5}) . \tag{6}$$

where G is the generation counter, and max_GEN is the maximum generation
number in the evolution process.

3.3   Selection

The objective value of ith individual is denoted as O_i, then the distance of
objective value of the ith individual to the selected reference point is referred
as $Distance(O_i, O)$. When two adjacent points are compared, if they have the
pareto dominance relationship, the point which dominates the other has a higher
rank. Otherwise, the distances of objective values of these two individuals to the
selected reference point are measured. The point that has a smaller distance
ranks higher. In this way, all points in the population have a unique rank value.
After all the individuals in the population are ranked according to Algorithm 2,
the best μ out of λ individuals are selected as the parent to reproduce. The
ranking procedure is described as Algorithm 2.

Algorithm 2. Ranking Procedure

1 $I_j = j, \forall j \in \{1, ..., \lambda\}$;
2 **for** $i = 1$ *to* λ **do**
3 $\quad$ Compute $Distance(O_i, O)$;
4 **end**
5 **for** $i = 1$ *to* λ **do**
6 $\quad$ **for** $j = 1$ *to* $\lambda - 1 - i$ **do**
7 $\quad\quad$ **if** $I_{j+1} \prec I_j$ **then**
8 $\quad\quad\quad$ swap(I_j, I_{j+1});
9 $\quad\quad$ **end**
10 $\quad\quad$ **else if** $Distance(O_j, O) > Distance(O_{j+1}, O)$ **then**
11 $\quad\quad\quad$ swap(I_j, I_{j+1});
12 $\quad\quad$ **end**
13 $\quad$ **end**
14 **end**

3.4 Differential Mutation and Standard Mutation

The evolutionary strategy adopted in our approach is slightly different from the one used in stochastic ranking [13]. In [13], only $\mu - 1$ individuals are mutated in a differential way, whereas the remaining individuals are mutated according to the standard mutation strategy. The details of mutation in RPCH are presented in Algorithm 3. And $\tau' = \varphi/\sqrt{2n}$ and $\tau = \varphi/\sqrt{2\sqrt{n}}$. In the differential mutation, the base vector are randomly selected from the the best M of best μ individuals. The setting of M will be presented in Section 4.

Algorithm 3. Differential Mutation and Standard Mutation

1 $(\boldsymbol{x}_i, \boldsymbol{\sigma}_i) \leftarrow (\boldsymbol{x}'_{i;\lambda}, \boldsymbol{\sigma}'_{i;\lambda}), i = 1, ..., \mu$;
2 **for** $k := 1$ *to* λ **do**
3 $\quad i \leftarrow mod(k, \mu)$;
4 $\quad$ **if** $(k < 3.5\mu)$*(differential variation)* **then**
5 $\quad\quad \boldsymbol{\sigma}'_k \leftarrow \boldsymbol{\sigma}_k$;
6 $\quad\quad \boldsymbol{x}'_k \leftarrow \boldsymbol{x}_{best} + \gamma(\boldsymbol{x}_i - \boldsymbol{x}_{rand})$;
7 $\quad$ **end**
8 $\quad$ **else**
9 $\quad\quad \boldsymbol{\sigma}'_{k,j} \leftarrow \boldsymbol{\sigma}_{i,j} \exp(\tau' N(0, 1) + \tau N_j(0, 1)), j = 1, ..., n$;
10 $\quad\quad \boldsymbol{x}'_k \leftarrow \boldsymbol{x}_i + \boldsymbol{\sigma}'_k N(0, 1)$;
11 $\quad\quad \boldsymbol{\sigma}'_k \leftarrow \boldsymbol{\sigma}_i + \alpha(\boldsymbol{\sigma}'_k - \boldsymbol{\sigma}_i)$;
12 $\quad$ **end**
13 **end**

4   Experimental Study

4.1   Experimental Setting

The capability of our approach is demonstrated on 24 benchmark instances collected on the special session on real-parameter constrained optimization problems at CEC2006 [7]. We conduct 25 runs for each benchmark instance with 5×10^5 FES at maximum. The value of ε is set to 0.0001 just as that in the special session.

There are six parameters involved in our method: the population size λ, the parent number μ, the smoothing factor α, the step size γ, and the expected rate of convergence φ and the best vector of M. The setting of actual parameter values are as follows: $\lambda = 280, \mu = 40, \gamma = 0.88, \varphi = 1.8, M = 10$ for all instances. And in the standard mutation part, the exponential smoothing factor $\alpha = 0.4$ for μ individuals, $\alpha = 0.3$ for μ individuals, and $\alpha = 0.2$ for the other 1.5μ individuals.

4.2   General Performance of the Proposed Approach

An improved best known solution has been found in this paper for test function g22, which is a heavily constrained instance. The best known objective value reported at CEC2006 [7] for g22 is 236.4309755040, while a better objective value found by our RPCH is 236.3542569137 [1].

For instance g22, the probability keeps as one during the process. Because instance g22 is heavily constrained, which has 8 linear equality constraints and 11 nonlinear equality constraints. So the probability remain to be one for instance g22. With regard to test function g20, it is heavily constrained, which has 6 nonlinear inequality constraints, 2 linear equality constraints and 12 nonlinear equality constraints. And even the best known solution is a slight infeasible, and there is no feasible solution found so far. Therefore, we do not consider test function g20 in this paper.

From table 1, we can find that RPCH can achieve 100% feasible rate for all instances except for instance g22, and 100% success rate can be achieved for all instances except for instance g02 and g22 within 5×10^5 FES. For 14 out of 24 test functions (i.e., g03, g04, g05, g06, g07, g08, g09, g11, g12, g13, g15, g16, g18, g24) can find optimal values in every run by using 1×10^5 FES.

4.3   Comparison with State-of-the-Art Approaches

RPCH does not achieve 100% success rate in test function g02 and g22 for different reasons. RPCH can not jump out of local optimum because the search space

---

[1] Our improved solution is x^*={236.3542569137; 135.3429802710; 200.6125871166; 6461.338558449911; 3000000.1320832283; 4000001.7928515930; 3.2999998075187832E7; 130.0000013203; 170.0000192478; 299.9999986786; 399.9999820705; 330.0000192474; 184.7159051072; 249.2367160447; 127.6825141442; 269.9998836856; 159.9999505442; 5.2982628170; 5.1358507338; 5.5984855308; 5.4379953630; 5.0750856623 } with f(x^*)=236.3542569137.

Table 1. The optimal values of 24 instances are given in table 1. The best, mean, worst, standard deviation values, Feasible Rate (FR), Success Rate (SR), and Success Performance (SP) are listed in table 1. Feasible Rate is computed as $N(feasible\ runs)/N(total\ runs)$. Success Rate is computed as $N(successful\ runs)/N(total\ runs)$. Success Performance is computed as $mean(FES\ for\ successful\ runs) \times N(total\ runs)/N(successful\ runs)$.

Prob.	opt	best	mean	worst	std	FR	SR	SP
g01	-15.0000	-15.0000	-15.0000	-15.0000	0.0E00	100%	100%	109256
g02	-0.8036	-0.8036	-0.7826	-0.7094	2.2E-02	100%	24%	1143625
g03	-1.0005	-1.0005	-1.0005	-1.0005	2.1E-10	100%	100%	67827
g04	-30665.5387	-30665.5387	-30665.5387	-30665.5387	7.2E-12	100%	100%	73024
g05	5126.4967	5126.4967	5126.4967	5126.4967	3.2E-12	100%	100%	78982
g06	-6961.8139	-6961.8139	-6961.8139	-6961.8139	0.0E00	100%	100%	26510
g07	24.3062	24.3062	24.3062	24.3062	1.3E-14	100%	100%	95132
g08	-0.0958	-0.0958	-0.0958	-0.0958	9.2E-18	100%	100%	3304
g09	680.6301	680.6301	680.6301	680.6301	3.5E-13	100%	100%	85030
g10	7049.2480	7049.2480	7049.2480	7049.2480	3.9E-12	100%	100%	112156
g11	0.7499	0.7499	0.7499	0.7499	1.1E-16	100%	100%	15691
g12	-1.0000	-1.0000	-1.0000	-1.0000	0.0E00	100%	100%	7762
g13	0.0539	0.0539	0.0539	0.0539	2.0E-17	100%	100%	44497
g14	-47.7649	-47.7649	-47.7649	-47.7649	3.8E-15	100%	100%	126022
g15	961.7150	961.7150	961.7150	961.7150	5.7E-13	100%	100%	39771
g16	-1.9052	-1.9052	-1.9052	-1.9052	7.8E-16	100%	100%	63313
g17	8853.5339	8853.5339	8853.5339	8853.5339	8.9E-11	100%	100%	129640
g18	-0.8660	-0.8660	-0.8660	-0.8660	3.1E-17	100%	100%	72452
g19	32.6556	32.6556	32.6556	32.6556	4.4E-06	100%	100%	216238
g21	193.7245	193.7245	193.7245	193.7245	1.1E-02	100%	100%	108196
g22	236.4310	236.3578	245.8468	250.8468	3.8E01	80%	4%	9961000
g23	-400.0551	-400.0551	-400.0551	-400.0551	4.8E-13	100%	100%	131152
g24	-5.5080	-5.5080	-5.5080	-5.5080	8.9E-16	100%	100%	33857

of g02 highly consists of feasible space. RPCH can not achieve 100% success rate for function g22 because g22 is highly constrained. However, most algorithms even can not find feasible solutions for g22 in the literature [15], while our algorithm can find the feasible solution in most runs and an improved solution has been found by our algorithm. What's more, our algorithm can reach 100% success rate for all the rest test functions.

5   Conclusion and Future Work

We have presented a reference point based constraint handling method to solve constrained optimization problems. We first compute the sum of constraint violation value of each solution after all the constraint violation values have been normalized. Then a reference point is selected by an introduced mechanism. Finally, in each iteration individuals are selected either according to the pareto dominance relationship or the distances of objective value of each individual to the reference point.

We have evaluated RPCH on 24 benchmark, the experimental results reveal that RPCH can solve most of the problems efficiently. However, the experiments suggest that, when a problem is heavily constrained or it has a large feasible space, RPCH is not able to solve it in the number of FES required in the experiments, and how to overcome this problem will be studied in our future work.

We may also apply our reference point based constraint handling idea to other evolutionary algorithms such as differential evolution or genetic algorithms.

References

1. Ben Hamida, S., Schoenauer, M.: Aschea: new results using adaptive segregational constraint handling. In: Proceedings of the 2002 Congress on Evolutionary Computation, CEC 2002, vol. 1, pp. 884–889. IEEE (2002)
2. Cai, Z., Wang, Y.: A multiobjective optimization-based evolutionary algorithm for constrained optimization. IEEE Transactions on Evolutionary Computation 10(6), 658–675 (2006)
3. Coello, C.A.C.: Use of a self-adaptive penalty approach for engineering optimization problems. Computers in Industry 41(2), 113–127 (2000)
4. Deb, K.: An efficient constraint handling method for genetic algorithms. Computer Methods in Applied Mechanics and Engineering 186(2), 311–338 (2000)
5. Homaifar, A., Qi, C.X., Lai, S.H.: Constrained optimization via genetic algorithms. Simulation 62(4), 242–253 (1994)
6. Koziel, S., Michalewicz, Z.: A decoder-based evolutionary algorithm for constrained parameter optimization problems. In: Eiben, A.E., Bäck, T., Schoenauer, M., Schwefel, H.-P. (eds.) PPSN 1998. LNCS, vol. 1498, pp. 231–240. Springer, Heidelberg (1998)
7. Liang, J., Runarsson, T., Mezura-Montes, E., Clerc, M., Suganthan, P., Coello, C.C., Deb, K.: Problem definitions and evaluation criteria for the cec 2006. Special Session on Constrained Real-parameter Optimization, Technical Report (2006)
8. Mallipeddi, R., Suganthan, P.N.: Ensemble of constraint handling techniques. IEEE Transactions on Evolutionary Computation 14(4), 561–579 (2010)
9. Mezura Montes, E., Coello, C.A.C.: A simple multimembered evolution strategy to solve constrained optimization problems. IEEE Transactions on Evolutionary Computation 9(1), 1–17 (2005)
10. Mezura-Montes, E., Coello, C.A.C.: Constraint-handling in nature-inspired numerical optimization: past, present and future. Swarm and Evolutionary Computation 1(4), 173–194 (2011)
11. Michalewicz, Z., Attia, N.: Evolutionary optimization of constrained problems. In: Proceedings of the 3rd Annual Conference on Evolutionary Programming, pp. 98–108. Citeseer (1994)
12. Runarsson, T.P., Yao, X.: Stochastic ranking for constrained evolutionary optimization. IEEE Transactions on Evolutionary Computation 4(3), 284–294 (2000)
13. Runarsson, T.P., Yao, X.: Search biases in constrained evolutionary optimization. IEEE Transactions on Systems, Man, and Cybernetics, Part C: Applications and Reviews 35(2), 233–243 (2005)
14. Takahama, T., Sakai, S.: Constrained optimization by the ε constrained differential evolution with gradient-based mutation and feasible elites. In: IEEE Congress on Evolutionary Computation, CEC 2006, pp. 1–8. IEEE (2006)
15. Wang, Y., Cai, Z.: IEEE Transactions on Combining multiobjective optimization with differential evolution to solve constrained optimization problems. Evolutionary Computation 16(1), 117–134 (2012)

New Interactive-Generative Design System: Hybrid of Shape Grammar and Evolutionary Design - An Application of Jewelry Design

Somlak Wannarumon Kielarova[1](✉), Prapasson Pradujphongphet[1], and Erik L.J. Bohez[2]

[1] Faculty of Engineering, Naresuan University, Phitsanulok, Thailand
somlakw@nu.ac.th, prapassonp@gmail.com
[2] School of Engineering and Technology, Asian Institute of Technology, Pathumthani, Thailand
bohez@ait.ac.th

Abstract. This paper proposes a new methodology for developing a computer-based design system. It places designers at the centre of design process to perform their tasks collaboratively with the design system. The proposed system is developed based on interactive shape grammar and evolutionary design algorithm, which is able to increase the creativity and productivity of design activity. Designers can utilize the generated designs to initialize their conceptual design process more easily and rapidly. The source of form diversity is derived from genetic operators. Subjective user preference is used for design evaluation. The system can be integrated with computer-controlled model-making machines to automatically build physical artifacts. As a result, designers can easily start their conceptual design process through obtaining the desired designs and the resulting physical artifacts in line. The human-computer synergy is illustrated for the design of jewelry, but it is applicable to other industrial product design problems.

Keywords: Evolutionary design · Evolutionary strategy · Shape grammar · Generative design · Parametric design · Jewelry design

1   Introduction

In conceptual design stage, designers generally carry out activities such as generating design ideas, recording them, and making decisions whether to continue to generate more ideas or instead to explore more possibilities of the preferred existing ones [1]. Therefore, at this stage designers generate a number of various alternatives [2].

Currently, there are various computer-aided design (CAD) packages available in the market. Nevertheless the available computer-aided design packages are mostly used in detailed design stage rather than in conceptual design, because most of CAD systems are not able to suitably support designers in conceptual design activities.

Generative design (GD) system is a computer-based design system that can support designers in divergent thinking throughout design exploration and design generation.

© Springer International Publishing Switzerland 2015
Y. Tan et al. (Eds.): ICSI-CCI 2015, Part I, LNCS 9140, pp. 302–313, 2015.
DOI: 10.1007/978-3-319-20466-6_33

It enables designers to explore a large design space, which provides a larger range of design possibilities than manual design process [3]. Furthermore, a generative design system can also generate designs that might have not been predicted by designers themselves [4]. Despite design exploration and design generation being crucial in conceptual design stage, most of available CAD systems cannot properly support designers in this stage. The proposed design system was therefore developed to address these issues.

This paper aims to bring the benefits of computer-based design systems into the early stage of design process, which is important for creative design. This paper proposes a generative evolutionary design system, which combines shape grammar and evolutionary strategy algorithm for achieving the above purposes. The system is an interactive design system, which is used for jewelry design application. It is able to increase creativity and productivity of design activity. Designers can apply the generated designs to initialize their conceptual designs more easily and rapidly through obtaining the desired designs and the resulting physical artifacts in line.

The paper is organized in five main sections. The related theories and research papers focusing on topics such as generative design, shape grammar and evolutionary strategy are presented in Section 2. The development of hybrid shape grammar and evolutionary strategy is introduced in Section 3. The experimental results are discussed in Section 4. Finally, the research work is concluded in Section 5.

2 Related Research

2.1 Generative Design

Generative design (GD) is a process that employs computational capabilities to support designers working in design process and/or automate some parts of the design process [5]. GD is expected to help designers in divergent thinking. Singh and Gu had investigated five commonly used GD techniques in architecture [5]. Those techniques are cellular automata, shape grammars, L-systems, genetic algorithms, and swarm intelligence. Their review of the existing literature indicates that most of the available GD systems are developed based on one of the above GD techniques.

Several researchers have studied various issues of GD. One of the most important issues in developing a GD system is user-system interaction. There are different models of user-system interaction depending on the objectives of the uses of GD systems. Chase [4] summarized the user-system interaction paradigms in the following ways: minimal user interaction using optimization techniques; tight control over rule application; and computer generation with user selection. He also explained the possible interaction scenarios, which are categorized into full control, partial control, and no control. Prats et al. [6] recommended that an effective design system should provide a communication channel between designer and the system. An generative design tool that applies partial control interaction model to guide conceptual design process was developed by Orbay and Kara [7]. Tapia [8] developed a shape grammar-based design system, which allows user to define and select rules. Additionally, this system also offers user the selection of control mode such as user control and system control.

There are only a few publications on generative design systems in jewelry applications. The GD system developed by Kumar *et al.* [9] is a geometrical modeling method to generate 3D patterns for traditional Indian Kundan jewelry. Another system was developed by Sharma *et al.* [10]. In this case the relevant CAD system for ornament design industry in India was studied and a GD system named Estampa was implemented. The system can generate ornamental patterns by transforming primitive motifs. A generative design system for jewelry design application was also developed by Wannarumon S. *et al.* [11]. The first iteration of this system was based on evolutionary algorithm. Later on, Shape grammar method was used to create a shape generation mechanism to generate jewelry ring design without considerations of gemstone setting [12].

It can be seen, based on the literature examples discussed above, that shape grammar integrated into generative design systems have a good potential to support design applications in art and design domains.

2.2   Shape Grammar

Shape grammars (SG) were first introduced by Stiny and Gips in 1972 [13]. Four basic components of an SG are defined by Stiny [14]:

• an initial shape
• set of shapes
• set of shape rules
• set of symbols

To generate a design, shape grammar needs sets of shape rules and to perform calculations with shapes in the following order: recognition of a given shape and its possible replacements; compiling rules; and exploring the shape grammar [8, 13]. Shape grammars generate designs by formalizing the spatial relations between their elements [6]. Shape transformations can be explicitly described and used for systematic generation of design alternatives using shape grammars. Alternating the sequence of shape rules can generate different designs. A new type of rule named 'piecewise line-rule' used in conjunction with the decomposition rule to describe and transform curved outlines was presented by Prat [15]. His technique can be used to systematically generate design alternatives.

In this paper, we are interested in the shape grammar technique rather than other techniques, because SG is specifically suitable for form and style generation [5]. Furthermore it can be easily modified to create new design languages with small changes in grammar rules [16]. Several researchers have developed shape grammar-based GD systems for different applications. Stiny and Gips [13] had developed shape grammars for generating paintings and sculptures. In architecture applications, Stiny and Mitchell [17] proposed parametric shape grammar for generating the ground plan of Palladio's villas as Palladian style. The parametric shape grammar that generates the compositional forms and identifies the function zones of Frank Lloyd Wright's prairie-style houses was developed by Koning and Eizenberg [18]. Shape grammars were also implemented in engineering design problems such as roof truss design by Shea and Cagan [19]. Shape grammars were applied in industrial design applications and are illustrated in the following examples. Agarwal and Cagan [20] developed a function-based grammar to design coffee makers. Partial control protocol was applied to set up the interaction between designer and the system by allowing user to select design rules. Pugliese and Cagan [21] proposed a two-dimensional Harley-Davidson motorcycle shape grammar for capturing brand identity. Two-level shape grammars based on shape decomposition method were proposed for Zhuang ethnic embroidery design exploration by Cui and Tang [22]. Kielarova et al. [12] developed a shape grammar-based design system for jewelry design applications. The authors have studied shape transformations in jewelry ring design process to identify transformations of shapes from one state to another, and to develop shape grammar and shape rules.

2.3   Evolutionary Strategy

Schwefel was the first to investigate the evolutionary strategy (ES) and develop the (1+1)-ES system. Later on the scope of ES was expanded by Rechenberg [23]. ES was developed based on the concept of natural evolution. ES was initially applied for parameter optimization. Instead of binary strings real values are used in ES to code parameters. Contrary to genetic algorithm that includes both crossover and mutation, ES uses only mutation. The basic implementation of ES was a two member (1+1)-ES system, where one parent generates one offspring using mutation and the better of the two is selected, while the other is eliminated. In ES, each individual is represented by its genotype and strategy parameters that are both evolved. In order to improve the algorithm for parallel processing with respect to local optima, two general forms $(\mu+\lambda)$-ES and (μ,λ)-ES were suggested [24]. We are interested in ES's abilities to increase diversity of design alternatives.

3   Development of Interactive-Generative Design System: Hybrid of Shape Grammar and Evolutionary Design

We propose an interactive-generative design system based on a hybrid of shape grammar and evolutionary strategy. This research work is extension from the previous research [12], which has studied shape transformations in jewelry design and developed shape rules to be applied in shape grammar to generate new design alternatives. The previously published system was limited in its shape generating capabilities. Therefore a new system is proposed herein. This new system was developed by integrating ES into the shape grammar-based generative design system to generate more preferred alternatives. The outline of this system is shown in Fig.1.

3.1   Shape Grammar for Jewelry Design

The major issue for developing shape grammar systems for jewelry application is the variety of shapes of jewelry items. It is, therefore, necessary to define scope and limitations of the shape grammar to be developed. In this paper, we demonstrate the development of shape grammars, which can generate design elements for designing earrings. Gemstone earrings, shown in Fig. 2, were studied in terms of their characters and shapes for developing shape grammar. The color of the gemstone was not considered. After the analysis, the features of the earrings were abstracted as follows:

- The earrings are set with main gemstones and decorated with minor ones on the top as shown in Fig. 3. This constitutes a strong spatial relationship.
- Choosing different gemstone cuts, as shown in Fig. 4, can vary shapes of earrings.
- Design parameters are size of earring, size of gemstones, gemstone cuts, and shapes of decorative items.

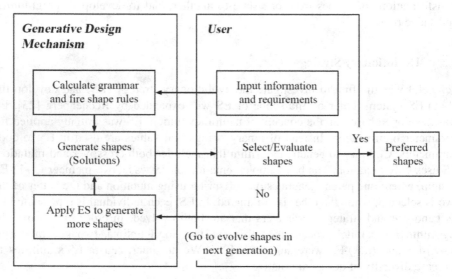

Fig. 1. Workflow of the proposed generative design system

The features described above indicate that it is possible to apply shape grammar rules to represent the gemstone earring design process.

The earring shape grammar consists of sets as follows:

- Shape grammar (SG) = {S_m, S_l, S_d, S_c, R, L, I}
- Set of main gemstone cuts (S_m) = {Round, Oval, Pear, Square, Heart Shape, etc.}
- Set of minor gemstone cuts (S_l) = {Round, Oval, Pear, Square, Heart Shape, etc.}
- Set of shapes of decorative elements (S_d) = {circle, oval, rectangle, square, triangle, other freeform items}
- Set of shapes of connectors (S_c) = {circle, oval, rectangle, square, triangle, other geometric shapes}
- Set of shape rule (R) = {Initial Rules, Shape Transformation Rules, Termination Rules}
- Set of symbol or label (L) = {•1, •2, •3, ... }
- Initial shape (I) shown in Fig 3. (b)

The initial shape consists of three main elements: main gemstone; minor gemstone; and connector. The shapes of main and minor gemstones can be varied within the set of standard gemstone cuts, i.e. round, oval, pear, square, etc. Transforming a set of geometric shapes and other freeform shapes can vary the shapes of decorative element items. Shape of the connector is limited in geometric shapes such as circle, square, etc. Shape grammar rules consist of initial rules, shape transformation rules, and termination rules. The initial rules are used in the starting step of the generative design process to generate a main gemstone, a minor gemstone or decorative item, and a connector. Furthermore, initial rules also establish their positions.

Fig. 2. Examples of earrings [25-28] used for studying their shape grammars

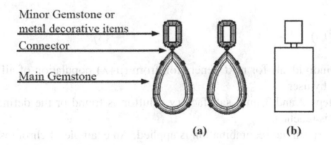

Fig. 3. (a) schematic outline of studied earrings; (b) initial shape

Shape transformation rules were obtained from [12], and are used for adaptation and transition of design elements. Termination rules are used to create and to position gemstone, decorative items, and connectors, when all tasks are finished. After the application of these rules the shape grammar process will stop. Label set is represented by dots in sequence, to signify shape-manipulating sequences along with spatial relationship.

Fig. 4. Examples of gemstone cuts

3.2   Modified Evolutionary Strategy

The concept of evolutionary strategy (μ+λ)-ES is employed to create an evolutionary design algorithm. The process begins by taking the generated shapes (solutions) from the shape grammar process as a set of initial parents. A new evolutionary design algorithm is then implemented in the following order:

1. Choose generated design items as μ parents that contain m design parameters

$$X=(x_1,x_2,x_3,...,x_m) \tag{1}$$

2. Create λ new offspring by mutation.

$$X'=(x'_1,x'_2,x'_3,...,x'_m) \tag{2}$$

Applying a random vector of size X with normal distribution performs the mutation

$$X'=X+N(0,s) \tag{3}$$

3. Select μ individuals for next generation from (μ+λ) population of all parents and offspring by user.
4. Repeat steps 2 and 3 until satisfactory solution is found or the defined computation time is reached.
   In the algorithm, no recombination is applied. An example of chromosome used in this process is shown in Fig. 5.

4   Experimental Results and Discussions

The prototype system was developed using Visual Basic script in Rhinoceros 5.0 software [29]. Using the hybrid shape grammar and evolutionary design algorithm method, the proposed generative design system can generate a large number of earring designs within the defined shape grammar depending on the process parameter setting. The generative design system works as shown in Fig.6.

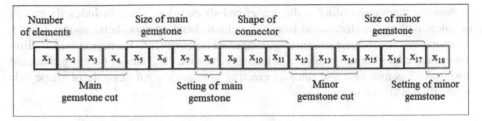

Fig. 5. Example of chromosome used in the evolutionary process

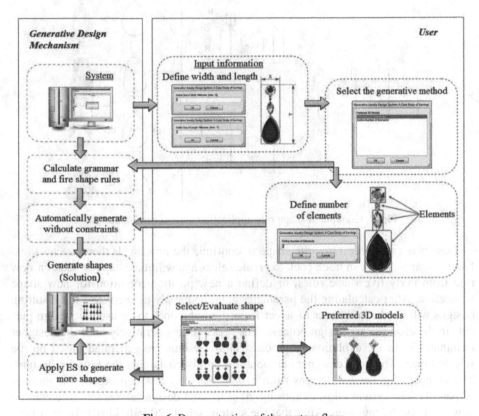

Fig. 6. Demonstration of the system flow

In the experiment, the process parameters set consists of the number of elements, number of gemstone cuts, types of setting, number of connector's shapes, number of outline transformation rules, and number of structural transformation rules. Consequently, the number of all possibilities of earring designs is 1,180,096. Designers can control direction of design generation through the input parameters that are size of earrings, number and size of gemstones, gemstone cuts, and shapes of connectors. The generative design system then will generate initial shapes. The system calculates shape transformations for all possibilities in the design space, calculating from the input parameters. The system applies the rules to the shapes and then generates a set of possible shapes according to the predefined shape rules, which includes six groups of rules: outline transformation rules (straighten, bend convex, bend concave, etc.); structure transformation rules (mirror, rotate, add element, etc.); safety rules (to eliminate sharp edges/corners); rules of selecting gemstone cut; rules of defining sizes and positions of gemstones; and rules of creating connectors. An example of shape rule application is shown in Fig. 7.

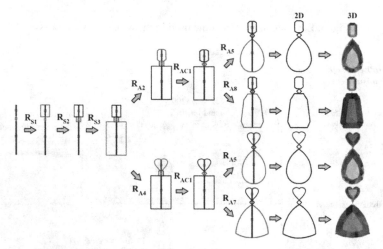

Fig. 7. Example of shape rule application used in earring design

Designers can select a favorite shape to continue the process. If there are no satisfactory shapes, they can track back to create/select a new initial shape or select a new rule from thirty-five shape rules, or define a new position/condition for new shape generation. After calculating the parameter data, and applying the rules, the resulting designs will be shown to user to select from them the ones to be used as design parents in the evolutionary design process. The process runs iteratively until the designer terminates it. In the evolutionary process, more designs are generated based on the selected ones. Thus, the designer can obtain a family containing the favorite designs. The number of generated designs from evolutionary process depends on the number of parents. The system allows choosing 20 generated designs as parents to create 60 new offspring by mutation. After that the system selects 20 individuals for next generation. In this evolutionary process, the designer works as the fitness function. Therefore different user can generate different set of designs. In other words, the system can generate designs according to user's style. The generative design system can work as an external working memory for imaginary process, which can solve overload of visuospatial working memory of human designers. The system can also be integrated with computer-controlled model-making machines to automatically build physical artifacts. As a result, designers can easily start their conceptual designs through obtaining the desired designs and the resulting physical artifacts in line.

Figure 8 (on the left) shows some of the resulting designs generated by the system in an example where a designer defines the input parameter of main gemstone as pear cut. On the other hand designs obtained when no inputs were defined, and the system ran automatically, are shown in Fig. 8 (on the right).

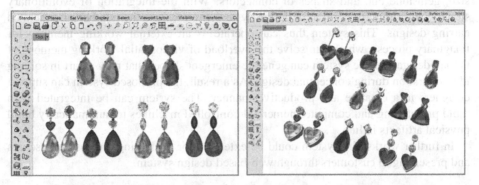

Fig. 8. Example of the resulting earring designs generated by the system

Furthermore, the design elements generated by the system can be used for designing a set of earrings and pendant as shown in Fig. 9.

Fig. 9. Example of a set of earrings and pendant generated by the system

A workshop was organized for five participants who were given the task of designing jewelry earrings using the generative design system. All participants were satisfied with the performance of the system. They remarked that the system is easy and intuitive to use. However, they required a short amount of time to become familiar with the system.

5   Conclusions

This paper presents an interactive-generative design system, which is a hybrid of shape grammar and evolutionary strategy. The system is designed to allow user to work with it semi-interactively in both shape grammar process and evolutionary design process.

The generative design system can support designers to automatically generate gemstone earring shapes based on input parameters such as earring size, gemstone size, gemstone cut, and shapes of connectors. With the integration of evolutionary strategy algorithm, the proposed design system can increase the number of generated earring designs. This system thus can operate as an external working memory for imaginary process, which can solve the overload of visuospatial working memory of human designers. The system can generate emergent shapes that play a part in solving idea saturation during conceptual design. As a result, the proposed system can support designers in a creative and productive manner. The system can be integrated with rapid prototyping and computer numerical controlled machines to automatically build physical artifacts in line.

In further works, the system could be extended for evaluating the aesthetic scores and presenting to customers through web-based design system.

Acknowledgements. The research has been carried out as part of the research projects funded by National Research Council of Thailand and Naresaun University with Contract No. R2558B026. The authors would like to gratefully thank all participants for their collaborations in this research including Generative Jewelry Design Lab. team, Department of Industrial Engineering, Naresuan University for their hard work. Finally, the authors would like to thank Dr. Filip Kielar for correcting the manuscript.

References

1. Kolli, R., Pasman, G.J., Hennessey, J.M.: Some considerations for designing a user environment for creative ideation. In: Interface 1993, pp. 72–77. Human Factors and Ergonomics Society (1993)
2. French, M.J.: Conceptual Design for Engineers. Design Council, London (1985)
3. Krish, S.: A Practical Generative Design Method. Computer-Aided Design **43**, 88–100 (2011)
4. Chase, S.C.: A Model for User Interaction in Grammar-based Design Systems. Automation in Construction **11**, 161–172 (2002)
5. Singh, V., Gu, N.: Towards an Integrated Generative Design Framework. Design Studies **33**, 185–207 (2012)
6. Prats, M., Lim, S., Jowers, I., Garner, S.W., Chase, S.: Transforming Shape in Design: Observations from Studies of Sketching. Design Studies **30**, 503–520 (2009)
7. Orbay, G., Kara, L.B.: Learning geometric design knowledge from conceptual sketches and its utilization in shape creation and optimization. In: ASME Conference, pp. 683–691 (2009)
8. Tapia, M.: A Visual Implementation of a Shape Grammar System. Environment and Planning B: Planning and Design **26**, 59–73 (1999)
9. Kumar, G.V., Garg, T.K., Puneet, T.: Geometrical Modeling of 3D Patterns for Traditional Indian Kundan Jewelry. Int. J. of Engineering Science and Technology **3**, 5666–5670 (2011)
10. Sharma, S., Singh, Y., Virk, G.S.: Study on Designing and Development of Ornamental Products with Special Reference to the Role of CAD in it. Int. J. of Current Engineering and Technology **2**, 443–447 (2012)

11. Wannarumon, S., Bohez, E.L.J., Annanon, K.: Aesthetic Evolutionary Algorithm for Fractal-based User-centered Jewelry Design. AIEDAM **22**, 19–39 (2008)
12. Kielarova, S.W., Pradujphongphet, P., Bohez, E.L.J.: An Approach of Generative Design System: Jewelry Design Application. IEEM **2013**, 1329–1333 (2013)
13. Stiny, G., Gips, J.: Shape grammars and the generative specification of painting and sculpture. In: Information Processing 1972, pp. 1460–1465. Amsterdam, Holland (1972)
14. Stiny, G.: Introduction to Shape Srammars. Environment and Planning B: Planning and Design **7**, 343–351 (1980)
15. Prats, M.: Shape Exploration in Product Design. Ph.D. dissertations. Open University, Milton Keynes, UK (2007)
16. Alcaide-Marzal, J., Diego-Más, J.A., Asensio-Cuesta, S., Piqueras-Fiszman, B.: An Exploratory Study on the Use of Digital Sculpting in Conceptual Product Design. Design Studies **34**, 264–284 (2013)
17. Stiny, G., Mitchell, W.J.: The palladian grammar. Environment and Planning B: Planning and Design **5**, 5–18 (1978)
18. Koning, H., Eizenberg, J.: The Language of the Prairie: Frank Lloyd Wright's Prairie Houses. Environment and Planning B: Planning and Design **8**, 295–323 (1981)
19. Shea, K., Cagan, J.: The Design of Novel Roof Trusses with Shape Annealing: Assessing the Ability of a Computational Method in Aiding Structural Designers with Varying Design Intent. Design Studies **20**, 3–23 (1999)
20. Agarwal, M., Cagan, J.: A Blend of Different Tastes: the Language of Coffeemakers. Environment and Planning B: Planning and Design **25**, 205–226 (1998)
21. Pugliese, M.J., Cagan, J.: Capturing a rebel: modeling the Harley-Davidson brand through a Motorcycle Shape Grammar. Res. Eng. Design **13**, 139–156 (2002)
22. Cui, J., Tang, M.-X.: Integrating Shape Grammars into a Generative System for Zhuang Ethnic Embroidery Design Exploration. Computer-Aided Design **45**, 591–604 (2013)
23. Beyer, H.-G., Schwefel, H.-P.: Evolution Strategies. Natural Computing **1**, 3–52 (2002)
24. Dianati, M., Song, I., Treiber, M.: An Introduction to Genetic Algorithms and Evolution Strategies. Technical Report. University of Waterloo, Ontario (2002)
25. Right Honghua Development Ltd. http://www.finejewellerymaking.com
26. Goldies Jewelry Inc. http://www.goldiesjewelry.com
27. Charles Jewelry Inc. http://charlesjewelry.com
28. C Jewelry Fashion Inc. http://www.cjewelryfashion.com
29. Rhinoceros Software. http://www.rhino3d.com

Differential Evolution

Differential Evolution with Novel Local Search Operation for Large Scale Optimization Problems

Changshou Deng[✉], Xiaogang Dong, Yanlin Yang, Yucheng Tan, and Xujie Tan

School of Information Science and Technology, JiuJiang University, Jiujiang 332005, China
{csdeng,xx_dongxiaogang,yangyanling,yctan}@jju.edu.cn

Abstract. Many real-world optimization problems have a large number of decision variables. In order to enhance the ability of DE for these problems, a novel local search operation was proposed. This operation combines orthogonal crossover and opposition-based learning strategy. During the evolution of DE, one individual was randomly chosen to undergo this operation. Thus it does not need much computing time, but can improve the search ability of DE. The performance of the proposed method is compared with two other competitive algorithms with benchmark problems. The compared results show the new method's effectiveness and efficiency.

Keywords: Large scale optimization · Differential evolution · Orthogonal crossover · Quasi-opposition learning

1 Introduction

Differential evolution (DE), proposed by Storn and Price, is a simple yet efficient algorithm for global optimization problems in continuous domain [1].It has been widely used in various applications [2].However, DE still suffers from the "curse of dimensionality", which implies that the performance of DE will deteriorate rapidly while the scale of the search space increases [3]. Thus DE usually fails to find the optimal solutions to large scale optimization problems. Much work have been tried to enhance the performance of DE for large scale optimization problems. One way to improve the performance of DE is by using new crossover operators. Noman et al. proposed a crossover-based adaptive local search operation to enhance the performa-nce of standard DE algorithm [2]. Wang et al. used an orthogonal crossover to enhance the search ability of DE [4]. These works have improved the performance of DE for low dimensional problems. However, when the scale size of problem grows up to 1000 or even more, they can not avoid being trapped into local minimum. Another promising approach to deal with large scale problems is opposition-based learning (OBL) [5, 6]. OBL has been successfully applied to enhance the performance of DE. The key concept of OBL is to evaluate the current solutions and their opposite ones simultaneously. And the central opposition theorem has proved that the probability that the opposite of solution is closer to the global optimum is higher than the probability of a second random guess [7]. In recent years, one paradigm that received much attention is

© Springer International Publishing Switzerland 2015
Y. Tan et al. (Eds.): ICSI-CCI 2015, Part I, LNCS 9140, pp. 317–325, 2015.
DOI: 10.1007/978-3-319-20466-6_34

cooperative co-evolution [8] . Cooperative co-evolution is a divide-and-conquer strategy where a large scale problem is grouped into several lower-dimensional sub-problems. Each sub-problem is usually easier to be solved in a round-robin way. However, how to effectively group the sub-problems are still challengeable [3].

This paper is inspired by the orthogonal crossover [4] and OBL [5]. We combine orthogonal crossover and OBL to propose DE with novel local search operation (DENLS) for solving large scale optimization problems.

2   Differential Evolution

DE has four main steps: Generation of Initial Population, Mutation operation, Crossover operation and Selection operation. One of the most promising schemes, DE/Rand/1/Bin, is presented in great detail. It is supposed that we are going to find the minimization of the objective function $f(x)$.

2.1   Generation of Initial Population

The DE algorithm starts with the initial population $X = (x_{ij})_{NP \times D}$ with the size of NP and the dimension size of D , which is generated by:

$$x_{ij}(G) = x_j^l + rand(0,1)(x_j^u - x_j^l) \tag{1}$$

where $G = 0, i = 1,2,\ldots,NP$, $j = 1,2,\ldots,D$, x_j^u and x_j^l denotes the upper constraints and the lower constraints respectively.

2.2   Mutation Operation

For each target vector $x_i (i = 1,2,\ldots NP)$, a mutant vector is produced by

$$v_i(G+1) = x_{r_1} + F(x_{r_2} - x_{r_3}) \tag{2}$$

where $i, r_1, r_2, r_3 \in \{1,2,\ldots NP\}$ are randomly chosen and must be different from each other. And F is the scaling factor for the difference between the individual x_{r_2} and individual x_{r_3} .

2.3   Crossover Operation

DE employs the crossover operation to add the diversity of the population. The approach is given as follows.

$$u_i(G+1) = \begin{cases} v_i(G+1), & if \quad rand \leq CR \quad or \quad j = rand(i) \\ x_i(G), & otherwise \end{cases} \tag{3}$$

where $i = 1,2,\ldots,NP$, $j = 1,2,\ldots,D$, $CR \in [0,1]$ is the crossover probability and $rand(i) \in (1,2,\ldots,NP)$ is the randomly selected number. The crossover operation can ensure that at least one component of the trial individual comes from the mutation vector.

2.4 Selection Operation

Selection operation decides whether the trial individual $u_i(G+1)$ should be a member of the next generation, it is compared to the corresponding $x_i(G)$. The selection operation is based on the survival of the fitness between the trial individual and the corresponding one such that:

$$x_i(G+1) = \begin{cases} u_i(G+1), & if\ f(u_i(G+1) \le f(x_i(G))) \\ x_i(G), otherwise \end{cases} \tag{4}$$

3 DE with Novel Local Search Operation

In the proposed DENLS, the dynamic model in [10] is adopted and a novel local search operation was proposed to balance the exploration ability of DE. In the local search operation, the orthogonal crossover is used and further the quasi-opposition learning strategy is adopted in the orthogonal crossover to enhance the exploitation ability.

3.1 Orthogonal Crossover

Orthogonal crossover operation was originally proposed to improve the performance of genetic algorithm and a quantization technique called QOX was introduced to deal with continuous optimization [11]. To enhance the ability of orthogonal crossover, opposition-based learning (OBL) [5, 6] is embedded into the orthogonal crossover.

QOX [11] is based on orthogonal array. An orthogonal array for K factors with Q levels and M combinations can be denoted to $L_M(Q^K)$. For example, $L_9(3^4)$ is used in DENSL as follows:

$$L_9(3^4) = \begin{bmatrix} 1 & 1 & 1 & 1 \\ 1 & 2 & 2 & 2 \\ 1 & 3 & 3 & 3 \\ 2 & 1 & 2 & 3 \\ 2 & 2 & 3 & 1 \\ 2 & 3 & 1 & 2 \\ 3 & 1 & 3 & 2 \\ 3 & 2 & 1 & 3 \\ 3 & 3 & 2 & 1 \end{bmatrix} \tag{5}$$

Given two parent individuals $x = (x_1, x_2, ..., x_D)$ and $y = (y_1, y_2, ..., y_D)$, x and y can define a search interval with $[\min(x_i, y_i), \max(x_i, y_i)]$ for the *ith* decision variable. QOX defines Q levels $l_{i1}, l_{i2}, ..., l_{iQ}$ for the *ith* decision variable as follows:

$$l_{ij} = \min(x_i, y_i) + \frac{j-1}{Q-1}(\max(x_i, y_i) - \min(x_i, y_i)) \quad j = 1, ..., Q \tag{6}$$

According to Eq. (5) and Eq. (6), given two parent individuals x and y, nine children can be generated. Details can refer to [4].

3.2   Quasi-Opposition Learning

Rahnamayan et al. proposed quasi-opposition-based learning and proved that a quasi-opposite point is more likely to be closer to the solution of the optimization problem than the opposite one [12]. Opposition as defined by [12] is given in Eq. (7).

Definition 1 let $x \in [a, b]$ be any real number. Its opposite, ox is defined as

$$ox = a + b - x \tag{7}$$

Definition 2 let $x \in [a, b]$ be any real number. Its quasi-opposite point, qox is defined as

$$qox = rand(c, ox) \tag{8}$$

where c is the center of the interval $[a, b]$ and $rand(c, ox)$ is a random number uniformly distributed between c and ox.

3.3   Proposed Method

The crossover operation only generates one single trial vector $u_{i,G}$, which is a vertex of the hyper-rectangle space defined by the mutation vector $v_{i,G}$ and the individual $x_{i,G}$. Therefore, DE can not search the hyper-rectangle sufficiently. To overcome this limitation, we use orthogonal crossover and quasi-opposition learning to build a local search. This local search can generate eighteen trial vectors to locate more space. This local search operation needs to evaluate eighteen new solutions, and thus, it can not be used for each pair of mutation vector and individual vector. In order to achieve the advantage of this local search operation while not consuming too much computing time, only one individual in current population was randomly chosen to undergo this local search operation. In addition, the mutation operation for this chosen individual is also changed into Eq. (9).

$$v_i(G+1) = x_{r_1} + rand(0,1) \cdot (x_{r_2} - x_{r_3}) \tag{9}$$

where $rand(0,1)$ is a uniformly random number between zero and one.

The Pseudo code of DENLS is presented in figure 1.

DENLS pesudocode:

Step 1.Parameters setting: NP(population size),F(scaling factor),CR(crossover control parameter),Orthogonal Array, FES_{max} (maximum number of function evaluations(FES)).

Step 2. Generation of Initial Population P

Step 3. Evaluation of the Population P

Step 4. i=1

Step 5. if FES> FESmax then goto Step 15 else goto Step 6.

Step 6. K=rand [1, NP]

Step 7. if (K equals to i) then goto Step 8 else goto Step 12

Step 8. Using Eq. (9) to generate mutation vector

Step 9. Using orthogonal crossover to generate nine trial vectors

Step10. To generate another nine trial vectors with quasi-opposition learning

Step11. Choose the best one from the above eighteen trial vectors, FES=FES+18, then goto   Step 13

Step12. Undergo mutation and crossover operation, according to Eq. (2) and Eq.(3) without generation G, FES=FES+1

Step13. if $f(u_i) \leq f(x_i)$, then update P with u_i replacing x_i

Step14. if (i equals to NP) goto Step 5, else i=i+1;

Step15. Stop and output the vector with the smallest fitness value in P.

Fig. 1. The Pseudo code of DENLS

4   Numerical Experiment

4.1   Experimental Settings

In order to evaluate the performance of the proposed DENLS, thirteen well-studied optimization problems are chosen as test bed   [3]. Among the thirteen problems, problems $f_1 - f_7$ are unimodal and problems $f_8 - f_{13}$ are multimodal. The dimensional sizes of these problems are 1000. The compared algorithms include OXDE [2] and a competitive cooperative co-evolution DECC-G [3].

For all experiments, the FES is set to 5E6. The size of population is set to 100 for all algorithms. And F is set to 0.9, CR is set to 0.9. Twenty-five independent runs are carried out for each method in each instance.

Table 1. Compared Results between DENLS with OXDE, DECC-G

fun	Algo	Best	Worst	Mean	Std	P Value
f_1	OXDE	2.37e+002	8.91e+002	4.76e+002 −	1.67e+002	9.72e-011
	DECC-G	4.45e-029	1.58e-028	9.61e-029 −	3.11e-029	9.73e-011
	DENLS	0.00e+000	0.00e+000	0.00e+000	0.00e+000	
f_2	OXDE	3.71e+001	6.89e+001	5.27e+001 −	6.89e+000	1.42e-009
	DECC-G	3.68e-015	8.16e-014	1.70e-014 −	1.77e-014	1.42e-009
	DENLS	2.64e-266	6.66e-262	3.95e-263	0.00e+00	
f_3	OXDE	1.17e+006	1.98e+006	1.54e+006 +	2.33e+005	1.42e-009
	DECC-G	2.25e-004	2.80e-003	1.20e-003 +	6.70e-004	1.42e-009
	DENLS	2.04e+006	6.50e+006	4.64e+006	1.18e+006	
f_4	OXDE	2.28e+001	2.91e+001	2.59e+001 −	1.84e+000	1.42e-009
	DECC-G	2.22e-002	4.09e-002	3.19e-002 −	4.72e-003	1.42e-009
	DENLS	1.33e-200	2.76e-197	2.51e-198	0.00e+000	
f_5	OXDE	1.16e+004	3.80e+004	2.28e+004 −	7.08e+003	1.42e-009
	DECC-G	9.87e+002	9.85e+002	9.86e+002 +	4.11e-001	1.42e-009
	DENLS	9.89e+002	9.89e+002	9.89e+002	7.99e-003	
f_6	OXDE	6.52e+003	9.36e+003	7.86e+003 −	8.86e+002	9.73e-011
	DECC-G	0.00e+000	0.00e+000	0.00e+000 ≈	0.00e+000	NaN
	DENLS	0.00e+000	0.00e+000	0.00e+000	0.00e+000	
f_7	OXDE	4.24e+000	7.35e+000	5.68e+000 −	7.25e-001	1.42e-009
	DECC-G	1.50e-003	3.70e-003	2.62e-003 −	6.68e-004	1.42e-009
	DENLS	1.01e-005	1.71e-004	6.91e-005	4.29e-005	
f_8	OXDE	-4.18e+005	-4.15e+005	-4.17e+005 −	7.27e+002	9.73e-011
	DECC-G	-4.19e+005	-4.19e+005	-4.19e+005 −	9.28e-011	2.77e-012
	DENLS	-4.19e+005	-4.19e+005	-4.19e+005	1.19e-010	
f_9	OXDE	3.63e+002	5.81e+002	4.94e+002 −	5.32e+001	9.73e-011
	DECC-G	3.55e-014	0.00e+000	1.25e-014 −	7.49e-015	3.51e-010
	DENLS	0.00e+000	0.00e+000	0.00e+000	0.00e+000	
f_{10}	OXDE	5.95e+000	6.93e+000	6.49e+000 −	2.79e-001	9.73e-011
	DECC-G	1.15e-013	1.39e-013	1.27e-013 −	7.11e-015	9.01e-011
	DENLS	8.88e-016	8.88e-016	8.88e-016	0.00e+000	
f_{11}	OXDE	3.46e+000	9.13e+000	5.02e+000 −	1.19e+000	9.73e-011
	DECC-G	6.66e-016	1.33e-015	9.50e-016 −	1.44e-016	8.05e-011
	DENLS	0.00e+000	0.00e+000	0.00e+000	0.00e+000	
f_{12}	OXDE	1.85e+000	4.53e+000	3.01e+000 −	7.11e-001	9.73e-011
	DECC-G	6.20e-028	1.95e-027	1.24e-027 −	3.30e-028	9.73e-011
	DENLS	4.71e-034	4.71e-034	4.71e-034	1.75e-049	
f_{13}	OXDE	1.33e+003	2.78e+003	1.94e+003 −	3.40e+002	9.73e-011
	DECC-G	5.27e-024	1.10e-002	2.64e-003 −	4.79e-003	9.34e-011
	DENLS	1.35e-032	1.35e-032	1.35e-032	5.59e-048	

Notes: NaN means that both algorithms can locate optimum each time.

4.2   Compared Results

The Best, Worst, Mean, and Std (standard deviation) values of three algorithms were recorded in table 1. And Wilcoxon rank-sum test with significance level 0.05 have been conducted to compare the three algorithms. The p values between DENSL and the compared algorithms are recorded as well. '-' indicates the p value is less than 0.05 and the performance of DENSL is better than that of compared algorithms (Win). '≈' indicates that there is no significant difference between two compared algorithms (Tie). '+' indicates the p value is less than 0.05 and the performance of DENSL is worse than that of the compared algorithms (Lose). The Win-Tie-Lose results of the three algorithms in solving thirteen problems are summarized in table 2.

Table 2. Win–Tie–Lose

	OXDE	DECC-G
DENSL	12-0-1	10-1-2

The results in tables 1-2 show us that the performance of DENSL is better than that of OXDE in solving twelve problems except problem 3 and the performance of DENSL is not worse than that of DECC-G in solving eleven problems except problems 3 and 5.To present more details about the convergence performance of the three algorithms, the evolution process of the algorithms in solving several problems are given in Figs.2-7.

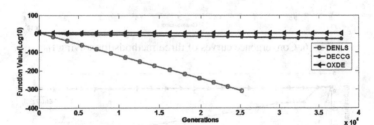

Fig. 2. Convergence curves of three methods for solving f_1

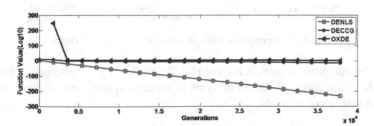

Fig. 3. Convergence curves of three methods for solving f_2

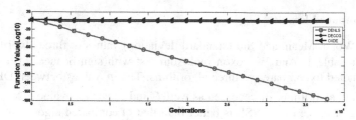

Fig. 4. Convergence curves of three methods for solving f_4

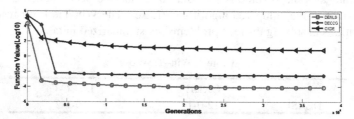

Fig. 5. Convergence curves of three methods for solving f_7

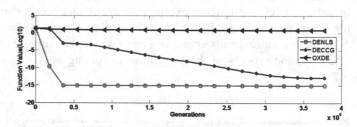

Fig. 6. Convergence curves of three methods for solving f_{10}

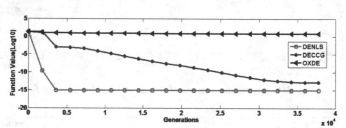

Fig. 7. Convergence curves of three methods for solving f_{13}

From the compared results, it is clear that DENLS is a good alternative for solving large scale optimization problems in terms of solution quality and convergence speed compared with other methods.

5　　Conclusion

DE can be easily trapped into local optima when solving large scale optimization problems. In order to enhance the ability of DE for solving large scale optimization problems, DE with a novel local search operation was proposed. In the novel local search operation, quasi-opposition based learning was adopted to enhance the performance of orthogonal crossover operation. The proposed DENLS is a hybrid algorithm of DE/rand/bin/1 with this novel local search operation. The compared results in solving the commonly used thirteen large scale optimization problems show that the DENLS is effective and efficient in tackling most of the large optimization problems.

Acknowledgments. This work is partially supported by Natural Science Foundation of China under grant No. 61364025, State Key Laboratory of Software Engineering under grant No. SKLSE2012-09-39 and the Science and Technology Foundation of Jiangxi Province, China under grant No. GJJ13729 and No. GJJ14742, and Science Foundation of Jiujiang University Under grant no.2013KJ27.

References

1. Storn, R., Price, K.V.: Differential evolution–A simple and efficient heuristic for global optimization over continuous spaces. J. Glob. Optim. **11**(4), 341–359 (1997)
2. Noman, N., Hitoshi, I.: Accelerating differential evolution using an adaptive local search. IEEE Transactions on Evolutionary Computation **12**(1), 107–125 (2008)
3. Yang, Z., Tang, K., Yao, X.: Large scale evolutionary optimization using cooperative coevolution. Information Sciences **178**(15), 2985–2999 (2008)
4. Wang, Y., Cai, Z., Zhang, Q.: Enhancing the search ability of differential evolution through orthogonal crossover. Information Sciences **185**(1), 153–177 (2012)
5. Rahnamayan, S., Gary, W.: Solving large scale optimization problems by opposition-based differential evolution (ODE). WSEAS Transactions on Computers **7**(10), 1792–1804 (2008)
6. Hui, W., Shahryar, R., Zhijian, W.: Parallel differential evolution with self-adapting control parameters and generalized opposition-based learning for solving high-dimensional optimization problems. J. Parallel Distrib. Comput. **73**(1), 62–73 (2013)
7. Rahnamayan, S., Tizhoosh, R., Salama, M.: Opposition versus randomness in soft computing techniques. Applied Soft Computing **8**(2), 906–918 (2008)
8. Potter, M.A., De Jong, K.A.: A cooperative coevolutionary approach to function optimization. In: Davidor, Y., Schwefel, H.-P., Männer, R. (eds.) Parallel Problem Solving from Nature, PPSN III. LNCS, vol. 866, pp. 249–257. Springer, Heidelberg (1994)
9. Tagawa, K., Ishimizu, T.: Concurrent differential evolution based on MapReduce. International Journal of Computers **4**(4), 161–168 (2010)
10. Qing, A.: Dynamic differential evolution strategy and applications in electromagnetic inverse scattering problems. IEEE Transactions on Geoscience and Remote Sensing **44**(1), 116–125 (2006)
11. Leung, Y.W., Wang, Y.: An orthogonal genetic algorithm with quantization for global numerical optimization. IEEE Transactions on Evolutionary Computation **5**(1), 41–53 (2001)
12. Rahnamayan, S., Tizhoosh, R., Salama, M.A.: Quasi-oppositional differential evolution. In: IEEE Congress on Evolutionary Computation, CEC 2007, pp. 2229–2236. IEEE (2007)

Clustering Time-Evolving Data Using an Efficient Differential Evolution

Gang Chen and Wenjian Luo[(⊠)]

Anhui Province key Laboratory of Software Engineering in Computing
and Communication, School of Computer Science and Technology,
University of Science and Technology of China, Hefei, Anhui, China
jeken@mail.ustc.edu.cn, wjluo@ustc.edu.cn

Abstract. The previous evolutionary clustering methods for time-evolving data usually adopt the temporal smoothness framework, which controls the balance between temporal noise and true concept drift of clusters. They, however, have two major drawbacks: (1) assuming a fixed number of clusters over time; (2) the penalty term may reduce the accuracy of the clustering. In this paper, a Multimodal Evolutionary Clustering (MEC) based on Differential Evolution (DE) is presented to cope with these problems. With an existing chromosome representation of the ACDE, the MEC automatically determines the cluster number at each time step. Moreover, instead of adopting the temporal smoothness framework, we try to deal with the problem from view of the multimodal optimization. That is, the species-based DE (SDE) for multimodal optimization is adopted in the MEC. Thus the MEC is a hybrid of the ACDE and the SDE, and designed for time-evolving data clustering. Experimental evaluation demonstrates the MEC achieves good results.

Keywords: Time-evolving data · Differential evolution · Multimodal optimization

1 Introduction

Time-evolving data refers to a collection of data that evolves over the time, which is ubiquitous in many dynamic scenarios, such as daily news, web browsing, e-mail, stocks markets and dynamic social network. Usually, a promise clustering result is desired at each time step. The key challenge to traditional clustering algorithms is that the data objects to be clustered evolve over the time, both as a result of long term trend due to changes in statistical properties and short term variations due to noise.

Chakrabarti et al. [1] first addressed the problem of clustering time-evolving data, called *evolutionary clustering*, and proposed a framework called temporal smoothness. Their framework considers the frequently shift of clustering results in a very short time is unexpected, and so, it tries to smooth the clusters over time. It divides the objective function into two terms: *snapshot quality* (Sq) measuring the clustering quality on the current data, and *history quality* (Hq) verifying how similar the current clustering result is with the previous one. Actually, the *history quality* is a

© Springer International Publishing Switzerland 2015
Y. Tan et al. (Eds.): ICSI-CCI 2015, Part I, LNCS 9140, pp. 326–338, 2015.
DOI: 10.1007/978-3-319-20466-6_35

penalty term which penalizes the deviation of the current clustering result from the previous. Thus their framework trades off the benefits of maintaining a consistent clustering over time with the cost of deviating from an accurate representation of the current data.

Based on the temporal smoothness framework, several evolutionary clustering methods have been proposed [2-4]. These methods control the balance between temporal noise and true concept drift, and achieve promise results. However, they have two major drawbacks:

(1) Assuming a fixed number of clusters over time. Limited to the static clustering model, most of the existing methods required pre-fixed cluster number k at each time step. And so, they are inadequate to handle the very typical scenario where the number of clusters at different times is unknown and varies with the time.

(2) The penalty term may reduce the accuracy of the clustering. Usually, the *history quality* is scaled by a smoothing parameter which reflects the reference to the history model. However, it is difficult to decide how much weight should be assigned to the history quality term. Though some existing works propose some strategies to adaptively estimate the optimal smoothing parameter [5, 6], it still in a way pulls down the accuracy of the clustering on the current data.

In this paper, an efficient evolutionary clustering method based on DE (differential evolution) is presented to deal with the problem of time-evolving data clustering. Evolutionary algorithms are widely used for static clustering. However, the work that clusters time-evolving data based on an evolutionary algorithm is little [6, 7]. We believe that the DE has advantages to the time-evolving data clustering.

On one hand, with the special chromosome representation scheme, the DE for clustering could automatically determine the number of clusters. In this paper, we adopt the chromosome representation in the ACDE [8], which is detailed in Section 3.

On the other hand, compared to the typical k-means, the DE could perform a global search in the solution space and provides robust and adaptive solutions. Therefore, instead of adopting the temporal smoothness framework, we try to deal with the problem from view of the multimodal optimization. The *temporal smoothness framework* assumes that the data to be clustered at different time steps should be identical, i.e., the data of time steps are "snapshots" of the same set of objects at different time. In this paper, the data of different time steps could be arbitrary I.I.D. samples from different underlying distributions. We adopt the SDE [9] to search the global and local optimal solutions. In order to ensure the current clustering does not deviate much from historical clustering results, the global/local optimum that is most similar to the historical clustering result is screened out as the best result at the current time step.

Thus, the DE used in this paper is a hybrid of ACDE [8] and SDE [9]. By adopting the chromosome representation in the ACDE, the proposed method could deal with the situation where the number of clusters is unknown and varies over time. Meanwhile, the best result at the current time step could be selected from the global and local optimal clusters obtained by the SDE, which is considered the most consistent with the historical clustering results.

2   Background

2.1   Evolutionary Clustering

The topic of evolutionary clustering has attracted significant attention in recent years. Chakrabarti et al. [1] first formulated the problem and proposed the temporal smoothness framework, where a history cost is added to the cost function to ensure the temporal smoothness. Chi et al. [2] extended the spectral clustering to the evolutionary setting and proposed two frameworks PCQ and PCM, by incorporating the temporal smoothness to restructure a cost function. Both the frameworks are to optimize the following cost function:

$$C_{total} = \alpha \cdot Sq + (1 - \alpha) \cdot Hq .   \tag{1}$$

Where Sq measures the clustering quality when the solution is applied to the current data and Hq is the measure of history quality. These methods significantly advance the evolutionary clustering literature. However, as stated in section 1, these methods assume the number of clusters is pre-fixed at each time step, which limits the applications in real world.

2.2   Evolutionary Algorithms for Static Clustering

Evolutionary algorithms for static clustering have been widely studied, and some of them have been put into practice [8]. For example, Bandyopadhyay et al. [10] applied evolutionary algorithms for clustering to distinguish landscape regions like mountains, rivers, vegetation areas and habitations in satellite images.

Evolutionary algorithms basically evolve the clustering solutions through operators that use probabilistic rules to process data partitions sampled from the search space. Roughly speaking, the solutions that more fitted the objective functions have higher probabilities to be sampled. Thus, the evolutionary search is biased towards more promising clustering solutions and tends to perform a more computationally efficient exploration of the search space than traditional randomized approaches.

2.3   The ACDE

In this paper, we adopt the special chromosome representation scheme of the ACDE, proposed by Das et al [8]. The ACDE is used to automatically cluster large unlabeled data sets. Different from most of the existing clustering techniques, the ACDE requires no prior knowledge of the data to be classified.

In the ACDE, for the data points of d dimensions at each time step, and for a user-specified maximum number of clusters K_{max}, a chromosome is a vector of real numbers of $K_{max} + K_{max} \times d$ dimensions. The first K_{max} positions are the control genes, each of which is positive floating-point number in [0, 1] and controls whether the corresponding cluster is to be activated. The remaining positions are reserved for

K_{max} cluster centers, where each center has d dimensions. For example, the i-th chromosome vector in the population is shown as follows:

$x_i =$	$T_{i,1}$	$T_{i,2}$	$\cdots$	$T_{i,K_{ma}}$	$m_{i,1}$	$m_{i,2}$	$\cdots$	$m_{i,K_{ma}}$

In the first K_{max} positions, if $T_{i,j} > 0.5$, the j-th cluster in the i-th chromosome is active and selected for partitioning the associated data set. Otherwise, if $T_{i,j} < 0.5$, the particular j-th cluster is inactive. Suppose $K_{max} = 4$, and we have a chromosome $x = $ (0.3, 0.7, 0.1, 0.9, 6.1, 3.2, 2.1, 6, 4.4, 7, 5, 8, 4.6, 8, 4, 4) for a 3-D data set. According to the rule, the second center (6, 4.4, 7) and the fourth one (8, 4, 4) are active for partitioning the data set.

The fitness function corresponds to the statistical mathematical function to evaluate the results of the clustering algorithm on a quantitative basis. The ACDE uses the CS measure [12], which is outlined as:

$$CS(x) = \frac{\sum_{i=1}^{K} \left[\frac{1}{N_i} \sum_{s_i \in C_i} \max_{s_q \in C_i} \{ d(s_i, s_q) \} \right]}{\sum_{i=1}^{K} \left[\min_{j \in K, j \neq i} \{ d(m_i, m_j) \} \right]}. \tag{2}$$

Where x is an individual having K clusters ($C_1, ..., C_K$), and m_i ($i=1,...,K$) is the centroid of C_i.

Thus the fitness function of an individual x can be described as:

$$f(x) = \frac{1}{CS(x) + e}. \tag{3}$$

Where e is the small constant and set to be 0.002.

In addition, in the ACDE [8], to avoid erroneous chromosomes, a cluster center is probably recalculated. That is, the centroid m_i is computed by averaging the data vectors that belong to cluster C_i by the following equation.

$$m_i = \frac{1}{N_i} \sum_{s_j \in C_i} s_j. \tag{4}$$

2.4   Multimodal Optimization and Species-Based DE

Multimodal optimization refers to finding multiple global and local optima of an objective functions, so that the user can have a better knowledge about different optimal solutions in the search space. Evolutionary algorithms have a clear advantage over the classical optimization techniques to deal with multimodal optimization problems, due to their population-based approaches are able to detect multiple solutions within a population in a single simulation run. Thus, numerous evolutionary optimization techniques have been developed since late 1970s for locating multiple optima.

Detection and maintenance of multiple solutions are the challenge of using EAs for multimodal optimization. Niching is a common technique used in multimodal optimization, which refers to the technique of finding and preserving multiple stable niches, or favorable parts of the solution space possibly around multiple solutions, so as to prevent convergence to a single solution. In this paper, a kind of niching techniques named speciation is used and a species-based DE (SDE) by Li [9] is adopted.

The SDE [9] is a speciation method, which classifies a DE population into groups according to their similarity based on Euclidean distance. The definition of a speciation depends on a parameter R_s, which denotes the radius measured in Euclidean distance from the center of a species to its boundary. The center of a species, called species seed, is considered the fittest individual in the species. All individuals whose distance from the species seed is no larger than the predefined parameter R_s are merged into the same species.

3   The Proposed Methods

3.1   Algorithm Design Principle

Let f^{t-1}, f^t be the two underlying data distribution at time step $t-1$ and t, respectively. And p^{t-1}, p^t are the corresponding model distributions. In evolutionary clustering setting, f^{t-1} and f^t are assumed to be close to each other. Zhang et al. [5] addressed the evolutionary clustering problem from view of statistical models, and proposed two general on-line frameworks, in which the *historical data dependent* (HDD) aims to minimize the loss function:

$$\zeta_{hdd} = -\int \left[(1-\lambda)f^t(x) + \lambda f^{t-1}(x)\right] log\ p^t(x)\,dx \ . \tag{5}$$

Where λ is a parameter that reflects the reference to historical data. In equation (5), $(1-\lambda)f^t(x) + \lambda f^{t-1}(x)$ induces another distribution. Zhang et al. employed an EFM to estimate the density of the deduced distribution.

In equation (5), we can observe that the parameter λ plays a role in trading of the reference between current data and historical data. So it essentially belongs to *temporal smoothness framework*. From multimodal optimization point of view, evolutionary clustering problem aims to find global optimal solutions that could reflect the current data accurately, and not deviate from the historical models simultaneously. Therefore, we could assume that the data of time step $t-1$ and t are arbitrary I.I.D. samples from different underlying distributions, and consider them simultaneously in the course of evolution. Thus the task is regarded as finding multiple global and local optima, and then screening out the best solution that fit the goals.

In multimodal optimization, if multiple solutions are known, the implementation could be quickly switched to another solution without much interrupting the optimal system performance. Evolutionary algorithms (EAs), have a clear natural advantage over classical optimization techniques. That is because the EAs maintain a population of possible solutions. And on termination of the algorithm, we will have multiple good solutions rather than only the best solution.

In order to screen out the best solution that fit the evolutionary properties, we adopt the NMI metric [13] to compare the clustering solutions:

$$NMI = \frac{\sum_{h=1}^{c} \sum_{l=1}^{c} n_{h,l} \, log(\frac{N \cdot n_{h,l}}{n_h \cdot n_l})}{\sqrt{\left(\sum_{h=1}^{c} n_h \, log \frac{n_h}{N}\right)\left(\sum_{l=1}^{c} n_l \, log \frac{n_l}{N}\right)}}. \tag{6}$$

Where $n_{h,l}$ denotes the number of agreements between clusters h and l, n_h and n_l is the number of data points in cluster h and l, respectively, and N represents the number of data points in the whole dataset. NMI usually lies in [0, 1]. A higher NMI value indicates that the current cluster solution more similar to the historical ones.

We further illustrate how to pick out the best solution. Assuming at time step t, SDE could finally find out three species seeds $x_{t,1}$, $x_{t,2}$ and $x_{t,3}$, corresponding to the three peeks in fig. 1, respectively. Apparently, $f(x_{t,2}) > f(x_{t,1}) > f(x_{t,3})$, and the partitions yielded by these three chromosomes are considered the three best solutions. However, because the NMI between $x_{t,1}$ and the historical solution x_{t-1} are the highest, $x_{t,1}$ is screened out as the best solution. Because the partition yielded by $x_{t,1}$ is most similar with the historical result.

Overall, we consider that each global/local optimum could be a reasonable partition for current data, while the global/local optimum that best clusters the historical data is selected for the final solution.

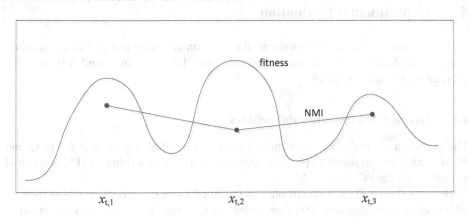

Fig. 1. Illustration on how to pick out the best solution

3.2   Algorithm Description

Here, a Multimodal Evolutionary Clustering (MEC) method based on SDE is presented, which produces multiple solutions at each time step, and compares the solutions to determine the best one, which is most similar with the historical clustering results. The procedure can be summarized as follows.

```
Input: data set 𝑺ₜ₋₁, 𝑺ₜ
        the radius Rₛ
        the solution of time step t-1, Pₜ₋₁
Output: the best solution xₜ
1.   Randomly generate an initial population with N
         individuals according to section 2.3.
2.   While not reach termination criteria
3.     Evaluate all individual in the population.
4.     Sort all individuals in decreasing order of their
           fitness values.
5.     Determine the species seeds for the current
           population according to SDE.
6.     For each species as identified via its species seed
7.       Run the DE with F = 0.8, Cr = 0.9, and a local
             search by one step k-means described in[6].
8.       On the termination of DE, compare the best
             solution xₛ with Pₜ₋₁ using NMI, if NMI(xₛ)>
           NMI(xₜ),then update the best solution.
9.     Keep only the N fitter individuals from the
           combined population.
10.  End while
11. Return the best solution xₜ.
```

4   Experimental Evaluation

In this section, we report the experimental results on synthetic and real world data set. Unless stated otherwise, all experiment are run 10 times independently and the average performance is given.

4.1   Baseline and Experimental Settings

The proposed method MEC is compared with two baseline algorithms: the PCQ and PCM, which are proposed in [2]. The spectral clustering algorithm in [14] is adopted to implement the PCQ and PCM.

Besides, the ACDE is implemented as the third baseline algorithm, which directly adopts the *temporal smoothness framework*. In fact, the fitness function presented in equation (3) could be easily modified to fit the *temporal smoothness framework*. According to equation (1), the *temporal smoothness* could be expressed as the adaptability of an individual in the old environment (the past time step). In other words, the history quality Hq is about how well the partitions, yielded by the individual x, when is applied to cluster the history data. Thus, formally, the history quality could be defined as:

$$Hq(x,t) = f_t(x) = \frac{1}{CS_t(x)+e} . \tag{7}$$

Where $CS_t(x)$ is the CS measure which is evaluated on the partitions yielded by the chromosome x at time step t, and e is the same as equation (3).

Finally, following the *temporal smoothness framework*, the fitness function of individual x is as follows.

$$F(x,t) = \alpha \cdot f_t(x) + (1-\alpha) \cdot Hq(x,t-1)$$
$$= \alpha \cdot \frac{1}{CS_t(x)+e} + (1-\alpha) \cdot \frac{1}{CS_{t-1}(x)+e} . \tag{8}$$

Where the timestamp $t-1$ indicates that the CS measure is evaluated on the history data (in this paper, we only consider the adjacent time step t and $t-1$).

The comparisons are conducted using the fitness value and the NMI. Both the values of metrics are the higher, the better cluster performance is. For fair comparison, we calculate the fitness values of cluster solutions produced by PCQ and PCM according to equation (3).

The parameter α in PCQ, PCM and ACDE is set to be 0.8 in all the experiments. For ACDE and MEC, the population size is set to be 10 times of the dimension of the data objects. For example, the synthetic data set is 2-dimenstional, thus the population size in ACDE and MEC is 20. The crossover probability is 0.9 and the scale factor F is set to be 0.8. The maximum and minimum number of clusters is 20 and 2, respectively.

4.2   Experiments on Synthetic Data

The data sets are generated by means of the similar generation algorithm in [2]. Two hundred 2-D data objects are initially generated as shown in fig. 1(a), by two Gaussian distributions generating 100 data objects with mean values of (3, 3) and (3, 7), respectively. In order to simulate the data evolving process, the means of the two Gaussian distributions are moved slowly towards a pre-fixed evolving direction, which is set as $\Delta_1 = (0.5, -0.5)$ for the upper cluster and $\Delta_2 = (-0.5, 0.5)$ for the lower cluster. The overall data objects of the whole 6 time steps are shown in fig. 1(b).

We use two experiment on synthetic data sets to investigate the performance of the proposed method MEC. In the first experiment, we verify our methods in the situation where there are concept drifts, but the number of clusters keeps the same over the time. In the second experiment, we demonstrate the proposed methods could deal with the scenario where the number of clusters at different times varies.

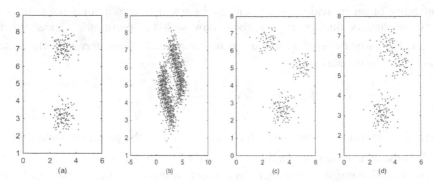

Fig. 2. The synthetic data set

Table 1. Experimental results on synthetic data set

methods	criteria	1	2	3	4	5	6
MEC	*Sq*	**1.983**	**1.5522**	**1.4458**	1.5206	1.4960	**1.5679**
	Hq	0	**1.7533**	**1.4336**	**2.0668**	**1.7529**	**1.7529**
	no. of clusters found	2	2	2	2	2	2
ACDE	*Sq*	**1.983**	**1.5522**	1.395	1.5413	**1.983**	**1.5679**
	Hq	0	1.5522	1.3555	1.5417	1.395	1.5679
	no. of clusters found	2	2	2	2	2	2
PCQ	*Sq*	**1.983**	1.5383	1.40429	**1.5658**	**1.983**	**1.5679**
	Hq	0	1.5649	1.3549	1.431	1.395	1.5679
	no. of clusters found	2	2	2	2	2	2
PCM	*Sq*	**1.983**	1.5383	1.4042	**1.5658**	**1.983**	**1.5679**
	Hq	0	1.5649	1.3549	1.431	1.395	1.5679
	no. of clusters found	2	2	2	2	2	2

The first experimental results on synthetic data are reported in Table 1. The bold values in each column denote the best values in terms of the fitness values or NMI values. *Sq* is the abbreviation of *snapshot quality*, and *Hq* is *history quality*. For the PCQ and PCM, the number of clusters is pre-fixed, resulting in 2 clusters at each time step. From Table 1, we can see that ACDE and MEC correctly partition the data set into 2 clusters all the time. Though the *snapshot quality* are not always better than PCQ and PCM over all the time steps (actually they are the same at time step 1 and 6), however, the *history quality* of MEC are better than the others. This could be attributed to the ability of MEC to provide multiple solutions and picks out the one that most similar to the historical solutions. The NMI values shown in Table 2 indicate that, the cluster accuracy of each method has its advantage at different time steps.

In the second experiment, we generate another data set, described in fig. 1(c, d). The lower cluster keeps static over the time, while the two upper clusters moves towards each other. At time step 4, the means of these two clusters overlap together at location (4, 6), thus the total number of clusters turn into 2, see fig. 1(d). We then examine how the ACDE and MEC response to the change of number of clusters.

Table 2. The NMI values on synthetic data

methods	criteria	1	2	3	4	5	6
MEC	*NMI*	**0.9077**	0.8501	0.9293	0.8301	0.919	**0.9596**
ACDE	*NMI*	0.8926	0.8331	**0.9513**	0.7994	**0.9596**	0.9204
PCQ	*NMI*	0.8907	0.8544	0.8177	0.822	0.9133	0.8217
PCM	*NMI*	**0.9077**	**0.8612**	0.8153	**0.8379**	0.9120	0.8379

We do not compare the proposed methods with PCQ and PCM, because they are incapable to deal with changeable number of clusters. Table 3 displays the cluster performance of ACDE and MEC in the second experiment. We can see that ACDE and MEC correctly partition the data into 2 clusters at time step 4. Moreover, in order to keep smooth to the previous cluster solution, ACDE and MEC still partition the data into 2 clusters at time step 5. However, ACDE partition the data at time step 3 into 2 clusters, while MEC successfully partitions them into 2. From the second experiment, we can see that ACDE and MEC could automatically determine the number of clusters at each time step, and maintain a consistent clustering over time.

Table 3. The cluster performance of ACDE and MEC

methods	criteria	1	2	3	4	5	6
MEC	*Sq*	**1.5519**	1.4288	**1.3608**	**1.4603**	**1.4159**	1.3600
	Hq	0	**1.805**	**1.4385**	**1.4603**	**1.4159**	**1.5267**
	NMI	**0.9772**	**0.9772**	**0.9033**	**0.9624**	**0.7771**	**0.9772**
	no. of clusters found	3	3	3	2	2	3
ACDE	*Sq*	**1.5519**	1.5882	1.2823	1.3467	1.3467	**1.4788**
	Hq	0	1.4964	1.0248	1.3467	1.3467	1.5032
	NMI	**0.9772**	0.9056	0.7771	**0.9624**	0.7771	**0.9772**
	no. of clusters found	3	3	2	2	2	3

4.3   Experiments on Real World Data set

In this section, we present outcomes on the UCI-benchmarking data sets. We use the *Iris* and *breast cancer* data sets to test the performance of MEC, compared with ACDE, PCQ and PCM. Both the data sets are available in the UCI depository.

Each cluster in the datasets is set a pre-fixed direction for adding the Gaussian distribution to simulate the evolving behavior of the data. The experimental results are displayed in the Table 4 to 5. Since the data attributes change randomly, it is difficult to determine each data point's belonging clusters. Thus we do not use the NMI to verify the performance.

From Table 4 and 5, we can see that the MEC outperforms the other methods in terms of the *snapshot quality* and *history quality*. The ACDE and MEC correctly partition the breast cancer data set into 2 clusters. While on the Iris data set, MEC partitions the data into 2 clusters at time step 2 to 5. The ACDE, however, is not satisfying on finding the optimal number of clusters. Note that in most of the literatures, the Iris data set are often partitioned into 2 clusters, which is considered acceptable.

Table 4. The experimental results on Iris data set

methods	criteria	1	2	3	4	5	6
MEC	Sq	**1.4556**	**1.1932**	1.1082	**1.122**	**1.1208**	**1.1555**
	Hq	0	**1.5011**	**1.1385**	**1.1336**	**1.1959**	**1.1907**
	no. of clusters found	3	2	2	2	2	3
ACDE	Sq	1.4117	0.9515	**1.1421**	0.8914	0.8146	1.1253
	Hq	0	1.4527	1.1099	0.8663	0.8219	1.1168
	no. of clusters found	2	4	2	3	4	2
PCQ	Sq	0.9521	0.8403	0.8681	0.7911	0.7841	0.8055
	Hq	0	1.2772	0.8381	0.8112	0.8083	0.7727
	no. of clusters found	3	3	3	3	3	3
PCM	Sq	0.9671	0.8541	0.9018	0.8253	0.815	0.8754
	Hq	0	1.2976	0.8669	0.8482	0.8355	0.8462
	no. of clusters found	3	3	3	3	3	3

Table 5. The experimental results on breast cancer data set

methods	criteria	1	2	3	4	5	6
MEC	Sq	**0.9891**	**0.9573**	0.9046	**0.9182**	**0.9753**	**1.0046**
	Hq	0	**1.0242**	**0.9089**	**0.9282**	**1.0026**	**1.0098**
	no. of clusters found	2	2	2	2	2	2
ACDE	Sq	0.9301	0.9144	**0.9202**	0.8981	0.9036	0.9701
	Hq	0	0.9417	0.9099	0.9102	0.9045	0.9636
	no. of clusters found	2	2	2	2	2	2
PCQ	Sq	0.9218	0.9180	0.8871	0.8719	0.8702	0.9029
	Hq	0	0.88929	0.8692	0.8868	0.8776	0.8714
	no. of clusters found	2	2	2	2	2	2
PCM	Sq	0.9035	0.9149	0.9079	0.9036	0.9073	0.9045
	Hq	0	0.9389	0.9022	0.9177	0.9076	0.9044
	no. of clusters found	2	2	2	2	2	2

5   Conclusion

In this paper, we deal with evolutionary clustering problem from view of multimodal optimization. By adopting the chromosome representation in the ACDE, the proposed method could deal with the situation where the number of clusters is unknown and varies over time. Meanwhile, the best result at the current time step could be selected from the global and local optimal clusters found by the SDE, which is considered the most consistent with the historical clustering results. Compared with the existing work, the experimental results of the proposed method are not bad. Although there is still much room for improvement, especially the details on controlling the search process of multimodal optimization for clustering time-evolving data, we think that the basic framework of the proposed method is promising.

Acknowledgements. This work is partly supported by Anhui Provincial Natural Science Foundation (No. 1408085MKL07).

References

1. Chakrabarti, D., Kumar, R., Tomkins, A.: Evolutionary clustering. In: 12th ACM SIGKDD, pp. 554–560 (2006)
2. Chi, Y., et al.: Evolutionary spectral clustering by incorporating temporal smoothness. In: 13th ACM International Conference on Knowledge Discovery and Data Mining, pp. 153–162 (2007)
3. Wang, L., et al.: Low-Rank Kernel Matrix Factorization for Large-Scale Evolutionary Clustering. IEEE Transactions on Knowledge and Data Engineering **24**(6), 1036–1050 (2012)
4. Xu, K.S., Kliger, M., Hero III, A.O.: Adaptive evolutionary clustering. Data Mining and Knowledge Discovery **28**(2), 304–336 (2014)
5. Zhang, J., et al.: On-line Evolutionary Exponential Family Mixture. In: IJCAI, pp. 1610–1615 (2009)
6. Chen, G., Luo, W., Zhu, T.: Evolutionary clustering with differential evolution. In: 2014 IEEE Congress on Evolutionary Computation (CEC), pp. 1382–1389. IEEE (2014)
7. Ma, J., et al.: Spatio-temporal data evolutionary clustering based on MOEA/D. In: 13th Annual Conference Companion on Genetic and Evolutionary Computation, pp. 85–86. ACM (2011)
8. Das, S., Abraham, A., Konar, A.: Automatic clustering using an improved differential evolution algorithm. IEEE Transactions on Systems, Man and Cybernetics, Part A **38**(1), 218–237 (2008)
9. Li, X.: Efficient differential evolution using speciation for multimodal function optimization. In: 7th Annual Conference on Genetic and Evolutionary Computation, pp. 873–880. ACM (2005)
10. Bandyopadhyay, S., Maulik, U.: An evolutionary technique based on K-means algorithm for optimal clustering in RN. Information Sciences **146**(1), 221–237 (2002)

11. Chou, C.-H., Su, M.-C., Lai, E.: A new cluster validity measure and its application to image compression. Pattern Analysis and Applications 7(2), 205–220 (2004)
12. Strehl, A., Ghosh, J.: Cluster ensembles—a knowledge reuse framework for combining multiple partitions. The Journal of Machine Learning Research 2, 583–617 (2003)
13. Chen, W.-Y., et al.: Parallel spectral clustering in distributed systems. IEEE Transactions on Pattern Analysis and Machine Intelligence 33(3), 568–586 (2011)

Multi-objective Differential Evolution Algorithm for Multi-label Feature Selection in Classification

Yong Zhang[✉], Dun-Wei Gong, and Miao Rong

School of Information and Electrical Engineering,
China University of Mining and Technology, Xuzhou 221116, China
yongzh401@126.com

Abstract. Multi-label feature selection is a multi-objective optimization problem in nature, which has two conflicting objectives, i.e., the classification performance and the number of features. However, most of existing approaches treat the task as a single objective problem. In order to meet different requirements of decision-makers in real-world applications, this paper presents an effective multi-objective differential evolution for multi-label feature selection. The proposed algorithm applies the ideas of efficient non-dominated sort, the crowding distance and the Pareto dominance relationship to differential evolution to find a Pareto solution set. The proposed algorithm was applied to several multi-label classification problems, and experimental results show it can obtain better performance than two conventional methods. *abstract* environment.

Keywords: Classification · Multi-label feature selection · Multi-objective · Differential evolution

1 Introduction

Multi-label classification is a challenging problem that emerges in many modern real applications [1,2]. By removing irrelevant or redundant features, feature selection can effectively reduce data dimensionality, speeding up the training time, simplify the learned classifiers, and/or improve the classification performance [3]. However, this problem has not received much attention yet. In the few existing literature, a main way is to convert multi-label problems into traditional single-label multi-class ones, and then each feature is evaluated by new transformed single-label approach [4-6]. This way provides a connection between single-label learning and multi-label learning. However, since a new created label maybe contain too many classes, this way may increase the difficulty of learning, and reduce the classification accuracy.

Differential evolution (DE) has been applied to single-label feature selection [7,8], because of population-based characteristic and good global search capability. However, the use of DE for multi-label feature selection has not been investigated. Compared with single-label classification learning [9], since there can

© Springer International Publishing Switzerland 2015
Y. Tan et al. (Eds.): ICSI-CCI 2015, Part I, LNCS 9140, pp. 339–345, 2015.
DOI: 10.1007/978-3-319-20466-6_36

be complex interaction among features, and these labels are usually correlated, multi-label feature selection becomes more difficult. Furthermore, multi-label feature selection has two conflicting objectives: maximizing the classification performance and minimizing the number of features. Therefore, in this paper, we study an effective multi-objective approach for multi-label feature selection based on DE.

2   Problem Formulation

We use a binary string to represent solutions of the problem. Taking a set of data with D features as an example, a solution of the problem can be represented as follows:

$$X = (x_1, x_2, \cdots, x_D), \ x_i \in \{0,1\}, \ i = 1, 2, \cdots, D. \tag{1}$$

Selecting hamming loss [5] to evaluate the classification performance of classifier which is decided by feature subsets, a multi-label feature selection problem is formulated as a combinatorial multi-objective optimization one with discrete variables:

$$\min \ F(X) = (Hloss(X), \ |X|) \tag{2}$$

Where $|X|$ represents the number of features, $Hloss(X)$ is the hamming loss in terms of the feature subset X.

3   Proposed Algorithm

3.1   Encoding

In DE, an individual refers to a possible solution of the optimized problem, thus it is very important to define a suitable encoding strategy first. This paper adopts the probability-based encoding strategy proposed in our previous work [10]. In this strategy, an individual is represented as a vector of probability,

$$P_i = (p_{i,1}, p_{i.2}, \cdots, p_{i.D}), \ p_{i,j} \in [0,1] \tag{3}$$

Where the probability $p_{i,j} > 0.5$ means that the j-th feature will be selected into the i-th feature subset.

3.2   Improved Randomized Localization Mutation

For mutation operator in DE, the traditional approach chooses the base vector at random within three vectors [11]. This approach has an exploratory effect but it slows down the convergence of DE. Randomized localization mutation (RLM) is first introduced to deal with single objective optimization in [12]. Since it can get a good balance between the global exploratory and convergence capabilities, this paper extends it to the multi-objective case by incorporating the Pareto domination relationship. The improved mutation is described as follows:

$$V_i(t) = P_{i,best}(t) + F \cdot (P_{r2}(t) - P_{r3}(t)) \tag{4}$$

Where $P_{i,best}(t)$ is the non-dominated one among the three random vectors, $P_{r2}(t)$ and $P_{r3}(t)$ are the rest two vectors.

3.3   Selection Based on Efficient Non-dominated Sort

For selection operator, fast non-dominated sorting (FNS) proposed in [13] were often used to finding Pareto-optimal individuals in DE. Efficient non-dominated sort (ENS) [14] is a new, computationally efficient comparison technique. Theoretical analysis shows that it has a space complexity of $O(1)$, which is smaller than FNS. Based on the advantage above, this paper uses a variation of the ENS, together with the crowding distance, to update the external archive.

Supposing that the parent population at generation t is S_t, the set of trial vectors produced by crossover and mutation is Q_t , first all the individuals among $R_t = S_t \cup Q_t$ are classed into different rank sets according to ENS. Herein, a solution to be assigned to the Pareto front needs to be compared only with those that have already been assigned to a front, thereby avoiding many unnecessary dominance comparisons. Individuals belonging to the first rank set are of best ones in R_t. And then, the new population is selected from subsequent rank sets in the order of their ranking. If the number of individuals selected goes beyond the population size, then individuals that have high rank and crowding distance values are deleted.

3.4   Implement of the Proposed Algorithm

Based on these operators above and some established operators, detailed steps of the proposed algorithm are described as follows:

Step 1: Initialize. First, set relative parameters, including the population size N, the scale factor F, the crossover probability CR, and the maximal generation times T_{max}. Then, initialize the positions of individuals in the search space.

Step 2: Implement the mutation proposed in subsection 3.2.

Step 3: Implement the uniform crossover technique introduced in [15] to generate a trail vector, i.e., a new offspring;

Step 4: Select the new population. First, evaluate the fitness of each offspring by the method introduced in subsection 3.1; then, combine these offsprings and the parent population, and generate new population by using the method proposed in subsection 3.3;

Step 5: Judge whether the algorithm meets termination criterion. If yes, stop the algorithm, and output the individuals with the first rank as finial result; otherwise, go to step 2.

Furthermore, Figure 1 shows the flowchart of the proposed multi-objective feature selection algorithm.

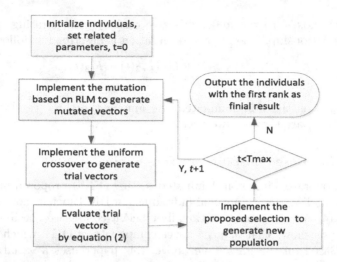

Fig. 1. Flowchart of the proposed DE-based multi-objective feature selection algorithm

3.5   Complexity Analysis

Since ENS and the crowding distance both need $O\,(MNlogN)$ basic operation, the mutation operator needs $O\,(3N)$ basic operation, and the crossover operator needs $O\,(N)$ basic operation, the time complexity of the proposed algorithm can be simplified as $O\,(MNlogN)$.

4   Experiments and Results

We compared the proposed algorithm with two conventional multi-label feature selection methods, ReliefF based on the binary relevance (RF-BR) [4] and mutual information approach based on pruned problem transformation (MI-PPT) [6].

Table 1. Data sets used in experiments

Data sets	Patterns	Features	Labels
Emotions	593	72	6
Yeast	2417	103	14
Scene	2407	294	6

Table 2. Best hamming loss obtained by the two algorithms

	Proposed algorithm		RF-BR	
Datasets	HLoss	No. of features	HLoss	No. of features
Emotions	0.18	17.89	0.22	17.2
Yeast	0.193	39.29	0.24	40.58
Scene	0.087	140.24	0.12	56.3

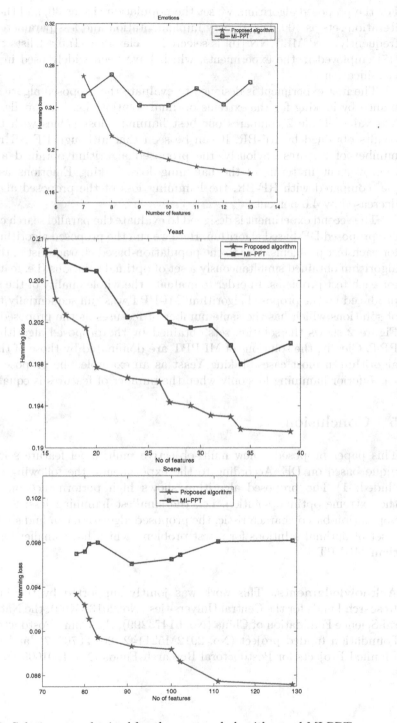

Fig. 2. Solution sets obtained by the proposed algorithm and MI-PPT

For the proposed algorithm, we set the population size as 30, and the maximum iteration sets as 100. Due to easy implementation and less parameter, the most frequently used ML-KNN [16] is selected as classifier. Table 1 lists the datasets [17] employed in the experiments, which have been widely used in multi-label classification.

The first experiment is designed to evaluate the proposed algorithms performance by looking for the extreme optimal solution, i.e., the smallest hamming loss value. Table 2 compares our best hamming loss values with the existing results obtained by RF-BR. It can be seen that although RF-BR reduced the number of features obviously, the proposed algorithm obtained a significant improvement in terms of the hamming loss. Taking Emotions as an example, compared with RF-BR, the hamming loss of the proposed algorithm has decreased by 4.0 percent.

The second experiment is designed to evaluate the parallel search capability of the proposed DE-based algorithm. Here we run the proposed algorithm only once for each test problems. Due to the population-based characteristic, the proposed algorithm obtained simultaneously a set of optimal solutions (i.e, feature subset) for each test problems. In order to evaluate the whole quality of the solution set produced by the proposed algorithm, MI-PPT was run sequentially to find a set of solutions which has the same number of features as the proposed algorithm. Figure 2 shows the solution sets obtained by the proposed algorithm and MI-PPT. Clearly, the solutions of MI-PPT are dominated by those of the proposed algorithm in most cases. Taking Yeast as an example, the proposed algorithm has inferior hamming loss only when the number of features is equal to 15.

5 Conclusion

This paper proposed a new multi-objective multi-label feature selection technique based on DE. According to the experiments, the following can be concluded: 1) The proposed algorithm shows high performance on looking for the extreme optimal solution, i.e., the smallest hamming loss; 2) Due to the population-based characteristic, the proposed algorithm can find simultaneously a set of optimal solutions for a test problem, which have smaller hamming loss than MI-PPT.

Acknowledgments. This work was jointly supported by the Fundamental Research Funds for the Central Universities (No. 2013XK09), the National Natural Science Foundation of China (No. 61473299), the China Postdoctoral Science Foundation funded project (No. 2012M521142, 2014T70557), and the Jiangsu Planned Projects for Postdoctoral Research Funds (No. 1301009B).

References

1. Schapire, R., Singer, Y.: Boostexter: A boosting-based system for text categorization. Machine Learning. **39**, 135–168 (2000)
2. Sun, F.M., Tang, J.H., Li, H.J., Qi, G.J., Huang, S.: Multi-label image categorization with sparse factor representation. IEEE Transactions on Image Processing. **23**, 1028–1037 (2014)
3. Xue, B., Zhang, M.J., Browne, W.N., Huang, S.: Particle swarm optimization for feature selection in classification: a multi-objective approach. IEEE Transactions on cybernetic. **43**, 1656–1671 (2013)
4. Spolaor, N., Alvares Cherman, E., Carolina Monard, M., Lee, H.D.: A comparison of multi-label feature selection methods using the problem transformation approach. Electronic Notes in Theoretical Computer Science. **292**, 135–151 (2013)
5. Chen, W., Yan, J., Zhang, B., Chen, Z., Yang, Q.: Document transformation for multi-label feature selection in text categorization. In: 7th IEEE International on Data Mining, pp. 451–456. IEEE Press, Omaha NE (2007)
6. Doquire, Gauthier, Verleysen, Michel: Feature selection for multi-label classification problems. In: Cabestany, Joan, Rojas, Ignacio, Joya, Gonzalo (eds.) IWANN 2011, Part I. LNCS, vol. 6691, pp. 9–16. Springer, Heidelberg (2011)
7. Sikdar, U.K., Ekbal, A., Saha, S., Uryupina, O., Poesio, M.: Differential evolution-based feature selection technique for anaphora resolution. Soft Computing (2014). doi:10.1007/s00500-014-1397-3
8. Ani, A., Alsukker, A., Khushaba, R.N.: Feature subset selection using differential evolution and a wheel based search strategy. Swarm and Evolutionary Computation **9**, 15–26 (2013)
9. Unler, A., Murat, A.: A discrete particle swarm optimization method for feature selection in binary classification problems. European Journal of Operational Research **206**, 528–539 (2010)
10. Zhang, Y., Gong, D.W.: Feature selection algorithm based on bare bones particle swarm optimization. Neurocomputing **148**, 150–157 (2015)
11. Fieldsend, J., Everson, R.: Using unconstrained elite archives for multi-objective optimization. IEEE Transactions on Evolutionary Computation **7**, 305–323 (2003)
12. Kumar, P., Pant, M.: Enhanced mutation strategy for differential evolution. In: IEEE World Congress on Computational Intelligence, pp. 1–6. IEEE Press, Brisbane (2012)
13. Zhang, X., Tian, Y., Cheng, R., Jin, Y.C.: An efficient approach to non-dominated sorting for evolutionary multi-objective optimization. IEEE Transactions On Evolutionary Computation. (2015). doi:10.1109/TEVC.2014.2308305
14. Deb, K., Pratap, A., Agarwal, S., Meyarivan, T.: A fast and elitist multiobjective genetic algorithm: NSGA-II. IEEE Transactions On Evolutionary Computation **6**, 182–197 (2002)
15. Englebrecht, A.P.: Computational Intelligence: An Intorduction(second edition). John Wiley and Sons, West Sussex (2007)
16. Zhang, M.L., Zhou, Z.H.: ML-KNN: A lazy learning approach to multi-label learning. Pattern Recognition **40**, 2038–2048 (2007)
17. A Java Library for Multi-Label Learning. http://mulan.sourceforge.net/datasets.html

Research on Network Coding Optimization Using Differential Evolution Based on Simulated Annealing

Liying Zhang, Xinjian Zhuo, and Xinchao Zhao[(✉)]

School of Science, Beijing University of Posts and Telecommunications,
Beijing 100876, China
zhaoxc@bupt.edu.cn

Abstract. Network coding can reduce the data transmission time and improves the throughput and transmission efficiency. However, network coding technique increases the complexity and overhead of network because of extra coding operation for information from different links. Therefore, network coding optimization problem becomes more and more important. In this paper, a differential evolution algorithm based on simulated annealing (SDE) is proposed to solve the network coding optimization problem. SDE introduces individual acceptance mechanism based on simulated annealing into canonical differential evolution algorithm. SDE finds out the optimal solution and keeps the population diversity during the process of evolution and avoids falling into local optimum as far as possible. Simulation experiments show that SDE can improve the local optimum of DE and finds network coding scheme with less coding edges.

Keywords: Network coding · Differential evolution · Simulated annealing · SDE

1   Introduction

Network coding [1] is a type of data transfer mode, which is an extension on the concept of routing. In a conventional network transmission policy, the intermediate nodes in the network only simply forward or copy forward the information of upstream link. Network coding allows intermediate nodes mixed coding the information from different incoming links to generate new information and forwards [2]. Network coding combines the information exchange technologies of routing and coding. Its core idea is linear or nonlinear process the information received from each channel on each node and forward to downstream nodes. The intermediate nodes play the role of encoder or signal processor. Network coding shows its superiority by improving network throughout, improving load balancing, reducing transmission delay, saving energy consumption of nods and increasing network robustness. Network coding also could be widely used in Ad Hoc networks, sensor networks, P2P content distribution, distribution file storage, network security and many other fields. The theory and application of network coding has become a hot research field [3].

However, introduction of coding operation and transferring of encoding vector brings additional calculations and control overhead for nodes in network. The core

© Springer International Publishing Switzerland 2015
Y. Tan et al. (Eds.): ICSI-CCI 2015, Part I, LNCS 9140, pp. 346–353, 2015.
DOI: 10.1007/978-3-319-20466-6_37

optimization problem of network coding is how to seek an optimal network coding scheme when network resources are consumed as less as possible and only the essential link of network coding is retained [2]. At the same time, it should ensure that network speeds reach the theoretical maximum of the premise. Kim et al. [4] proved that network coding optimization problem is NP-hard problem, and then they are the first to use genetic algorithm (GA) to solve the network coding optimization problem. With the increase of the problem size, the single cycle time of genetic algorithm increases dramatically and the running time of the algorithm is also unacceptable. The quality of the obtained results also significantly decline. This method does not consider as far as possible to reduce the unnecessary links.

Differential evolution algorithm (DE) was first proposed in 1995 by Storn et al [7]. The principle of differential algorithm is to randomly select multiple individuals to construct difference vector to update individual and diversify the search direction of the individual. It aims to decline the function value. It has shown a clear advantage on solving continuous optimization problems [5,6]. Similar with generic algorithm (GA), DE is free of gradient information and includes mutation, crossover and selection operations. It updates population with one-elimination mechanism. DE is essentially a kind of greedy GA with elitism. On the other hand, DE algorithm takes into account correlation between variables, so it has a great advantage on solving variable coupling problem when comparing with GA. The greedy selection operation of DE easily results in premature convergence. In order to maintain population diversity well and search routes of DE, simulated annealing strategy is introduced into DE to select the better network coding scheme according to the acceptance probability distribution.

This paper presents a hybrid DE based on simulated annealing idea for network coding optimization problem. Section 2 describes DE algorithm, simulated annealing principle and network coding optimization; Section 3 combines simulated annealing with differential evolution for network coding optimization problem; Section 4 presents the simulation results and analysis; and finally concluding remarks.

2   Background Knowledge

2.1   Differential Evolution Algorithm

DE firstly generates an initial population and then produces new population through mutation, crossover and selection operation, where $i = 1,...,NP$, NP population size; $j = 1,...,D$, D is problem dimension, g is current iteration variable.

Initialization. Initialize population $Xi = (x_{i,1}, x_{i,2},..., x_{i,D})$ as follows [8]:

$$x_{i,j} = randj \times (U_j - L_j) + L_j \tag{1}$$

Where *randj* is a random number in (0, 1), U_j and L_j denote the upper and lower bounds for the *j*-dimension.

Mutation. Difference vector of different individuals is scaled and is added to another different individual, then a trial solution is obtained, where $r1 \neq r2 \neq r3 \neq i$, F denotes scaling factor, the most common mutation operation, DE / rand / 1, is following:

$$v_{i,g} = x_{r1,g} + F(x_{r2,g} - x_{r3,g}) \tag{2}$$

Crossover. DE usually uses binomial crossover operation:

$$u_{i,j,g} = \begin{cases} v_{i,j,g} & rand(0,1) \leq CR \| j = j_{rand} \\ x_{i,j,g} & other \end{cases} \tag{3}$$

CR is crossover operation probability, which is usually assigned in [0.4, 0.9], j_{rand} is a constant which locates in $[1, D]$.

Selection. DE adopts greedy selection strategy according to the fitness value f of target vector and test vector. For minimization problem, selection operation is as follows:

$$x_{i,g+1} = \begin{cases} v_{i,g} & f(u_{i,g}) < f(x_{i,g}) \\ x_{i,g} & other \end{cases} \tag{4}$$

$x_{i,g+1}$ is a solution vector for next generation, DE produces new population through differential mutation, crossover and selection operations.

2.2   Simulated Annealing

Simulated annealing [9] is a stochastic optimization algorithm with Monte Carlo iterative solution strategy. It is originated from solid's annealing process, which needs setting an initial temperature and an annealing control parameter in order to adapt to a certain probability to accept a lower value of the solution. It makes the algorithm always have the ability to jump out of local optimum. SA has proven to be an optimal global algorithm theoretically, but the prerequisite is to choose the appropriate annealing control parameters. The slower the cooling schedule, or rate of decrease, the more likely the algorithm is to find an optimal or near-optimal solution.

1. Give an initial solution x and an initial temperature T, then compute the function value;
2. Generate random perturbations Δx and get a new solution $x' = x + \Delta x$, then compute function perturbations $\Delta f = f(x') - f(x)$;
3. Accept new solution or remain unchanged according to Metropolis criteria:

$$x = \begin{cases} x' & if \ (\Delta f < 0) \\ x' & if \ [p(\Delta f) \geq rand] \\ x & else \end{cases} \tag{5}$$

$p(\Delta f) = e^{-\frac{\Delta f}{T}}$ is the acceptance probability and *rand* is a uniform random in [0,1];

4. Decrease temperature. $T = T \times \lambda$, λ is parameter of temperature decreasing;

5. If terminate then output final solution; else Return to (2).

2.3   Network Coding Optimization

Network coding optimization aims at reducing computing complexity and other expenses of network as far as possible resulting from network coding under the condition that the maximum information transmission speed reaches the max multicast speed. Network coding optimization problems can be divided into following four categories [10]: minimum cost multicast, maximum undirected network throughput, minimize coding nodes and edges and the network topology design based on network coding. The first two topics also exist in the routing network, but they are NP-complete in the routing network. The latter two topics were proposed after network coding problem appeared. Research on these two issues has great influence in practical applications for the promotion of network coding.

Figure 1 is the simplest butterfly diagram network. S is source node. t1 and t2 are receiving nodes. The information sent from S should arrive at two receiving nodes simultaneously with certain propagation rate R. However, due to the limitation of channel capacity, such as C-D channel, coding is necessary if information x and y is required to transfer to receiving nodes simultaneously [10]. When data is sent to multiple recipients (t1, t2) at the same time, network coding can significantly enhance the multicast rate of networks.

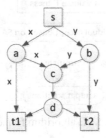

Fig. 1. Simple butterfly diagram network

Kim et al. [11] applied GA to solving network coding link optimization problem firstly. Simulation experiments indicate that GA has better performance than traditional greedy algorithm. After that, chromosome gene block operation method is proposed for crossover and mutation step and distributed algorithm is given. However, the evolution process of GA is easy to fall into local optimum. Reference [10] shows that standard DE has more advantages than GA on solving network coding optimization problem. Based on these observations, this paper introduces selection strategy of simulated annealing to solve the network coding optimization problem.

3   Hybrid Differential Evolution Based on Annealing Strategy

3.1   Algorithm Framework

In theory, differential evolution algorithm achieves optimization by swarm evolution operation based on "survival of the fittest" ideology. The key operation of DE is its unique mutation, but the essential feature of the operation also determines that algorithm converges slowly in the late period and is possible to fall into local optimum. Simulated annealing (SA) strategy can effectively avoid local optimum when appropriate control parameter is chosen[12]. SA always has a probability of escaping from current local optimum. This probability changes over time and tends to zero with the progress of algorithm. Therefore, the idea of simulated annealing is introduced to DE algorithm to avoid falling into local optimum and enhance the robustness, while the advantages of DE are maintained. SDE framework is as follows.

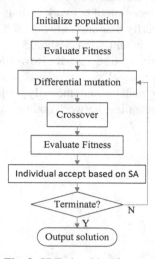

Fig. 2. SDE algorithm framework

3.2   New Individual Accept Strategy Based on Simulated Annealing

Fitness function f() is used to evaluate new individual generated by differential operations. Simulated annealing-based acceptance strategy is used to accept the new or the original individual with probability p, which is defined as follows [13]:

$$p = \begin{cases} 1, & f(x_{new}^g) < f(x_{old}^g) \\ \exp\left(-\dfrac{f(x_{new}^g) - f(x_{old}^g)}{T^g}\right), & f(x_{new}^g) \geq f(x_{old}^g) \end{cases} \tag{6}$$

$f(x_{old}^g)$ is fitness value for the current solution, T^g is the current annealing temperature and $T^{g+1} = \lambda T^g$, $0 < \lambda < 1$. The new individual acceptance strategy based on simulated annealing utilizes a certain probability to balance the new or the original individual.

It is also possible to accept the new solution even it is unsatisfactory. So is always possible to escape from the current local optimum. Of course, new individual is directly sent to next generation if it is an even better.

4   Simulation Experiment

4.1   Network Coding Examples and Parameter Settings

Network coding optimization problem is researched to verify the superiority of SDE algorithm with the following classic network coding instances. Four network models are used: three butterfly diagram network (F3), seven butterfly diagram network (F7), random network (Fr) and fifteen butterfly diagram network (F15). The goal of algorithm is to find the network coding optimization model with the minimal number of coding edges. F3 is showed in Figure 3. Random network model consists of 24 nodes including a sending node and 3 receiving node, as shown in Figure 4.

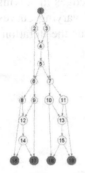

Fig. 3. Three butterfly diagram network model   **Fig. 4.** Random network model

In this experiment, population size is 50, maximal iteration is 200, scale factor F is 0.25, crossover probability is 0.7. Initial temperature of simulated annealing is 10000 and the cool factor is 0.9 [16]. Both algorithms were independently run50 times on F3, F7 and Fr and on 30 times F15 due to the large-scale model.

4.2   Numerical Comparison and Analysis

The final computing results of two algorithms were shown in Table 1. "Best", "Average" and "STD" denote the minimum, average and standard deviation respectively of the coding edges in multiple independent runs on the network models.

Table 1. Comparison between algorithms on butterfly network models

models	SDE			DE		
	Best	Average	STD	Best	Average	STD
F3	3.0	3.3	0.63	3.0	4.4	1.27
Fr	5.0	7.9	1.23	7.0	8.3	1.47
F7	14.0	14.7	0.71	15.0	15.3	1.41
F15	37.0	38.0	0.85	38.0	38.4	0.56

Observed from Table 1, the "Best" and "Average" items of SDE are all better than DE except for the small scale network F3. It indicates the even better exploration ability and the stability of SDE. Both algorithms found the optimal coding edges on relatively simple example F3 (3 coding edges) in multiple runs. The simulation results show that SDE is better than DE to reduce coding edges and has better optimization results especially on medium and large-scale network models.

Standard deviation of SDE are all better than those of DE except for F15. This phenomenon is not difficult to understand because the acceptance poclicy of SDE is based on simulated annealing strategy and has a certain probability accept worse individual. This strategy can maintain population diversity well and avoid local convergence. Thus even better result is able to obtain. However, it is also possible to obtain final solutions with great difference for large-scale complex problems.

4.3   The Curve of Dynamic Convergence Behavior

In order to investigate the online changes of coding edges in solving network coding optimization problems, Figure 5 - Figure 8 show the varying curves of mean coding edges at every same iteration in multiple runs versus the iteration variable on four network optimization model.

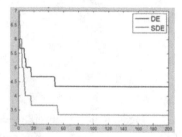

Fig. 5. F3 three butterfly diagram network

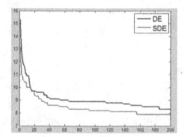

Fig. 6. Fr random network

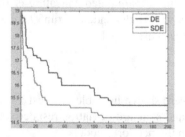

Fig. 7. F7 seven butterfly diagram network

Fig. 8. F15 fifteen butterfly diagram network

The red curve in figures 5 to 8 are the convergence process of SDE, and the black curve represents convergence process of DE. It can be seen that SDE converges faster than DE in the early period which means that SDE can detect the excellent solution faster. The average values of SDE are always less than those of DE for all network models which mean that SDE can always find even less coding edges than DE.

5 Conclusion

A hybrid differential evolution algorithm with simulated annealing strategy (SDE) is proposed for network coding optimization in this paper. Individual acceptance strategy based on simulated annealing ensures the population diversity and avoids falling into local optimal. Simulation results show that SDE can significantly improve the loscally optimal network coding optimization problems. The comparison between SDE and other techniques such as simulated annealing genetic algorithm will also be made in next work. Our future work is how to further improve the network coding optimization on large-scale problems and improve the robustness of SDE algorithm.

Acknowledgement. This research is supported by Natural Science Foundation of China (61375066, 61105127).

References

1. Ahlswede, R., Cai, N., Li, S.Y.R., Yeung, R.W.: Network information flow. IEEE Trans. on Information Theory **46**, 1204–1206 (2000)
2. Deng, Z.L., Zhao, J., Wang, X.: Genetic Algorithm Solution of Network Coding Optimization. Journal of Software **20**, 2269–2279 (2009)
3. Yang, L., Zheng, G., Hu, X.H.: Research on Network Coding: A Survey. Journal of Computer Research and Development **45**, 400–407 (2008)
4. Kim, M., Ahn, C.W., Medard, M., Effros, M.: On minimizing network coding resources: an evolutionary approach. In: Proc. Of the NetCod (2006)
5. Ye, H.T., Luo, F., Xu, Y.G.: Differential evolution for solving multi-objective optimization problems: a survey of the state-of-the-art. Control Theory & Applications **30**, 922–928 (2013)
6. Li, Y.H., Mo, L., Zuo, J.: Shuffled differential evolution algorithm based on optimal scheduling of cascade hydropower stations. Computer Engineering and Application **48**, 228–231 (2012)
7. Storn, R., Price, K.: Differential evolution – a simple and efficient heuristic for global optimization over continuous spaces. Journal of Global Optimization **11**, 341–359 (1997)
8. Hu, Z.B., Xiong, C.W.: Study of hybrid differential evolution based on simulated annealing. Computer Engineering and Design **28**, 1989–1992 (2007)
9. Kirkpatrick, S., Vecchi, M.P.: Optimization by simulated annealing. Science **220**, 671–680 (1983)
10. Cao, X.J.: Two types of network optimization problems. Master thesis of Beijing University of Posts and Telecommunications (2014)
11. Kim, M., Medard, M., Aggarwal, V.: Evolutionary approaches to minimizing network coding resources. In: Proc. of the IEEE INFOCOM 2007, pp. 1991-1999 (2007)
12. Shao, X., Wang, R.C., Huang, H.P., Sun, L.J.: Research of network coding optimization based on simulated annealing genetic algorithm. Journal of Nanjing University of Posts and Telecommunications (Natural Science) **33**, 80–85 (2013)
13. Zhao, X., Lin, W., Yu, C., et al.: A new hybrid differential evolution with simulated annealing and self-adaptive immune operation. Computers & Mathematics with Applications **66**, 1948–1960 (2013)

Brain Storm Optimization Algorithm

Random Grouping Brain Storm Optimization Algorithm with a New Dynamically Changing Step Size

Zijian Cao[1(✉)], Yuhui Shi[2], Xiaofeng Rong[1], Baolong Liu[1],
Zhiqiang Du[1], and Bo Yang[1]

[1] Xi'an Technological University, Xi'an 710021, China
bosscao@163.com
[2] Xi'an Jiaotong-Liverpool University, Suzhou 215123, China

Abstract. Finding the global optima of a complex real-world problem has become much more challenging task for evolutionary computation and swarm intelligence. Brain storm optimization (BSO) is a swarm intelligence algorithm inspired by human being's behavior of brainstorming for solving global optimization problems. In this paper, we propose a Random Grouping BSO algorithm termed RGBSO by improving the creating operation of the original BSO. To reduce the load of parameter settings and balance exploration and exploitation at different searching generations, the proposed RGBSO adopts a new dynamic step-size parameter control strategy in the idea generation step. Moreover, to decrease the time complexity of the original BSO algorithm, the improved RGBSO replaces the clustering method with a random grouping strategy. To examine the effectiveness of the proposed algorithm, it is tested on 14 benchmark functions of CEC2005. Experimental results show that RGBSO is an effective method to optimize complex shifted and rotated functions, and performs significantly better than the original BSO algorithm.

Keywords: Brain storm optimization · Dynamic step size · Random grouping

1 Introduction

Global optimization problems have become more and more complex, from simple unimodal functions to hybrid rotated shifted multimodal functions [1]. Effective and efficient optimization algorithms are always needed to tackle increasingly complex real-world optimization problems in different fields of engineering, social sciences and physical sciences. It has led the researchers to develop various optimization techniques founded on evolutionary computation and swarm intelligence. In the past few decades, a lot of swarm intelligence optimization algorithms, such as particle swarm optimization (PSO) [2,3], ant colony optimization (ACO) [4], bee colony optimization (BCO) [5], firefly optimization algorithms (FFO)[6], bacterial forging optimization (BFO)[7], artificial raindrop algorithm (ARA)[8], have been proposed to tackle those challenging complex real-world problems.

© Springer International Publishing Switzerland 2015
Y. Tan et al. (Eds.): ICSI-CCI 2015, Part I, LNCS 9140, pp. 357–364, 2015.
DOI: 10.1007/978-3-319-20466-6_38

Among those existing meta-heuristics, brain storm optimization (BSO) was proposed to act as a global optimization technique by emulating the collective behavior of human beings in the problem solving process [9, 10]. Like other swarm intelligence algorithms, BSO has achieved successful applications in areas such as optimal satellite formation reconfiguration [11], the design of DC Brushless Motor [12], economic dispatch considering wind power [13], and multi-objective optimization[14,15]. However, in evolutionary computation research, there have always been attempts to further improve any given findings. In this paper, we present an improved variant of the BSO algorithm named Random Grouping BSO (RGBSO). On the one hand, the RGBSO decreases the time complexity of the algorithm with random group strategy instead of clustering method, and on the other hand, reduces the load of parameter settings and balances exploration and exploitation at different searching generations with a new dynamically changing step size. The proposed algorithm is tested on 14 benchmark functions of CEC2005, and the results show that the RGBSO algorithm significantly improves the performance of BSO on complex global optimization problems.

The rest of this paper is organized as follows. Section 2 describes the improved RGBSO. Section 3 presents the test benchmark functions, the experimental setting for each algorithm, and the experimental results. Conclusions are given in Section 4.

2   The Improved Algorithm-RGBSO

2.1   Random Grouping Strategy

In the original BSO, a k-means clustering method was used in the grouping operator. As we all known, the k-means clustering method needs heavier time computational burden. During the evolutionary process, BSO executes the k-means clustering in every generation to group ideas.

However, it is not necessary to use much accurate k-means clustering method to group the ideas into different groups. In our improved algorithm, the random grouping strategy is used to replace the k-means clustering method. The strategy of random grouping stems from the following two ideas. On the one hand, in many facilitating creative thinking, the exchange and discussion in the scheme of random grouping may be propitious to increase the chance of creativity. To the contrary, the well-regulated or precise grouping may be easy to fall into the fixed mindset. On the other hand, similar to other swarm intelligence algorithms, BSO is also a stochastic optimization technique, and random grouping will increase the chance to seek good solutions in a heuristic search mode.

Based on the above idea, we propose the random grouping strategy. The random grouping strategy is implemented as the following steps.

1) Step 1: Randomly divide the N ideas into m groups based on a group size s (assuming $N = s * m$)), i.e., $G = \{G_1, G_2, ..., G_m\}$.

2) Step 2: For each group $G_{i(1 \leq i \leq m)}$, compare the function fitness value of each idea in each group.

3) Step 3: Choose the group center idea of each group from which has minimal function fitness value in each group.

The Pseudo-code for the random grouping is summarized in Algorithm 1.

Algorithm 1. Random Grouping( )

1   Begin
2   G = { };   %grouping set
3   rand_ints =randperm(N); %generating N different random integer between 1 to N
4   s=N/m; %N is the total ideas, and m is the group number
5   for (i=1:s:N)
6   index = rand_ints (i:i+s-1);
7   G = {G{1:end} index};
8   end
9   fit_values = Inf*ones(m,1); %storing the best fitness value
10   best = ones(m,1); %storing number value of the best idea
11   for (i=1:m)
12   for (j=1:s)
13   if fit_values(i,1) > fitness_popu(group{i}(j),1)
14   fit_values(i,1) = fitness_popu(group{i}(j),1);
15   best(i,1) = group{i}(j);
16   end
17   end
18   centers(i,:) = popu(best(i,1),:); % popu is all ideas in the population
19   end
20   end

2.2   A New Dynamic Step Size Parameter Control

To maintain the diversity of population, a new individual in the original BSO is updated according to as

$$X_{new} = X_{select} + \xi * normrnd(0,1) \tag{1}$$

where *normrnd* is the Gaussian random with mean 0 and variance 1, ξ is an adjusting factor   which is expressed as

$$\xi = rand * \log sig((0.5 * \max\_iteration - current\_iteration) / k) \tag{2}$$

where *rand* is a random value between 0 and 1. The *max_iteration* and *current_iteration* denote the maximum number of iteration and current number of iteration respectively. The *logsig* is a logarithmic sigmoid transfer function, and *k* is a predefined parameter for changing slopes of the *logsig* function.

This approach to control the size of step can also balance exploration and exploitation at different searching generations. However, it just takes effect only for very short interval which is shown in Fig.1 (1). Hence, we present a simple dynamic step size strategy, and the dynamic function is described as the following

$$\xi = rand * \exp(1\text{-}(\max\_iteration)/(\max\_iteration - current\_iteration+1)) \qquad (3)$$

where *rand* is a random value between 0 and 1. The *max_iteration* and *current_iteration* are the same to formula 2. Fig.1 (2) shows the adjusting factor which controls the scale of step.

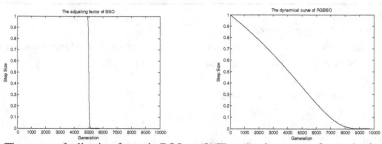

(1) The curve of adjusting factor in BSO   (2) The adaptive curve of step size in RGBSO

Fig. 1. The comparison chart of adjusting factor between BSO and RGBSO

2.3   Pseudo-Code of the RGBSO

As described above, the pseudo-code of the RGBSO is summarized in Algorithm 2.

Algorithm2. RGBSO( )

```
1    Begin
2       Randomly initialize N ideas and evaluate their fitness
3       Initialize m cluster centers (m<N)
4       while(stopping condition not met)
5          Execute Algorithm 1: Random Grouping( )
6          for (i=1 to N)                        % creating operation
7            if rand()<p_one
8              if rand()<p_one_center
9                 Give the group center idea to X_selected
10             else
11                Randomly select an idea in a group to X_selected
12             endif
13           else
14             if rand()<p_two_center
15                Combine the two groups' center ideas to X_selected
16             else
17                Combine two random ideas from the two groups to X_selected
18             endif
19           endif
20           Create X_new using X_selected according to formula (1) and (3)
21           Accept X_new if f(X_new) is better than f(X_i)
22         endfor
23       endwhile
24    end
```

2.4   The Compared Analysis of Computational Complexity

Due to the simplicity of modification in RGBSO compared with the original BSO, the analysis of the computational complexity about RGBSO and BSO is fairly straightforward. The difference of the computational complexity of RGBSO and BSO mainly lies in the grouping of ideas. So, the analysis of the computational complexity of the grouping operator depends on the k-means clustering method of BSO and the random grouping strategy of RGBSO.

In our experiments, the run time of the k-means in BSO is 42.592810s in 10000 runs, and the random grouping in RGBSO is only 0.455166s. The k-means method takes much longer run time than the random grouping strategy does.

3   Benchmark Tests and Experimental Results

For a fair comparison, all the experiments are conducted on the same machine with an Intel 3.4 GHz CPU, 4GB memory. The operating system is Windows 7 with MATLAB 8.0 (R2012b).

3.1   Benchmark Functions

In this paper, we choose 14 widely known rotated and shifted benchmark functions in CEC2005 which are given in [1]. All functions are tested on 30 dimensions. Because the BSO algorithm is analogous with the PSO algorithm, we specially compare the performance of RGBSO with the PSO, the variants of PSO, DE, and the original BSO. We choose two PSO variants, i.e. CLPSO [16].The parameter settings of all the algorithms are given in Table 1.

Table 1. Algorithms Parameter Setting

No.	Algorithm	Parameter Setting
1	RGBSO	M=5,p_one=0.8, p_one_center=0.4, p_two_center=0.5
2	BSO	M=5,p_replace=0.2, p_one=0.8, p_one_center=0.4,
3	DE	F=0.5, CR=0.9
4	PSO	ω: 0.9~0.4, c1= c2=1.49445, global version
5	CLPSO	ω: 0.9~0.4, c1= c2=1.49445, vmaxd=0.2*Range

3.2   Comparison Results for 30-D

1. Comparisons on Solution Accuracy. The results of solution accuracy are given in Table 2 in terms of the mean optimum solution and the standard deviation of the solutions obtained in the 25 independent runs by each algorithm over 300,000 FEs on 14 benchmark functions. In each row of the table, the mean values are listed in the first part, and the standard deviations are listed in the last part, and the two parts are divided with a symbol "±". The best results among the algorithms are shown in boldface.

Table 2. The results of solution accuracy

Fun	RGBSO	BSO	DE	PSO	CLPSO
F1	3.41E-17±3.41E-17	2.65E-03±3.03E-03	5.46E-01±2.59E+00	4.67E+03±2.86E+03	**0.00E+00±0.00E+00**
F2	**1.08E-05±9.27E-06**	2.76E+00±8.16E-01	1.19E-01±5.88E-01	6.51E+03±8.15E+03	3.59E+02±9.69E+01
F3	4.46E+05±2.21E+05	1.33E+06±3.58E+05	**3.40E+05±1.69E+05**	2.29E+07±4.93E+07	1.42E+07±3.99E+06
F4	2.08E+03±4.33E+03	1.29E+04±9.06E+03	**1.25E+02±5.87E+02**	4.56E+03±6.54E+03	5.31E+03±1.13E+03
F5	2.91E+04±1.19E+02	3.12E+04±6.02E+02	**2.90E+04±2.17E-01**	3.43E+04±2.28E+03	2.92E+04±7.19E+01
F6	4.69E+02±7.96E+02	1.52E+03±2.71E+03	5.43E+06±1.51E+07	8.15E+08±9.30E+08	**7.28E+00±1.02E+01**
F7	**2.10E-02±1.69E-02**	5.35E+03±2.32E+02	4.70E+03±9.47E-13	5.36E+03±5.34E+02	4.70E+03±1.52E-12
F8	**2.02E+01±7.42E-02**	2.05E+01±1.17E-01	2.09E+01±7.37E-02	2.09E+01±6.56E-02	2.09E+01±4.57E-02
F9	3.38E+01±7.35E+00	4.97E+01±1.02E+01	2.18E+01±5.63E+00	6.90E+01±2.28E+01	**0.00E+00±0.00E+00**
F10	**2.82E+01±7.26E+00**	4.59E+01±1.19E+01	4.28E+01±1.59E+01	1.26E+02±3.82E+01	9.33E+01±1.75E+01
F11	**9.99E+00±3.48E+00**	2.43E+01±2.58E+00	2.30E+01±6.15E+00	2.02E+01±2.59E+00	2.57E+01±1.69E+00
F12	**3.02E+03±2.85E+03**	3.16E+04±1.70E+04	2.08E+05±8.19E+04	4.83E+03±5.14E+03	5.76E+04±9.67E+03
F13	3.17E+00±6.60E-01	5.56E+00±1.35E+00	3.24E+00±5.77E-01	3.65E+00±1.14E+00	**1.89E+00±2.34E-01**
F14	1.31E+01±3.50E-01	1.32E+01±3.92E-01	1.30E+01±2.66E-01	1.21E+01±5.27E-01	1.27E+01±2.54E-01
w/t/l	-	14/0/0	9/0/5	13/0/1	9/0/5

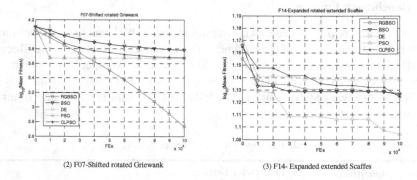

(2) F07-Shifted rotated Griewank   (3) F14- Expanded extended Scaffes

Fig. 2. Convergence performance of the 6 algorithms on parts of functions

The results among RGBSO and other algorithms are summarized as "*w/t/l*" in the last row of the Table, which means that RGBSO wins in *w* functions, ties in *t* function and loses in *l* functions. From the Table 2 it can be observed that the mean value and the standard deviation value of the RGBSO performs better for all 14 function than the original BSO, worse only for 5 functions F3, F4, F5, F9 and F14 than DE, and worse for only one function F14 than PSO. The mean value and the standard deviation value of the RGBSO is performs better for 9 functions F2, F3, F4, F5, F7, F8, F10, F11 and F12 than the CLPSO.

2. The Comparison Results of Convergence Speed. The Fig.2 presents the convergence graphs in terms of the mean fitness values achieved by each of 6 algorithms for 25 runs. From the Fig.2 we can observe that ABSO has fast or similar convergence speed than the other five algorithms.

3. The Comparison Results of Mean CPU Time. The mean CPU time of 5 algorithms are shown in Table 3. Table 3 shows that the total mean CPU time of RGBSO is shorter than BSO, DE and PSO for functions F1 to F14. In all, the total mean CPU time of RGBSO is ranked in the second, and only worse than CLPSO.

Table 3. The mean CPU times for 30-D problems (Time(s))

Fun	RGBSO	BSO	DE	PSO	CLPSO
F1	2.64E+02 (2)	3.93E+02 (6)	3.69E+02 (4)	3.48E+02 (3)	3.13E+01 (1)
F2	2.81E+02 (2)	4.16E+02 (6)	3.78E+02 (4)	3.52E+02 (3)	3.21E+01 (1)
F3	3.84E+02 (2)	5.72E+02 (5)	5.22E+02 (4)	4.97E+02 (3)	4.78E+01 (1)
F4	2.88E+02 (2)	4.10E+02 (5)	3.80E+02 (4)	3.67E+02 (3)	3.23E+01 (1)
F5	2.75E+02 (2)	3.90E+02 (5)	3.63E+02 (4)	3.55E+02 (3)	2.85E+01 (1)
F6	2.67E+02 (2)	3.82E+02 (5)	3.56E+02 (4)	3.41E+02 (3)	3.16E+01 (1)
F7	3.84E+02 (2)	5.39E+02 (4)	5.43E+02 (5)	5.14E+02 (3)	4.51E+01 (1)
F8	3.91E+02 (2)	5.90E+02 (6)	5.45E+02 (4)	5.05E+02 (3)	4.78E+01 (1)
F9	2.72E+02 (2)	4.70E+02 (6)	3.91E+02 (4)	3.38E+02 (3)	3.10E+01 (1)
F10	3.88E+02 (2)	6.56E+02 (6)	5.93E+02 (5)	4.93E+02 (3)	4.87E+01 (1)
F11	5.27E+02 (2)	8.15E+02 (6)	7.74E+02 (5)	6.42E+02 (3)	1.01E+02 (1)
F12	3.23E+02 (2)	5.05E+02 (6)	4.28E+02 (5)	3.91E+02 (4)	4.62E+01 (1)
F13	3.06E+02 (2)	4.93E+02 (6)	4.11E+02 (5)	3.78E+02 (3)	3.37E+01 (1)
F14	4.32E+02 (2)	7.08E+02 (6)	6.00E+02 (4)	5.64E+02 (3)	5.05E+01 (1)
Total	4.78E+03	7.34E+03	6.65E+03	6.08E+03	**6.08E+02**
Rank	2	6	4	3	1

4 Conclusion

In this paper, a random grouping BSO named RGBSO is proposed for solving complex shifted and rotated global optimization problems. Experiments on the 14 chosen test problems were carried out in this paper. From the experimental results, it is observed that the random grouping strategy greatly reduces the run time of RGBSO by compared with the original BSO algorithm. From the analysis of experimental results, we can conclude that RGBSO significantly improves the performance of the original BSO. We are considering applying RGBSO to solve some real-world global optimization problems. Further work also includes research into dynamic clustering to make the algorithm more efficient.

Acknowledgments. This research was partially supported by National Natural Science Foundation of China (Grant No. 61273367), the Project of Department of Education Science Research of Shaanxi Province, China (Grant No. 14JK1360) and the Project of Science Research Plan in Xi'an of China (Grant No. CXY1437(4)).

References

1. Suganthan, P.N., Hansen, N.J, Liang, J.J. et al.: Problem definitions and evaluation criteria for the CEC 2005 special session on real-parameter optimization, Nanyang Technological University, Singapore, Technical Report (2005)
2. Kennedy, J., Eberhart, R.C.: Particle swarm optimization. In: Proceedings of IEEE International Conference on Neural Network, pp. 1942–1948 (1995)
3. Shi, Y.H., Eberhart, R.C.: A modified particle swarm optimizer. In: IEEE International Conference on Evolutionary Computation, Anchorage, Alaska, USA (1998)
4. Dorigo, M., Maniezzo, V., Colorni, A.: Ant system: Optimization by a colony of cooperating agents. IEEE Transactions on Systems, Man, Cybernetics B 26(2), 29–41 (1996)
5. Karaboga, D., Basturk, B.: A powerful and efficient algorithm for numerical function optimization: Artificial bee colony (ABC) algorithm. Journal of Global Optimization 39(3), 459–471 (2007)
6. Yang, X.: Nature-inspired metaheuristic algorithms, Beckington. Luniver Press, UK (2008)
7. Passino, K.M.: Bacterial foraging optimization. International Journal of Swarm Intelligence Research 1(1), 1–16 (2010)
8. Jiang, Q.Y., Wang, L., Hei, X.H. et al.: Optimal approximation of stable linear systems with a novel and efficient optimization algorithm. In: IEEE Congress on Evolutionary Computation, pp. 840–844 (2014)
9. Shi, Y.: Brain storm optimization algorithm. In: Tan, Y., Shi, Y., Chai, Y., Wang, G. (eds.) ICSI 2011, Part I. LNCS, vol. 6728, pp. 303–309. Springer, Heidelberg (2011)
10. Shi, Y.H.: An optimization algorithm based on brainstorming process. International Journal of Swarm Intelligence Research 2(4), 35–62 (2011)
11. Sun, C.H., Duan, H.B., Shi, Y.H.: Optimal satellite formation reconfiguration based on closed-loop brain storm optimization. IEEE Computational Intelligence Magazine 8(4), 39–51 (2013)
12. Duan, H.B., Li, S.T., Shi, Y.H.: Predator-prey based brain storm optimization for DC brushless motor. IEEE Transactions on Magnetics 49(10), 5336–5340 (2013)
13. Jadhav, H.T., Sharma, U., Patel, J., et al.: Brain storm optimization algorithm based economic dispatch considering wind power. In: IEEE International Conference on Power and Energy, pp. 588–593 (2012)
14. Shi, Y.H.: Multi-objective optimization based on brain storm optimization algorithm. International Journal of Swarm Intelligence Research 4(3), 1–21 (2013)
15. Xue, J.Q., Wu, Y.L., Shi, Y.H., Cheng, S.: Brain storm optimization algorithm for multi-objective optimization problems. In: The Third International Conference on Swarm Intelligence, pp. 513–519 (2012)
16. Liang, J., Qin, A.K., Suganthan, P.N., Baskar, S.: Comprehensive learning particle swarm optimizer for global optimization of multimodal functions. IEEE Transactions on Evolutionary Computation 10(3), 281–295 (2006)

An Adaptive Brain Storm Optimization Algorithm for Multiobjective Optimization Problems

Xiaoping Guo[1], Yali Wu[1(✉)], Lixia Xie[1], Shi Cheng[2], and Jing Xin[1]

[1] Xi'an University of Technology, Xi'an, Shaanxi, China
yliwu@xaut.edu.cn
[2] Division of Computer Science, University of Nottingham Ningbo, Ningbo, China
shi.cheng@nottingham.edu.cn

Abstract. Brain Storm Optimization (BSO) algorithm is a new swarm intelligence method that arising from the process of human beings problem-solving. It has been well validated and applied in solving the single objective problem. In order to extend the wide applications of BSO algorithm, a modified Self-adaptive Multiobjective Brain Storm Optimization (SMOBSO) algorithm is proposed in this paper. Instead of the k-means clustering of the traditional algorithm, the algorithm adopts the simple clustering operation to increase the searching speed. At the same time, the open probability is introduced to avoid the algorithm trapping into local optimum, and an adaptive mutation method is used to give an uneven distribution on solutions. The proposed algorithm is tested on five benchmark functions; and the simulation results showed that the modified algorithm increase the diversity as well as the convergence successfully. The conclusions could be made that the SMOBSO algorithm is an effective BSO variant for multiobjective optimization problems.

Keywords: Brain storm optimization · Multiobjective optimization · Clustering operation · Mutation method

1 Introduction

Optimization technique has been a significant and successful tool in solving various problems. Multiobjective optimization problems (MOPs) have gained much attention in recent years. Unlike single objective problems, each objective of MOPs usually conflicts with each other [7]. Due to the characteristic of a MOP, its optimum solution is usually not unique, but consists of a set of candidate solutions among which no one solution is better than other solutions with regards to all objectives.

This work is partially supported by National Natural Science Foundation of China under Grant Number 61203345 and 61273367, and by Ningbo Science & Technology Bureau (Science and Technology Project Number 2012B10055).

© Springer International Publishing Switzerland 2015
Y. Tan et al. (Eds.): ICSI-CCI 2015, Part I, LNCS 9140, pp. 365–372, 2015.
DOI: 10.1007/978-3-319-20466-6_39

In the multiobjective optimization algorithm, the Non-dominated Sorting Genetic Algorithm (NSGA) is the most widely used algorithm, in which the idea of a non-dominated sorting was introduced to the genetic algorithm that transform the computation of the objective function into multiple virtual fitness. NSGA-II is the improved NSGA-based non-dominated sorting genetic algorithm [3]. It uses a fast non-dominated sorting process, elitist and non-operating parameters to niche operator that overcomes the shortcomings of traditional NSGA which has high computational complexity, non-elite maintaining strategies and without specifying shared radius. Speed constrained multiobjective particle swarm optimization (SMPSO) has used crowding distance to maintain external archive collection which is used in NSGA-II [4]. The SMPSO algorithm introduced the binomial variation in the population space, which can be a good solution to the multimodal problem. Two solutions are randomly selected from the archive and choose the one that has a larger crowd distance as the global best [4].

Brain Storm Optimization (BSO) algorithm, inspired by human idea generation process, is originally proposed solving single objective optimization problems [5]. As a new swarm intelligence algorithm, BSO has received much attention since proposed in 2011. Currently, the main works of BSO research could be categorized into three class: 1) the analysis of BSO algorithm, such as solution clustering analysis [1], and population diversity maintenance [2]; 2) the new variants of BSO algorithms, e.g. BSO Algorithm for multiobjective optimization problems [6,7]; 3) the application of BSO algorithm.

The BSO algorithm, which used the k-means clustering and Gaussian/ Cauchy mutation, has been utilized to solve multiobjective problems [6,7]. In this paper, a self-adaptive multiobjective BSO (SMOBSO) algorithm with clustering strategy and mutation is proposed to solve multiobjective optimization problems. Instead of the k-means method, the algorithm adopts a simple grouping method in the cluster operator to reduce the algorithm computational burden. Moreover, a parameter named open probability is introduced in mutation, which is dynamically changed with the increasing of iterations. The proposed algorithm is tested on the ZDT benchmark functions with different dimensions [3]. The simulation results showed that SMOBSO would be a promising algorithm in solving multiobjective optimization problems.

The remaining of the paper is organized as follows. The proposed algorithm is described in Section 2. The parameter settings and results are given in Section 3. Finally, the conclusions and further research are detailed in Section 4.

2   Self-adaptive Brain Storm Optimization for Multiobjective Optimization Problems

Compared with the original multiobjective BSO algorithm [7], this paper makes improvements about clustering, mutation, and global archive operations. Three modifications are: a simple clustering operator has replaced the original k-means clustering, an open probability in the mutation operation, and a new updating strategy of the archive set. These three parts improve the performance of the proposed algorithm together. The SMOBSO algorithm is described in Algorithm 1:

Algorithm 1. Procedure of SMOBSO algorithm

---

1 Initialize number of the population with the size of N_p, maximum iteration, probability parameter p_1, p_2, p_3, p_4, size of archive is max A, number of cluster and cluster center;
2 Initialize the individuals of the population p and calculate their fitness values. Set a null archive Rep;
3 **while** *the maximum iteration has not reached* **do**
4   Calculate the non-dominated solutions and store them in Rep and then crowing distance of each individual in Rep;
5   Cluster individuals using a new clustering strategy in 2.1;
6   Get the elite and general clusters according to clustering results and if there are non-dominated individuals or not;
7   Select the mutation that will be mutated according to 2.2;
8   Update to generate new individuals;
9   Put non-dominated individuals into archive one by one and update archive using 2.3;
10 Output the archive;

2.1 Grouping Operator

In the original BSO, k-means clustering was used in the grouping operator. However, there is no strict requirement for grouping operator in BSO algorithm, only individuals should be divided into different classes. Although originally k-means clustering method is accurate. But it needs many computational costs. In this paper, a new simple clustering method is proposed which randomly select M (M is the number of clustering) different individuals as the centers throughout the search area to make the algorithm more simple. Each individual is clustered to its nearest class after calculated the Euclidean distance between each individual and the center of all classes. The detail steps of clustering strategy are given in Algorithm 2.

Algorithm 2. The clustering strategy in grouping operator

---

1 Randomly select M different ideas from the current generation as the centers of M groups, denoted as S_j, $1 \leq j \leq M$;
2 **while** *The clustering of all individuals has not completed* **do**
3   Calculate Euclidean distance between each individual $\mathbf{x}_i$ ($1 \leq i \leq N_p$) in the current generation and every center;
4   For each individual, comparing the M distance values, the individual will be clustered in the cluster which has the smallest M value. In this process, the cluster centers do not always change;

2.2   Mutation Operator

Individuals will be mutated after $\mathbf{x}_{select}$ has been selected ($\mathbf{x}_{select}$ is selected according to [6]). Gaussian mutation was used in original BSO algorithm. However, there are two drawbacks: 1) Due to behavior of BSO search is a random process, there is no feedback about the complex transfer function of S-type and cannot get a good search information. This defect will be more obvious in dealing with different optimization problems; 2) $\log sig()$ function and random value $rand()$ is all in the range of $(0,1)$. ξ multiplied by the Gaussian random value. Such random noise may have little effect on the global search when facing a large search range.

The open probability is introduced to avoid the algorithm trapping into local optimum. An adaptive mutation method is used to give an uneven distribution on solutions. In order to ensure the convergence, open probability of p_r should be small at the beginning of the search. The effect of the p_r value became smaller with the increasing of iterations. In order to generate new ideas and to avoid falling into local best solutions, p_r should be increased with the increasing of iterations. Based on this principle, the equation of formula variation is as follows:

$$x_{new,d} = \begin{cases} L_d + (H_d - L_d) \times rand() & rand() < p_r \\ x_{select,d} + (x_{1,d} - x_{2,d}) \times rand() & \text{otherwise} \end{cases} \qquad (1)$$

Where L_d and H_d are lower and upper bounds of dth dimension, and $x_{1,d}$ and $x_{2,d}$ are two selected unequal dth dimension values in the population.

2.3   Global Archive

Circulation crowded distance is used to maintain global archive. Crowding distance is used to estimate the intensity of a solution and other solutions around it. Firstly, set all individual crowding distance to 0, and then calculate objective values of individuals in the archive. Then sort each objective function values in ascending order. Crowding distance of the first and last individual will be set to infinity, then calculate the other crowding distance following the formula:

$$distance(i) = distance(i) + \frac{f_m(i+1) - f_m(i-1)}{f_m^{\max} - f_m^{\min}} \qquad (2)$$

where $f_m(i)$ represents the ith individual function value on the mth target, $f_m^{\max}$ and $f_m^{\min}$ respectively denote the maximum and minimum of all individuals in the function value of the mth target. The average length of cube edge that formed by the solution $i+1$ and $i-1$ will be calculated. The specific method for the archive set updating shows as follows: 1) Non-dominated individuals in the population will be put into an archive one by one. If an individual is dominated by individuals in external archive, then the individual will be deleted from the archive. Otherwise the individual will join the archive. 2) If number of individuals in archive is less than the maximum number, then there is no need to delete, otherwise crowding distance of all individuals in the current archive will be calculated and individual that has the smallest crowded distance will be deleted to keep the number of individuals is less than or equal to the maximum capacity. This method is different from the NSGA-II which will sort crowding distance between all non-dominated individuals that were newly generated and in archive. Then choose individuals that have far crowded distances into the next generation. It is benefit for the individuals to be distributed more even in the Pareto front.

3 Results and Discussions

3.1 Parameter Settings

During the test, the population size is set to be 200 and the maximum size of the Pareto set is fixed at 100. The pre-determined probability values p_1, p_2 and p_3 are set to 0.8, 0.8 and 0.2, respectively. All algorithms are implemented in MATLAB using a real-number representation for decision variables. For each experiment, 30 independent runs were conducted to collect statistical results. The number of the dimensions is set to 5, 10, 20, and 30 for each test problem, respectively.

3.2 Results

The ZDT benchmark functions [3] are used in this paper to evaluate the performance of the SMOBSO algorithm. In ZDT functions, the ZDT1 and ZDT3 have a convex of the Pareto frontier, and ZDT3 is discontinuous. The high-dimensional space ratio HR box, which is obtained by repeating for 30 runs, is shown in Figure 1 and 2. The Figure 1 shows the comparisons among different variant of BSO algorithms, which include SMOBSO, MMBSO, MBSO_G, and MBSO_C. While the Figure 2 shows the comparison among different swarm intelligence algorithms, which include SMOBSO, NSGA-II [3], SMPSO [4].

All the result in Figure 1 shows that SMOBSO for all test functions has the most concentrated distribution, and dirty data is clearly smaller than the other algorithms. According to definition of HR, the closer of HR value to 1, the better performance algorithm has and the closer to the real front, the more uniformly distributed. It is clearly showed that HR value of SMOBSO is almost around 1. The SMOBSO algorithm is the best in comparison of other similar algorithms.

To compare the SMOBSO algorithm with the other multiobjective optimization algorithm, the SMPSO and NSGA-II algorithm are used as our comparison algorithm in this paper. The comparisons of high-dimensional space ratio HR among SMOBSO, NSGA-II, and SMPSO solving ZDT benchmark problems are shown in Figure 2. From the figure 2, we can see that SMOBSO is slightly worse than the other two algorithms in high-dimensional space ratio for ZDT1, ZDT2, and ZDT3 problem. For ZDT4 and ZDT6 problem, with the number of dimensions increasing, the dirty data of NSGA-II are significantly increased; and SMOBSO and SMPSO results are similar. In conclusion, the proposed algorithm shows strong potential for multiobjective optimization algorithm.

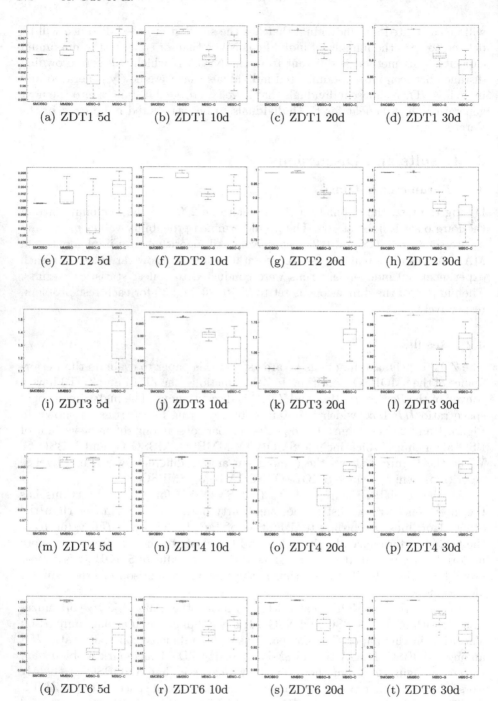

Fig. 1. The comparisons of high-dimensional space ratio HR among SMOBSO, MMBSO, MBSO_G, MBSO_C solving ZDT benchmark problems. The statistics box-plots are derived from 30 independent runs.

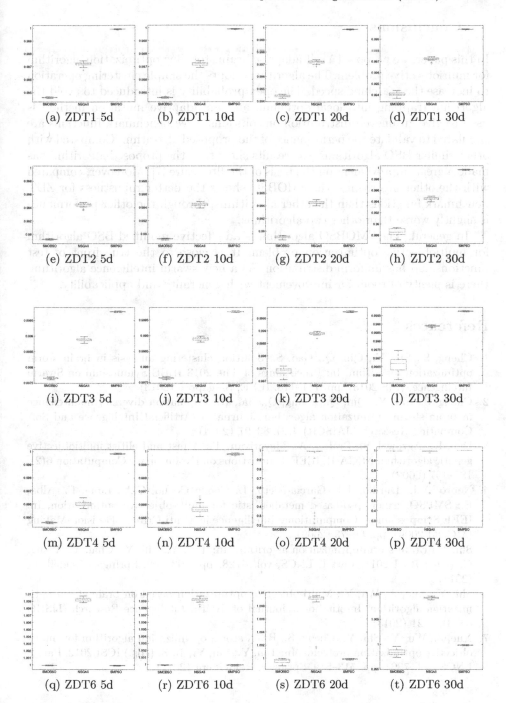

Fig. 2. The comparisons of high-dimensional space ratio HR among SMOBSO, NSGA-II, and SMPSO solving ZDT benchmark problems. The statistics boxplots are derived from 30 independent runs.

4 Conclusions

In this paper, we proposed a self-adaptive brain storming optimization algorithm for multiobjective problem. The algorithm adopts the simple clustering operation to increase the searching speed. The open probability is introduced to avoid the algorithm trapping into local optimum; and an adaptive mutation method is used to give an uneven distribution on solutions. Five benchmark functions are simulated to validate the performance of the proposed algorithm. Compared with other similar BSO algorithms, the results show that the proposed algorithm has made a great improvement on the basis of the literature [7]. Moreover, compared with the other algorithms, the SMOBSO shows the better robustness for ZDT benchmark functions than the other algorithms, through the other performance is slightly worse than other two algorithms.

In general, The SMOBSO algorithm is an effective modified BSO algorithm for multiobjective optimization problem. It is close to the true front of test functions and has uniform distribution. As a new swarm intelligence algorithm, there is plenty of room for improvement with generality and applicability.

References

1. Cheng, S., Shi, Y., Qin, Q., Gao, S.: Solution clustering analysis in brain storm optimization algorithm. In: Proceedings of The 2013 IEEE Symposium on Swarm Intelligence, (SIS 2013), pp. 111–118. IEEE, Singapore (2013)
2. Cheng, S., Shi, Y., Qin, Q., Zhang, Q., Bai, R.: Population diversity maintenance in brain storm optimization algorithm. Journal of Artificial Intelligence and Soft Computing Research (JAISCR) 4(2), 83–97 (2014)
3. Deb, K., Agrawal, S., Pratap, A., Meyarivan, T.: A fast and elitist multiobjective genetic algorithm: NSGA-II. IEEE Transactions on Evolutionary Computation 6(2), 182–197 (2002)
4. Nebro, A.J., Durillo, J.J., García-Nieto, J., Coello Coello, C.A., Luna, F., Alba, E.: SMPSO: a new pso-based metaheuristic for multi-objective optimization. In: IEEE Symposium on Computational Intelligence in Multi-Criteria Decision-Making (MCDM 2009), pp. 66–73 (2009)
5. Shi, Y.: Brain storm optimization algorithm. In: Tan, Y., Shi, Y., Chai, Y., Wang, G. (eds.) ICSI 2011, Part I. LNCS, vol. 6728, pp. 303–309. Springer, Heidelberg (2011)
6. Shi, Y., Xue, J., Wu, Y.: Multi-objective optimization based on brain storm optimization algorithm. International Journal of Swarm Intelligence Research (IJSIR) 43(3), 1–21 (2013)
7. Xue, J., Wu, Y., Shi, Y., Cheng, S.: Brain storm optimization algorithm for multi-objective optimization problems. In: Tan, Y., Shi, Y., Ji, Z. (eds.) ICSI 2012, Part I. LNCS, vol. 7331, pp. 513–519. Springer, Heidelberg (2012)

Enhanced Brain Storm Optimization Algorithm for Wireless Sensor Networks Deployment

Junfeng Chen[1(✉)], Shi Cheng[2], Yang Chen[3], Yingjuan Xie[1], and Yuhui Shi[4]

[1] College of IOT Engineering, Hohai University, Changzhou 213022, China
chen-1997@163.com
[2] Division of Computer Science, University of Nottingham Ningbo, Ningbo, China
shi.cheng@nottingham.edu.cn
[3] College of Science and Technology, Ningbo University, Ningbo, China
[4] Department of Electrical and Electronic Engineering,
Xi'an Jiaotong-Liverpool University, Suzhou, China
yuhui.shi@xjtlu.edu.cn

Abstract. Brain storm optimization is a young and promising swarm intelligence algorithm, which simulates the human brainstorming process. The convergent operation and divergent operation are two basic operators of the brain storm optimization. The k means clustering is utilized in the original brain storm optimization, which needs to define the k value before the search. To adaptively change the number of clusters during the search, a modified Affinity Propagation (AP) clustering method and an enhanced creating strategy are proposed on account of the structure information of single or multiple clusters. In addition, the modified brain storm optimization is applied to optimize the dynamic deployments of two different wireless sensor networks (WSN). Experimental results show that the proposed algorithm achieves satisfactory results and guarantees a high coverage rate.

Keywords: Brain Storm Optimization · Affinity propagation · Structure information · Wireless sensor networks

1 Introduction

Brain Storm Optimization (BSO) is a new swarm intelligence algorithm that simulates the problem-solving process of human brainstorming. The basic framework of BSO was introduced by Shi [7–9], who designed the clustering and creating operators by modelling and abstracting brainstorming process based on Osborn's four rules in 2011.

As a young and promising algorithm, BSO can be further improved by developing various search strategies. Zhan *et al.* [13] proposed a simple grouping method to reduce the algorithm computational burden. Chen *et al.* [1] introduced affinity propagation clustering into BSO by analyzing the clusters' variations over iterations. Xue *et al.* [10] designed a new creating operator of BSO with combinations of Gaussian mutation and Cauchy mutation. Zhou et al. [14]

© Springer International Publishing Switzerland 2015
Y. Tan et al. (Eds.): ICSI-CCI 2015, Part I, LNCS 9140, pp. 373–381, 2015.
DOI: 10.1007/978-3-319-20466-6_40

created new individuals in a batch-mode and used a mutative step-size according to the dynamic range of individuals on each dimension. Zhan *et al.* [12] investigated the influence on the performance of BSO with different control parameters. Duan *et al.* [5] introduced Predator-prey concept into BSO for the purpose of utilizing the global information and improving the swarm diversity. Yang *et al.* [11] proposed a new creating mechanism with inter-group discussion and intra-group discussion to get a tradeoff between the capability of global search and local search. Cheng *et al.* [2–4] analyzed and discussed the solution clustering, and other properties of BSO. In addition, the basic BSO and its variants have been applied successfully to global numerical optimization and optimizing design variables for a DC brushless motor [5].

In this paper, an enhanced BSO algorithm is proposed for optimizing Wireless Sensor Networks (WSN) deployment. The remaining paper is organized as follows. In Section 2, the original BSO algorithm is introduced and the performance of basic operators is discussed. In Section 3, an enhanced BSO method is proposed based on affinity propagation clustering and an improved creating operator. In Section 4, the proposed BSO algorithm is applied to solve the problem of WSN coverage and the experimental results are analyzed. Our concluding remarks are made in Section 5.

2   Original Brain Storm Optimization

Consider an unconstrained minimization function with finite dimensions in the form of $\min_{x \in S} f(x)$, where S is an n-dimensional search space, x denotes a candidate solution corresponding to a point in S, and $f(x)$ is an evaluation function corresponding to x.

The original BSO optimizes a problem by having a population of candidate solutions, and iteratively trying to improve candidate solutions with regards to a given measure of quality. Generally speaking, BSO has three main operators: clustering, creating, and selecting. The original BSO and most of variants employ k-means or k-medoids clustering algorithms, which require the number of clusters to be determined before running the algorithm. However, the proper number of clusters in BSO would not be available in advance. Moreover, the individuals should be grouped into diverse number of clusters as the BSO algorithm evolves. To be specific, candidate solutions can be partitioned into several groups in the initial stage of iteration, while they may be gathered into different number of clusters in the later iterations. In other words, the number of clusters might vary with the iteration. Affinity Propagation (AP) clustering algorithm can be a better clustering algorithm for a BSO than k-means clustering does at least under some scenarios [1]. AP can organize various individuals into proper groups without knowing the number of clusters in advance. The number of clusters is determined adaptively over iterations. For creating operator, the basic BSO generates new individuals in four patterns.

1. a random vector is added to a selected cluster center to develop a new individual.

2. a random vector is added to a random individual to form a new individual.
3. a random vector is added to a combination of two cluster centers to generate a new individual.
4. a random vector is is added to a combination of two random individuals to create a new individual.

We can see that the basic BSO takes the advantages of fine-grained search and coarse-grained search. Specifically, updating pattern 1 and pattern 2 are the local search strategies which can yield small variation, while updating pattern 3 and pattern 4 are the global search strategies that create large variation.

For the original BSO, there are some valuable aspects which need to be further discussed. Firstly, the cluster center can be viewed as a special individual, the fitness of which is not generally desired best individual. In some cases, it might be a good choice to specify the best individual of the cluster as the cluster center. Secondly, new individuals are generated one by one in the original version, which is a time consuming process. The mass-produced individuals are worth considering. Thirdly, the global strategies which only combine the information of two selected clusters might go against exploration within the whole search space.

3   Enhanced Brain Storm Optimization (EBSO)

3.1   AP Clustering Strategy

Affinity Propagation (AP) is a clustering algorithm based on the concept of "message passing" between data points [6]. Unlike k-means or k-medoids clustering algorithms, AP does not require knowing the number of clusters. Like k-medoids, AP finds "exemplars", members of the input set that are representative of clusters.

Let $\{\xi_1, \xi_2, \cdots, \xi_n\}$ be a set of data points, with no assumptions made about their internal structure, and let $\rho(\cdot)$ be a function that quantifies the similarity between any two points, that is $\rho(i, j) > \rho(i, k)$ iff ξ_j is more similar to ξ_i than ξ_k is.

The algorithm proceeds by alternating two message passing steps, to update two matrices: responsibility matrix and availability matrix. The value $\gamma(i, k)$ of responsibility matrix quantifies how well-suited ξ_k is to serve as the exemplar for ξ_i. The value $\lambda(i, k)$ of availability matrix represents how appropriate it would be for ξ_i to pick ξ_k as its exemplar [6].

First, responsibility updates are set

$$\gamma(i, k) \leftarrow \rho(i, k) - \max_{k' \neq k}\{\lambda(i, k') + \rho(i, k')\} \qquad (1)$$

Then, availability $\lambda(i, k)$ is updated to the self responsibility $\lambda(k, k)$ add the sum of the positive responsibilities candidate exemplar k receives from other samples [6].

$$\lambda(i, k) \leftarrow \min\left\{0, \gamma(k, k) + \sum_{i' \notin \{i,k\}} \max\{0, \rho(i', k)\}\right\} \qquad (2)$$

The self-availability $\lambda(k, k)$ is updated as follows:

$$\lambda(k, k) \leftarrow \sum_{i' \neq k} \max\{0, \gamma(i', k)\} \tag{3}$$

It is generally known that an appropriate clustering algorithm depends on data set and intended usage of the clustering results. However, the individuals might be over-clustered by AP algorithm in BSO. A single example of over-segmentation is shown in Fig. 1.

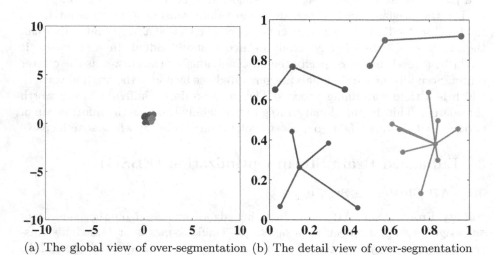

(a) The global view of over-segmentation (b) The detail view of over-segmentation

Fig. 1. The over-segmentation of candidate solutions

From Fig. 1 (a), it can be seen that the distribution of candidate solutions is concentrated on the central part of search space. So, generally speaking, these individuals are expected to be a group for BSO. However, AP algorithm distinguishes the subtle differences among the candidate solutions, which results in the over-segmentation, as shown in Fig. 1 (b). In this paper, we solve the issue by specifying the smallest distance between individuals.

3.2   Enhanced Creating Strategy

In this section, a new creating operator is proposed to strengthen the ability of exploration and exploitation. The enhanced creating strategy has the characteristics of synthesizing structure information of one or more clusters.

First, the fitness values of candidate solutions in each cluster are mapped into the uniform confidence interval for easily extracting information.Second, the structure information of each cluster C is extracted and denoted as a set of vectors $C\{u, v, w\}$, where u is the kernel of the cluster, v and w are the coverage and dispersion of candidate solutions in the cluster, respectively. The detailed procedure is listed as Algorithm 1.

Algorithm 1. Extracting structure information of each cluster.

---

1. Select a cluster with K individuals in M dimension.;
2. Calculate the fitness value of each individual $f(\boldsymbol{x})$, $i = 1, 2, \cdots, K$;
3. Mapping the fitness values $f(\boldsymbol{x})$ to the confidence interval $[0, 1]$;
4. Determine the best individual as the kernel in the selected cluster
   $\boldsymbol{u} = [u_1, u_1, \cdots, u_M] = \check{\boldsymbol{x}}$;
5. Calculate the coverage of the selected sparks
   $\boldsymbol{v} = [v_1, v_1, \cdots, v_M] = \frac{1}{L-1} \sum_{i=1}^{L} (\boldsymbol{x}_i - \boldsymbol{u})^2$;
6. **for** *each couple of individuals and their fitness values* $(\boldsymbol{x}_i, f(\boldsymbol{x}_i))$ **do**
7.   Calculate $\boldsymbol{o}_i = \sqrt{\frac{-(\boldsymbol{x}_i - \boldsymbol{u})^2}{2 \ln f(\boldsymbol{x}_i)}}$
8. Calculate the mean of $\boldsymbol{o}_i$ as $\bar{\boldsymbol{o}}$, *i.e.*, $\bar{\boldsymbol{o}} = \frac{1}{L} \sum_{i=1}^{L} \boldsymbol{o}_i$;
9. Calculate the dispersion of the selected cluster
   $\boldsymbol{w} = [w_1, w_1, \cdots, w_M] = \frac{1}{L-1} \sum_{i=1}^{L} (\boldsymbol{o}_i - \bar{\boldsymbol{o}})^2$;

Third, the structure information of multiple clusters can be calculated on the basis of the information of single clusters. For example, we have the cluster information $C_1\{\boldsymbol{u}_1, \boldsymbol{v}_1, \boldsymbol{w}_1\}$ and $C_2\{\boldsymbol{u}_2, \boldsymbol{v}_2, \boldsymbol{w}_2\}$. The synthesized information of two clusters is calculated as follow.

$$u_{12} = \frac{u_1 v_1' + u_2 v_2'}{v_1' + v_2'}, \quad v_{12} = v_1' + v_2', \quad w_{12} = \frac{w_1 v_1' + w_2 v_2'}{v_1' + v_2'}$$

where v_1', v_2' are the expectation of the clusters C_1 and C_2, respectively.

Similarly, the synthesized information of multiple clusters can be obtained in the same way. Finally, the new individuals can be generated based on diverse structure information. The detailed procedure is presented as Algorithm 2.

Algorithm 2. Creating new individuals with structure information $C(\boldsymbol{u}, \boldsymbol{v}, \boldsymbol{w})$.

---

1. Select a cluster with K individuals in M dimension.;
2. Create the G new individuals based on the structure information $C(\boldsymbol{u}, \boldsymbol{v}, \boldsymbol{w})$;
3. **for** $j = 1$ *to* G **do**
4.   Generate a normally distributed random vector $\boldsymbol{p}$ with the kernel $\boldsymbol{v}$ and the coverage $\boldsymbol{w}$, *i.e.*, $\boldsymbol{p} = \text{NormRand}(\boldsymbol{v}, \boldsymbol{w})$.;
5.   Generate a normally distributed random individuals $\boldsymbol{x}_i$ with the kernel $\boldsymbol{u}$ and the coverage $\boldsymbol{p}$, *i.e.*, $\boldsymbol{x}_i = \text{NormRand}(\boldsymbol{u}, \boldsymbol{p})$.;

---

4   WSN Deployment Based on Enhanced BSO

For better investigating the performance of the proposed BSO, dynamic deployments of WSNs are performed. First, the mathematical expressions of WSN Deployment are described, then EBSO is used to optimize the dynamic deployments of two different WSNs.

4.1   Model Representations of WSN Deployment

Assume that there are a set of wireless sensors, expressed as $\zeta = \{\zeta_1, \zeta_2, \cdots, \zeta_N\}$. The position of a wireless sensor ζ_i is recorded as (a_i, b_i). If the target point φ_j is located at (a_j, b_j), then the Euclidean distance between the wireless sensor ζ_i and the target point φ_j is calculated as follows.

$$D(\zeta_i, \varphi_j) = \sqrt{(a_i - a_j)^2 + (b_i - b_j)^2} \tag{4}$$

The perceived radius of the sensor node is set as R_i. We take the binary detection model for sensor ζ_i and target φ_j, expressed as

$$P(\zeta_i, \varphi_j) = \begin{cases} 1 & D(\zeta_i, \varphi_j) \leq R_i \\ 0 & D(\zeta_i, \varphi_j) > R_i \end{cases} \tag{5}$$

Then we have the joint perception probability of K sensors with target point φ_j.

$$J(\varphi_j) = 1 - \prod_{i=1}^{K}(1 - P(\zeta_i, \varphi_j)) \tag{6}$$

In the enhanced BSO, a $2N$-dimensional individual $\boldsymbol{x}_i$ represents all the N sensor nodes in two-dimensional space: $\boldsymbol{x}_i = [x_{i1}^H, x_{i1}^V, x_{i2}^H, x_{i2}^V, \cdots, x_{iN}^H, x_{iN}^V]$, where x_{iN}^H and x_{iN}^V present the positions of the N-th mobile sensor node in horizontal axis and vertical axis, respectively.

The fitness function of EBSO is the effective coverage, expressed as

$$f(\boldsymbol{x}) = \frac{\sum J(\varphi_j)}{X \times Y} \tag{7}$$

where the X and Y represent the rows or columns within a grid environment.

4.2   Experimental Results and Analysis

Experiments are carried out in a square area $S = 500 \times 500$, i.e., 0.25 square kilometres. In Experiment I, there are 15 stationary nodes (the green squares in Fig. 2) and 40 mobile nodes (the red circles in Fig. 2). The detection radius of each sensor is 40 metre. In Experiment II, there are 15 stationary nodes (the green squares in Fig. 3) and 70 mobile nodes (the red circles in Fig. 3). The detection radius of each stationary node is 40 metre, and the detection radius of each mobile node is 30 metre.

The EBSO has a population size 50 and the number of maximum iteration is set to 200 for one execution. The experiments are implemented by Matlab 8.0 and the simulations are run on the PC with Intel Core $i3$ 2350M 2.3GHz, 2 GB memory capacity and the Windows 7 operating system.

The simulation results of the Experiment I and Experiment II are illustrated as Fig. 2 and Fig. 3, respectively. Fig. 4 shows the variation of coverage rate over iterations.

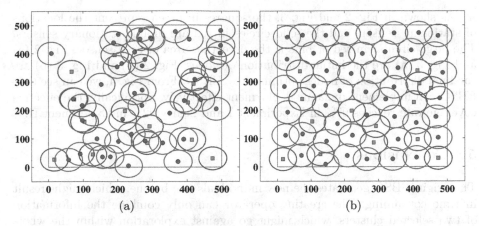

Fig. 2. The simulation results of Experiment I

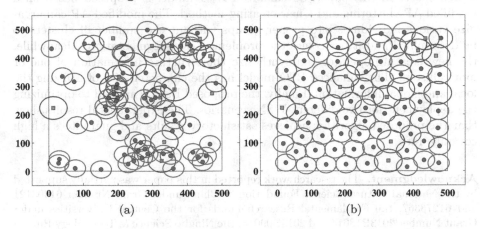

Fig. 3. The simulation results of Experiment II

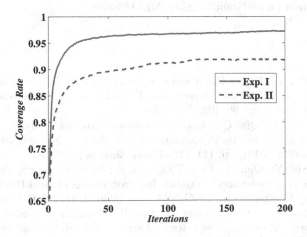

Fig. 4. The variations of coverage rate with iterations

As shown in Fig. 2 and Fig. 3, the squares in green represent the locations of stationary sensors and the solid circles in red are mobile stationary sensors. The large circular in red or blue is the detection area of each sensor. Fig. 2 (a) and Fig. 3 (a) are the initial distribution of nodes. Fig. 2 (b) and Fig. 3 (b) are the final distribution of nodes optimized by EBSO. From Fig. 4, we can see that EBSO can implement dynamic deployment effectively. Specifically, the effective coverage rate reaches 97% in Experiment I and nearly 92% in Experiment II.

5   Conclusions

The original BSO generates the new individuals one by one, which might result in time consuming. The creating operator can only combine the information of two selected clusters, which might go against exploration within the whole search space. In this paper, an enhanced BSO model is proposed based on a modified AP clustering and a new creating method. The modified AP can organize various individuals into proper groups without the number of clusters to be determined and it overcomes the problem of over-segmentation. Meanwhile, the new creating operator can synthesize the structure information of one or more clusters. The enhanced BSO model has the characteristics of utilizing the local and global information. Finally, the proposed BSO is applied to optimizing the deployment for two different WSN scenarios. The experimental results show that the proposed algorithm achieves satisfactory results and guarantees a high coverage rate.

Acknowledgment. The research work reported in this paper was partially supported by the National Natural Science Foundation of China under Grant Number 61403121 and 61273367, the Fundamental Research Funds for the Central Universities under Grant Number 2013B18614 and 2013B09014, the Ningbo Science & Technology Bureau with the Science and Technology Project Number 2012B10055, and the Ningbo Natural Science Foundation under Grant Number 2013A610148.

References

1. Chen, J., Xie, Y., Ni, J.: Brain storm optimization model based on uncertainty information. In: 2014 Tenth International Conference on Computational Intelligence and Security, pp. 99–103, November 2014
2. Cheng, S., Shi, Y., Qin, Q., Gao, S.: Solution clustering analysis in brain storm optimization algorithm. In: Proceedings of The 2013 IEEE Symposium on Swarm Intelligence, (SIS 2013), pp. 111–118. IEEE, Singapore (2013)
3. Cheng, S., Shi, Y., Qin, Q., Ting, T.O., Bai, R.: Maintaining population diversity in brain storm optimization algorithm. In: Proceedings of 2014 IEEE Congress on Evolutionary Computation, (CEC 2014), pp. 3230–3237. IEEE, Beijing (2014)
4. Cheng, S., Shi, Y., Qin, Q., Zhang, Q., Bai, R.: Population diversity maintenance in brain storm optimization algorithm. Journal of Artificial Intelligence and Soft Computing Research (JAISCR) 4(2), 83–97 (2014)

5. Duan, H., Li, S., Shi, Y.: Predator-prey brain storm optimization for dc brushless motor. IEEE Transactions on Magnetics **49**(10), 5336–5340 (2013)
6. Frey, B.J., Dueck, D.: Clustering by passing messages between data points. Science **315**, 972–976 (2014)
7. Shi, Y.: Brain storm optimization algorithm. In: Tan, Y., Shi, Y., Chai, Y., Wang, G. (eds.) ICSI 2011, Part I. LNCS, vol. 6728, pp. 303–309. Springer, Heidelberg (2011)
8. Shi, Y.: An optimization algorithm based on brainstorming process. International Journal of Swarm Intelligence Research (IJSIR) **2**(4), 35–62 (2011)
9. Shi, Y.: Developmental swarm intelligence: Developmental learning perspective of swarm intelligence algorithms. International Journal of Swarm Intelligence Research (IJSIR) **51**(1), 36–54 (2014)
10. Xue, J., Wu, Y., Shi, Y., Cheng, S.: Brain storm optimization algorithm for multi-objective optimization problems. In: Tan, Y., Shi, Y., Ji, Z. (eds.) ICSI 2012, Part I. LNCS, vol. 7331, pp. 513–519. Springer, Heidelberg (2012)
11. Yang, Y., Shi, Y., Xia, S.: Discussion mechanism based brain storm optimization algorithm. Journal of Zhejiang University (Engineering Science) **47**(10), 1705–1711 (2013)
12. Zhan, Z.H., Chen, W.N., Lin, Y., Gong, Y.J., long Li, Y., Zhang, J.: Parameter investigation in brain storm optimization. In: 2013 IEEE Symposium on Swarm Intelligence (SIS), pp. 103–110, April 2013
13. Zhan, Z.h., Zhang, J., Shi, Y.h., Liu, H.l.: A modified brain storm optimization. In: 2012 IEEE Congress on Evolutionary Computation (CEC), pp. 1–8, June 2012
14. Zhou, D., Shi, Y., Cheng, S.: Brain storm optimization algorithm with modified step-size and individual generation. In: Tan, Y., Shi, Y., Ji, Z. (eds.) ICSI 2012, Part I. LNCS, vol. 7331, pp. 243–252. Springer, Heidelberg (2012)

Biogeography Based Optimization

Biogeography Optimization Algorithm
for DC Motor PID Control

Hongwei Mo[1(✉)] and Lifang Xu[2]

[1] Automation College, Harbin Engineering University,
Harbin 150001, China
honwei2004@126.com
[2] Engineering Training Center, Harbin Engineering University,
Harbin 150001, China
mxlfang@163.com

Abstract. Biogeography optimization algorithm (BBO) is a new optimization algorithm based on biogeography. Unique migration pattern of BBO makes good habitat feature information can be widely distributed among multiple habitats, showing a diversity of solutions. It is applied to the DC motor PID control problems and compared with genetic algorithms (GA), differential evolution (DE), particle swarm optimization (PSO). Experimental results show that BBO has the ability of searching optimal solution in a small local neighborhood space. The output of PID control system of DC motor optimized under BBO has no overshoot, no steady-state error and has the shortest system dynamic response time.

Keywords:: Biogeography based optimization algorithm · Direct current motor · PID control

1 Introduction

Biogeography Based Optimization Algorithm (BBO) is a new optimization algorithm was proposed by Simon in 2008 [1]. The biogeography of species migration process is realized by adjusting the immigration rate and migration rate. Migration strategies are used to achieve information sharing and improve habitat suitability for solving global optimal solution [1]. BBO algorithm has been improved and applied for different areas [2-3].

BBO was applied to solve reactive power optimization problem. Studies have shown that BBO algorithm is not strongly dependent on the parameters for solving reactive power optimization [4]. BBO and DE are hybrid for solving mixed robot path planning in a static environment to determine the required number of points by the number of obstacles in the path [5].Panchal used BBO to classify satellite images and have achieved good results [6]. A novel high performance low-complex EA that combines the advantages of both BBO and ABC algorithms for optimization problems in both continuous and discrete domains was proposed in[7]. This algorithm has shown higher performance in comparison to other EAs when applied to some optimization

© Springer International Publishing Switzerland 2015
Y. Tan et al. (Eds.): ICSI-CCI 2015, Part I, LNCS 9140, pp. 385–394, 2015.
DOI: 10.1007/978-3-319-20466-6_41

problems. In[8], BBO was combined with fuzzy C-means clustering algorithm to be BBO-FCM and used for image segmentation. Experimental results show that BBO-FCM was superior to FCM and some other compared bio-inspired algorithms. In [9], BBO is combined with chaos theory and used to train the neural network architecture for fault diagnosis of pumping.

DC motors are widely used as actuator component in engineering applications. The selection of PID control parameter directly affects the performance of DC motor control systems. For a DC motor, overshoot, errors and response speed of control system are three main objectives to optimize without mutual conflict at the same time. To improve the performance of some other objectives will cause performance degradation. We use BBO to optimize PID parameters of DC motor in order to obtain overshoot, errors and response speed between a reasonable configuration and good dynamic response of control system.

2   Biogeography Based Optimization

The fitness function of BBO is generally defined as living suitability index (habitat suitability index, HSI). BBO model is based on a group of suitable habitats. For species with higher habitat adaptation index (Habitat Suitability Index, HSI), they have more species to move out. They often have a larger emigration rates and smaller immigration rate. Habitat with lower HSI has fewer species. Higher HIS solution will share their good characteristics with lower HSI solution to complete the evolution.

2.1   BBO Migration Operation

Fig.1 shows the migration of species habitats on earth by a nonlinear model.

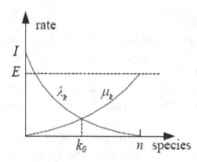

Fig. 1. Biogeography migration model

The lateral axis represents the number of species of habitats. And the vertical axis represents mobility,which represents immigration rate. I is the biggest immigration rate and E the maximum emigration rate. Immigration rate and emigration rate are equal to the equilibrium point for the habitat to accommodate the maximum number of species.

In this model, immigration rate and emigration rate are calculated as follows, respectively:

$$\begin{cases} \lambda_k = I\left(\dfrac{k}{n}-1\right)^2 \\ \mu_k = E\left(\dfrac{k}{n}\right)^2 \end{cases} \tag{1}$$

When population size of habitats increases, immigration rate decreases rapidly, while emigration rate increases slowly. When the number of the population almost saturated, immigration rate decreases slowly move out rate increases rapidly.

The migration operation is as follows.

```
 1: for  i =1 to  NP  do
 2:   Selected  X_i  with probability ∝ λ_i
 3:       if rndreal (0,1) <  λ_i then
 4:         for  j =1 to  NP  do
 5:             Selected  X_j  with probability ∝ μ_j
 6:               if rndreal (0,1) <  μ_j then
 7:                   Randomly select a variable  σ    from  X_j
 8:                   Replace a random variable in  X_i with  σ
 9:               end if
10:         end for
11:       end if
12: end for
```

2.2   Mutation Operation

BBO simulate changes in habitat environment by mutation operation. The number of species is inversely proportional to the probability and mutation probability. Mutation operation makes the solutions with low HSI have more opportunities to be improved. The mutation rate is caculated as follows:

$$m_k = m_{\max}\left(1-\dfrac{P_k}{P_{\max}}\right). \tag{2}$$

where $m_{\max}$ is an user-defined parameter, $P_{\max} = \arg\max P_i$, $i = 1, 2, ..., NP$. P_i is the probability of the corresponding number of species.

The mutation operator is as follows:

```
1: for  i =1 to  NP   do
2:   Compute the probability ∝ Pᵢ
3:     Select SIV  Xᵢ(j)   with probability  Pᵢ
4:       if rndreal (0, 1) < mᵢ, then
5:           Replace  Xᵢ(j) with a randomly generated SIV
6:         end if
7: end for.
```

3   Model of DC Motor Control System

3.1   Transfer Function of the DC motor

In Fig.2,the circuit of DC motor is shown in Fig.2.Dynamic mathematical model of the DC motor can be expressed by the following equations [10]:

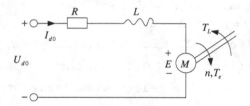

Fig. 2. Circuit of DC motor

$$U_{d0} = RI_d + L\frac{dI_d}{dt} + E \ . \tag{3}$$

where U_{d0} is the armature voltage, L the armature circuit inductance, I_d the armature current, R the total resistance of the armature circuit.

Ignore viscous friction and elasticity torque motor shaft kinetic equation is:

$$T_e - T_L = \frac{GD^2}{375}\frac{dn}{dt} \ . \tag{4}$$

where $T_L(Nm)$ is the motor load torque including no-load torquet, $GD^2(Nm^2)$ the torque of electric drive system moving parts converted to motor shaft.

Induced electromotive force and torque under rated excitation are:

$$E = C_e n \ . \tag{5}$$

$$T_e = C_m I_d \ . \tag{6}$$

where $C_m = \dfrac{30}{\pi} C_e (Nm/A)$ is the ratio of torque and current of motor under rated excitation.

Define the following time constants:

$T_l = \dfrac{L}{R}$ - electronmagnetic time constant of armature circuit with unit s;

$T_m = \dfrac{GD^2 R}{375 C_e C_m}$ - electromechanical time constant of electric drive system with unit s.

Substituting $T_l = \dfrac{L}{R}$ into equations (4) and (5),we get

$$U_{do} - E = R(I_d + T_l \frac{dI_d}{dt}) \quad . \tag{7}$$

$$I_d - I_{dL} = \frac{T_m}{R} \frac{dE}{dt} \quad . \tag{8}$$

where $I_{dL} = \dfrac{T_L}{C_m}$ - load current(A).

Under zero initial condition, Laplace transform is operated on both sides of the equation (7) and equation (8). The transfer function between voltage and current is:

$$\frac{I_d(s)}{U_{do}(s) - E(s)} = \frac{\frac{1}{R}}{T_l s + 1} \quad . \tag{9}$$

The transfer function between the current and the electromotive force is:

$$\frac{E(s)}{I_d(s) - I_{dL}(s)} = \frac{R}{T_m s} \quad . \tag{10}$$

Based on formula (10) and equation (11), the final finishing speed relative to the input DC voltage transfer function model is :

$$H(s) = \frac{N(s)}{U_{d0}(s)} = \frac{1/C_e}{T_m T_l s^2 + T_m s + 1} \quad . \tag{11}$$

A DC motor nameplate parameters include Inertia Rated voltage V, rated current A, rated speed, rated output power KW, motor shaft. According to [11], the parameters of DC motor are motor electromechanical time constant = 0.3023s, electromagnetic time constant = 0.0215s, motor potential coefficient = 0.612. The final transfer function of the motor speed and the input voltage model is:

$$H(s) = \frac{\Omega(s)}{U_a(s)} = \frac{1.634}{0.00649945s^2 + 0.3023s + 1}. \tag{12}$$

3.2   Objective Function

For optimizing the DC motor control system, the objective function is defined as:

$$J = \int_0^\infty (w_1 \mid e(t) \mid + w_2 u^2(t))dt + w_3 \cdot t_u. \tag{13}$$

In the function, system error, the controller output and the rise time are used as weights. Absolute error and the control input of the time integral of the squared term, as well as key performance indicators Functions, respectively, as the rise time of entry. To avoid overshoot, also uses a punitive function, which once produced overshoot, exceed the amount transferred as an optimal index, in which case the best indicators:

$$if \quad ey(t) < 0$$

$$J = \int_0^\infty (w_1 \mid e(t) \mid + w_2 u^2(t) + w_4 \mid ey(t) \mid)dt + w_3 \cdot t_u. \tag{14}$$

Formula(14), the weights, and, for the target output is pulled.Right value w_1 =0.999, w_2 =0.001, w_4 =100, w_3 =2.0。 where w_3 is weight and $w_4 \gg w_1$. $ey(t) = y(t) - y(t-1)$. $y(t)$ is the system output.

4   PID Control System of DC Motor Optimized by BBO

For using the BBO algorithm to optimize the PID parameters of the DC motor, we need to map the K_p, K_i, K_d into habitat feature vector SIVs, SIV = { K_p, K_i, K_d }. The optimization problem of DC motor PID control system becomes the minimum optimization problem of searching SIVs. In the evolutionary process of the algorithm, the solutions with higher HSI will share their good characteristics with solutions which have lower HSI, that is, share the better PID control parameter values. The number of samples used in the algorithm is 30. The number of evolution generation is G = 30. The parameter ranges [0,300], the range [0,10], the range [0,50]. The process of optimizing PID parameters by BBO is shown in Fig.3.

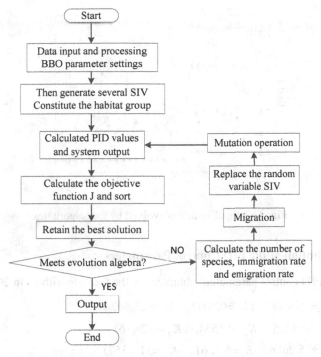

Fig. 3. Optimization process of DC motor PID parameters

5 Simulation and Analysis

The simulation is realized by Matlab/Simulink7.12.0. BBO is compared with Genetic Algorithm (GA), Differential Evolution (DE) and Particle Swarm Optimization (PSO).

5.1 Optimal Solutions Evolution

The best optimal objective functions of the four algorithms were obtained in 20 tests: BBO (4.2933), GA (4.4144), DE (4.7650), PSO (4.6444). Evolutionary process is shown in Figure 1.

Seen from Fig.4, BBO gets the optimal solution with least generation G = 11, followed by GA, PSO. The worst is DE. The best values of BBO and GA are close to the solution of the objective function and were significantly better than the other two algorithms. So for the four algorithms, BBO is the most effective for solving the problem.

It can be seen from Fig.4 that BBO has the best diverse distribution in neighborhood of optimal solutions. It can maintain the diversity of the solutions to some extent avoid the local minimum.

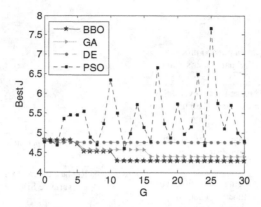

Fig. 4. Optimal solutions evolved by four algorithms

5.2   Distribution of the PID Control Parameters

The optimal PID control parameters obtained by the four algorithms in 20 tests were:

BBO (K_p=228.432, K_i=0.1045, K_d=25.8424)

GA (K_p=40.9285, K_i=0.1534, K_d=6.3416)

DE (K_p=75.2666, K_i=6.3661, K_d=31.1357)

PSO (K_p=96.6429, K_i=3.7513, K_d=30.5624)

The distribution is shown in Fig. 5 in a three-dimensional numerical space. It can be seen from Fig.5 that the search results of BBO and GA, whose values are distributed in [0,1]. The value is relatively small.

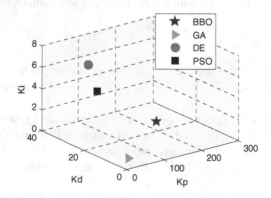

Fig. 5. Control parameters obtained by four algorithms

In Fig.5, the distribution of the PID control parameters obtained by BBO in 20 experiments is shown in 3D space. Fig.5 and Fig. 6 show that BBO tends to obtain the control parameters in the vicinity of the optimal solution Best J , while the solutions of PSO and

DE are relatively far from the optimal solution Best J. To some extent, BBO has the characteristics of searching the solutions within the periphery of the optimal solution.

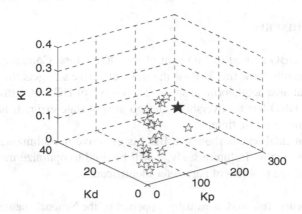

Fig. 6. Control parameters obtained by BBO in 20 experiments

In Fig. 6, we can see that the control parameters obtained by BBO in 20 experiments are distributed near the optimal solution.

5.3   DC Motor Speed Control

For the rated speed of the DC motor speed tracking, it can be seen from Figure 7 that the DC motor speed control under the BBO PID algorithm obtains the response without overshoot, but there is a certain steady-state tracking error of DE and PSO for DC motor speed control. DE control has the maximum steady-state error. Under the BBO control of the DC motor system its output has no steady-state error.

It can be seen from Fig. 7 that BBO control has the fastest dynamic response, followed by GA, DE, and PSO. DC motor under BBO control obtains steady-state output in 0.3s.

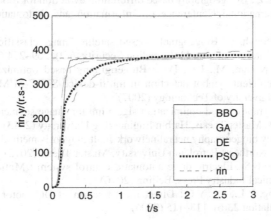

Fig. 7. DC motor speed control

For DC motor speed control, BBO gets optimal results of PID control system output. DE has the worst results.

6   Conclusions

In this paper, BBO is used to PID control for DC motors. Compared with GA, PSO and DE, the results show that BBO is the most effective and gets the optimal solution in a small local area neighborhood space. The output of DC motor PID control system optimized by BBO has no overshoot and no steady-state error. It has the minimum system dynamic response time.

The experimental results show that BBO is effective in optimizing PID control parameters of DC motor. In future study, it will be used to optimize more complex control system and the problem of parameter identification.

Acknowledgments. This work is partially supported by the National Natural Science Foundation of China under Grant No.61075113, the Excellent Youth Foundation of Heilongjiang Province of China under Grant No. JC201212, the Fundamental Research Funds for the Central Universities No.HEUCFX041306.

References

1. Simon, D.: Biogeography based optimization. IEEE Transaction on Evolutionary Computation. **12**, 702–713 (2008)
2. Xu, Z.D., Mo, H.W.: Disturbance multi-objective biogeography optimization algorithm. Control and Decision. **29**(2), 231–235 (2014)
3. Xu, Z.D., Mo, H.W.: Biogeography information optimization algorithm to improve the operator's migration. Pattern Recognition and Artificial Intelligence **25**(3), 544–548 (2012)
4. Sun, J., Gao, Y.H., Wang, C.: Biogeography based optimization algorithm for reactive power optimization. Nanchang University (Engineering & Technology) 35(4), 380–384, 391 (2013)
5. Mo, H.W., Li, Z.Z.: Bio-geography based differential evolution for robot path planning. In: 2012 IEEE International Conference on Information and Automation, ICIA, pp. 1–6 (2012)
6. Panchal, V.K., Singh, P.: Biogeography based satellite image classification. International Journal of Computer Science and Information Security. **6**(2), 269–274 (2009)
7. Ashrafinia, S., Naeem, M., Lee, D.C.: Biogeography based optimization algorithm for computational efficient symbol detection in multi-device STBC-MIMO systems. Master Thesis, Sharif University of Technology (2007)
8. Lee, B.: Biogeography based optimization algorithm for image segmentation technologies and applications. Master Thesis, Harbin Engineering University (2013)
9. Zhao, Y.J.: Study biogeography neural network fault diagnosis method based on optimization algorithms. Northeast Petroleum University. Master Thesis (2013)
10. Ruan, Y., Chen, B.S.: Electric drive automatic control system : Motion control systems, 4th edn. Mechanical Industry Press, Beijing (2010)
11. Ru, Z.X., Zhang, Z.L., Qi, Y.C.: Direct identification of DC motor model parameters. Computer Simulation **23**(6), 113–115 (2006)

Biogeography Based Optimization for Tuning FLC Controller of PMSM

Salam Waley[1], Chengxiong Mao[1], and Nasseer K. Bachache[1,2(✉)]

[1] Huazhong University of Science and Technology, Wuhan, China
[2] College University of Humanity Studies, Najaf, Iraq
tech_n2008@yahoo.com

Abstract. In this paper we embody the simulation of the Fuzzy Logic Controller. The controller governs the speed of a permanent magnet synchronous motor PMSM, which is employed in an elevator with different loads. This work aims to obtain the optimal parameters of FLC. Biogeography-Based-Optimization (BBO) is a new intelligent technique for optimization; it can be used to tune the parameters in different fields. The main contribution of this work is to show the ability of BBO to design the parameters of FLC by shaping the triangle memberships of the two inputs and the output. The results show the optimal controller (BBO-FLC) compared with the other controllers designed by genetic algorithm (GA). GA is a powerful method that has been found to solve the optimization problems. The implementation of the BBO algorithm has been done by M-file/MATLAB. The complete mathematical model of PMSM system has carried out using SIMULINK/MATLAB. The calculation of finesses function can be done by SIMULINK, and it linked with M-file/MATLAB to complete all steps of BBO. The results show the excellent performance of BBO-FLC compared with the GA-FLC and PI controller; also, the proposed method was very fast and needed only a few iterations.

Keywords: Biogeography Based Optimization (BBO) · Genetic Algorithm (GA) · Fuzzy Logic Controller (FLC) · Permanent Magnet Synchronous Motor (PMSM)

1 Introduction

In a new high-rise building, an elevator becomes the essential service facility. With the continuous improvement of running speed, the elevator's dynamic performance is closely related with human comfort, which is increasingly a cause for concern. Improving the elevators' comfort, and reducing vibration and noise during operation has become a hot research topic at home and abroad [1]. The development in microprocessor schemes and semiconductor technologies makes the AC drive give a high performance of speed control. This system is an excellent opportunity to use AC motors [2]. In the last few years, PMSM has become popular in the medium range of an AC machine and its drive. Nowadays, this technology has become the first choice because of its inherent advantages. These advantages include high torque to current

© Springer International Publishing Switzerland 2015
Y. Tan et al. (Eds.): ICSI-CCI 2015, Part I, LNCS 9140, pp. 395–402, 2015.
DOI: 10.1007/978-3-319-20466-6_42

ratio, large power to weight ratio, higher efficiency and robustness. There are many applications of PMSM in elevators, wind energy, EV drive, etc., because it allows an enlarged speed range with an inverter size that is lower than in a conventional flux-oriented induction motor drive [3]. In [4], the adaptive dynamic surface control (DSC) has been presented for the feedback controller of PMSM. Also, some control methods have been studied to stabilize the PMSM systems, such as the sliding mode control (SMC) [7], deferential geometry method [8], and passivity control [9, 10]. The tangible benefit of choosing the controller is its simplicity in implementation. It is not easy to find another controller with such a simple structure that is comparable in performance. Fuzzy rule-based models are easy to comprehend all applications of PMSM because it uses linguistic terms as well as the structure of if-then rules [11]. A very important step in the use of controllers is the controller parameters and tuning process [12]. Unfortunately, in spite of this, a large range of tuning techniques and the optimum performance cannot be achieved. In recent years many intelligent optimization techniques have emerged and have received much attention from researchers concerning genetic algorithm (GA), particle swarm optimization (PSO) techniques bee colony optimization (BCO), ant colony optimization (ACO), simulated annealing (SA), and bacterial foraging (BF) [13]. GA was the most used in the control field, such as in the search for optimal parameters of an FLC controller. But it still requires an enormous computational effort. In this paper we suggest a new computational theory named Biogeography-Based Optimization (BBO) to tune parameters of the FLC controller. This controller can govern a non-linear system.

2   PMSM Mathematical Models and the Vector Control

The basic idea of vector control is to manage the analog DC motor torque. Also, the control law is used in the ordinary three-phase AC motor. For magnetic field directional coordinates, we break down the current vector into the exciting current component, which produces the magnetic flux and torque.

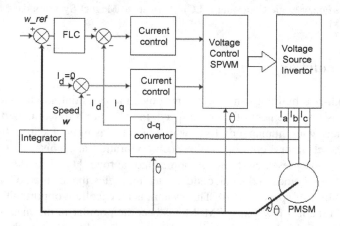

Fig. 1. Schematic diagram of the PMSM vector control

After coordinate transformation, the three-phase stator coordinate system (a-b-c coordinate) is varied to d-q coordinate. The two components (d-q coordinate) are perpendicular to each other and independent of each other. They are then adjusted respectively. Thus, the torque control of the AC motor is similar to the DC motor regarding their principles and characteristics. Therefore, the key for vector control is still both magnitude and the spatial location control of the current vector. Figure 1 shows the schematic diagram of the PMSM vector control.

3   Fuzzy Logic Controllers

Fuzzy logic controllers have the following advantages over the conventional controllers: they are cheaper to develop, they cover a wide range of operating conditions, and they are more readily customizable in natural language terms. In Mamdani type FIS, the crisp result is obtained by defuzzification [14]; the Mamdani FIS can be used for both multiple input and single output and a multiple inputs/multiple outputs system, as shown in Figure 2.

The usefulness of the fuzzy controller is adopted particularly in a complex and nonlinear system. The rules of conventional FLC produced depend on the operator's experience or general knowledge of the system in a heuristic way. The thresholds of the fuzzy linguistic variables are usually chosen arbitrarily in the design process. An improper controller value leads to an adverse consequence, unstable mode, collapse and separation. This work proposed BBO to design an optimal fuzzy logic controller (OFLC), where the optimized criteria are how to minimize the transient state.

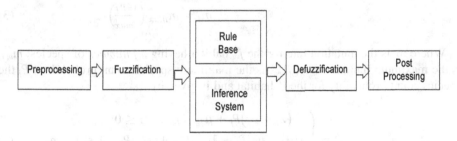

Fig. 2. Arrangement of fuzzy logic controller

4   Biogeography Based Optimization

Inspired by biogeography, Simon developed a new approach called Biogeography-Based-Optimization (BBO) in 2008. This algorithm is an example of how a natural process can be modeled to solve optimization [15]. In BBO, each possible solution is an island; the features that describe habitability are named the habitat suitability index (HSI). The goodness of each solution is named Suitability Index Variables (SIV). For example, concerning the natural process, why do some islands tend to accumulate many more species than others? Because they posses certain environmental features that are more suitable to sustaining them than other islands with fewer species. It is axiomatic that the habitats with high HSI have large populations, a high immigration rate, and a large number of species that migrate to other habitats. The rate of immigration will be lower if these habitats are already saturated with species. On the other hand, habitats with low HSI have a high immigration and low immigration rate, because of the sparse population.

The fitness function FF is associated with each solution of Biogeography Based Optimization BBO, which is analogous to the HSI of a habitat. A good solution is analogous to a habitat having a igh HSI, and a poor solution represents a habitat having a low HSI. The best solutions share their geographies of the lowest solutions throw migration. The best solutions have very small change compared with the lowest solutions, while the lowest solutions have a large change from time to time and accept many new features from the best solutions.

The immigration rate and emigration rate of the jth island may be formulated as follows, in equations 1 & 2.

$$\lambda_j = I\left(1 - \frac{j}{n}\right) \tag{1}$$

$$\mu_j = \frac{E.j}{n} \tag{2}$$

is Where μ_j, λ_j are the immigration rate and the emigration rate of jth individual; I is the maximum possible immigration rate; E is the maximum possible emigration rate; j is the number of species of jth individual; and n is the maximum number of species [16].

In BBO, the mutation is used to increase the diversity of the population to get the best solutions. The mutation operator modifies a habitat's SIV, randomly based on the mutation rate. The mutation rate m_j is expressed in (3).

$$m_j = m_{max}\left(\frac{1 - P_j}{P_{max}}\right) \tag{3}$$

Where m_j is the mutation rate for the jth habitat having *a j number* of species; m_{max} is the maximum mutation rate; P_{max} is the maximum species count probability; P_j the species count probability for the jth habitat and is given by:

$$\dot{P} = \begin{cases} -(\lambda_j + \mu_j)P_j + \mu_{j+1}P_{j+1}, & j \le 0 \\ -(\lambda_j + \mu_j)P_j + \lambda_{j-1}P_{j-1} + \mu_{j+1}P_{j+1} & 1 \le j \le n \\ -(\lambda_j + \mu_j)P_j + \lambda_{j-1}P_{j-1}, & j = n \end{cases} \tag{4}$$

Where μ_{j+1}, λ_{j+1} are the immigration and emigration rate for the jth habitat contain j+1 species; μ_{j-1}, λ_{j-1} are the immigration and emigration rate for the jth habitat contains j-1 species.

5   Implementing BBO Tuning for FLC Parameters

The implementation of BBO in this work is somewhat complex, because the performance of the system must be examined for all habitats during of the each iteration. Therefore, the optimization algorithm is implemented by the MATLAB m-file program and linked with the system simulation program in MATLAB SIMULINK, to check the system performance in the each iteration. In this paper, the problem is summarized in optimizing three variables (X1, X2 and X3 shown in Figure 6), they are: one output and two inputs (speed and the change in speed); they are represented as three dimensional spaces including the prams of the triangle memberships of FLC. A random 20 habitats were assumed and an algorithm of 100 iterations is used to estimate the optimal values of the FLC controller parameters. The fitness function FF, illustrated in equation (5), can be calculated by SIMULINK, as shown in Figure 3.

$$FF = ITSE = \int_0^t t * e^2(t)\, dt \tag{5}$$

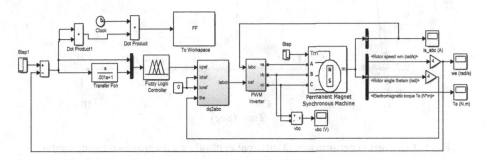

Fig. 3. System model implemented by SIMULINK/MATLAB

6   Results

Figure 4 shows the convergence of fitness function in 100 iterations and the comparison between GA and BBO. Figure 5 shows the step response with load and no

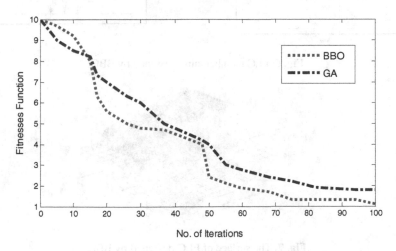

Fig. 4. The convergence of fitness function in 100 iterations

load using the proposed controller and GA-FLC, and the PI-controller tuned by conventional methods of trial and error. Figure 6 shows FLC designed by BBO, and Figure 7 shows the surface of FLC.

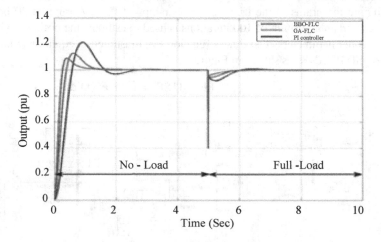

Fig. 5. Comparison performance of different controllers with proposed tuning methods

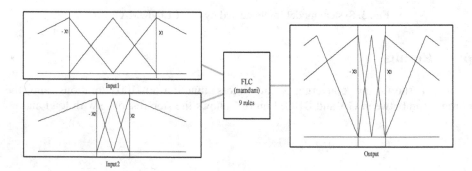

Fig. 6. FLC memberships designed by BBO

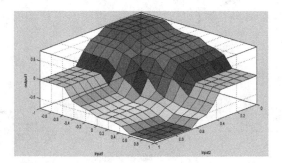

Fig. 7. The surface of FLC designed by BBO

7   Conclusions

The system responses of different tuning methods are illustrated in the simulation result, and a comparable performance between the three controllers in this research (PI controller, GA-FLC and BBO-FLC) is shown in Figure 5. We can obtain the following conclusions through simulation analysis:

1- This paper designs fuzzy logic control by a computational algorithm; it interjects control concepts of trial and error in fuzzy control and the conventional GA-FLC method, and then the control velocity modulation of an elevator.
2- Obviously, the BBO tuning of the FLC is the best intelligent method, which gives an excellent system performance, and the GA gives a good response with respect to the traditional trial and error method.
3- In addition to improving the system response, the BBO and GA can use a higher order system in the tuning process, which avoids the error of system order reduction. It gives a satisfactory solution during the first 50 iterations as shown in Figure 6.
4- The proposed method gives the control system strong flexibility, instantaneity and reliability because of the advanced prediction of the FLC predicting controller.
5- It makes the control system have a stronger real-time controllability because optimal fuzzy parameters have predicted a possible interference source. The lower the interference frequency, the more the BBO algorithm is controllable.
6- The elevator speed control system using BBO-FLC can be used as an effective complement to traditional control methods, thus, it can further enhance and improve the regulating quality for a control system with a different load torque.

References

1. Morrison, L.J., Angelini, M.P., Vermeulen, M.J., Schwartz, B.: Measuring the EMS patient access time interval and the impact of responding to high-rise buildings. Prehospital Emergency Care 9(1), 14–18 (2005)
2. Tezic, B., Jadric, M.: Design and implementation of the Extended Kalman Filter for the Speed and Rotor Position Estimation of BLDC. IEEE Transactions on Industrial Elect. 48, 1065–1073 (2001)
3. Dehkordi, A.B., Gole, A.A., Maguire, T.L.: Permanent Magnet Synchronous Machine Model for Real-Time Simulation. In: Conference on Power System Transients, IPST 2005, pp.159–164 (2005)
4. Wei, D.Q., Luo, X.S., Wang, B.H., Fang, J.Q.: Robust adaptive dynamic surface control of chaos in permanent magnet synchronous motor. Physics Letters, Section A 363(1–2), 71–77 (2007)
5. Luo, Y.: Current rate feedback control of chaos in permanent magnet synchronous motor. Proceedings of the CSU-EPSA 18(6), 31–34 (2006)
6. Ren, H., Liu, D.: Nonlinear feedback control of chaos in permanent magnet synchronous motor. IEEE Transactions on Circuits and Systems II 53(1), 45–50 (2006)

7. Reichhartinger, M., Horn, M.: Sliding-mode control of a permanent-magnet synchronous motor with uncertainty estimate on. International Journal of Mechanical and Materials Engineering1 2, 121–124 (2010)

8. Wei, D.Q., Luo, X.S., Fang, J.Q., Wang, B.H.: Controlling chaos in permanent magnet synchronous motor based on the deferential geometry method. Acta Physica Sinica 55(1), 54–59 (2006)

9. Qi, D.L., Wang, J.J., Zhao, G.Z.: Passive control of permanent magnet synchronous motor chaoticsystems. Journal of Zhejiang University 6(7), 728–732 (2005)

10. Wu, Z.Q., Tan, F.X.: Passivity control of permanent-magnet synchronous motors chaotic system. Proceedings of the Chinese Society of Electrical Engineering 26(18), 159–163 (2006)

11. Hirulkar, S., Damle, M., Rathee, V., Hardas, B.: Design of Automatic Car Breaking System Using Fuzzy Logic and PID Controller. In: 21st International Conference on Electronics Circuits and Systems, pp. 413–418 (2014).

12. Castillo, O., Patricia, M.: A review on interval type-2 fuzzy logic applications in intelligent control. Information Sciences 279, 615–631 (2014)

13. Bachache, N.K., Wen, J.Y.: PSO and GA designed Pareto of Fuzzy Controller in AC Motor Drive. International Journal of Control and Automation 6(5), 149–158 (2013)

14. Chen, C.-W.: Applications of neural-network-based fuzzy logic control to a nonlinear time-delay chaotic system. Journal of Vibration and Control 20(4), 589–605 (2014)

15. Roy, P.K., Mandal, D.: Optimal reactive power dispatch using quasi-oppositional biogeography-based optimization. International Journal of Energy Optimization and Engineering 1(4), 38–55 (2012)

16. Simon, D.: Biogeography-based optimization. IEEE Transactions on Evolutionary Computation 12(6), 702–713 (2008)

Enhancing the Performance of Biogeography-Based Optimization in Discrete Domain

Qingzheng Xu[✉], Na Wang, Jianhang Zhang, and Xiang Gu

Xi'an Communications Institute, Xi'an 710106, China
xuqingzheng@hotmail.com

Abstract. Oppositional Biogeography-Based Optimization using the Current Optimum (COOBBO) has been recently developed to solve combinatorial problems and outperforms other heuristic algorithms. The objective of this paper is mainly to ascertain various modified methods which can significantly enhance the performance and efficiency of COOBBO algorithm. The improvement measures include crossover approach, local optimization approach and greedy approach. Experiment results illustrate that, the combination model of "inver-over crossover + 2-opt local optimization + all greedy" may be the best choice of all when considering both the overall algorithm performance and computation overhead.

Keywords: Biogeography-Based Optimization · Traveling Salesman Problem · Crossover approach · Local optimization approach · Greedy approach

1 Introduction

Biogeography-Based Optimization (BBO), which was first introduced in 2008 by Simon [1], is inspired by the science of biogeography which studies the temporal and spatial distribution of species amongst islands. Based on previous research results in recent years, BBO has been successful applied not only to continuous optimization problems, but also to discrete domain problems [2-7]. Recently, we proposed a novel Oppositional Biogeography-Based Optimization using the Current Optimum, called COOBBO algorithm, for solving combinatorial problems [8]. Simulation results illustrated clearly that, it outperformed both standard BBO and Oppositional Biogeography-Based Optimization (OBBO) by Ergezer in 2011 [6].

While an excellent algorithm may be very critical for sophisticated problem solving, various modified techniques are also definitely an issue, particularly for real-world applications. Many experiment results proved that, using these improvement measures correctly can really help us increase population diversity, accelerate the convergence speed and reduce time complexity greatly. Here for a chance, we emphasize again that, we do not put forward a new algorithm, but focus on the entire impact of various skills that will give us a chance to enhance the performance and efficiency of COOBBO.

The remainder of this paper is organized as follows. In the following section, we provide a general description of our COOBBO algorithm, and then investigate various

© Springer International Publishing Switzerland 2015
Y. Tan et al. (Eds.): ICSI-CCI 2015, Part I, LNCS 9140, pp. 403–414, 2015.
DOI: 10.1007/978-3-319-20466-6_43

modified techniques briefly. Section 3 shows the simulation results to ascertain their effect and influence on basic algorithm. Finally, some concluding remarks and suggestions for further research are presented in Section 4.

2   Background

2.1   COOBBO Algorithm

As two hybrid soft computing algorithms, OBBO and COOBBO are both created by incorporating the idea of opposition-based learning into original BBO algorithm and suffer the same basic produce as shown in Figure 1. Actually, the only difference between original OBBO and COOBBO consists in that the definition of opposite path. In OBBO algorithm, an opposite path is defined as a candidate path that maximizes the distance between the adjacent vertices in the original path [6].

1:**procedure** OBBO (or COOBBO) algorithm (*Problem*)
2:     Randomly generate initial population, *P*
3:     Generate the opposite of initial population, *OP*
4:     Maintain the fittest amongst *P* and *OP*
5:     **while** *Generation ≤ gen limit* **do**
6:          Perform BBO Migration operator
7:          Calculate the fitness of *P*
8:          **if** *random ≤ Opposition Jumping Rate* **then**
9:               Create the opposite population, *OP*
10:               Calculate the fitness of *OP*
11:               Maintain the fittest amongst *P* and *OP*
12:          **end if**
13:          Restore *Elite Individuals*
14:     **end while**
15:     **return** *Best Individual*
16:**end procedure**

Fig. 1. General flowchart of OBBO and COOBBO algorithms

Opposition-Based Learning using the Current Optimum (COOBL) was original introduced to accelerate differential evolution in continuous domain by Xu [9, 10]. As a successful extension of COOBL from continuous domain to discrete domain, we developed a novel definition of opposite path, which could lead to a better solution efficiently [8]. Its core feature was that the sequence of candidate paths and the distances between adjacent nodes in the tour were considered simultaneously. In a sense, the candidate path and its corresponding opposite path had the same (or similar at least) distance from the optimal path in the current population. Experiment results also showed that, when compared with standard BBO and OBBO, the excellent performance of COOBBO algorithm was mainly attributed to the distinct definition of opposite path.

2.2   Modified Method

So far, many technical modifications have been invented for the purpose of performance improvement. In this paper, we mainly investigate three types of modified approaches that are widely studied and utilized for classical Traveling Salesman Problem (TSP).

(1) Crossover Approach. It is well known that, migration is the method of combining selected parent individuals and creating a child individual from them in the next generation. Hence, crossover operator is also believed to be the most important and useful component of COOBBO algorithm. Three crossover methods will be discussed later in this paper: matrix crossover [11], clycle crossover [12], and inver-over crossover [13].

The detailed procedure of matrix crossover is as follows [7]. (1) First, for an n-city problem, we need to convert the ordering information of all individuals to an n by n matrix. Each row in the matrix expression provides the position information of a city in the trip, and each column in each row represents the ordering relationship between the column city and row city. Based on this method, we convert all the individuals in the population to the matrix expression. (2) Second, based on the selection methods, we select individuals to perform migration operator. Once the parents are selected, we perform AND logic on two matrices and then we obtain one child matrix. (3) Third, we randomly fill in necessary information to create a valid child, which represent a TSP tour, and then the child matrix is complete. (4) In the last step, we transform the child individual from matrix expression to sequential representation.

Alternative crossover operator for TSP is cycle crossover as follows [7]. (1) We randomly select a city as the starting point in parent 1, and record its position. (2) In parent 2, find the city at the position we recorded in parent 1 and then record this city. Go back to parent 1, search for the city we found in parent 2 and then record its position in parent 1. (3) Repeat step 2 until we obtain a closed cycle, which means we have returned to the starting city. (4) We copy the cities from the closed cycle in parent 2, and the cities that are not in the closed cycle in parent 1, to obtain child 1. Similarly, we copy the cities from the closed cycle in parent 1, and the cities that are not in the closed cycle in parent 2, to obtain child 2.

Another crossover operator investigated in this paper is inver-over crossover to yield better individuals for the next generation [7]. (1) Two parents are utilized to generate a child. Randomly select a city in parent 1 as the starting point, city s. (2) Find s in parent 2 and choose the city next to it as the ending point, city e. Then find this ending point city in parent 1. (3) Reverse the cities between $s+1$ (the city next to the starting point city) and e in parent 1. That is the child created by inver-over crossover.

(2) Local Optimization Approach. As a complement to migration in COOBBO, local search optimization can find the optimal solutions by modifying the candidate solutions. We introduce two local optimizations measures, which have been successfully implemented in TSP: 2-opt [14] and 3-opt [15].

It is certainty that, 2-opt is a very simple but effective method for local research. For 2-opt method, we break two links in a random individual, and then connect the cities which only have one link connected, with the constraint that the resulting path includes all cities. Instead of replacing two links in the individual as in 2-opt, the 3-opt technique breaks three links and then randomly reconnects the cities that have broken links.

(3) Greedy Approach. As a general algorithm design paradigm, greedy methods have been applied to solve a wide variety of continuous and discrete problems. The definition of a greedy method is just as its name implies: always choose the immediate benefit, and refuse to take any losses [16]. Based on practice experiences, Greedy method does not always yield optimal solutions, but for many problems they do.

3   Experimental Results and Discussions

3.1   Experimental Setup

In the rest of this section, we will ascertain and compare the impact of all techniques mentioned in Section 2. Since these modified methods can not be exclusively performed solving TSP problem, they should be embedded in a general framework. In this paper, COOBBO is selected as the basic algorithm for its compact configuration and excellent performance. Then various hybrid COOBBO algorithms, as shown in Figure 2, are set up well for investigating the value measures implicit in hybrid algorithms.

```
-------------------------------------------------------------------------
1:procedure    hybrid    COOBBO    algorithm    (Problem,    CrossoverMethod,
LocalOptimizationMethod, GreedyMethod)
2:      Randomly generate initial population, P
3:      Generate the opposite of initial population, OP
4:      Maintain the fittest amongst P and OP
5:      while Generation ≤ gen limit do
6:            Perform crossover operator based on CrossoverMethod            //new
7:            Perform greedy operator based on GreedyMethod                  //new
8:            Perform local optimization operator based on LocalOptimizationMethod
                                                                            //new
9:            Perform greedy operator based on GreedyMethod                  //new
10:           Calculate the fitness of P
11:           if random ≤ Opposition Jumping Rate then
12:                 Create the opposite population, OP
13:                 Calculate the fitness of OP
14:                 Maintain the fittest amongst P and OP
15:           end if
16:           Restore Elite Individuals
17:      end while
18:      return Best Individual
19:end procedure
-------------------------------------------------------------------------
```

Fig. 2. General flowchart of hybrid COOBBO algorithms

By comparing the Figures 1 and 2, there are four distinctions between OBBO (COOBBO) algorithm and hybrid COOBBO algorithms, which are labeled in Lines 6-9 in Figure 2. Each improvement approaches can be chosen to run independently and manually on algorithm parameters, such as CrossoverMethod, LocalOptimization-Method, and GreedyMethod. Hence we can protect individual approach from the interference by other approaches.

In this paper, eight different well-known TSP problems are employed for performance verification of all modified methods. All the benchmarks are selected from TSPLIB [17], and their sizes vary from small problem to extra large problem. In order to minimize the effect of the stochastic nature of heuristic algorithms, all algorithms are conducted 100 independent experiments, both utilizing the same set of algorithm parameters as [8]. It is noted that, unless a change is stated, hybrid algorithms run inver-over crossover, and do not perform local optimization operator and greedy operator in all experiments. In this paper, tested algorithms will be stopped when the number of evaluation of cost function has reached the maximum number. Obviously, it is in accord with the termination criterion described in [8], which seems more fair and advisable, especially when compared with that employed by Ergezer [6].

As utilized in [8], four comparison criteria are also recoded and compared to analyze the contribution of different modified methods. They include Best Solution found at the end of the computation (BS), Computation Time (CT), Utilization Rate of opposite paths (UR) and Population Diversity (PD). Thereinto, UR and PD are defined respectively as following.

$$UR = \frac{OP_{res}}{N_p} \tag{1}$$

$$PD = -\sum_{e \in X} \frac{F(e)}{N_p} \log(\frac{F(e)}{N_p}) \tag{2}$$

where OP_{res} is the number of opposite paths reserved as offspring in the next generation, N_p is the size of the population including all candidate paths and their corresponding opposite paths, X is a set of edges included in the current population, and $F(e)$ is the number of edges e in the current population. For further details of comparison criteria, please carefully read the reference [8].

3.2   Crossover Method

Firstly, in order to check the performance of different crossover methods in the migration of COOBBO, we design three crossover strategies in this paper: matrix crossover (*CrossoverMethod* = 1), cycle crossover (*CrossoverMethod* = 2) and inver-over crossover (*CrossoverMethod* = 3). The simulation results on eight TSP benchmarks are shown in Table 1. The mean (left) and variance (right) of 100 independent experiments for each case are both listed and the best results in each row are highlighted in **boldface** font.

Table 1. Performances for different crossover methods

TSP problems		Crossover methods		
and criteria		Inver-over crossover	Cycle crossover	Matrix crossover
berlin52	BS	**10585.6729**±1106.8232	13962.2258±1820.3901	19090.7778±**543.2546**
	CT	**15.7656**±0.5733	20.7691±0.6810	33.8837±**0.4033**
	UR	3.4087±**0.6340**	5.1758±1.0683	42.4384±0.7741
	PD	6.0483±0.7837	8.6661±2.7154	**68.3919**±**0.0735**
bier127	BS	**280361.0292**±38778.3636	335819.6482±47120.0356	516662.9791±**8447.7298**
	CT	**70.9550**±2.7362	85.5037±3.1586	168.4036±**1.9537**
	UR	3.8659±1.0352	5.5124±1.302	74.4264±**0.8640**
	PD	14.7046±2.1019	27.3031±9.3704	**204.9903**±**0.2294**
ch130	BS	**17089.9077**±2858.6011	22660.3878±3255.9918	37579.2059±**910.5536**
	CT	**75.2717**±2.9781	89.2527±2.9303	172.6000±**2.3052**
	UR	4.3072±1.0183	6.3204±1.6010	54.4476±**0.7465**
	PD	15.3682±2.3285	29.6966±9.3633	**210.3570**±**0.2744**
kroA150	BS	**85580.4285**±15715.5394	120467.1556±19866.5277	203524.9152±**7368.3192**
	CT	**96.3781**±3.7218	114.2763±3.7809	225.2339±**2.7192**
	UR	4.8259±1.2510	6.6926±1.9878	50.4902±**0.5534**
	PD	19.5021±2.8978	36.0452±12.3256	**248.3056**±**0.3054**
kroA200	BS	**120597.2931**±19113.1388	163060.5333±22350.7237	279234.5228±**8166.8980**
	CT	**163.4809**±6.3209	189.698±6.7125	395.7475±**4.3098**
	UR	5.0262±1.3217	6.9101±1.5923	52.8035±**0.6579**
	PD	26.0689±3.4488	51.2904±14.3457	**344.4834**±**0.4392**
kroC100	BS	**48296.9643**±8517.3143	75028.9277±12709.5846	127590.1662±**3035.9217**
	CT	**46.8898**±1.9041	57.1769±1.6063	107.2795±**1.3750**
	UR	4.5670±0.8556	6.2929±1.6824	52.9462±**0.6007**
	PD	12.9923±1.7231	20.5180±6.9798	**154.1704**±**0.1985**
lin105	BS	**33257.0403**±5520.1857	53617.8518±8962.8354	77434.0851±**4526.5639**
	CT	**51.3555**±1.7913	62.7614±2.0238	114.6811±**1.4163**
	UR	4.2091±**0.8561**	6.0932±1.4705	44.3913±1.0255
	PD	14.3067±1.9913	22.1774±6.734	**161.6842**±**0.2329**
lin318	BS	**235023.9477**±33480.0785	318788.4121±35103.0976	454171.3861±**26468.5845**
	CT	**393.0959**±15.4872	444.2844±16.7241	1061.0963±**14.2891**
	UR	5.3975±1.5245	6.2498±1.5583	29.9671±**0.9732**
	PD	40.4241±5.3767	77.0247±24.9730	**569.3946**±**0.75272**

Based on the simulation results in Table 1, we have some good reason to believe that inver-over crossover dominates over the other methods in both BS and CT. As mentioned above, matrix crossover needs both expression transformation and valid child creation, which spends a great deal of time. So the CPU time required by matrix crossover becomes very long when the benchmark size increases.

As is known to all, the appropriate randomness can efficiently improve population diversity and avoid premature convergence for population-based soft algorithms. At the same time, intelligent algorithm should also circumscribe the randomness of population within narrow bounds and then seek a proper balance of exploitation and exploration. But, in the process of valid child creation of matrix crossover, some necessary cities are randomly filled to create a valid child and then the balance of exploitation and exploration is destroyed badly. As a result, matrix crossover is ranked the worst on most of comparison criteria among three methods. Besides, it is surprising that matrix crossover shows the best stability for its smallest variance of all. Then the attempt at explaining this phenomenon is our top task in the near future.

3.3   Local Optimization Method

The second test is to compare the performance of different local optimization approaches. In this paper, three improvement methods are discussed: no local optimization (*LocalOptimizationMethod* = 1), 2-opt (*LocalOptimizationMethod* = 2), and 3-opt (*LocalOptimizationMethod* = 3). The improvement performances are shown in Table 2 and the best results in each row are also highlighted in **boldface** font.

Table 2. Performances for different local optimization methods

TSP problems and criteria		No local optimization	2-opt local optimization	3-opt local optimization
		Local optimization methods		
berlin52	BS	10585.6729±1106.8232	**8894.8842±332.7840**	9927.7220±409.1705
	CT	**15.7656±0.5733**	18.6747±**0.5060**	21.2614±0.6892
	UR	3.4087±0.6340	3.4867±**0.2141**	**11.5389**±0.5142
	PD	6.0483±0.7837	30.7186±0.3279	**48.3719±0.2507**
bier127	BS	280361.0292±38778.3636	225532.4117±10162.9802	257053.2436±**9689.5476**
	CT	**70.9550±2.7362**	75.1322±2.9792	79.8512±3.1949
	UR	**3.8659**±1.0352	2.0488±**0.1222**	2.1109±0.1289
	PD	14.7046±2.1019	38.3343±0.9401	**67.6701±0.6155**
ch130	BS	17089.9077±2858.6011	**14171.3376**±671.7278	17122.4857±**661.7640**
	CT	**75.2717±2.9781**	77.7820±**2.8153**	79.9014±3.3327
	UR	**4.3072**±1.0183	2.0716±**0.1354**	2.1518±0.1382
	PD	15.3682±2.3285	37.8605±0.7931	**67.5397±0.6990**

Table 2. (*Continued*)

TSP problems and criteria		No local optimization	2-opt local optimization	3-opt local optimization
		Local optimization methods		
kroA150	BS	85580.4285±15715.5394	**73505.8046±3166.0688**	87485.8476±**3156.0360**
	CT	**96.3781±3.7218**	101.2262±3.9108	107.5780±4.2153
	UR	**4.8259**±1.2510	2.4676±**0.1497**	3.0251±0.2378
	PD	19.5021±2.8978	41.6502±1.4121	**72.8408±1.1334**
kroA200	BS	120597.2931±19113.1388	**108649.8755±4617.973**	124554.3216±**3898.1169**
	CT	**163.4809±6.3209**	168.1614±6.3264	170.0575±**5.8185**
	UR	**5.0262**±1.3217	2.3738±**0.1440**	2.5207±0.1738
	PD	26.0689±3.4488	45.6952±1.5906	**77.6007±1.4078**
kroC100	BS	48296.9643±8517.3143	**40843.2221±2075.6276**	50617.9084±2421.8124
	CT	**46.8898**±1.9041	50.1209±**1.8377**	54.7880±1.9410
	UR	4.5670±0.8556	2.9232±**0.2163**	**6.9964**±0.5579
	PD	12.9923±1.7231	37.6503±**0.9068**	**69.9279**±1.1320
lin105	BS	33257.0403±5520.1857	**28994.3855±1559.5959**	35921.3267±1743.4190
	CT	**51.3555±1.7913**	54.1039±1.8861	56.3952±2.0881
	UR	4.2091±0.8561	2.7982±**0.1823**	**6.3792**±0.4469
	PD	14.3067±1.9913	38.2931±**0.8673**	**70.7354**±1.0536
lin318	BS	**235023.9477±33480.0785**	240604.5007±10818.7375	267160.5702±**7109.9995**
	CT	**393.0959±15.4872**	404.3264±**14.4934**	426.0155±15.7691
	UR	**5.3975**±1.5245	1.9481±**0.1091**	1.9371±0.1342
	PD	40.4241±5.3767	54.5339±3.0000	**85.2502±2.0748**

Without any question, COOBBO algorithm with no local optimization requires the least CPU time among all tested methods. However with small increases in CPU time, the performance improvement may be very significant at a lower cost, except for lin318 problem, when utilizing 2-opt local optimization. In this particular case, the average BS of no local optimization is slightly less than that of 2-opt local optimization apparently. With the help of statistical analysis software, Statistical Product and Service Solutions (SPSS) 14.0 version, we can find that there is no significant difference in statistics (sig = 0.114, $p = 0.05$) between no local optimization and 2-opt local optimization. In conclusion, when considering both algorithm performance and computation time, 2-opt local optimization may be the best choice for COOBBO algorithm. Another fact we can observe from Table 2 is that, the performance improvement will be steadily decline, and even degeneration, especially for large scale TSP problems. Of course, the plucky inference should be proved or falsified by more extra experiments to cover a wide range of benchmark size in the future.

According to our observation of the process of 2-opt local optimization and 3-opt local optimization, we have found that their superior performances derive from high population diversity. The original order in a tour is confused and then realigned in order to seek the better solution, which brings in randomness and increases population diversity inevitably. And the intuitive influence in 3-opt local optimization is more distinct than that in 2-opt local optimization. As noted above, randomness and diversity can badly destroy the balance of exploitation and exploration of population-based algorithm. In our opinion, it is the power principle in algorithm design that leaves 3-opt local optimization in an awkward position.

Despite of the worst algorithm performance among three methods, no local optimization has the highest utilization rate of opposite paths for most of benchmark problems. Probably it seems implausible, because the high UR leads usually to a wide search space and a good optimization result. In fact, this conclusion applies only to the same algorithm without opposition-based learning (BBO algorithm in this paper). In our previous works [8], experiment results had already illustrated that COOBBO outperformed standard BBO and its excellent performance was attributed to the distinct definition of opposite path. Thus we may infer that the performance gap between standard BBO and COOBBO with local optimization must be much more remarkable than now as shown in Table 2.

3.4   Greedy Method

Finally, we test various greedy methods in similar way: no greedy (*GreedyMethod* = 1), half greedy (greedy method used in half of population, *GreedyMethod* = 2), and all greedy (greedy method used in entire population, *GreedyMethod* = 3). The greedy method is applied at both two steps (crossover operator and local optimization operator) of the COOBBO algorithm in each generation. The experimental results for different greedy method are shown in Table 3, and similarly the best results in each row are highlighted in **boldface** font.

Table 3. Performances for different greedy methods

TSP problems and criteria		Greedy methods		
		No greedy	Half greedy	All greedy
berlin52	BS	10585.6729±1106.8232	**8129.2240±237.4142**	8412.0144±385.9223
	CT	**15.7656±0.5733**	18.5217±0.5972	17.9673±0.5939
	UR	3.4087±0.6340	**9.3809**±3.4942	1.8345±**0.3447**
	PD	6.0483±0.7837	**11.0890±0.6490**	4.4932±**0.3429**
bier127	BS	280361.0292±38778.3636	170839.6973±6631.1940	**154413.8013±6596.8629**
	CT	**70.9550±2.7362**	74.0825±**2.6817**	74.3248±2.7873
	UR	**3.8659±1.0352**	2.5370±**0.2490**	3.0678±0.4246
	PD	14.7046±2.1019	**23.1495**±1.0181	14.1370±**0.7819**

Table 3. (*Continued*)

TSP problems and criteria		Greedy methods		
		No greedy	Half greedy	All greedy
ch130	BS	17089.9077±2858.6011	10047.0577±**455.0622**	**8882.2923**±512.9533
	CT	**75.2717**±2.9781	77.9531±**2.8970**	76.8778±2.9550
	UR	**4.3072**±1.0183	2.6275±**0.2998**	3.2770±0.3907
	PD	15.3682±2.3285	**22.7481**±1.1188	14.2228±**0.8272**
kroA150	BS	85580.4285±15715.5394	51588.1972±2706.9540	**43546.5763**±**2638.0468**
	CT	**96.3781**±3.7218	99.2130±3.6143	98.7319±**3.5164**
	UR	**4.8259**±1.2510	2.7287±**0.2461**	3.517±0.4314
	PD	19.5021±2.8978	**26.3202**±1.3272	17.3495±**1.2022**
kroA200	BS	120597.2931±19113.1388	81249.7957±4577.1808	**67858.3109**±**3900.7881**
	CT	**163.4809**±**6.3209**	168.3205±6.5453	171.4500±6.3965
	UR	**5.0262**±1.3217	2.4959±**0.1710**	3.1838±0.3630
	PD	26.0689±3.4488	**31.5255**±1.7917	22.8020±**1.6054**
kroC100	BS	48296.9643±8517.3143	27258.1630±**1297.3684**	**26887.4022**±2224.6761
	CT	**46.8898**±1.9041	49.8877±1.9892	48.6789±**1.8294**
	UR	**4.5670**±0.8556	3.6354±0.3905	3.5355±**0.5597**
	PD	12.9923±1.7231	**20.4741**±0.9838	11.0619±**0.6984**
lin105	BS	33257.0403±5520.1857	19600.8860±**1003.3149**	**18571.1600**±1406.4064
	CT	**51.3555**±**1.7913**	53.6700±1.8817	53.8862±1.8962
	UR	**4.2091**±0.8561	3.2621±**0.3396**	2.9983±0.4262
	PD	14.3067±1.9913	**21.8738**±1.0320	12.2363±**0.7913**
lin318	BS	235023.9477±33480.078	5191804.1596±10477.6094	**164103.4505**±**8397.6502**
	CT	**393.0959**±**15.4872**	399.2556±15.6145	394.9992±18.4682
	UR	**5.3975**±1.5245	2.0946±**0.1337**	2.6318±0.2295
	PD	40.4241±5.3767	**42.7564**±**2.5368**	35.7467±2.7509

Similarly with local optimization, COOBBO algorithm with no greedy requires the least CPU time among all tested methods and the performance improvement of other measures will also be very significant at a lower cost (small increases in CPU time in this paper). Except berlin52 problem, all greedy can get an advantage over half greedy on the degree of improvement. These data fully demonstrate that, greedy method is very helpful to improve algorithm performance in discrete domain, and the overall trend increases rapidly when more candidate solutions survive and reproduce over generations by greedy technique.

For the point of CPU time required, the distinction between all greedy and half greedy is not much at all, and even there are four benchmark problems with no significant difference in statistics ((bier127, kroA150, lin105 and lin318; highlighted in **boldface** font) as seen from Table 4. In brief, hybrid COOBBO algorithms can be improved efficiently via embedding all greedy approach, if algorithm performance and computation time are considered simultaneously.

Table 4. Performance comparison on CT of different greedy methods

	Half greedy	All greedy	Total
Half greedy	-	0.000; **0.532**; 0.010; **0.341**; 0.001; 0.000; **0.419**; **0.080**	4+4
All greedy	*	-	

For most of benchmark problems, no greedy has the highest UR and the worst algorithm performance among three methods, for the same reason as mentioned before. However the utilization rate of opposite paths of all greedy is higher than that of half greedy, which may be helpful to improve its performance.

Surprising though, half greedy has the highest population diversity among three measures. For this improvement technique, candidate solutions are derived from two sources. The first half evolves in the normal way and the second half suffers from extra greedy process. Hybrid COOBBO algorithm may obtain a better balance of exploitation and exploration with the help of this extra operation. At the same time, for all greedy, the whole population of candidate solutions suffers from extra greedy process and then its population diversity decrease instead of increase like it should.

4   Conclusion

In this paper, we focus on the impact and influence of various modified measures to enhance the performance and efficiency of COOBBO algorithm in discrete domain. These methods include crossover approach, local optimization approach and greedy approach. Experiment results illustrate clearly that, compared to other sets of algorithm parameters, the combination model of "inver-over crossover + 2-opt local optimization + all greedy" can improve much of the overall performance without increasing complexity and overhead.

As we all known, combinatorial problems are challenging benchmarks for heuristic algorithms. In near future research, we should introduce and test several new methods to create hybrid soft algorithms and then extend our research to real world applications, such as vehicle routing problems. For population-based intelligent algorithms, population diversity is undoubtedly one of the most important indicators of a population state. With the exception of edge entropy used in this paper, various measures should be proposed to evaluate population diversity.

Acknowledgments. This work was supported in part by the National Natural Science Foundation of China (Nos. 61375089 and 61305083).

References

1. Simon, D.: Biogeography-based optimization. IEEE Trans. Evolutionary Computation **12**, 702–713 (2008)
2. Du, D.: Biogeography-based optimization for combinatorial problems and complex systems. Ph.D Dissertation, Cleveland State University, USA (2014)
3. Song, Y., Liu, M., Wang, Z.: Biogeography-based optimization for the traveling salesman problems. In: International Joint Conference on Computational Science and Optimization, Huangshan, China, pp. 295–299 (2010)
4. Mo, H.M., Xu, L.F.: Biogeography based optimization for traveling salesman problem. In: International Conference on Natural Computation, Yantai, China, pp. 3143–3147 (2010)
5. Mo, H.M., Xu, L.F.: Biogeography migration algorithm for traveling salesman problem. International Journal of Intelligent Computing and Cybernetics **4**, 311–330 (2011)
6. Ergezer, M., Simon, D.: Oppositional biogeography-based optimization for combinatorial problems. In: IEEE Congress on Evolutionary Computation, New Orleans, USA, pp. 1496–1503 (2011)
7. Du, D., Simon, D.: Biogeography-based optimization for large scale combinatorial problems. In: Igelnik, B.; Zurada, J. (eds.) Efficiency and Scalability Methods for Computational Intellect, pp. 197–217 (2013)
8. Xu, Q.Z., Guo, L.M., Wang, N., Pan, J., Wang, L.: A novel oppositional biogeography-based optimization for combinatorial problems. In: International Conference on Natural Computation, Xiamen, China, pp. 414–420 (2014)
9. Xu, Q.Z., Wang, L., He, B.M., Wang, N.: Modified opposition-based differential evolution for function optimization. Journal of Computational Information Systems **7**, 1582–1591 (2011)
10. Xu, Q.Z., Wang, L., He, B.M., Wang, N.: Opposition-based differential evolution using the current optimum for function optimization. Journal of Applied Sciences **29**, 308–315 (2011). (in Chinese)
11. Fox, B., McMahon, M.: Genetic operators for sequencing problem. In: Rawlins, G. (ed.) Foundations of Genetic Algorithms, pp. 284–300 (1991)
12. Oliver, I., Smith, D., Holland, J.: A study of permutation crossover operators on the traveling salesman problem. In: International Conference on Genetic Algorithm and their Application, Mahwah, USA, pp. 224–230 (1987)
13. Tao, G., Michalewicz, Z.: Inver-over operator for the TSP. In: Eiben, A.E., Bäck, T., Schoenauer, M., Schwefel, H.-P. (eds.) PPSN 1998. LNCS, vol. 1498, pp. 803–812. Springer, Heidelberg (1998)
14. Croes, G.A.: A method for solving traveling-salesman problems. Operations Research **6**, 791–812 (1958)
15. Lin, S.: Computer solutions of the traveling salesman problem. Bell System Technical Journal **44**, 2245–2269 (1965)
16. Gutin, G., Yeo, A., Zverovich, A.: Traveling salesman should not be greedy: Domination analysis of greedy-type heuristics for the TSP. Discrete Applied Mathematics **117**, 81–86 (2002)
17. Reinelt, G.: TSPLIB — A traveling salesman problem library. ORSA Journal on Computing **3**, 376–384 (1991)

Motor Imagery Electroencephalograph Classification Based on Optimized Support Vector Machine by Magnetic Bacteria Optimization Algorithm

Hongwei Mo[✉] and Yanyan Zhao

Automation College, Harbin Engineering University, Harbin 150001, China
honwei2004@126.com, ayanyan2011@163.com

Abstract. In this paper, an analysis method of electroencephalograph (EEG) based on the motor imagery is proposed. Butterworth band-pass filter and artifact removal technique are combined to extract the feature of frequency band of ERD/ERS. Common spatial pattern (CSP) is used to extract feature vector. Support Vector Machine (SVM) is used for signal classification of motor imagery EEG. To improve classification performance, the parameters of SVM are optimized by a new bio-inspired method called Magnetic Bacteria Optimization Algorithm (MBOA). Experimental results show that MBOA has good performance on the problem of SVM optimization and obtain good classification results on EEG signals.

Keywords: Magnetic bacteria optimization algorithm · Support vector machine · Classification · EEG · Motor imagery

1 Introduction

As one kind of the brain computer interface (BCI) control signal, sensorimotor rhythms have been investigated extensively in BCI research[1]. Well-known BCI systems such as Wadsworth[2], Berlin[3], or Graz[4] BCIs employ sensorimotor rhythms as control signals. Sensorimotor rhythms are related to motor imagery without any actual movement. They comprise mu and beta rhythms, which are oscillations in the brain activity localized in the mu band (7–13 Hz), also known as the Rolandic band, and beta band (13–30 Hz), respectively. The amplitude of the sensorimotor rhythms varies when cerebral activity is related to any motor task although actual movement is not required to modulate the amplitude of sensorimotor rhythms.

In BCI system, given the inter and intrapersonal variations in EEG, to obtain satisfactory performance, the design of the classifier is often critical.

Support vector machine (SVM) is an extensively used machine learning method with many biomedical signal classification applications. Indeed, the SVM classifier exhibits a promising generalization capability. Several works have introduced SVM into the EEG classification application [5-10]. Meanwhile, the optimization mechanism involves kernel parameter setting in the SVM training procedure, which significantly influences the classification accuracy. The optimization of SVM classifier based on

© Springer International Publishing Switzerland 2015
Y. Tan et al. (Eds.): ICSI-CCI 2015, Part I, LNCS 9140, pp. 415–424, 2015.
DOI: 10.1007/978-3-319-20466-6_44

bio-inspired optimization techniques has proved successful in a number of different application fields. More recently, Subasi proposed the PSO optimized SVM in an attempt to classify the EMG signals for diagnosis of neuromuscular disorders [11]. Aydin proposed a multi-objective artificial immune algorithm to optimize the kernel and penalize parameters of SVM[12]. Fei used the particle swarm optimization-based SVM to study the diagnosis of arrhythmia cordis[13]. Fernandez applied the genetic algorithm-optimized SVM method in drug design quantitative structure–activity relationships (QSAR) modeling [14]. Nevertheless, in some of the studies conducted before, there was no similar optimization of SVM applied to motor imagery EEG classification. So our study attempts to increase the EEG signal classification accuracy rate by utilizing a novel approach called Magnetic Bacteria Optimization Algorithm (MBOA) to optimize the parameters of SVM.

Magnetic Bacteria Optimization Algorithm (MBOA) is a new intelligent optimi-zation algorithm [15]. The algorithm is inspired by magnetic bacteria, simulating magnetic bacteria mechanism which can move along the magnetic field lines. Lots of experiment results show that the MBOA can effectively solve optimization problems [16-21].

We adopted adaptive EOG artifact removal and Common Spatial Patterns (CSP) [22] for the EEG preprocessing and feature extraction process, respectively. All the algorithms were applied to the datasets collected from 2008 BCI Competition which are consisted of four different motor imagery tasks. Finally we apply our novel classifiers to the feature to evaluate the performance. For comparison, the other bio-inspired optimization techniques, such as Genetic Algorithm (GA) [23], Particle Swarm Algorithm (PSO) [24], Artificial Bee Colony(ABC)[25], Biogeography Based Optimization(BBO)[26] are also used to optimize the parameters of SVM and to test the classification accuracy.

2   Data Description

Experiment data comes from the 2008 International BCI competition dataset (Graz data A) [27].This data set consists of EEG data from 9 subjects. The BCI paradigm consisted of four different motor imagery tasks, namely the imagination of movement of the left hand (class 1), right hand (class 2), both feet (class 3), and tongue (class 4). Two sessions on different days were recorded for each subject. Each session is comprised of 6 runs separated by short breaks. One run consists of 48 trials (12 for each of the four possible classes), yielding a total of 288 trials per session.

At the beginning of each session, a recording of approximately 5 minutes was performed to estimate the EOG influence. The recording was divided into 3 blocks: (1) two minutes with eyes open (looking at a fixation cross on the screen), (2) one minute with eyes closed, and (3) one minute with eye movements.

In addition to the 22 EEG channels, 3 monopolar EOG channels were recorded and also sampled with 250 Hz. The EOG channels are provided for the subsequent application of artifact processing methods. These channels were bandpass filtered between 0.5 Hz and 100 Hz (with the 50 Hz notch filter enabled), and the sensitivity of the amplifier was set to 1 mV.

All data sets are stored in the General Data Format for biomedical signals(GDF). The GDF files can be loaded using the open-source toolbox BioSig.

Digital Filter is a kind of digital signal processing device, which has transmission characteristic. For motor imagery ERD / ERS phenomena, this study uses 10-27Hz Butterworth band-pass filter to filter each channel of EEG data.

3   SVM Optimized by MBOA

3.1   SVM and MBOA

Support vector machine is a popular classifier used in many classification problems. Given the training sample of instance-label pairs $(\mathbf{x}_i, y_i)$, $i = 1,...,l$, $\mathbf{x}_i \in \mathbf{R}^n$, $y_i \in \{1,-1\}$, support vector machines require the solution of the following primal problem[16]:

$$\min_{w,b,\xi} \frac{1}{2}\mathbf{w}^T\mathbf{w} + C\sum_{i=1}^{l}\xi_i \ . \tag{1}$$

Subject to $y_i(\mathbf{w}^T\mathbf{x}_i + b) \geq 1 - \xi_i$, $\xi_i \geq 0$, $i = 1,...,l$, where $C > 0$ is the penalty parameter of the error term.

The parameters of support vector machines with Gaussian kernel refer to the error penalty parameter C and the Gaussian kernel parameter γ, namely which is parameters (C, γ).

Magnetic bacteria optimization algorithm (MBOA) is a new optimization algorithm, inspired by the behavior of magnetic bacteria. MBOA obtains the optimal solution by regulating the moments of cells continually by the process of MTS generation, MTS expanding and MTS replacement. When the MBOA obtains the optimal solution, it corresponds to the state that when the moments of all cells are oriented in the geomagnetic field.

The MBOA includes the following steps:

Initialization. The initial population is filled with N number of randomly generated n-dimensional real-valued vectors. Let $X_i^0 = (x_{i1}^0, x_{i2}^0,..., x_{in}^0)$ represents the ith cell (for t=0) initialized randomly. Then each MTS x_{ij}^0 in a cell X_i^0 is generated as follows:

$$x_{ij}^0 = x_{\min j} + rand \times (x_{\max j} - x_{\min j}) \ . \tag{2}$$

where $i = 1, 2,..., N$, $j = 1, 2,..., n$. $x_{\max j}$ and $x_{\min j}$ are upper and lower bounds for the dimension j, respectively. $rand$ is a random number between 0 and 1.

Interaction Energy Calculation. In this step, for enhancing the search ability of the algorithm and producing diverse solutions in the algorithm, it is not necessary to follow the biology strictly. So, at first we randomly select a cell X_r ($r \in [1, N]$) in the population. t is the number of generation. The distance of two cells X_i and X_r, $D_i^t = \left(d_{i1}^t, d_{i2}^t, \ldots, d_{in}^t \right)$, is calculated as follows:

$$D_i^t = X_i^t - X_r^t . \tag{3}$$

From (3), we can obtain a distance vector matrix $D^t = (D_1^t, D_2^t, \ldots D_i^t, \ldots, D_N^t)' = \begin{bmatrix} d_{11}^t & d_{12}^t & \cdots & d_{1n}^t \\ d_{21}^t & d_{22}^t & \cdots & d_{2n}^t \\ \vdots & \vdots & \cdots & \vdots \\ d_{N1}^t & d_{N2}^t & \cdots & d_{Nn}^t \end{bmatrix}$ from the population. Then the

interaction energy $E_i^t = (e_{i1}^t, e_{i2}^t, \ldots, e_{ij}^t, \ldots, e_{in}^t)$ is calculated as follows[15]:

$$e_{ij}^t = \left(\frac{d_{ij}^t}{1 + c_1 \times d_{p'q'}^t + c_2 \times d_{pq}^t} \right)^3 . \tag{4}$$

Where c_1 and c_2 are constants. d_{pq}^t and $d_{p'q'}^t$ stand for randomly selected variables from D^t. $p, p' \in [1, N]$, $q, q' \in [1, n]$, $p \neq p'$, $q \neq q'$.

MTSs generation, moments are generated as follows[15]:

$$M_i^t = \frac{E_i^t}{B} . \tag{5}$$

where B is the magnetic field strength. Suppose $M_i^t = (m_{i1}^t, m_{i2}^t, \ldots, m_{ij}^t, \ldots, m_{in}^t)$, we can obtain a moment vector matrix $M^t = (M_1^t, M_2^t, \ldots M_i^t, \ldots, M_N^t)' = \begin{bmatrix} m_{11}^t & m_{12}^t & \cdots & m_{1n}^t \\ m_{21}^t & m_{22}^t & \cdots & m_{2n}^t \\ \vdots & \vdots & \cdots & \vdots \\ m_{N1}^t & m_{N2}^t & \cdots & m_{Nn}^t \end{bmatrix}$.

The MTSs are generated as follows:

$$v_{ij}^t = x_{ij}^t + m_{ls}^t \times rand . \tag{6}$$

where m_{ls}^t stands for the moment of a randomly selected MTS from M^t. $l \in [1, N]$, $s \in [1, n]$.

MTSs Generation. After MTSs generation, the MTSs in the cell are regulated as follows:

We set a magnetic field strength probability as 0.5.

If $rand > 0.5$, the MTSs in the cell are regulated and their moments are as follows:

$$u_{ij}^t = v_{cbestj}^t + (v_{cbestj}^t - v_{ij}^t) \times rand . \tag{7}$$

Otherwise, they are regulated as follows.

$$u_{ij}^t = v_{ij}^t + (v_{cbestj}^t - v_{ij}^t) \times rand \quad . \tag{8}$$

where v_{cbestj}^t stands for the jth dimension of current best cell V_{cbest}^t in the current generation.

MTSs Replacement. After the MTSs regulation, we set a replacement probability 0.5, some worse cells with worse moments are replaced by the following way :
   If rand>0.5,

$$x_{ij}^{t+1} = m_{l'j}^t \times rand \quad . \tag{9}$$

where l' is a random number between 1 and N. $m_{l'j}^t$ stands for the moment of a randomly selected MTS from $M_{l'}^t$.

   In general, we replace the last 1/5 of population by new generated MTSs.

   From the process of MBOA, it can be seen that the MBOA has different steps from the other popular natural computing mentioned above. The distance matrix is used to generate diverse MTSs with good moments. MTSs regulation guides the generated MTSs to the better MTS in current generation. This step not only makes the other relative worse MTSs close to the better one and thus the algorithm has good global search ability, but also enhance the local search ability. MTSs replacement is used to enhance the diversity of solutions in one generation.

3.2   Procedures of MBOA Optimizing SVM

The procedure of optimizing the parameters of SVM based on MBOA is as follows:
- Select support vectors from sample vectors to construct sample training set X.
- Use each support vector in sample training set to obtain a set of SVM parameters, and obtain the population X of cells.
- Calculate the distance matrix D of X.
- Calculate the interaction energy E of cells and obtain the moments of cells.
- MTS generation according to Eq.(4) and Eq.(5).
- Use the classification accuracy $F(x) = \frac{1}{m} \sum_{i=1}^{\infty} (f_i - y_i)^2$ as the fitness of a cell and calculate the fitness of each cell.
- MTS regulation according to Eq.(7),(8) and calculate the fitness of each cell.
- MTS replacment according to Eq.(9) and calculate the fitness of each cell.
- If termination criteria is met, the algorithms stops, output the results. It not, it returns step three.

4   Experiments and Results

In this section, we use MBOA to optimize the SVM penalty factor C and kernel parameter g. The parameters of MBOA are given below:

MBOA setting: In the MBOA, only the magnetic field B needs to be set up as a parameter,the earth magnetic field strength $B = 3$, and parameters $C_1 = 50$, $C_2 = 0.0003$.

In order to show the performance of MBOA, we compare MBOA with some other popular optimization algorithms including GA, PSO, ABC and BBO. A GA is a method that is based on natural selection in the theory of biological evolution. PSO is based on the swarming behavior of birds, fish, and other creatures. ABC is an optimization algorithm inspired by the behavior of bee colony. BBO is an optimization method based on biogeography.

The parameters of them are set as follows:

GA Settings: In our experiments, we employ a real number coded standard GA having evaluation, fitness scaling, seeded selection, random selection, crossover, mutation and elite units. Single point crossover operation with the rate of 0.8 is employed. Mutation operation restores genetic diversity lost during the application of reproduction and crossover. Mutation rate in our experiments is 0.01.

PSO Settings: In our experiments cognitive and social components are both set to 2. Inertia weight, which determines how the previous velocity of the particle influences the velocity in the next iteration, is 0.8.

ABC Settings: Limit=100, which is a control parameter in order to abandon the food source.

BBO Settings: For BBO, we use the following parameters: habitat modification probability is 1, immigration probability bounds per gene are [0,1] , step size for numerical integration of probabilities, maximum immigration and migration rates for each island are 1 , and mutation probability is 0.

For all the algorithms, set the size of the population 20, the number of iterations 20.Search range of optimize parameters C and g is set as [0.1, 1000].

We selects 2008 BCI competition four categories motor imagery EEG data A01, A03, A07, A09 as training and test sets. The tendency results of classification are shown in Fig.1, 2, 3 and 4. The accuracy results of classification are shown in Table1, 2, 3 and 4 respectively.

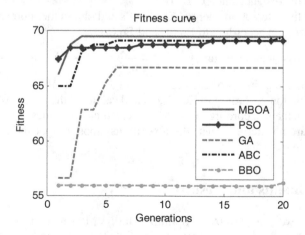

Fig. 1. Fitness curve corresponding to each algorithm of EEG dataset A01

Table 1. The optimal value of EEG dataset A01(%)

Algorithm	MBOA	PSO	GA	ABC	BBO
Accuracy	69.4444	60.4167	66.6667	69.4444	56.2500

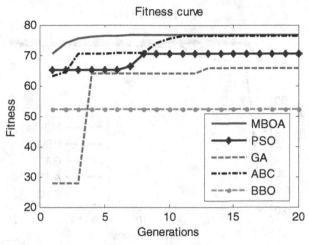

Fig. 2. Fitness curve corresponding to each algorithm of EEG dataset A03

Table 2. The optimal value of EEG dataset A03(%)

Algorithm	MBOA	PSO	GA	ABC	BBO
Accuracy	76.7361	70.4861	65.6250	76.3889	52.0833

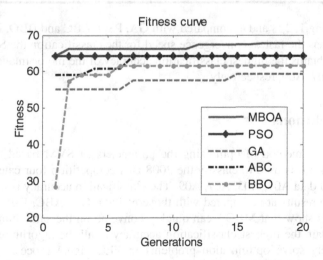

Fig. 3. Fitness curve corresponding to each algorithm of EEG dataset A07

Table 3. The optimal value of EEG dataset A07(%)

Algorithm	MBOA	PSO	GA	ABC	BBO
Accuracy	67.7083	64.2361	59.3750	64.2361	61.4583

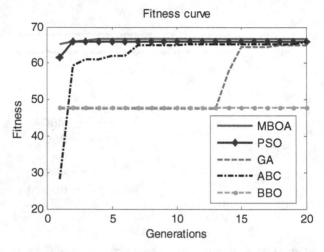

Fig. 4. Fitness curve corresponding to each algorithm of EEG dataset A09

Table 4. The optimal value of EEG dataset A09(%)

Algorithm	MBOA	PSO	GA	ABC	BBO
Accuracy	66.6667	65.9722	64.9306	65.9722	47.9167

From the Fig.1, 2, 3 and 4, compared with GA, PSO, ABC and BBO, it can be seen that MBOA has the fastest convergence speed for the classification the SVM parameters. And it obtains the highest classification accuracy of the motor imagery EEG data A01, A03, A07, A09, respectively.

5   Conclusions

In this paper, a method of optimizing the parameters of SVM based on MBOA is proposed. Then it is used to classify the 2008 BCI competition four categories motor imagery EEG data A01, A03, A07, A09. The classification accuracy is used as fitness function. The results are compared with those of PSO, GA, ABC, BBO. The experimental results show that MBOA can quickly converge on the same number of iterations and obtain the highest classification accuracy in all the algorithms. So MBOA can effectively solve optimization problems in EEG signal processing. In future, MBOA can be improved in further to obtain better results.

References

1. Nicolas-Alonso, L.F., Gomez-Gil, J.: Brain Computer Interfaces: a Review. Sensors **12**, 1211–1279 (2012)
2. Wolpaw, J.R., McFarland, D.J., Vaughan, T.M.: Brain-computer Interface Research at the Wadsworth Center. IEEE Trans. Rehabil. Eng. **8**, 222–226 (2000)
3. Blankertz, B., Losch, F., Krauledat, M., Dornhege, G., Curio, G., Müller, K.-R.: The Berlin Brain-Computer Interface: Accurate Performance from First-session in BCI-naïve Subjects. IEEE Trans. Biomed. Eng. **55**(10), 2452–2462 (2008)
4. Pfurtscheller, G., Neuper, C., Muller, G.R., Obermaier, B., Krausz, G., Schlogl, A., Scherer, R., Graimann, B., Keinrath, C., Skliris, D.: Graz-BCI: State of the Art and Clinical Applications. IEEE Trans. Neural Sys. Rehabil. Eng. **11**, 1–4 (2003)
5. Hortal, E., Planelles, D., Costa, A., Iáñez, E., Úbeda, A., Azorín, J.M., Fernández, E.: SVM–based Brain-Machine Interface for Controlling a Robot Arm Through Four Mental Tasks. Neurocomputing **151**, 116–121 (2015)
6. Siuly, L.: Y.: A Novel Statistical Algorithm for Multiclass EEG Signal Classification. Engineering Applications of Artificial Intelligence **34**, 154–167 (2014)
7. Jrad, N., Congedo, M., Phlypo, R., Rousseau, S., Flamary, R., Yger, F., Rakotomamonjy, A.: Sw-svm: Sensor Weighting Support Vector Machines for EEG-based Brain–computer Interfaces. J. Neural Eng. **8**, 056004 (2011)
8. Fu, K., Qu, J., Chai, Y., Dong, Y.: Classification of Seizure Based on the Time-frequency Image of EEG Signals Using HHT and SVM. Biomedical Signal Processing and Control **13**, 15–22 (2014)
9. Joshi, V., Pachori, R.B., Vijesh, A.: Classification of Ictal and Seizure-free EEG Signals Using Fractional Linear Prediction. Biomedical Signal Processing and Control **9**, 1–5 (2014)
10. Ianez, E., Ubeda, A., Hortal, E., Azorin, J.M.: Mental tasks selection method for a SVM-based BCI system. In: IEEE International Systems Conference (2013)
11. Subasi, A.: Classification of EMG Signals Using PSO Optimized SVM for Diagnosis of Neuromuscular Disorders. Computers in Biology and Medicine **43**, 576–586 (2013)
12. Aydin, I., Karakose, M., Akin, E.: A Multi-objective Artificial Immune Algorithm for Parameter Optimization in Support Vector Machine. Applied Soft Computing **11**, 120–129 (2011)
13. Fei, S.: Diagnostic Study on Arrhythmia Cordis Based on Particle Swarm Optimization-based Support Vector Machine. Expert Systems with Applications **37**, 6748–6752 (2010)
14. Fernandez, M., Caballero, J., Fernandez, L., Sarai, A.: Genetic Algorithm Optimization in Drug Design QSAR: Bayesian-regularized Genetic Neural Networks (BRGNN) and Genetic Algorithm-optimized Support Vectors Machines. Comprehensive Review Mol Divers **15**, 269–289 (2011)
15. Mo, H.W.: Research on magnetotactic bacteria optimization algorithm. In: The Fifth International Conference on Advanced Computational Intelligence (ICACI 2012), Nanjing (2012)
16. Mo, H.W., Xu, L.F.: Magnetotactic bacteria optimization algorithm for multimodal optimization. In: IEEE Symposium on Swarm Intelligence (SIS), Singapore, pp. 240–247 (2013)
17. Mo, H., Liu, L., Xu, L., Zhao, Y.: Research on magnetotactic bacteria optimization algorithm based on the best individual. In: Pan, L., Păun, G., Pérez-Jiménez, M.J., Song, T. (eds.) BIC-TA 2014. CCIS, vol. 472, pp. 318–322. Springer, Heidelberg (2014)

18. Mo, H.W., Geng, M.J.: Magnetotactic bacteria optimization algorithm based on best-rand scheme. In: 6th Naturei and Biologically Inspired Computing, Porto Portugal, pp. 59–64 (2014)
19. Mo, H.W., Liu, L.L.: Magnetotactic bacteria optimization algorithm based on best-target scheme. In: International Conference on Nature Computing and Fuzzy Knowledge, Xiamen, China, pp. 103–114 (2014)
20. Mo, H.W., Liu, L.L., Xu, L.F.: A Power Spectrum Optimization Algorithm Inspired by Magnetotactic Bacteria. Neural Computing and Applications 25(7–8), 1823–1844 (2014)
21. Mo, H., Liu, L., Geng, M.: A new magnetotactic bacteria optimization algorithm based on moment migration. In: Tan, Y., Shi, Y., Coello, C.A. (eds.) ICSI 2014, Part I. LNCS, vol. 8794, pp. 103–114. Springer, Heidelberg (2014)
22. Nasihatkon, B., Boostani, R., Jahromi, M.Z.: An Efficient Hybrid Linear and Kernel CSP Approach for EEG Feature Extraction. Neurocomputing 73, 432–437 (2009)
23. Goldberg, D.: Genetic Algorithms in Search Optimization and Machine Learning. Addison-Wesley, Reading (1989)
24. Kennedy, J., Eberhart, R.: Particle swarm optimization. In: IEEE Int. Conf. on Neural Networks, Piscataway, NJ, pp. 1942–1948 (1995)
25. Tereshko, V.: Reaction–diffusion model of a honeybee colony's foraging behaviour. In: Deb, K., Rudolph, G., Lutton, E., Merelo, J.J., Schoenauer, M., Schwefel, H.-P., Yao, X. (eds.) PPSN 2000. LNCS, vol. 1917, pp. 807–816. Springer, Heidelberg (2000)
26. Simon, D.: Biogeography-based Optimization. IEEE Trans on Evolutionary Computation. 12, 702–713 (2008)
27. BCI Competition IV. http://www.bbci.de/competition/iv/

Cuckoo Search

Adaptive Cuckoo Search Algorithm with Two-Parent Crossover for Solving Optimization Problems

Pauline Ong[1(✉)], Zarita Zainuddin[2], Chee Kiong Sia[1], and Badrul Aisham Md. Zain[1]

[1] Faculty of Mechanical and Manufacturing Engineering,
Universiti Tun Hussein Onn Malaysia (UTHM), 86400 Parit Raja, Batu Pahat, Johor, Malaysia
{ongp,sia,aisham}@uthm.edu.my
[2] School of Mathematical Sciences, Universiti Sains Malaysia (USM),
11800 Pulau Pinang, Malaysia
zarita@cs.usm.my

Abstract. Cuckoo search algorithm (CSA) experiences an upsurge in popularity since its invention due to its effectiveness in solving optimization problems. In this paper, a new CSA was proposed, in which the two-parent crossover operator was integrated in order to alleviate the deficiency of lack of information exchange. In addition, an adaptive step size strategy was introduced. The resultant algorithm was validated on optimizing benchmarking functions and a real-world problem. The experimental analysis highlighted the faster convergence ability of the proposed algorithm to the optimal solution.

Keywords: Cuckoo search algorithm · Two-parent crossover · Swarm intelligence · Numerical optimization

1 Introduction

Swarm intelligence (SI) - an emerging research field of artificial intelligence, consists of particles that are capable of accomplishing autonomous tasks by means of self-organization and cooperation principles among the particles within an environment [1]. Inspired by how nature adapts to challenging circumstances, SI has roots in multitudinous domains, particularly for problems with optimization at the heart. Particle swarm optimization (PSO), perhaps, is the most notable representative of SI based optimization technique. Each particle in PSO contributes its individual best experience to the swarm, leading to the convergence towards the optimality [2]. Other bio-inspired optimizers with similar searching strategy as in PSO are artificial bee colony [3], and ant colony optimization [4], to name but a few.

Cuckoo search algorithm (CSA), a new SI optimization technique, has been added to the pool recently [5]. Comprising of Lévy flight and brood parasitism behavior of certain cuckoo species, CSA has shown an obvious predominance in solving problems in the presence of high dimensionality and non-linearity, ranging from fault diagnosis, pattern recognition, job scheduling, software testing, data fusion, network design to image processing problems [6]. Despite its great promise of faster convergence but

© Springer International Publishing Switzerland 2015
Y. Tan et al. (Eds.): ICSI-CCI 2015, Part I, LNCS 9140, pp. 427–435, 2015.
DOI: 10.1007/978-3-319-20466-6_45

with fewer control parameters, a further challenge in CSA is to improve its efficiency, in terms of accelerating the convergence speed and avoiding the local optima.

In this regard, Ong enumerated considerations for CSA, which, the adaptive step size adjustment strategy was formed, for fast convergence purpose [7]. Walton approached the CSA through the addition of information exchange among the elites in order to generate a better offspring solution [8]. As well, Li *et al.* presented an orthogonal learning design framework in balancing the exploitation and exploration in CSA [9]. Wang *et al.* added a harmony search based mutation operator to CSA [10], for expediting the convergence rate of CSA. Zhang *et al.* developed dimensional entropy gain method to CSA; where a punishment was employed to the inferior solution in their topology in order to improve the quality of offspring [11].

While prior researches have identified possible contributions to the existing CSA framework, this work has attached its improvement from two aspects: (i) adaptive search strategy, where the step size of Lévy flight is updated adaptively. The step size should neither extremely narrow nor extremely wide in preventing premature or slow convergence; (ii) two-parent crossover operator, which, the information exchange between good solutions is allowed, considering the possibility that a high quality of potential solution can be generated. The modified algorithm was then tested through the optimization of benchmark functions, as well as a real world application problem.

The paper is organized as follows. Section 2 presents the framework of the standard CSA. Its limitations are then discussed in Section 3 and subsequently, the proposed modified CSA, namely, adaptive cuckoo search algorithm with two-parent crossover operator (ACSAC), is described. In Section 4, the analysis of comparative results for both benchmark functions and real-world problem are performed and lastly, conclusions are summarized in Section 5.

2   Cuckoo Search Algorithm

CSA was developed by Yang and Deb based on the obligate brood parasitism engaged by some cuckoo species [5]. Cuckoos, such as the *ani* and *Guira* cuckoos, employ unique reproduction strategy in which the female cuckoos lay their fertilized eggs in the nest of other species. The unwitting host birds are fooled due to the high resemblance between the cuckoo eggs and the host eggs. The host birds, somehow, evict the parasitic egg if it is spotted, and this incites the cuckoos to evolve better mimicry. The ongoing arms race between cuckoo egg mimicry opposed to host adaption triggers the formulation of CSA [5].

Such simulation, similarly to other SI based optimization techniques, starts with of a pool of randomly generated initial potential solutions (the host nests), which is characterized by:

$$\mathbf{x}_{i,j} = \mathbf{x}_{\min,j} + rand(0,1)(\mathbf{x}_{\max,j} - \mathbf{x}_{\min,j}) , \tag{1}$$

where $i = 1, 2, ..., n$, $j = 1, 2, ..., d$, while $\mathbf{x}_{\max,j}$ and $\mathbf{x}_{\min,j}$ denote the upper bound and lower bound of dimension j, respectively. The number of potential solutions is given by n while d represents the dimension of underlying problem.

The fitness of each possible solution is then evaluated. The solution with high fitness value is considered as showing high similarity with the host egg and thus, it is more likely to be passed on to next iteration. The host birds, anyhow, detect the cuckoo eggs with a discovery probability $p_a \in [0,1]$. If this is the case, the host nests are destroyed and new nests are built in other places, which is characterized by:

$$\mathbf{x}_{i,j}^{t+1} = \mathbf{x}_{i,j}^{t} + \alpha L(\lambda) , \tag{2}$$

$$L(\lambda) = \left| \frac{\Gamma(1+\lambda) \times \sin(\pi\lambda/2)}{\Gamma[(1+\lambda)/2] \times \lambda \times 2^{(\lambda-1)/2}} \right|^{1/\lambda} , \tag{3}$$

Here, t represents the current iteration number, Γ is the gamma function while λ is a constant ($1 < \lambda \leq 3$). The step size $\alpha > 0$ controls the scale of Lévy flight search patterns $L(\lambda)$, in which it ensures that the distribution of new solution is neither too narrow nor too wide. The fixed step size $\alpha = 1$ is used in CSA [5].

3   Adaptive Cuckoo Search Algorithm with Two-Parent Crossover

Scrutinizing the CSA revealed that there is lack of information exchange in its searching process. Each cuckoo moves independently without interaction. The cuckoo neither memorizes its individual or global best location nor shares its best experience with others. In this regard, the two-parent crossover operator that allows the information exchange among the potentially good solutions is proposed, considering the possibility that a better new solution might be produced from this group of elites. In the proposed ACSAC, 25% of the solutions with highest fitness values are considered as elite. The crossover among two randomly chosen top eggs from this group is performed by [12]:

$$\begin{aligned}
\mathbf{x}_{i,1}^{t+1} &= rand \times top\_egg_1^{t} + (1 - rand) \times top\_egg_2^{t} \\
\mathbf{x}_{i,2}^{t+1} &= rand \times top\_egg_2^{t} + (1 - rand) \times top\_egg_1^{t}
\end{aligned} \tag{4}$$

where $rand$ is a random value from a uniform distribution over the interval $[0,1]$.

On the other hand, the step-length of Lévy flight α should be assigned judiciously to maintain an appropriate balance between global and local searching. If the generated new solutions are distributed widely, the search may experience slow convergence, as the new solutions may be located outside the search space. If the new solutions are generated in narrower regions than previous, it may lose its diversity. An adjustable step size α, in general, is preferable. Intensive search around the regions with high survivability should be performed, since those regions, probably, contain the optimal solutions. In contrast, more aggressive search approach seems reasonable if the current habitat quality is poor, in order to improve its diversity. Thus, the second modification in the ACSAC is to update the step size adaptively according to:

Algorithm 1. Adaptive Cuckoo Search Algorithm with Two-Parent Crossover

Begin

Generate initial population of n host nest $\mathbf{x}_i, i = 1, 2, ..., n$

Define minimum Lévy flight step size $\alpha_{\min}$ and initial Lévy flight step size α_0

Evaluate the fitness function $F_i = f(\mathbf{x}_i)$

    while (iteration < Max Generation)

        for all top eggs

            Generate a cuckoo egg $\mathbf{x}_j$ using (4)

            Evaluate the fitness $F_j = f(\mathbf{x}_j)$

        end for

        for all non-top eggs

            Generate a cuckoo egg $\mathbf{x}_j$ using (2) and (5)

            Evaluate the fitness $F_j = f(\mathbf{x}_j)$

        end for

        Choose a nest i among n host nest randomly

        if ($F_j < F_i$) (for minimization problem)

            Replace $\mathbf{x}_i$ with $\mathbf{x}_j$

            Replace F_i with F_j

        end if

        Abandon a fraction p_a of the worst nests

        Generate new nests randomly to replace nests lost

        Evaluate the fitness of new nests

    end while

End

$$\alpha = \alpha_L \left(1 + \alpha_0 \tanh\left(\gamma / F_{best}^t\right) / \sqrt{t}\right), \tag{5}$$

where α_L, α_0 and t are the predefined minimum step size, initial step size, and t-th iteration number, respectively. $\tanh(\cdot)$ is the hyperbolic tangent, γ is the best fitness value in the initial population, and F_{best}^t represents the best fitness in the t-th iteration.

As shown in (5), the rate of fitness improvement over the best solution in the initial population is denoted by γ / F_{best}^t, which controls the scale of Lévy flight. A more intense exploitation around the current solution is undertaken whenever an improvement in terms of fitness is seen. The new cuckoo egg is distributed widely if the current best fitness value is comparative poorer than the best fitness value in initial population, in order to further diversify the distribution. Assuming that the cuckoo eggs are located far from the optimal solutions initially, the term $\alpha_0 / \sqrt{t}$ is included in (5) to prompt more thorough exploration at the beginning. As the number of generation t increases, the value of $\alpha_0 / \sqrt{t}$ is gradually decreases, assuming that the cuckoo eggs are now approaching optimal solution. In this case, more localized search should be performed. The steps involved in the ACSAC are shown in Algorithm 1.

4   Numerical Simulations

4.1   Benchmark Problems

Benchmark functions taken from the optimization literature were used to analyze the feasibility of the ACSAC, as described in Table 1. The population size was chosen as 20. The simulations were repeated for 30 trials. The searching for global optimum continued unless the best fitness value was below the threshold value of $\xi \leq 10^{-5}$. At each run, the Euclidean distance between the obtained best solution coordinates and the known global optima was measured. The results from 30 trials were then averaged and compared against the standard CSA. In addition, the two-tailed t-test was applied in order to validate the statistical significance of the obtained performances. Table 2 summarizes the performances of CSA and ACSAC. The iteration curves were depicted in Fig. 1 and Fig. 2, for the average results from 30 independent runs.

As shown in Fig.1, both algorithms demonstrated the behavior of exponential-like decrease of the distance error as the computation continued; however, the CSA needed more iteration steps for convergence, as presented in Table 2. More encouragingly than the CSA, the ACSAC reduced the number of iterations needed in converging to optimality from 1753 to 1222 on average, exhibited an improvement of 30%. It can be noticed that both algorithms took short iterations in convergence, attributed to that the De Jong's function is one of the simplest unimodal function.

For the Rosenbrock's function, the comparative result reaffirmed the superior convergence characteristic of the ACSAC, demonstrated an improvement of 30% in convergence rate as compared to CSA. The fastest allowable value needed by CSA in converging to optimality was 23218 iterations, and conversely, the slowest allowable value needed by ACSAC in reaching the known optimal was 22701 iterations, which was 8000 less iterations than the standard CSA.

Table 1. Problem description and parameter setting of benchmark function

Function	Definition	Dimension d	Search Space	Global Optimum $f(x_*)$	Optimum Point x_*
De Jong	$f(\mathbf{x}) = \sum_{i=1}^{d} x_i^2$	50	[-5.12,5.12]	0	(0,0,...,0)
Rosenbrock	$f(\mathbf{x}) = \sum_{i=1}^{d-1}[(1-x_i)^2 + 100(x_{i+1} - x_i^2)^2]$	10	[-100,100]	0	(1,1,...,1)

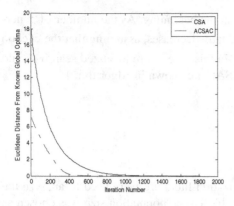

Fig. 1. De Jong's function: Convergence characteristic performance of standard CSA and the proposed ACSAC

Fig. 2. Rosenbrock's function: Convergence characteristic performance of standard CSA and the proposed ACSAC

Table 2. Performance comparison of CSA and the proposed ACSAC (in terms of number of iterations needed in converging to optimality)

Function	CSA				ACSAC				Statistically
	Best	Worst	Average	Time,s	Best	Worst	Average	Time, s	Significant?
De Jong	1673	1815	1753	2.14	1104	1317	1222	2.46	Yes
Rosenbrock	23218	34916	27635	34.46	11328	22701	19687	31.77	Yes

Table 3. Performance comparison of ACSAC with other optimization methods

Function	Generation	Average Best Fitness Value		
		GA	PSO	ACSAC
De Jong	1	334.35	334.12	337.01
	500	17.79	0.27	0.0050
	1000	7.65	0.27	5.85e-05
	1500	4.52	0.27	9.97e-07
	2000	2.96	0.27	1.80e-08
Rosenbrock	1	5.54e+09	1.79e+10	2.37e+09
	500	7.34e+04	7.93e+09	5.06e+07
	1000	7.31e+04	4.30e+08	9.93e+06
	1500	7.31e+04	3.02e+05	1.88e+06
	2000	7.31e+04	8.97e+04	9.90e+04

Apart from the performance assessment under fixed tolerance rate, another commonly used approach – comparison of best fitness value for a fixed number of iteration, was adopted. The convergence characteristics of genetic algorithm (GA) and PSO were compared against with the ACSAC, too. Table 3 presents the average best fitness values from 30 independent trials at iteration number of 1, 500, 1000, 1500 and 2000 for all considered optimization methods.

It is pertinent to note that for De Jong's function, both GA and PSO showed fast convergence initially; however, merely marginal or no improvement in terms of the best fitness values can be noticed as iteration proceeded, which might be due to these algorithms were trapped in local optima. Only the ACSAC has reached the optimality at iteration number of 1000, where the GA and PSO were failed in this regard. Furthermore, the best fitness values of both GA and PSO were not below the threshold value of $\xi \leq 10^{-5}$, although maximum number of iteration is set to 100,000.

For the Rosenbrock's function, both GA and PSO attained better best fitness value than ACSAC at iteration number of 2000; however, both failed to converge to optimality in all 30 independent runs, even the maximum number of generation is set to 100,000. In contrast, the ACSAC was able to reach the optimal solution after 19,687 iterations averagely.

4.2   Real-World Problem – Optimization of Pulp and Paper Properties

Throughout the pulping process in pulp and paper industry, a marginal change in the pulping variables, for instance, the temperature and pressure, may lead to high variation in pulp and paper properties, for instance, the kappa number and tensile strength. The optimization of pulping condition is often a challenging task due to the fact that each response has its own optimal experimental condition, and, it is often conflicting with each other. For instance, due to fiber degradation, a severe experimental milieu should be avoided in order to obtain a satisfying tensile strength, but severe pulping condition is preferable in getting a high tear index.

The data from 27 experimental trials which study the effect of four types of pulping variables (sodium hydroxide (NaOH), ethanol (EtOH) concentration, temperature and time) on the properties of pulp and paper (screened yield, kappa number, tensile index and tear index) are presented in Table 4 [13]. The proposed ACSAC is used to determine the optimal experimental milieu, in which it aims to maximize the screened yield, tensile index and tear index while simultaneously minimize the kappa number. The quadratic models which correlate the pulping variables to the response variables are formed initially, which are given as [13]:

$$Y_{ScreenedYield} = 28.51 - 4.66X_A - 0.30X_{Et} + 0.44X_T - 0.19X_t - 2.69X_A X_T - 1.27X_A^2 \quad (6)$$

$$Y_{KappaNumber} = 32.2667 - 36.63X_A - 7.60X_{Et} - 7.67X_T - 3.72X_t + 4.20X_A X_{Et} \quad (7)$$
$$+ 3.40X_A X_T + 4.10X_{Et} X_T + 31.60X_A^2 + 7.40X_{Et}^2 + 8.10X_T^2 + 3.10X_t^2$$

$$Y_{TensileIndex} = 59.50 + 14.18X_A - 2.34X_{Et} + 2.88X_T + 2.06X_t - 12.50X_A X_{Et} \quad (8)$$
$$- 16.21X_A X_T - 20.84X_A^2 - 8.36X_{Et}^2 - 6.05X_T^2 - 7.06X_t^2$$

$$Y_{TearIndex} = 4.10 + 0.64X_A - 0.18X_{Et} + 0.07X_T - 0.06X_t - 0.60X_A X_{Et} \quad (9)$$
$$- 0.80X_A X_T - 0.56X_T X_t - 0.90X_A^2 - 0.56X_{Et}^2$$

where X_A, X_{Et}, X_T and X_t are the wt% of NaOH, EtOH, temperature and time, respectively. The multi-objectives optimization problem was transformed into a single-objective maximization problem, in which the objective function was formed as:

$$Objective\_Function = Y_{ScreenedYield} + Y_{TensileIndex} + Y_{TearIndex} - Y_{KappaNumber} \qquad (10)$$

The population size was chosen as 20 and the predefined threshold value of $\xi \le 10^{-10}$ is selected. In addition, the obtained best fitness value was compared against the results obtained by the wavelet neural networks (WNNs) and the response surface methodology (RSM) on the same data.

The obtained fitness values for the corresponding optimal pulping conditions predicted by RSM (NaOH, 18.15%; EtOH, 38.62%; T, 165°C; t, 170min), WNNs (NaOH, 18.15%; EtOH, 38.12%; T, 165.23°C; t, 167.71min) and ACSAC (NaOH, 26.20%; EtOH, 19.87%; T, 187.18°C; t, 153.09min) are 295.33, 294.77 and 356.27, respectively. It can be concluded that the proposed ACSAC outperformed the others, in which a higher fitness value indicates better optimization results, since the problem was modeled as maximization problem.

Table 4. Experimental values of pulping and response variables from 27 trials

No	Experimental Conditions				Experimental Values for the Pulp and Paper Properties			
	NaOH (%)	EtOH (%)	Temperature (°C)	Time (min)	Screened Yield (%)	Kappa Number	Tensile Index (N m/g)	Tear Index (mN N^2/g)
1	10	45	170	120	28.25	102.0	28.59	2.83
2	20	45	170	120	28.12	32.6	62.99	4.13
3	25	60	180	90	24.51	23.3	46.5	3.84
4	25	60	160	150	22.52	24.9	64.74	4.18
5	15	30	180	150	32.92	59.6	46.06	3.55
6	15	60	180	90	31.71	57.5	44.88	3.38
7	30	45	170	120	25.91	25.4	51.26	3.64
8	20	45	170	180	29.45	30.9	51.29	3.95
9	25	30	180	150	24.07	25.4	56.88	3.56
10	20	45	170	60	28.75	39.5	56.12	4.13
11	15	60	160	150	28.98	58.7	39.52	3.35
12	15	30	160	150	31.50	72.7	33.53	2.96
13	25	30	180	90	25.60	28.1	60.59	4.65
14	20	45	170	120	29.54	31.7	57.1	4.23
15	25	30	160	90	25.65	34.6	59.82	4.45
16	15	60	180	150	32.10	53.0	49.6	3.56
17	15	30	160	90	29.03	75.1	33.14	3.01
18	20	45	190	120	26.84	31.6	59.1	4.23
19	25	60	160	90	25.99	28.5	47.06	3.73
20	15	60	160	90	30.54	62.5	35.28	3.06
21	25	60	180	150	23.24	20.2	48.4	3.45
22	20	15	170	120	26.57	47.0	54.27	3.65
23	20	75	170	120	28.01	32.0	50.52	3.5
24	20	45	170	120	27.86	32.5	58.4	3.95
25	20	45	150	120	27.15	48.8	50.32	3.9
26	25	30	160	150	25.67	31.0	64.69	4.68
27	15	30	180	90	31.66	63.3	41.83	3.56

5   Conclusion

The proposed ACSAC, accelerates the convergence characteristic of the standard CSA at an acceptable error level, with the utilization of adaptive step size strategy and two-parent crossover. The experimental results on benchmark functions of De Jong's and Rosenbrock's functions are encouragingly, where improvements of 30% in convergence rate as compared to standard CSA are noticed. Performance comparison with GA and PSO at fixed iteration number demonstrated its superiority from the aspect of faster convergence characteristic and free of stagnation. On the other hand, the promising capability of the proposed ACSAS was shown through finding the optimal experimental conditions of pulping process using a real-world data.

Acknowledgments. Financial support from the Malaysian government with cooperation of Universiti Tun Hussein Onn (UTHM) in the form of FRGS Vot 1490 is gratefully acknowledged.

References

1. Coelho, L.D.S.: Gaussian quantum-behaved particle swarm optimization approaches for constrained engineering design problems. Expert Systems with Applications **37**, 1676–1683 (2010)
2. Kennedy, J., Eberhart, R.: Particle swarm optimization. In: Proceedings of IEEE International Conference on Neural Networks, pp. 1942–1948. IEEE Publisher (1995)
3. Karaboga, D., Basturk, B.: On the performance of artificial bee colony (ABC) algorithm. Applied Soft Computing **8**, 687–697 (2008)
4. Dorigo, M., Maniezzo, V., Colorni, A.: The Ant System: An Autocatalytic Optimizing Process (1991)
5. Yang, X.S., Deb, S.: Cuckoo search via levy flights. In: World Congress on Nature & Biologically Inspired Computing, NaBIC 2009, pp. 210–214 (2009)
6. Yang, X.S.: Cuckoo search and firefly algorithm: Overview and analysis. Studies in Computational Intelligence **516**, 1–26 (2014)
7. Ong, P.: Adaptive Cuckoo Search Algorithm for Unconstrained Optimization. The Scientific World Journal **2014**, 8 (2014)
8. Walton, S., Hassan, O., Morgan, K., Brown, M.R.: Modified cuckoo search: A new gradient free optimisation algorithm. Chaos, Solitons & Fractals **44**, 710–718 (2011)
9. Li, X., Wang, J., Yin, M.: Enhancing the performance of cuckoo search algorithm using orthogonal learning method. Neural Comput & Applic **24**, 1233–1247 (2014)
10. Wang, G.G., Gandomi, A.H., Zhao, X., Chu, H.C.E.: Hybridizing harmony search algorithm with cuckoo search for global numerical optimization. Soft Comput (2014)
11. Zhang, Q., Wang, L., Cheng, J., Pan, R.: Improved cuckoo search algorithm using dimensional entropy gain. Neural Comput & Applic (2014)
12. Higashi, N., Iba, H.: Particle swarm optimization with Gaussian mutation. In: Proceedings of the 2003 IEEE Swarm Intelligence Symposium, SIS 2003, pp. 72–79. IEEE Publisher (2003)
13. Zainuddin, Z., Wan Daud, W.R., Pauline, O., Shafie, A.: Wavelet neural networks applied to pulping of oil palm fronds. Bioresource technology **102**, 10978–10986 (2011)

Automated, Adaptive, and Optimized Search for CSPs via Cuckoo Search

Ricardo Soto[1,2,3], Broderick Crawford[1,4,5], Javier Flores[1], Felipe Mella[1],
Cristian Galleguillos[1(✉)], Franklin Johnson[6], and Fernando Paredes[7]

[1] Pontificia Universidad Católica de Valparaíso, Valparaíso, Chile
{ricardo.soto,broderick.crawford}@ucv.cl,
{javier.flores.v,felipe.mella.101,cristian.galleguillos.m}@mail.pucv.cl
[2] Universidad Autónoma de Chile, Santiago, Chile
[3] Universidad Científica del Sur, Lima, Perú
[4] Universidad Central de Chile, Santiago, Chile
[5] Universidad San Sebastián, Santiago, Chile
[6] Universidad de Playa Ancha, Valparaíso, Chile
franklin.johnson@upla.cl
[7] Escuela de Ingeniería Industrial, Universidad Diego Portales, Santiago, Chile
fernando.paredes@udp.cl

Abstract. Constraint Programing is a programming paradigm devoted
to the efficient solving of constraint satisfaction problems (CSPs). A
CSP is a formal problem representation mainly composed of variables
and constraints defining relations among those variables. The resolution
process of CSPs is commonly carried out by building and exploring a
search tree that holds the possibles solutions. Such a tree is dynamically
created by interleaving two different phases: enumeration and propagation. During enumeration, the variables and values are chosen to build
the possible solution, while propagation intend to delete the values having no chance to reach a feasible result. Autonomous Search is a new
technique that gives the ability to the resolution process to be adaptive
by re-configuring its enumeration strategy when poor performances are
detected. This technique has exhibited impressive results during the last
years. However, such a re-configuration is hard to achieve as parameters
are problem-dependent and their best configuration is not stable along
the search. In this paper, we introduce an Autonomous Search framework that incorporates a new optimizer based on Cuckoo Search able to
efficiently support the re-configuration phase. Our goal is to provide an
automated, adaptive, and optimized search system for CSPs. We report
encouraging results where our approach clearly improves the performance
of previously reported Autonomous Search approaches for CSPs.

Keywords: Swarm-based optimization · Nature-inspired algorithms ·
Cuckoo search · Constraint Programming · Autonomous search

1 Introduction

Constraint Programing (CP) is a programming paradigm focused on solving
constraint satisfaction and optimization problems. A main idea behind this

© Springer International Publishing Switzerland 2015
Y. Tan et al. (Eds.): ICSI-CCI 2015, Part I, LNCS 9140, pp. 436–447, 2015.
DOI: 10.1007/978-3-319-20466-6_46

paradigm is to model the problems by using variables and constraints. The variables are the unknowns of the problem and each one has a non-empty domain of possible values that the variable can take. Constraints represent the relations among the variables and can be considered as rules that must be followed to find the solution. A problem is solved when each variable has taken a value from its domain and no constraint is violated. The resolution of the CSP requires the exploration of possible values for each variable. This process is usually carried out by using a search engine called solver, which attempts to find a proper solution by building and exploring a search tree. The construction of the tree can be divided in two main phases: enumeration and propagation. The first one selects the order in which variables and values are chosen, while propagation tries to eliminate the values having no chance to reach a solution.

The enumeration is a key phase on the solving process as the performance is greatly influenced by the selection of an appropriate enumeration strategy. However, selecting the proper strategy is known to be hard as the performance of strategies are commonly unpredictable. Autonomous Search is a new technique that gives the ability to the search process to be adaptive by automatically re-configuring its enumeration strategy when poor performances are detected. The idea is to interleave a set of strategies during the search process, replacing underperforming strategies by more promising ones. However, such a strategy re-configuration is hard to achieve as parameters are problem-dependent and their best configuration is not stable along the search.

In this paper, we introduce an Autonomous Search framework that incorporates a new optimizer based on Cuckoo Search able to efficiently support the re-configuration phase. Cuckoo Search is a modern metaheuristic based on the breeding behavior of certain Cuckoo species that has successfully been used to solve complex optimization problems [13]. Our goal is to provide an automated, adaptive, and optimized search system for CSPs. We report encouraging results where our approach clearly improves the performance of previously reported Autonomous Search approaches for CSPs. The rest of this work is organized as follows: Section 2 presents the related work. Section 3 and 4 present the problem and the proposed solution, respectively. Finally, the experimental evaluation is illustrated followed by conclusions and future work.

2   Related Work

A preliminary work in Autonomous Search (AS) for CP is the one presented in [2]. This framework proposed an interesting architecture composed of 4 elements. The idea is to support the dynamic replacement of enumeration strategies. The strategies are evaluated via performance indicators, and better ones replace worse ones during solving time. This preliminary framework was used as basis of different related works. For instance, a more recent framework based on this idea is reported in [5]. This approach uses two layers, where an hyper-heuristic placed on the top-layer manages the selection of strategies of the search engine placed on the lower-layer. An hyper-heuristic can be seen as a method to choose heuristics [8]. In this approach, two different top-layers have been proposed, one using a genetic algorithm [10,4] and another using a particle swarm optimizer [6]. Similar approaches have also been implemented for solving optimization problems instead of pure CSPs [9]. In Section 5 we contrast the proposed approach with the best AS optimizers reported in the literature.

3   Constraint Satisfaction Problems and Autonomous Search

A constraint satisfaction problem $\mathcal{P}$ is formally defined by a triple $\mathcal{P} = \langle \mathcal{X}, D, C \rangle$ where $\mathcal{X}$ is an n-tuple of variables $\mathcal{X} = \langle x_1, x_2, \ldots, x_n \rangle$. $\mathcal{D}$ is the corresponding n-tuple of domains $\mathcal{D} = \langle d_1, d_2, \ldots, d_n \rangle$ such that $x_i \in d_i$, and d_i is a set of values, for $i = 1, \ldots, n$. $\mathcal{C}$ is an m-tuple of constraints $\mathcal{C} = \langle c_1, c_2, \ldots, c_m \rangle$, and a constraint c_j is defined as a subset of the Cartesian product of domains $d_{j_1} \times \cdots \times d_{j_{n_j}}$, for $j = 1, \ldots, m$. A solution to a CSP is an assignment $\{x_1 \to a_1, \ldots, x_n \to a_n\}$ such that $a_i \in d_i$ for $i = 1, \ldots, n$ and $(a_{j_1}, \ldots, a_{j_{n_j}}) \in c_j$, for $j = 1, \ldots, m$.

As previously mentioned, the enumeration strategy controls which variable x_i and which value from d_i is selected to build the potential solution. In this work, we aim at online controlling a set of enumeration strategies which are dynamically interleaved during solving time. Our purpose is to select the most promising one for each part of the search tree. In order to properly perform this selection, we employ a quality rank that is governed by a Choice Function (CF) [6]. The CF is composed of performance indicators and parameters that control the relevance of each indicator within the CF. Considering any enumeration strategy S_j, the CF f in step n for S_j is defined by equation 1.

$$f_n(S_j) = \sum_{i=1}^{l} \alpha_i f_{in}(S_j) \qquad (1)$$

where l is the number of indicators and α_i is the control parameter for indicator i. The idea is to assign a weight to each indicators in order to increase the relevance of a given indicator and decreasing another one. This allow the solver to adjust even more the process of selecting the best strategy.

There are many indicators that can be used, but no one of them is the best to evaluate the strategies on every problem [6]. We employ the following CF for the experiments: $\alpha_1 SB + \alpha_2 In1 + \alpha_3 In2$, where SB is the number of shallow backtracks [1] (SB), $In1 = CurrentMaximumDepth - PreviousMaximumDepth$, and $In2 = CurrentDepth - PreviousDepth$, where $Depth$ refers to the depth reached within the search tree.

A main problem in this context, is that α parameters are problem-dependent and their best configuration is not stable along the search. To this end, we incorporate a Cuckoo Search algorithm, which allow the solver to optimize the relevance of each indicator during the resolution, thus giving a suitable configuration for each problem. This is done by carrying out a sampling phase where the CSP is partially solved to a given cutoff. The performance information gathered in this phase via the indicators is used as input data of the optimizer, which attempt to determine the most successful α parameters for the CF. This tuning process is very important as the correct configuration of the CF may have essential effects on the ability of the solver to properly solve specific CSPs.

4   Cuckoo Search Algorithm

The Cuckoo Search Algorithm is a metaheuristic inspired in the aggressive reproduction strategy used by certain species. The problems are modeled by representing each possible solution as a nest/egg. The cuckoos deposit the eggs on the random nest and then leave, the best nests are the ones that will carry the next generation of cuckoos. Each new generation will try to find a better nest than the previous, but the best one found so far will always be remembered [13,3,11,12]. The algorithm can be described by three simple rules:

- Each Cuckoo lays one single egg on a random nest
- The better nest will carry the next generation
- There is a change P_a that an egg will be discovered and discarded, in which case the mother will lay an egg on a different nest

The algorithm can be separated in three different phases.

1. Initial Phase: During this phase the initial solutions are randomly generated and evaluated.
2. Improving Solutions Phase: A local search is done in an effort to improve the actual solutions, Lévy Flights are used to generate random walks and if a new solution found is better than the previous one, then it is replaced.
3. New Solutions Phase: During this phase a percentage defined by the parameter P_a of nests are discovered and destroyed, using random permutations new solutions are created to replace the ones discovered.
4. The second and the third phases are repeated until the best solution is found.

Algorithm 1 depicts the Cuckoo Search procedure employed. We model the objective function $f(x)$ according to the CF employed, where each unknown x_i of the objective function represents an α_i of indicator i. The population size n is the amount of cuckoos generated (every cuckoo represents a potential solution). The max generation needs to be defined as how many iterations the algorithm is going to execute. Finally the $p_a \in [0, 1]$ represents the chance for a nest to be discovered and replaced by a new random solution.

The fitness of a potential solution given by Cuckoo Search is tested by using the indicator $IN3$, which corresponds to the search space reduction achieved during last step of the CSP to be solved. A bigger reduction means a smaller amount of potential solutions to be explored, which in turn means improvement in search time.

Algorithm 1. Cuckoo Search Algorithm

1: *Objective function* $f(x)$, $x = (x_1, ..., x_d)$
2: *Generate the initial population of* n *nest/solutions* x_i ($i = 1, 2, ..., n$)
3: *Evaluate the fitness of solutions w.r.t IN3*
4: **while** (t < *MaxGeneration*) *or* (*Stop criterion*) **do**
5:    *Choose a cuckoo/nest/solution randomly amon* n (*say*, j)
6:    *Generate a new solution by Levy flights* (*say*, i)
7:    *Evaluate its quiality/fitness* F_i
8:    **if** ($F_i > F_j$) **then**
9:       *replace* j *with the new nest* i;
10:   **end if**
11:   *A fraction* (p_a) *of the worst nests*
12:   *are abandoned and new ones are built using random permutations.*
13:   *Evaluate the quality of the solutions and create a rank.*
14: **end while**
15: *Results visualization.*

---

5   Experimental Evaluation

The proposed solution was tested on different problems and also compared to previous work. We test several instances of well-known CSPs, mentioned as follows:

- N-Queens problem with $n = \{8, 10, 12, 20, 50, 75\}$
- Magic Square problem size $n = \{3, 4, 5, 6\}$
- Sudoku puzzle $n = \{2, 5, 7\}$
- Knights Tournament with $n = \{5, 6\}$
- Quasi Group with $n = \{3, 5, 6\}$
- Langford with size $n = \{2\}$ and $k = \{12, 16, 20, 23\}$

The Autonomous search system has been implemented using Java and the Ecl^ips^e constraint programming system. All tests were carried out on an Intel Core i3-2120 3.30 GHz with 4 GB RAM running Windows 7 32-bits. All problems are solved until a solution is found or until a maximum amount of step is reached (65535 steps). When the solver is not able to reach a solution before this bound it is set to t.o. (time-out). There are 8 variable selection heuristics and 3 value selection heuristics that combined form a portfolio of 24 enumeration strategies described in Table 1.

Tables 2, 3, and 4 illustrate the performance in terms of runtime required to find a solution for each enumeration strategy individually (S1 to S24) and the proposed approach based on Cuckoo Search (CS). The results greatly validate our proposal, which is the only one in solving all instances of all problems, taking the best average runtime. Tables 5, 6, and 7 depict the results in terms of backtracks, which are similar to the previous ones. This demonstrates the ability of the proposed approach to correctly select the strategy to each segment of the solving process. We also contrast the proposed approach with the two previously

Table 1. Portfolio used

Id	Variable ordering	Value ordering
S_1	First variable of the list	min. value in domain
S_2	The variable with the smallest domain	min. value in domain
S_3	The variable with the largest domain	min. value in domain
S_4	The variable with the smallest value of the domain	min. value in domain
S_5	The variable with the largest value of the domain	min. value in domain
S_6	The variable with the largest number of attached constraints	min. value in domain
S_7	The variable with the smallest domain. If are more than one, choose the variable with the bigger number of attached constraints.	min. value in domain
S_8	The variable with the biggest difference between the smallest value and the second more smallest of the domain	min. value in domain
S_9	First variable of the list	mid. value in domain
S_{10}	The variable with the smallest domain	mid. value in domain
S_{11}	The variable with the largest domain	mid. value in domain
S_{12}	The variable with the smallest value of the domain	mid. value in domain
S_{13}	The variable with the largest value of the domain	mid. value in domain
S_{14}	The variable with the largest number of attached constraints	mid. value in domain
S_{15}	The variable with the smallest domain. If are more than one, choose the variable with the bigger number of attached constraints.	mid. value in domain
S_{16}	The variable with the biggest difference between the smallest value and the second more smallest of the domain	mid. value in domain
S_{17}	First variable of the list	max. value in domain
S_{18}	The variable with the smallest domain	max. value in domain
S_{19}	The variable with the largest domain	max. value in domain
S_{20}	The variable with the smallest value of the domain	max. value in domain
S_{21}	The variable with the largest value of the domain	max. value in domain
S_{22}	The variable with the largest number of attached constraints	max. value in domain
S_{23}	The variable with the smallest domain. If are more than one, choose the variable with the bigger number of attached constraints	max. value in domain
S_{24}	The variable with the biggest difference between the smallest value and the second more smallest of the domain.	max. value in domain

reported AS systems for CP, one supported by a genetic algorithm (GA) [6] and the other one supported by a particle swarm optimizer (PSO) [7]. Table 8 depicts solving time and number of backtracks required by GA and PSO in contrast with our proposal. This comparison shows that CS and PSO outperforms GA, in terms of number of problems solved and backtracks needed to successfully reach a solution. Moreover, considering runtime, the CS algorithm based optimizer performs notably better than its competitors. A graphical comparison can be seen in Figures 1 and 2.

Table 2. Runtime in ms for strategies S1 to S8

Problem	Strategies							
	S1	S2	S3	S4	S5	S6	S7	S8
Q-8	5	5	5	4	2	4	4	2
Q-10	5	8	3	4	4	5	3	4
Q-12	12	11	11	11	13	14	11	10
Q-20	20405	4867	20529	20529	1294	26972	15	93
Q-50	t.o.	t.o.	532	t.o.	t.o.	t.o.	524	t.o.
Q-75	t.o.	t.o.	4280	t.o.	t.o.	t.o.	4217	t.o.
MS-3	1	5	1	1	1	4	1	1
MS-4	14	2340	6	21	21	1500	6	11
MS-5	1544	t.o.	296	6490	t.o.	t.o.	203	1669
MS-6	t.o.	t.o.	t.o.	t.o.	t.o.	t.o.	t.o.	t.o.
S-2	35	30515	10	50	225	1607	10	10
S-5	7453	t.o.	2181	8274	t.o.	t.o.	2247	897
S-7	26882	t.o.	2135	25486	t.o.	t.o.	2187	31732
K-5	1825	t.o	2499	t.o	t.o	t.o	t.o	t.o
K-6	90755	t.o	111200	89854	t.o	t.o	39728	t.o
QG-5	t.o.	t.o.	7510	t.o.	t.o.	t.o.	9465	t.o.
QG-6	45	t.o.	15	45	t.o.	3605	15	t.o.
QG-7	256	8020	10	307	943	16896	10	16
LF 2-12	20	242	4	29	43	32	4	22
LF 2-16	70	70526	231	115	1217	489	237	7
LF 2-20	191	t.o.	546	318	61944	11	553	240
LF 2-23	79	t.o.	286	140	68254	19	285	19
$\overline{x}$	8311	11653.9	7252	8922.3	11163.5	3935.3	2986.3	2315.6

Table 3. Runtime in ms for strategies S9 to S16

Problem	Strategies							
	S9	S10	S11	S12	S13	S14	S15	S16
Q-8	5	5	4	5	2	4	4	2
Q-10	5	8	7	5	5	5	3	4
Q-12	11	11	11	11	13	14	11	10
Q-20	20349	4780	18	23860	1250	36034	17	87
Q-50	t.o.	t.o.	532	t.o.	t.o.	t.o.	533	t.o.
Q-75	t.o.	t.o.	4336	t.o.	t.o.	t.o.	4195	t.o.
MS-3	1	4	1	1	1	4	1	1
MS-4	13	2366	6	21	21	1495	6	11
MS-5	1498	t.o.	297	6053	t.o.	t.o.	216	1690
MS-6	t.o.	t.o.	t.o.	t.o.	t.o.	t.o.	t.o.	t.o.
S-2	35	29797	10	50	225	1732	10	10
S-5	7521	t.o.	2394	9015	t.o.	t.o.	2310	972
S-7	26621	t.o.	2069	26573	t.o.	t.o.	2094	30767
K-5	1908	t.o	2625	t.o	t.o	t.o	t.o	t.o
K-6	93762	t.o	102387	109157	t.o	t.o	46673	t.o
QG-5	t.o.	t.o.	9219	t.o.	t.o.	t.o.	10010	t.o.
QG-6	40	t.o.	15	45	t.o.	3565	15	t.o.
QG-7	240	13481	10	348	1097	18205	11	15
LF 2-12	20	270	4	29	44	32	5	21
LF 2-16	69	55291	250	118	1273	530	235	8
LF 2-20	185	t.o.	538	312	61345	11	541	237
LF 2-23	79	t.o.	285	140	71209	19	278	19
$\overline{x}$	8464.6	10601.3	5953.3	10337.9	11373.8	4742.4	3358.4	2257

Table 4. Runtime in ms for strategies S17 to S24 and CS

Problem	Strategies								
	S17	S18	S19	S20	S21	S22	S23	S24	CS
Q-8	5	5	4	4	4	2	4	2	290
Q-10	4	7	2	4	5	4	3	5	310
Q-12	11	10	11	11	14	13	11	8	380
Q-20	22286	4547	16	13135	26515	1249	16	1528	685
Q-50	t.o.	t.o.	520	t.o.	t.o.	t.o.	521	t.o.	3269
Q-75	t.o.	t.o.	4334	t.o.	t.o.	t.o.	4187	t.o.	12981
MS-3	1	1	1	1	1	1	1	1	340
MS-4	88	37	99	42	147	37	102	79	485
MS-5	t.o.	t.o.	t.o.	165878	t.o.	153679	t.o.	t.o.	590
MS-6	t.o.	t.o.	t.o.	t.o.	t.o.	t.o.	t.o.	t.o.	1135
S-2	5	18836	30	5	100	1710	30	40	330
S-5	t.o.	t.o.	2590	t.o.	t.o.	t.o.	2670	t.o.	245
S-7	3725	t.o.	338	5350	t.o.	t.o.	378	9168	555
K-5	1827	t.o	2620	t.o	t.o	t.o	t.o	t.o	1429
K-6	96666	t.o	97388	90938	t.o	t.o	40997	t.o	7575
QG-5	9743	t.o.	20	10507	t.o.	t.o.	21	t.o.	295
QG-6	7075	t.o.	125	6945	t.o.	t.o.	130	t.o.	345
QG-7	9	1878	12	9	1705	9	12	14	315
LF 2-12	18	242	4	29	33	43	5	13	310
LF 2-16	66	55687	245	107	510	1297	240	584	440
LF 2-20	170	t.o.	562	294	11	58732	569	15437	560
LF 2-23	75	t.o.	272	126	20	73168	276	10	505
$\bar{x}$	8339.7	8125	5459.7	17258	2422.1	22303.4	2640.7	2068.4	1516.8

Table 5. Backtracks requires for strategies S1 to S8

Problem	Strategies							
	S1	S2	S3	S4	S5	S6	S7	S8
Q-8	10	11	10	10	3	9	10	3
Q-10	6	12	4	6	6	6	4	5
Q-12	15	11	16	15	17	16	16	12
Q-20	10026	2539	11	10026	862	15808	11	63
Q-50	>121277	>160845	177	>121277	>173869	>143472	177	>117616
Q-75	>118127	>152812	818	>118127	>186617	>137450	818	>133184
MS-3	0	4	0	0	0	4	0	0
MS-4	12	1191	3	10	22	992	3	13
MS-5	910	>191240	185	5231	>153410	>204361	193	854
MS-6	>177021	>247013	>173930	>187630	>178895	>250986	>202927	>190877
S-2	18	10439	4	18	155	764	4	2
S-5	4229	>89125	871	4229	>112170	>83735	871	308
S-7	10786	>59828	773	10786	>81994	>80786	773	10379
K-5	767	>179097	767	>97176	>228316	>178970	>73253	>190116
K-6	37695	>177103	37695	35059	>239427	>176668	14988	>194116
QG-5	>145662	>103603	8343	>145656	>92253	>114550	8343	>93315
QG-6	30	>176613	0	30	>83087	965	0	>96367
QG-7	349	3475	1	349	4417	4417	1	4
LF 2-12	16	223	1	16	29	22	1	12
LF 2-16	39	24310	97	39	599	210	97	0
LF 2-20	77	>158157	172	77	26314	1	172	64
LF 2-23	26	>157621	64	26	29805	3	64	7
$\bar{x}$	3611.8	4221.5	2381.6	3878.1	5185.8	1786	1327.3	781.8

Table 6. Backtracks requires for strategies S9 to S16

Problem	Strategies							
	S9	S10	S11	S12	S13	S14	S15	S16
Q-8	10	11	10	10	3	9	10	3
Q-10	6	12	12	6	6	6	4	5
Q-12	15	11	16	15	17	16	16	12
Q-20	10026	2539	11	10026	862	15808	11	63
Q-50	>121277	>160845	177	>121277	>173869	>143472	177	>117616
Q-75	>118127	>152812	818	>118127	>186617	>137450	818	>133184
MS-3	0	4	0	0	0	4	0	0
MS-4	12	1191	3	10	22	992	3	13
MS-5	910	>191240	185	5231	>153410	>204361	193	854
MS-6	>177174	>247013	>174068	>187777	>179026	>251193	>203089	>191042
S-2	18	10439	4	18	155	764	4	2
S-5	4229	>89125	871	4229	>112174	>83735	871	308
S-7	10786	>59828	773	10786	>81994	>80786	773	10379
K-5	767	>179126	767	>97176	>228316	>178970	>73253	>190116
K-6	37695	>177129	37695	35059	>239427	>176668	14998	>194116
QG-5	>145835	>103663	8343	>145830	>92355	>114550	8343	>93315
QG-6	30	>176613	0	30	>83087	965	0	>93820
QG-7	349	3475	1	349	583	4417	1	4
LF 2-12	16	223	1	16	29	22	1	12
LF 2-16	39	24310	97	39	599	210	97	0
LF 2-20	77	>158157	172	77	26314	1	172	64
LF 2-23	26	>157621	64	26	29805	3	64	7
$\bar{x}$	3611.8	4221.5	2382	3878.1	4866.3	1786	1327.8	781.8

Table 7. Backtracks requires for strategies S17 to S24 and CS

Problem	Strategies								
	S17	S18	S19	S20	S21	S22	S23	S24	CS
Q-8	10	11	10	10	9	3	10	2	3
Q-10	6	12	4	6	6	6	4	37	3
Q-12	15	11	16	15	16	17	16	13	3
Q-20	10026	2539	11	10026	15808	862	11	1129	10
Q-50	>121277	>160845	177	>121277	>173869	>143472	177	>117616	1
Q-75	>118127	>152812	818	>118127	>186617	>137450	818	>133184	818
MS-3	1	0	1	1	1	0	1	1	0
MS-4	51	42	3	29	95	46	96	47	1
MS-5	>204089	>176414	>197512	74063	>201698	74711	>190692	>183580	26
MS-6	>237428	>176535	>231600	>190822	>239305	>204425	>204119	>214287	257
S-2	2	6541	9	2	89	887	9	12	4
S-5	>104148	>80203	963	>104148	>78774	>101058	963	>92557	308
S-7	1865	>80295	187	1865	>93675	>91514	187	2626	93
K-5	767	>179126	767	>97178	>178970	>228316	>73253	>190116	4
K-6	37695	>177129	37695	35059	>176668	>239427	14998	>160789	2196
QG-5	7743	>130635	0	7763	>96083	>94426	0	>95406	0
QG-6	2009	>75475	89	2009	>108987	>124523	89	>89888	1
QG-7	3	845	1	3	773	1	1	1	0
LF 2-12	16	223	1	16	22	29	1	6	0
LF 2-16	39	24592	98	39	210	599	98	239	50
LF 2-20	77	>158028	172	77	1	26314	172	4521	47
LF 2-23	26	>157649	64	26	3	29805	64	0	19
$\bar{x}$	3550.1	3481.6	2054.3	7706.5	1419.5	10252.4	932.4	664.2	174.8

Table 8. Solving time in ms and Backtracks requires for optimizers CS, PSO and GA

Problem	CS Runtime	CS Backtracks	PSO Runtime	PSO Backtracks	GA Runtime	GA Backtracks
Q-8	290	3	4982	3	645	1
Q-10	310	3	7735	1	735	4
Q-12	380	3	24369	1	875	40
Q-20	685	10	52827	11	7520	3879
Q-50	3269	1	1480195	0	6530	15
Q-75	12981	818	t.o.	818	16069	17
MS-3	340	0	2745	0	735	0
MS-4	485	1	15986	0	1162	42
MS-5	590	26	565155	14	1087	198
MS-6	1135	257	t.o.	>47209	t.o.	>176518
S-2	330	4	10967	2	15638	6541
S-5	245	308	2679975	13	8202	4229
S-7	555	93	967014	256	25748	10786
K-5	1429	4	4563751	106	21089	50571
K-6	7575	2196	t.o.	12952	170325	21651
QG-5	295	0	59158	0	11862	7763
QG-6	345	1	44565	0	947	0
QG-7	315	0	28612	0	795	4
LF 2-12	310	0	10430	1	1212	223
LF 2-16	440	50	20548	0	1502	97
LF 2-20	560	47	28466	1	1409	64
LF 2-23	505	19	30468	3	1287	0
$\bar{x}$	1516.8	174.8	557786.8	675.4	14065.5	5053.6

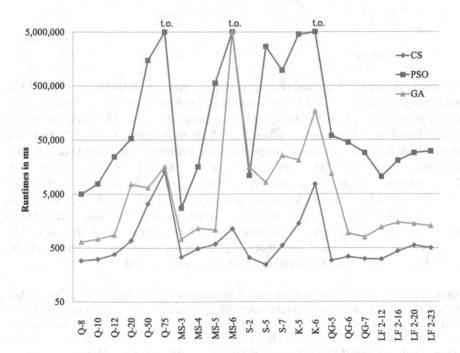

Fig. 1. Comparing runtimes of adaptive approaches

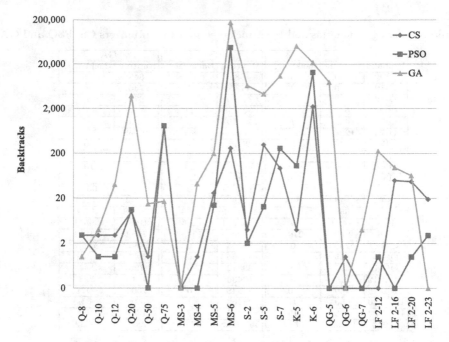

Fig. 2. Comparing backtracks of adaptive approaches

6   Conclusions

Autonomous search is a powerful approach to provide self-tuning capabilities to solvers in order to obtain faster resolution processes, but by maintaining the quality of solutions. In this paper, we have presented a new Autonomous Search system based on the modern Cuckoo Search metaheuristic. Cuckoo Search is a simple to implement algorithm for optimization problems with rapid convergence to optimal solutions. The experimental results have illustrated the efficiency of the proposed approach validating the adaptation capabilities to dynamic environments, which clearly improves performance of solvers during resolution. There exist different directions for future work, perhaps the most straightforward one is the integration of more recent metaheuristics to the framework. The incorporation of additional and more sophisticated strategies to the portfolio would be an interesting research direction to pursue as well.

Acknowledgments. Cristian Galleguillos is supported by Postgraduate Grant Pontificia Universidad Católica de Valparaíso 2015. Ricardo Soto is supported by Grant CONICYT / FONDECYT / INICIACION / 11130459. Broderick Crawford is supported by Grant CONICYT / FONDECYT / REGULAR / 1140897. Fernando Paredes is supported by Grant CONICYT / FONDECYT / REGULAR / 1130455.

References

1. Barták, R., Rudová, H.: Limited assignments: a new cutoff strategy for incomplete depth-first search. In: Proceedings of the 20th ACM Symposium on Applied Computing (SAC), pp. 388–392 (2005)
2. Castro, C., Monfroy, E., Figueroa, C., Meneses, R.: An approach for dynamic split strategies in constraint solving. In: Gelbukh, A., de Albornoz, A., Terashima-Marín, H. (eds.) MICAI 2005. LNCS (LNAI), vol. 3789, pp. 162–174. Springer, Heidelberg (2005)
3. Civicioglu, P., Besdok, E.: A conceptual comparison of the cuckoo-search, particle swarm optimization, differential evolution and artificial bee colony. Artificial Intelligence Review **39**(4), 315–346 (2013)
4. Crawford, B., Soto, R., Castro, C., Monfroy, E.: A hyperheuristic approach for dynamic enumeration strategy selection in constraint satisfaction. In: Ferrández, J.M., Sánchez, J.R.A., de la Paz, F., Toledo, F.J. (eds.) IWINAC 2011, Part II. LNCS, vol. 6687, pp. 295–304. Springer, Heidelberg (2011)
5. Crawford, B., Soto, R., Castro, C., Monfroy, E., Paredes, F.: An Extensible Autonomous Search Framework for Constraint Programming. Int. J. Phys. Sci. **6**(14), 3369–3376 (2011)
6. Crawford, B., Soto, R., Monfroy, E., Palma, W., Castro, C., Paredes, F.: Parameter tuning of a choice-function based hyperheuristic using Particle Swarm Optimization. Expert Systems with Applications **40**(5), 1690–1695 (2013)
7. Crawford, B., Soto, R., Monfroy, E., Palma, W., Castro, C., Paredes, F.: Parameter tuning of a choice-function based hyperheuristic using particle swarm optimization. Expert Syst. Appl. **40**(5), 1690–1695 (2013)
8. Hamadi, Y., Monfroy, E., Saubion, F.: Autonomous Search. Springer (2012)
9. Monfroy, E., Castro, C., Crawford, B., Soto, R., Paredes, F., Figueroa, C.: A reactive and hybrid constraint solver. Journal of Experimental and Theoretical Artificial Intelligence **25**(1), 1–22 (2013)
10. Soto, R., Crawford, B., Monfroy, E., Bustos, V.: Using autonomous search for generating good enumeration strategy blends in constraint programming. In: Murgante, B., Gervasi, O., Misra, S., Nedjah, N., Rocha, A.M.A.C., Taniar, D., Apduhan, B.O. (eds.) ICCSA 2012, Part III. LNCS, vol. 7335, pp. 607–617. Springer, Heidelberg (2012)
11. Soto, R., Crawford, B., Galleguillos, C., Monfroy, E., Paredes, F.: A pre-filtered cuckoo search algorithm with geometric operators for solving sudoku problems. The Scientific World Journal **2014**(465359), 12 (2014)
12. Yang, X.-S.: Nature Inspired Meta-heuristic Algorithms. University of Cambridge. Luniver Press, UK (2010)
13. Yang, X-S., Deb, S.: Cuckoo search via lévy flights. In: World Congress on Nature and Biologically Inspired Computing (NaBIC 2009), pp. 210–214 (2009)

Hybrid Methods

Instance Selection Approach for Self-Configuring Hybrid Fuzzy Evolutionary Algorithm for Imbalanced Datasets

Vladimir Stanovov[✉], Eugene Semenkin, and Olga Semenkina

Siberian State Aerospace University, Krasnoyarsk, Russia
{vladimirstanovov,eugenesemenkin}@yandex.ru,
semenkina.olga@mail.ru

Abstract. We propose an instance selection technique with subsample balancing for an evolutionary classification algorithm. The technique creates subsamples of the training sample in a way to guide the learning process towards problematic areas of the search space. For unbalanced datasets, the number of instances of different classes is artificially balanced to get better classification results. We apply this technique to a self-configured hybrid evolutionary fuzzy classification algorithm. We performed tests on 4 datasets to evaluate the accuracy as well as other classification quality measures for different parameters of the active instance selection procedure. The results shown by our algorithm are comparable or even better than other algorithms on the same classification problems.

Keywords: Instance selection · Fuzzy classification · Evolutionary algorithm · Genetics-based machine learning

1 Introduction

Recent advances in computer and internet technologies have led to the need to process, analyse and understand massive amounts of data. Data mining (DM) and machine learning methods have made this possible by creating accurate models for particular problems. Still, due to the growing size of data to be analysed, the problem of machine learning method scalability remains important for researchers.

Although this problem is often extensively solved by applying parallel computation methods and modern hardware solutions, sometimes they cannot be applied due to resource limitations. The data reduction (DR) methods represent an intelligent way to process large amounts of data with small resources. DR can be performed in different ways, depending on the problem to be solved, for example by selecting features or selecting instances. In this study we concentrate on instance selection methods (IS), or training set selection (TSS). IS focuses on preparing the dataset for the learning algorithm, i.e. creating a training subsample. IS methods include boosting, sampling, prototype selection and active learning.

As in instance selection we use only a part of the training set, and the problem of relevance of the subsample to the original sample must be considered. Although we achieve lower training time, removing instances may result in information loss.

Y. Tan et al. (Eds.): ICSI-CCI 2015, Part I, LNCS 9140, pp. 451–459, 2015.
DOI: 10.1007/978-3-319-20466-6_47

However, for large datasets removing instances does not necessarily lead to information loss, and moreover, may help to avoid over-fitting. That means that IS should be considered not only a method for saving computational resources, but also a way to increase the overall accuracy. So, the idea of developing an IS method that would not only decrease the computation time, but also increase accuracy is behind this study. Machine learning methods today often use evolutionary algorithms (EAs) as a powerful optimization technique, which allows the complex structures of the solution to be created. These algorithms may use specific genetic operators and population organization schemes. This field is called genetics-based machine learning (GBML) [1, 2].

In this study we used a self-configuring hybrid evolutionary fuzzy classification algorithm. This algorithm is our implementation of an algorithm developed by the H. Ishibuchi group [3], with some modifications. We applied our instance selection technique because this algorithm is very sample-dependent as it includes several heuristics in the learning process. They are the heuristic rule generation in initialization and the Michigan part, which uses instances to generate new rules; and the heuristic class label and rule weight specification, which uses confidence values.

The paper is organized as follows: Section 2 describes our machine learning method, Section 3 explains the IS method, Section 4 contains experimental setup and results, and Section 5 concludes the paper.

2   Hybrid Evolutionary Fuzzy Classification Algorithm

As a classification method, in this study we applied the algorithm that we have previously used in our works [4, 5]. The idea of the algorithm was developed by the H. Ishibuchi group in [3]. We made several modifications, including applying different genetic operators and self-configuration. We will give a very short description of the main features of this algorithm here.

In the Pittsburgh part, each individual consists of several rules, and the number of rules is not fixed. The compatibility grade for every pattern was calculated using a product operator. The winner-rule strategy was applied in the fuzzy inference procedure – the resulting class label for a pattern was equal to the label of the rule having the largest product of compatibility grade and rule weight. We used several fuzzy granulations for every variable – four partitions into 2, 3, 4 and 5 fuzzy sets, plus the "don't care" condition, i.e. there were 15 fuzzy sets used. The class label for each rule, as well as the class weight, was determined heuristically based on the confidence value. We applied the self-configuration scheme to choose the probabilities of selection, mutation and the Michigan part. The self-configuration scheme, introduced in [6, 7] has already been successfully used in our previous works [4, 5].

3   Instance Selection Method

The instance selection method that we propose creates subsamples of the original training sample to train the classifier on them. The size of the subsample was set to 5, 10, 15, 20, 25 and 30 percent of the original sample in our computational experiments.

Reducing the number of instances available to the classifier may or may not lead to lower accuracy on the test sample, depending on which instances are selected in the subsample. Similar approaches have already been studied, for example [8, 9].

The way to improve the classification quality is to use a selection procedure for instances. To be able to use a selection procedure, we have to define a fitness function for instances. For example, we may want to select those instances, which are difficult to classify for the learning algorithm, and design a fitness function based on classification results. To avoid over-fitting and to get more accurate results, we may change the training subsample during the learning process. For example, we may change the subsample every 50, 100, 200 or 400 generations, called the adaptation periods, during which the population adjusts itself to a new environment.

The instance selection scheme that we propose uses counters U_i for every instance. These values show the number of times that an instance i has been in the subsample and was classified correctly. If an instance was misclassified during the training process, its counter value is reset. Here we follow two main principles: firstly, we should select instances, which have not been used before (exploration), and secondly, we would like to keep misclassified instances in the training subset (exploitation). That means that our instance selection algorithm has to concentrate on problematic areas of the feature space, but it should also try to discover new areas. To follow these goals, at first we define all counter values $U_i = 1$, $i = 1...m$. Then, we select a training subset with uniform probabilities and launch the algorithm for one adaptation period. After that, we use the best current individual for the subsample, to classify the instances and recalculate U_i values. If an instance j was correctly classified, then $U_j = U_j + 1$, else $U_j = 1$. Note that we change U_i only for instances in the subset. Then we select a new training subsample in a way similar to proportional selection in GA, with probabilities equal to:

$$p_i = \frac{1/U_i}{\sum_{j=\overline{1,n}} 1/U_j} \tag{1}$$

The U_i values actually represent the inverse fitness values for the instances in this case. This means that instances with larger counters, i.e. those which have been previously correctly classified many times, will get lower probabilities. Unused or difficult-to-classify instances get large and equal probabilities to be selected.

After several adaptation periods, the counter values change in such a way that the algorithm concentrates on problematic areas where the different class instances are close to each other. At the same time, areas where the class number is clear get lower probabilities.

In fig. 1 we provide an example with two Gaussian samples, and show the instances selected (left side). As we can see, the algorithm focuses on the area where classes mix with each other. The reason why IS can provide better classification is that during the learning process the algorithm becomes more sensitive to misclassification, as it leads to a change in counter values U_i. At each generation of the algorithm, we take the current best individual for the subsample and test it on the whole training sample. If the accuracy improves, we save this new best individual separately and include it in the population on all generations. This allows good genetic information from previous generations to be saved and prevents over-fitting.

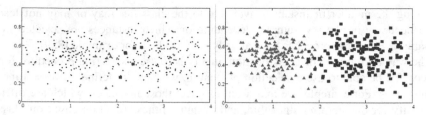

Fig. 1. Instance selection: 2 Gaussians example. The algorithm will select instances from the area of class mixture.

At the end of each adaptation period, we check all individuals in the population on the whole training set. This stage is important, as the population may contain good solutions for the whole dataset, although these solutions show average results for the subsample. The instance selection was performed separately for every class in the training sample. The number of instances to be selected for every class was predefined and depended on the number of instances available. For example, in case of imbalanced datasets, the number of instances of the minority class can be so small that if we use instance selection without taking the class label into consideration, the subsample may happen to contain only majority class instances, or vice versa. For this reason, we applied an artificial balancing strategy, which sets the number of instances to be selected into the subsample for every class as equal to each other as possible. This strategy leads to more balanced subsamples for training, which is important for the learning process.

4   Experimental Setup and Results

We performed a series of computational experiments to evaluate the effectiveness of the approach on a set of classification problems. The entire algorithm was developed in C++ using only standard libraries and ran on a four-core Intel Core-i7 2600K processor.

Table 1. Datasets used

Dataset	Number of instances	Number of features	Number of classes	Imbalance ratio
Magic	19020	10	2	1.843
Phoneme	5404	5	2	2.407
Page-blocks	5472	10	5	175.468
Satimage	6435	36	6	2.448

The population size was set to 100, the number of generations to 10000 and the maximum number of rules to 40 for all experiments. For the instance selection method we used the following settings: the subsample size was set to 5, 10, 15, 20, 25 and 30% of the training sample, and the adaptation period length was equal to 50, 100, 200 and 400 generations. We performed a 10-fold stratified cross-validation for every experiment. In this study we tested our approach on datasets from the KEEL repository [10]. The parameters of these datasets are provided in Table 1. First, we measured the algorithm efficiency without instance selection, i.e. with training on the whole sample. The results are shown in table 2.

Table 2. Results of the standard algorithm, test errors in percent

Dataset	Training error, %	Test error, %	Number of rules	Average rule length
Magic	16.044	16.606	11.1	8.71
Phoneme	15.659	17.099	17.4	2.96
Page-blocks	3.523	4.202	9.8	3.62
Satimage	12.926	14.353	16.6	15.99

In Table 3 we show the test sample error values for the Magic problem with instance selection for different sizes of the subsample and different adaptation periods.

Table 3. Results with instance selection for the Magic problem, test errors

Percent of training sample	Number of generations			
	50	100	200	400
5	15.788	15.383	15.594	16.119
10	15.457	15.299	15.589	15.683
15	15.252	15.263	15.489	15.404
20	15.263	15.368	15.199	15.194
25	15.457	15.410	15.420	15.252
30	15.221	15.284	**15.079**	15.163

The best result with instance selection was obtained with 30% of the training sample used and an adaption period of 200 generations. However, the results do not differ very much from 5% to 30%, which means that the data is homogenous. The instance selection was able to provide higher accuracy, although the Wilcoxon test did not show a significant change in accuracy. Table 4 contains the results for the Phoneme problem.

Table 4. Results with instance selection for the Phoneme problem, test errors

Percent of training sample	Number of generations			
	50	100	200	400
5	18.504	19.245	18.929	19.003
10	18.876	19.117	19.059	18.746
15	17.911	17.709	18.320	18.727
20	17.394	17.837	18.651	18.135
25	17.395	17.450	17.857	18.136
30	**16.875**	17.062	16.988	17.654

For the Phoneme problem, the use of instance selection mostly gave worse results, however, the Wicoxon test showed no significant difference in accuracy. One may see that a larger subsample usually gives better results; however, it increases the calculation time. Table 5 shows results for the Page-blocks dataset.

Table 5. Results with instance selection for the Page-blocks problem, test errors

Percent of training sample	Number of generations			
	50	100	200	400
5	*7.947*	*8.443*	*9.248*	*8.479*
10	5.810	6.031	6.670	7.274
15	4.002	4.496	4.897	5.775
20	3.252	3.399	3.929	4.405
25	3.783	3.782	3.691	4.148
30	**3.380**	3.690	3.782	4.057

For the Page-blocks dataset we observed a much bigger divergence in classification accuracy. This can be because this dataset is highly imbalanced, so that the change of subsample size influences the ratio in the subsample, as we used the artificial balancing strategy for the subsample. The best result in the sense of accuracy was received for 30% of the training sample size, and an adaptation period of 50 generations. According to the Wilcoxon test, this result was *not* significantly different from the original algorithm. With a training subsample size equal to 5% the accuracy values were much lower and the Wilcoxon test showed that they are significantly worse than those shown by the original algorithm. The results for the Satimage problem are in table 6.

Table 6. Results with instance selection for the Satimage problem, test errors

Percent of training sample	Number of generations			
	50	100	200	400
5	14.046	15.384	15.865	15.337
10	14.172	13.832	14.590	14.575
15	13.381	14.093	13.599	14.453
20	13.428	13.363	13.534	14.126
25	13.023	13.255	13.893	13.815
30	*12.928*	13.022	13.349	13.924

According to the Wilcoxon test, the best result for the Satimage dataset was significantly better than the result of the original algorithm. The same or even better results were obtained already with 10% of the size of the original sample.

Together with accuracy we calculated $Recall_\mu$ measure for datasets containing more than 2 classes according to [11]. This value shows the average accuracy among classes. In table 7 we provide the $(1-Recall_\mu)*100$ values for test sample for Page-blocks problem. The $1-Recall_\mu$ value for the original algorithm is equal to 39.9%, which means a huge imbalance in classification results. The accuracy value for this case is about 4.2%, but it should not be considered as a good result, as most of the test instances are classified into minority class.

Table 7. 1-Recall$_\mu$ values in percent for test sample, Page-blocks problem

Percent of training sample	Number of generations			
	50	100	200	400
5	**10.183**	12.915	15.180	12.060
10	12.918	13.713	14.589	17.509
15	20.508	19.625	18.355	19.263
20	23.045	21.226	22.719	22.359
25	28.351	26.137	23.048	22.239
30	28.037	27.731	28.423	29.511

The instance selection method with artificial balancing strategy significantly changes the 1-Recall$_\mu$ values. For example, when using only 5% of the training sample, the subset becomes much more balanced. This leads to the decrease in accuracy, which we observed before, however the solutions received with instance selection describe minority classes significantly better. For example, for 5% of the training sample and 50 generation adaptation period, the difference between accuracy and 1-Recall$_\mu$ is only 2.236%. For other datasets the difference in 1-Recall$_\mu$ is about 2-3% due to lower imbalance rates. We also compared the best instance selection results with other approaches for the same problems. The comparison is shown in table 8.

Table 8. Comparison to other approaches, test accuracy

Dataset	Our method	Parallel fuzzy GBML [4]	GP-Coach [12]	FARC-HD [13]
Magic	15.079±0.68	**14.89**	20.18	15.49
Phoneme	16.875±1.44	**15.96**	-	17.86
Page-blocks	**3.380±0.64**	3.62	8.77	4.99
Satimage	**12.928±0.83**	12.96	27.50	12.68

Presented results show that for two out of four problems our approach allowed us to get better results than the algorithm, which we used as a prototype. One should also mention that we used much less computational resources. For two of the problems we were able to get the best result compared to other approaches, although the difference is not significant.

5   Conclusion

In this paper we described an instance selection method for evolutionary classification machine learning algorithms. This method is based on the idea of applying a probabilistic mechanism to create subsamples of the original training sample for the classifier to learn for several generations/iterations. The presented instance selection technique allows artificial balancing of the subsample, which gives the opportunity for the classifier to learn more adequate models, describing minority classes better. The accuracy of the presented method is comparable to the classification method, used as prototype, and may even surpass it. The time complexity experiment has shown that applying instance selection results in better computation times, together with higher accuracy. The instance selection technique can be applied to other evolutionary classification methods. This method can also be used for larger datasets with some modifications.

Acknowledgements. Research is performed with the financial support of the Ministry of Education and Science of the Russian Federation within the federal R&D programme (project RFMEFI57414X0037).

References

1. Fernandez, A., Garcia, S., Luengo, J., Bernado-Mansilla, E., Herrera, F.: Genetics-Based Machine Learning for Rule Induction: State of the Art, Taxonomy, and Comparative Study. IEEE Transactions on Evolutionary Computation **14**(6), 913–941 (2010)
2. Booker, L.B., Goldberg, D.E., Holland, J.H.: Classifier systems and genetic algorithms. Artif. Intell. **40**(1–3), 235–282 (1989)
3. Ishibuchi, H., Mihara, S., Nojima, Y.: Parallel Distributed Hybrid Fuzzy GBML Models With Rule Set Migration and Training Data Rotation. IEEE Transactions on fuzzy systems **21**(2) (2013)
4. Stanovov, V., Semenkin, E.: Hybrid self-configuring evolutionary algorithm for automated design of fuzzy logic rule base. In: 11th International Conference on Fuzzy Systems and Knowledge Discovery, FSKD, Xaimen, China, pp. 317–321 (2014)
5. Stanovov, V., Semenkin, E.: Fuzzy rule bases automated design with self-configuring evolutionary algorithm. In: ICINCO 2014 – Proceedings of the 11th International Conference on Informatics in Control, Automation and Robotics, vol. 1, pp. 318–323 (2014)
6. Semenkina, M., Semenkin, E.: Hybrid self-configuring evolutionary algorithm for automated design of fuzzy classifier. In: Tan, Y., Shi, Y., Coello, C.A.C. (eds.) ICSI 2014, Part I. LNCS, vol. 8794, pp. 310–317. Springer, Heidelberg (2014)
7. Semenkin, E., Semenkina, M.: Self-configuring genetic algorithm with modified uniform crossover operator. In: Tan, Y., Shi, Y., Ji, Z. (eds.) ICSI 2012, Part I. LNCS, vol. 7331, pp. 414–421. Springer, Heidelberg (2012)
8. Cano, J.R., Herrera, F., Lozano, M.: Using Evolutionary Algorithms as Instance Selection for Data Reduction in KDD: An Experimental Study. IEEE Transactions on Evolutionary Computation **7**(6), 561–575 (2003)
9. Olvera-López, J.A., Carrasco-Ochoa, J.A., Martínez-Trinidad, J.F., Kittler, J.: A review of instance selection methods. Artif. Intell. Rev. **34**, 133–143 (2010)
10. Alcalá-Fdez, J., Sánchez, L., Garcia, S., del Jesus, M.J., Ventura, S., Garrell, J.M., Otero, J., Romero, C., Bacardit, J., Rivas, V.M., Fernández, J.C., Herrera, F.: KEEL: A software tool to assess evolutionary algorithms for data mining problems. Soft Comput. **13**(3), 307–318 (2009)
11. Sokolova, M., Lapalme, G.: A systematic analysis of performance measures for classification tasks. Information Processing and Management **45**, 427–437 (2009)

12. Berlanga, F.J., Rivera, A.J., del Jesus, M.J., Herrera, F.: GP-COACH: Genetic programming-based learning of compact and accurate fuzzy rulebased classification systems for high-dimensional problems. Inf. Sci. **180**(8), 1183–1200 (2010)

13. Alcala-Fdez, J., Alcala, R., Herrera, F.: A fuzzy association rulebased classification model for high-dimensional problems with genetic rule selection and lateral tuning. IEEE Trans. Fuzzy Syst. **19**(5), 857–872 (2011)

An Improved Hybrid PSO Based on ARPSO
and the Quasi-Newton Method

Fei Han[1(✉)] and Qing Liu[1,2]

[1] School of Computer Science and Communication Engineering,
Jiangsu University, Zhenjiang 212013, Jiangsu, China
hanfei@mail.ujs.edu.cn, clyqig2008@126.com
[2] School of Computer Science and Technology,
Nanjing University of Science and Technology, Nanjing 210094, Jiangsu, China

Abstract. Although attractive and repulsive particle swarm optimization (ARPSO) algorithm keeps the diversity of the swarm adaptively to avoid premature convergence, its search performance is still restricted because of its stochastic search mechanism. In this study, a new hybrid algorithm combining ARPSO with the Quasi-Newton method is proposed to improve the search ability of the swarm. In the proposed algorithm, ARPSO keeps the reasonable search space by controlling the swarm not to lose its diversity, while the Quasi-Newton method is used to perform local search efficiently. The Quasi-Newton method makes the hybrid algorithm converge to optimal solution accurately. The experimental results verify that the proposed hybrid PSO has better convergence performance than some classic PSO algorithms.

Keywords: Attractive and repulsive particle swarm optimization · Diversity · The Quasi-Newton method

1 Introduction

As an effective population-based stochastic optimization technique, particle swarm optimization (PSO) [1,2] has been widely used in large-scale, highly nonlinear, and complex optimization problems. Although PSO has shown good performance in solving many optimization problems, it suffers from the problem of premature convergence like most of the stochastic search techniques, particularly in multimodal optimization problems.

To improve the search ability of PSO, many improved PSO were proposed. Passive congregation PSO (PSOPC) introduced passive congregation for preserving swarm integrity to transfer information among individuals of the swarm [3]. PSO with a constriction (CPSO) defined a "no-hope" convergence criterion and a "rehope" method as well as one social/confidence parameter to re-initialize the swarm [4, 5]. Although these classical PSO and their improvements improve the search ability of the swarm, they still could not overcome the problem of premature convergence and thus converge to local minima with high probability.

© Springer International Publishing Switzerland 2015
Y. Tan et al. (Eds.): ICSI-CCI 2015, Part I, LNCS 9140, pp. 460–467, 2015.
DOI: 10.1007/978-3-319-20466-6_48

To overcome the problem of premature convergence, attractive and repulsive PSO (ARPSO) was proposed in [6]. It used a diversity measure to control the movement of the swarm, which the swarm alternated between phases of attraction and repulsion according to the diversity value of the swarm, which prevented premature convergence to a high degree [6]. Although ARPSO and its improvements [7] keep the diversity of the swarm adaptively, they still perform search with stochastic mechanism in essence and there still has much room to improve.

The algorithm referred to as GPSO in [8], is one of the most efficient gradient-based PSO. GPSO combined the standard PSO with second derivative information to perform global exploration and accurate local exploration, respectively [8]. Since the standard PSO could not control the diversity of the swarm effectively to keep the reasonable search space, this hybrid PSO may still converge to local minima in some cases.

To overcome the problems described above, we proposed an improved hybrid PSO called HARPSOGS [9] which combined ARPSO with the steepest gradient descent method. Different from HARPSOGS, ARPSO is combined with the Quasi-Newton method in this paper to further improve the search ability of PSO, In the new method, to avoid the swarm be trapped into local minima, firstly, ARPSO is used to perform stochastic and rough search. In the solution space obtained by ARPSO, the Quasi-Newton method is then used to perform further search. When the deterministic search converges to local minima, the proposed algorithm turns to ARPSO to adjust the search space by improving the diversity of the swarm. With the deterministic search method in a proper search space, the proposed PSO may converge to the global minima with a high likelihood.

2   Particle Swarm Optimization

PSO is an evolutionary computation technique in search for the best solution by simulating the movement of birds in a flock [1, 2]. The population of the birds is called swarm, and the members of the population are particles. Each particle represents a possible solution to the optimization problem. During each iteration, each particle flies independently in its own direction guided by its own previous best position as well as the global best position of all particles. Assume that the dimension of the search space is D, and the swarm is $S=(X_1, X_2, X_3, ..., X_{Np})$; each particle represents a position in the D dimension; the position of the i-th particle in the search space is denoted as $X_i=(x_{i1}, x_{i2}, ..., x_{iD})$, $i=1, 2, ..., N_p$, where N_p is the number of all particles. The own previous best position of the i-th particle is called $pbest$ which is expressed as $P_i=(p_{i1}, p_{i2}, ..., p_{iD})$. The global best position of all particles is called $gbest$ which is denoted as $P_g= (p_{g1}, p_{g2}, ..., p_{gD})$. The velocity of the i-th particle is expressed as $V_i= (v_{i1}, v_{i2}, ..., v_{iD})$. The adaptive PSO [10] was described as:

$$V_i(t+1) = W(t) \times V_i(t) + c1 \times rand() \times (P_i(t) - X_i(t)) + c2 \times rand() \times (P_g(t) - X_i(t)) \quad (1)$$

$$X_i(t+1) = X_i(t) + V_i(t+1) \quad (2)$$

where $c1$, $c2$ are the acceleration constants with positive values; $rand()$ is a random number ranged from 0 to 1; $W(t)$ is inertia weight.

Attractive and repulsive particle swarm optimization (ARPSO), a diversity-guided method, was proposed to avoid premature convergence effectively [6]. The movement of particles in ARPSO depends on the diversity value of the swarm, and the velocity update of particles was described as follows:

$$V_i(t+1) = W(t) \times V_i(t) + dir \times [c1 \times rand() \times (P_i(t) - X_i(t)) + c2 \times rand() \times (P_g(t) - X_i(t))] \quad (3)$$

where $dir = \begin{cases} -1 & diversity < d_{low} \\ 1 & diversity > d_{high} \end{cases}$.

In ARPSO, a function was proposed to calculate the diversity of the swarm as follows:

$$diversity\,(\ ) = \frac{1}{N_p \times |L|} \times \sum_{i=1}^{N_p} \sqrt{\sum_{j=1}^{D} (p_{ij} - \overline{p}_j)^2} \quad (4)$$

where $|L|$ is the length of the maximum radius of the search space; p_{ij} is the j-th component of the i-th particle and $\overline{p}_j$ is the j-th component of the average over all particles.

In the attraction phase ($dir=1$), the swarm is attracting, and consequently the diversity decreases. When the diversity drops below the lower bound, d_{low}, the swarm switches to the repulsion phase ($dir=-1$). When the diversity reaches the upper bound, d_{high}, the swarm switches back to the attraction phase. ARPSO alternates between phases of exploiting and exploring-attraction and repulsion-low diversity and high diversity and thus improve its search ability [6].

3   The Proposed Hybrid PSO

In this paper, an improved hybrid PSO, named HARPSOQN combining ARPSO with the Quasi-Newton method, is proposed, and the detailed steps are as follows:

Step 1: Initialize the velocities and positions of all particles randomly, the maximum iteration number and optimization target.

Step 2: Calculate the *pbest* of each particle, *gbest* of all particles and the diversity of the swarm, and update the *pbest* of each particle and the *gbest* of all particles.

Step 3: Each particle updates its position according to the following equation:

$$X' = X_i(t) + V_{arpso} \quad (5)$$

where V_{arpso} is the velocity update of the i-th particle obtained by Eq.(3).

Step 4: With the position obtained by Eq.(5), each particle searches in the direction guided by the BFGS method [11], one of most efficient quasi-Newton method, and the particle is updated by the following iterative equations:

$$X''(k+1) = X''(k) + \beta(k) \times Grad(k), \quad X''(0) = X' \quad (6)$$

$$B_{k+1} = B_k - \frac{B_h \cdot s_k \cdot s_k^T \cdot B_{k'}}{s_k^T \cdot B_k \cdot s_k} + \frac{y_k \cdot y_h^T}{y_k^T \cdot s_k}, \ B_0 = I \tag{7}$$

$$s_k = X''(k+1) - X''(k), \quad y_k = \frac{\partial f(X''(k+1))}{\partial X''(k+1)} - \frac{\partial f(X''(k))}{\partial X''(k)} \tag{8}$$

In Eq. (6), the parameter $\beta(k)$ is determined by one-dimensional search method such as binary chop.

Step 5: Compare the position obtained by ARPSO with the position obtained by the Quasi-Newton method, and select the one with the better fitness function value as the new position for the particle, which is described as follows:

$$X_i(t+1) = \begin{cases} X' & if \ f(X') < f(X'') \\ X'' & else \end{cases} \tag{9}$$

Step 6: Once the new population is generated, return to Step 2 until the goal is met or the predetermined maximum learning epochs are completed.

4   Experimental Results and Discussion

In this section, the performance of the proposed hybrid PSO is compared with some classical PSO including APSO, CPSO and PSOPC, ARPSO, GPSO and HARPSOGS on six functions in the De Jong test suite of benchmark optimization problems. For combing ARPSO with the steepest gradient descent method, HARPSOGS is renamed as HARPSOSGD in this section. Table 1 shows these classical test functions used in the experiments. The Sphere functions is convex and unimodal (single local minimum). The Rosenbrock test function has a single global minimum located in a long narrow parabolic shaped flat valley and tests the ability of an optimization algorithm to navigate flat regions with small gradients. The Rastrigin and Griewangk functions are highly multimodal and test the ability of an optimization algorithm to escape from local minima. The Schaffer and LevyNo.5 both are two-dimensional functions with many local minima, which the former has infinite minima and the latter has 760 local minima. All the programs are carried out in MATLAB 7.0 environment on an Intel Core 2 Duo 2.93 GHZ CPU.

The population size in PSOPC is 120 and the one in other PSO is 20 in all experiments. And the acceleration constants $c1$ and $c2$ in APSO, ARPSO, GPSO, HARPSOSGD and HARPSOQN all are set as 2.0. The constants $c1$ and $c2$ both are 2.05 in CPSO, and they both are 0.5 in PSOPC. The decaying inertia weight w starting at 0.9 and ending at 0.4 is set for APSO, CPSO, ARPSO, HARPSOSGD and HARPSOQN according to [10]. The initial inertial weight and the final one in PSOPC are 0.9 and 0.7, respectively. According to [7], the parameters, d_{low} and d_{high}, are set as 5e-6 and 0.25, respectively, in ARPSO, HARPSOSGD and HARPSOQN. All the results shown in this paper are the mean values of 20 trials.

Table 1. The specification of the six test functions

Test function	Equation	Search space	Global minima
Sphere (F1)	$\sum_{i=1}^{n} x_i^2$	$(-100,100)^n$	0
Rosenbrock (F2)	$\sum_{i=1}^{n-1}(100(x_{i+1}-x_i^2)^2+(1-x_i)^2)$	$(-100,100)^n$	0
Rastrigin (F3)	$10n+\sum_{i=1}^{n}(x_i^2-10\cos(2\pi x_i))$	$(-100,100)^n$	0
Griewangk (F4)	$1+\dfrac{1}{4000}\sum_{i=1}^{n}x_i^2-\prod_{i-1}^{n}\cos(\dfrac{x_i}{\sqrt{i}})$	$(-100,100)^n$	0
Schaffer (F5)	$-0.5+\dfrac{\sin\sqrt{x_1^2+x_2^2}-0.5}{(1+0.001(x_1^2+x_2^2))^2}$	$(-100,100)^n$	-1
LevyNo.5 (F6)	$\sum_{i=1}^{5}[i\cos((i-1)x_1+i)]\sum_{j=1}^{5}[j\cos((j-1)x_2+j)]$ $(x_1+1.42513)^2+(x_2+0.80032)^2$	$(-100,100)^n$	-176.1376

Table 2. Mean absolute error obtained by the seven PSO on the six test functions

Functions (Dimension)		APSO	CPSO	PSOPC	ARPSO	GPSO	HARPSOSGD	HARPSOQN
					Mean absolute error			
F1	10	2.80e-4	1.64	5.89e-9	3.3e-3	0	5.35e-15	0
	20	0.22	1.11e+3	8.42e-07	0.38	0	7.48e-13	0
	30	0.21	4.43e+3	5.33e-7	11.02	0	6.54e-12	0
F2	10	4.73	1.46e+4	7.98	2.73	7.87e-11	5.40e-10	0
	20	21.51	1.59e+5	18.63	191.19	2.20e-9	1.13e-9	0
	30	65.06	1.30e+6	60.31	352.68	2.14e-7	5.67e-7	0
F3	10	2.7e-2	107.54	0.058	79.48	0	2.21e-12	0
	20	3.29	433.15	0.42	14.91	0	6.88e-11	0
	30	4.11	3.18e+3	1.09	40.58	0.001	9.94e-10	0
F4	10	7.6e-2	0.36	5.32e-7	0.14	0	2.55e-14	0
	20	0.34	1.10	7.50e-5	0.67	0	1.09e-13	0
	30	1.68	1.38	7.21e-3	1.11	0	2.21e-9	0
F5	2	0.0091	0.0097	0.0077	0.0095	0.009	0	0
F6	2	6.4311	10.3421	1.1e-5	18.6752	1.21e-1	1.1e-5	1.1e-5

Table 2 shows mean absolute error for the six test functions on different dimensions by using the seven PSO. From Table 2, the proposed algorithm obtains better convergence accuracy than other PSO in all cases, which indicates that HARPSOQN has the best search ability among all PSO.

Table 3 lists CPU time of the seven PSO on the six functions with ten and thirty dimensions. From Table 3, some conclusions can be drawn. First, since ARPSO adaptively adjusts the diversity of the swarm and PSOPC uses much more particles to perform search, they require more time than APSO and CPSO. APSO and CPSO lose the diversity of the swarm and converge to local minima quickly, so they require less CPU time than other PSO in almost all cases. Second, although HARPSOSGD converges more accurately than APSO, CPSO, PSOPC and ARPSO, it spends less time than these stochastic PSO in most of cases. Finally, GPSO and HARPSOQN require most time in all cases except that they spend least time on the F1 and F4 functions. This is mainly because GPSO and HARPSOQN spend most of time to search the Qusai-Newton direction. Moreover, HARPSOQN spends less time than GPSO in all cases.

Table 3. CPU time versus the order magnitude of the absolute error obtained by the seven PSO on the six test functions with ten and thirty dimensions

Functions (Dimension)	APSO	CPSO	PSOPC	ARPSO	GPSO	HARPSOSGD	HARPSOQN
	CPU time (s) (the order of magnitude of the absolute error)						
F1 10	2.754 (10^{-1})	3.182 (10^{2})	8.14 (10^{-3})	9.135 (10^{1})	0.66 (0)	2.96 (10^{-8})	0.35 (0)
F1 30	8.491 (10^{1})	9.007 (10^{3})	8.47 (10^{1})	24.231 (10^{2})	0.67 (0)	6.336 (10^{-8})	0.67 (0)
F2 10	4.653 (10^{1})	5.331 (10^{3})	18.228 (10^{-11})	19.539 (10^{1})	780 (10^{-11})	6.512 (10^{-9})	418 (0)
F2 30	11.53 (10^{3})	12.201 (10^{7})	12.069 (10^{2})	27.199 (10^{3})	1985 (10^{-11})	11.387 (10^{-7})	1800 (10^{-6})
F3 10	6.719 (10^{2})	7.801 (10^{2})	20.34 (10^{1})	22.24 (10^{2})	869 (0)	18.103 (10^{-10})	300 (10^{-13})
F3 30	66.719 (10^{2})	67.801 (10^{3})	68.51 (10^{2})	85.24 (10^{2})	1614 (0)	76.103 (10^{-9})	1200 (0)
F4 10	8.199 (10^{-2})	9.458 (10^{-1})	8.76 (10^{-1})	10.023 (10^{-1})	7.65 (0)	4.177 (10^{-9})	2.568 (10^{-8})
F4 30	44.199 (10^{-2})	64.458 (10^{-1})	48 (10^{1})	76.023 (10^{-1})	128 (10^{-13})	38.177 (10^{-9})	32 (10^{-13})
F5 2	1.9390 (10^{-3})	1.698 (10^{-3})	1.26 (10^{-3})	2.343 (10^{-3})	330 (0)	2.170 (10^{-10})	120 (0)
F6 2	5.634 (10^{-1})	4.18 (10^{-1})	2.65 (10^{-5})	12.80 (10^{-1})	560 (10^{-1})	2.180 (10^{-5})	375 (0)

Fig. 1 shows the diversity values of the swarm in seven PSO on the six test functions with ten dimensions. Obviously, the swarms in the APSO, CPSO and PSOPC lose their diversity quickly, while ARPSO, GPSO, HARPSOSGD and HARPSOQN keep the diversity of the swarm adaptively in the whole search process. GPSO adjust the diversity of the swarm at higher levels than ARPSO, HARPSOSGD and HARPSOQN in all cases. HARPSOSGD and HARPSOQN adjust the diversity value of the swarm more frequently than ARPSO in almost all functions, which indicates that HARPSOSGD and HARPSOQN control the diversity of the swarm more efficiently than the ARPSO.

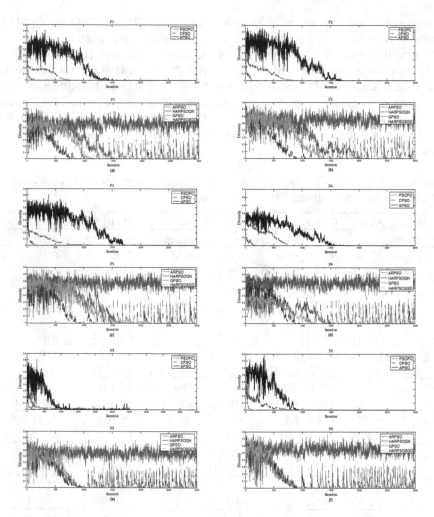

Fig. 1. The diversity values of the seven PSO on the six test functions with ten dimensions (a) Sphere (b) Rosenbrock (c) Rastrigin (d) Griewangk (e) Schaffer (f) LevyNo.5

5   Conclusions

To improve the search ability of PSO, a hybrid PSO combing ARPSO with the Quasi-Newton method was proposed in this paper. In the proposed PSO, ARPSO was used to keep the reasonable search space by adjusting the diversity of the swarm adaptively, and the Quasi-Newton method was used to perform local search efficiently. The experimental results verified the effectiveness and efficiency of the proposed hybrid PSO. Future work will include how to apply this hybrid PSO to solve discrete problems such as gene selection for microarray data.

Acknowledgements. This work was supported by the National Natural Science Foundation of China (Nos.61271385, 60702056).

References

1. Kennedy, J., Eberhart, R.: Particle swarm optimization. IEEE International Conference on Neural Networks **4**, 1942–1948 (1995)
2. Eberhart, R.C., Kennedy, J.: A new optimizer using particle swarm theory. In: The Sixth International Symposium on Micro Machines and Human Science, pp. 39–43 (1995)
3. He, S., Wu, Q.H., Wen, J.Y.: A particle swarm optimizer with passive congregation. Biosystems **78**, 135–147 (2004)
4. Clerc, M.: The swarm and the queen: towards a deterministic and adaptive particle swarm optimization. In: 1999 Congress on Evolutionary Computation, pp. 1951–1957 (1999)
5. Corne, D., Dorigo, M., Glover, F.: New ideas in optimization. McGraw Hill (1999)
6. Riget, J., Vesterstrom, J.S.: A diversity-guided particle swarm optimizer - the arPSO. Technical report. University of Aarhus, Department of Computer Science, Aarhus, Denmark (2002)
7. Han, F., Zhu, J.S.: Improved particle swarm optimization combined with backpropagation for feedforward neural networks. International Journal of Intelligent Systems **28**(3), 271–288 (2013)
8. Noel, M.M.: A new gradient based particle swarm optimization algorithm for accurate computation of global minimum. Applied Soft Computing **12**(1), 353–359 (2012)
9. Liu, Q., Han, F.: A hybrid attractive and repulsive particle swarm optimization based on gradient search. In: Huang, D.-S., Jo, K.-H., Zhou, Y.-Q., Han, K. (eds.) ICIC 2013. LNCS, vol. 7996, pp. 155–162. Springer, Heidelberg (2013)
10. Shi, Y., Eberhart, R.C.: A modified particle swarm optimizer. In: The 1998 IEEE International Conference on Evolutionary Computation, pp. 69–73 (1998)
11. Battiti, R., Masulli, F.: BFGS optimization for faster and automated supervised learning. In: Proceedings of International Neural Network Conference, pp. 757–760 (1990)

Multi-Objective Optimization

Cooperative Multi-objective Genetic Algorithm with Parallel Implementation

Christina Brester[(✉)] and Eugene Semenkin

Siberian State Aerospace University, Krasnoyarsk, Russia
christina.brester@gmail.com, eugenesemenkin@yandex.ru

Abstract. In this paper we introduce the multi-agent heuristic procedure to solve multi-objective optimization problems. To diminish the drawbacks of the evolutionary search, an island model is used to involve various genetic algorithms which are based on different concepts (NSGA-II, SPEA2, and PICEA-g). The main benefit of our proposal is that it does not require additional experiments to expose the most appropriate algorithm for the problem considered. For most of the test problems the effectiveness of the developed algorithmic scheme is comparable with (or even better than) the performance of its component which provides the best results separately. Owing to the parallel work of island model components we have managed to decrease computational time significantly (approximately by a factor of 2.7).

Keywords: Heuristic search · Multi-objective genetic algorithm · Multi-agent approach · Island model · Cooperation

1   Introduction

In recent times there has been a growing interest in the sphere of Evolutionary Ma-chine Learning: owing to a number of benefits which heuristic-based optimization methods have demonstrated, researchers have proposed several effective applications of Evolutionary Computation in the Machine Learning field [1], [2], [3]. This has become possible for several reasons: evolutionary algorithms are universal and might be used to find the optimal solution in both continuous and discrete search spaces; they could be applied in a dynamic environment; in most cases the effectiveness of evolutionary approaches is not lower than the performance of non-evolutionary ones [4].

However, some researchers highlight the negative sides of the Evolutionary Computation and Machine Learning integration. Firstly, it is always necessary to investigate a number of algorithms to define the most effective one for the problem considered because the performance of evolutionary algorithms varies significantly for different problems. Secondly, these methods require more computational resources compared with alternative non-evolutionary algorithms.

This paper is devoted to solving optimization problems with several criteria, and therefore, we attempt to develop a modified multi-objective genetic algorithm (MOGA) with these drawbacks removed.

© Springer International Publishing Switzerland 2015
Y. Tan et al. (Eds.): ICSI-CCI 2015, Part I, LNCS 9140, pp. 471–478, 2015.
DOI: 10.1007/978-3-319-20466-6_49

To overcome the disadvantages of the evolutionary search, an island model is used to involve genetic algorithms (GA) which are based on different concepts (NSGA-II, SPEA2, and PICEA-g). Moreover, this model allows us to parallelize calculations and, consequently, to reduce computational time.

As a result, we have managed to implement the multi-agent heuristic procedure to solve multi-objective optimization problems, which does not require additional experiments to expose the most appropriate algorithm for the problem considered. Besides, due to the parallel work of island model components we have achieved a significant decrease in computational time (roughly by a factor of 2.7). According to the results obtained, for most of the test problems the effectiveness of the developed algorithmic scheme is comparable with the performance of its component which provides the best results separately.

The rest of the paper is organized as follows: in Section 2 a description of the cooperative algorithm developed is presented. The test problems used to investigate the effectiveness of our proposal are introduced in Section 3. The experiments conducted, the results obtained, and the main inferences are included in Section 4. The conclusion and future work are presented in Section 5.

2   Developed Approach

2.1   Cooperative Multi-objective Genetic Algorithm

Designing a MOGA, researchers are faced with some issues which are referred to fitness assignment strategies, diversity preservation techniques, and ways of elitism implementation. However, the common scheme of any MOGA includes the same steps as any conventional one-criterion GA:

> *Generate the initial population*
> *Evaluate criteria values*
> *While (stop-criterion!=true), do:*
> *{Estimate fitness-values;*
> *Choose the most appropriate individuals with the mating selection operator based on their fitness-values;*
> *Produce new candidate solutions with recombination;*
> *Modify the obtained individuals with mutation;*
> *Compose the new population (environmental selection);*
> *}*

In contrast to one-criterion GAs, the outcome of MOGAs is the set of non-dominated points which form the Pareto set approximation.

To eliminate a number of questions which are raised while designing multi-criteria evolutionary methods, in this study we propose a cooperation of several GAs based on various heuristic mechanisms.

Generally speaking, an *island model* [5] of a GA implies the parallel work of several algorithms. A parallel implementation of GAs has shown not just an ability to preserve genetic diversity, since each island can potentially follow a different search trajectory, but also could be applied to separable problems. The initial number of individuals M is spread across L subpopulations: $M_i=M/L$, $i=1,...,L$. At each T-th generation algorithms exchange the best solutions (*migration*). There are two parameters: *migration size*, the number of candidates for migration, and *migration interval*, the number of generations between migrations. Moreover, it is necessary to define the island model topology, in other words, the scheme of migration. We use the fully connected topology that means each algorithm shares its best solutions with all other algorithms included in the island model. The multi-agent model is expected to preserve a higher level of genetic diversity. The benefits of the particular algorithm could be advantageous in different stages of optimization. In this study the Non-Sorting Genetic Algorithm II (NSGA-II) [6], the Preference-Inspired Co-Evolutionary Algorithm with goal vectors (PICEA-g) [7], and the Strength Pareto Evolutionary Algorithm 2 (SPEA2) [8] were used to be involved as parallel working islands (Figure 1).

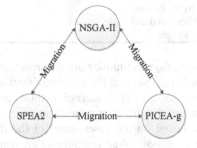

Fig. 1. The island model implemented

The next subsection provides a concise description of the algorithms included in the cooperation and their essential features.

2.2   Brief Description of Island Model Components

Several decades ago Goldberg suggested the usage of the Pareto-dominance idea as the main principle of fitness assignment in any evolutionary algorithm [9]. Since that time this strategy has proved its effectiveness and substituted other alternative proposals. Therefore, the chosen methods (NSGA-II, SPEA2, and PICEA-g) are based on the Pareto-dominance idea. However, there are various ways of its implementation [10]: some algorithms use the dominance rank (the amount of individuals by which the candidate-solution is dominated); in others the dominance depth is evaluated (this implies the division of a population into several fronts and determination of the front which an individual belongs to); the dominance count might also be taken into consideration (in other words, the amount of points dominated by a certain individual), and so on. Thus, the algorithms involved in the island model accomplish diverse fitness assignment strategies based on the Pareto-dominance idea (Table 1).

Table 1. Basic features of the MOGA used

MOGA	Fitness Assignment	Diversity Preservation	Elitism
NSGA-II	Pareto-dominance (niching mechanism) and diversity estimation (crowding distance)	Crowding distance	Combination of the previous population and the offspring
PICEA-g	Pareto-dominance (with generating goal vectors)	Nearest neighbour technique	The archive set and combination of the previous population and the offspring
SPEA2	Pareto-dominance (niching mechanism) and density estimation (the distance to the k-th nearest neighbour in the objective space)	Nearest neighbour technique	The archive set

Besides, diversity preservation techniques are incorporated in most of the MOGAs to maintain variety within Pareto Set and Front approximations. There are also several ways to implement these techniques [11]. Kernel methods estimate the density with a Kernel function, which takes the distance to another point as an argument. Nearest neighbour techniques are based on the assessment of the distance between a given point and its k-th nearest neighbour. And histograms present another class of density estimators that use a hypergrid to calculate neighbourhoods. In most cases, these approaches define the distance between points in the objective space.

Moreover, there is the problem of losing effective individuals during the optimization process due to stochastic effects, and to solve this problem the idea of elitism has been suggested. Generally, there are two ways to implement it. The first strategy to cope with the problem is to combine the parent population with the offspring and then to apply a deterministic selection procedure taking into account fitness values of individuals from the mating pool. Another strategy is based on the usage of a secondary population which is called archive to copy there promising solutions at each generation. Actually, one of these techniques might be implemented: in NSGA-II the first strategy is used, whereas in SPEA2 the second one is applied, but they might also be combined (as a case in point, PICEA-g).

In Table 1 a brief description of the MOGAs involved in the cooperative algorithm and their main features are summarized.

In the approach developed we have tried to engage various heuristic concepts to implement fitness assignment strategies, diversity preservation techniques, and the elitism idea. On the one hand, it is supposed to lead to an increase in the algorithm reliability. On the other hand, due to the parallel structure of the island model computational time might be decreased.

The next section includes a description of the test problems which have been used to investigate the effectiveness of the approach proposed.

3   Test Problems

To investigate the effectiveness of the approach proposed in comparison with its components, we have engaged a set of high-dimensional test problems designed by the international scientific community to compare the effectiveness of developed algorithmic schemes (the CEC 2009 competition [12]). There are problems with discrete and continuous, convex and non-convex Pareto Sets and Fronts.

In this study we use a number of these test instances which are unconstrained two- and three-objective optimization problems with real variables.

In the CEC 2009 competition the metric IGD was used to estimate the quality of obtained Pareto Front approximations:

$$IGD(A, P^*) = \frac{\sum_{v \in P^*} d(v, A)}{|P^*|}, \tag{1}$$

where P^* is a set of uniformly distributed points along the Pareto Front (in the objective space), A is an approximate set to the Pareto Front, $d(v, A)$ is the minimum Euclidean distance between v and the points in A. In short, the $IGD(A, P^*)$ value reflects the average distance from P^* to A.

These continuous multi-objective optimization test problems have been proposed in the past 25 years. In the CEC 2009 competition they were gathered to investigate the algorithms developed. Although in this study we do not compare our proposal with the winners of this competition, it is fair to notice that for most of the test problems the effectiveness of the approach developed is higher than the effectiveness of some methods from the list of winners (the list of thirteen best algorithms).

The next section provides a description of the experiments conducted, the results obtained and a brief discussion of them.

4   Experiments and Results

Firstly, conventional algorithmic schemes were applied to solve the problems introduced. All algorithms were provided with the same amount of resources: according to the rules of the CEC 2009 competition, the maximal number of function evaluations was equal to 300 000. The maximal number of solutions in the approximate set produced by each algorithm for computing the IGD metric was 100 and 150 for two-objective and three-objective problems respectively. For all of the test instances IGD values were averaged over 25 runs of each algorithm.

In the experiments conducted the following settings were defined: binary tournament selection, uniform recombination and the mutation probability $p_m = 1/n$, where n is the length of the chromosome. As usual, MOGAs (NSGA-II, SPEA2, and PICEA-g) operated with binary strings and therefore, we used standard binary coding to get real values of variables.

Secondly, a similar experiment was conducted for the developed cooperative multi-objective algorithm. The computational resources (300 000 function evaluations) were distributed to all of the components equally. The migration size was 50 (in total each island got 100 points from two others), and the migration interval was 25 generations. Again all results were averaged over 25 runs.

The main criterion which was used to compare the effectiveness of the algorithm proposed with the performance of its components was the IGD metric. However, we also measured computational time required in each case. The results obtained are presented in Table 2.

The first experiment revealed that there was no one MOGA which demonstrated the highest effectiveness (in the sense of the IGD metric) for all of the test problems. The best results provided with NSGA-II, SPEA2, and PICEA-g separately are highlighted with in bold.

Table 2. Experimental results

Test Func.	NSGA-II		PICEA-g		SPEA2		Cooperative algorithm		Result of t-test
	IGD	Time (sec.)	IGD	Time (sec.)	IGD	Time (sec.)	IGD	Time (sec.)	
UF1	**0.097**	196.060	0.107	42.327	0.010	236.677	0.068	56.566	**Outperforms the best value**
UF2	0.061	181.520	**0.060**	84.538	0.078	262.089	0.056	64.837	Corresponds to the **best** value
UF3	**0.191**	181.150	0.222	36.781	0.326	237.594	0.202	55.952	Corresponds to the **best** value
UF4	**0.055**	182.233	0.0570	75.837	0.083	243.208	0.058	60.271	Corresponds to the **best** value
UF5	**0.426**	181.509	0.498	33.844	0.518	240.198	0.338	56.391	**Outperforms the best value**
UF6	0.335	183.085	0.346	34.997	**0.319**	237.906	0.254	56.008	**Outperforms the best value**
UF7	**0.085**	181.039	0.091	75.556	0.125	245.891	0.084	60.269	**Outperforms the best value**
UF8	0.269	190.269	**0.191**	166.056	0.259	253.813	0.259	87.240	Corresponds to the **second** value
UF9	0.319	191.105	**0.290**	107.157	0.407	406.996	0.314	78.532	Corresponds to the **best** value
UF10	0.626	186.267	**0.421**	118.744	0.534	290.870	0.533	75.119	Corresponds to the **best** value

Then we compared these best IGD values obtained by MOGAs with the results of the cooperative algorithm. A t-test (with the significance level p=0.01) was used to expose the significant difference in the pairs of IGD values. As a result, it turned out that in seven cases there was no difference between the best results provided with MOGAs separately and the IGD values obtained with the cooperation of these MOGAs (in Table 2 'Corresponds to the best value' indicates these cases). Furthermore, the cooperative method outperformed the best MOGA twice (in Table 2 it is labeled 'Outperforms the best value') and only once its effectiveness corresponded to the second (in the sense of the IGD values) MOGA (this case is marked 'Corresponds to the second value'). This implies that our proposal is an effective alternative to the random choice of the appropriate MOGA for the problem considered.

Also the parallel implementation allows us to save computational time: the average number of seconds spent with conventional MOGAs (NSGA-II, SPEA2, and PICEA-g) is 176, whereas the cooperative algorithm requires 65 seconds. On average, it works faster than the fastest MOGA (PICEA-g) and much faster than two others. Certainly, these results depend on different characteristics of the computer used, but it might be roughly assessed that the computational time has been decreased essentially.

5   Conclusion

In this paper, we have proposed the multi-agent heuristic procedure to solve multi-objective optimization problems which does not require additional experiments to expose the most appropriate algorithm for the problem considered. This cooperative technique might be effectively used instead of any of its component. Moreover, the parallel work of island model components allows us to decrease the computational time significantly. For most of the test problems the effectiveness of the developed algorithmic scheme is comparable with (or even better than) the performance of its component which provides the best results separately.

The algorithm developed has already been applied to select informative features from data bases (two criteria were introduced – the Intra- and Inter-class distances). Also it has been successfully used to design neural network models taking into account two criteria (the computational complexity and the accuracy). All these applications will be presented in the next paper.

Thus, it might be concluded that due to advances in the algorithm proposed it might be effectively used in the *Machine Learning* field.

Acknowledgements. Research is performed with the financial support of the Ministry of Education and Science of the Russian Federation within the State Assignment for the Siberian State Aerospace University, project 2.1889.2014/K.

References

1. Khritonenko, D., Semenkin, E.: Application of artificial neural network ensembles for city ecology forecasting using air chemical composition information. In: Proceedings of the International Conference on Environment Engineering and Computer Application (ICEECA2014), Hong Kong, China, (2014) – In press
2. Stanovov, V., Semenkin, E.: Hybrid self-configuring evolutionary algorithm for automated design of fuzzy logic rule base. In: 11th International Conference on Fuzzy Systems and Knowledge Discovery, pp. 317–321 (2014)
3. Semenkina, M., Semenkin, E.: Hybrid self-configuring evolutionary algorithm for automated design of fuzzy classifier. In: Tan, Y., Shi, Y., Coello, C.A. (eds.) ICSI 2014, Part I. LNCS, vol. 8794, pp. 310–317. Springer, Heidelberg (2014)

4. Freitas, A.: Data Mining and Knowledge Discovery with Evolutionary Algorithms. Spinger-Verlag, Berlin (2002)
5. Whitley, D., Rana, S., Heckendorn, R.: Island model genetic algorithms and linearly separable problems. In: Corne, C., Shapiro, J.L. (eds.) Evolutionary Computing. LNCS, vol. 1305, pp. 109–125. Springer, Heidelberg (1997)
6. Deb, K., Pratap, A., Agarwal, S., Meyarivan, T.: A fast and elitist multiobjective genetic algorithm: NSGA-II. IEEE Transactions on Evolutionary Computation **6**(2), 182–197 (2002)
7. Wang, R.: Preference-Inspired Co-evolutionary Algorithms. A thesis submitted in partial fulfillment for the degree of the Doctor of Philosophy, University of Sheffield, p. 231 (2013)
8. Zitzler, E., Laumanns, M., Thiele, L.: SPEA2: Improving the Strength Pareto Evolutionary Algorithm for Multiobjective Optimization. Evolutionary Methods for Design Optimisation and Control with Application to Industrial Problems EUROGEN 2001 **3242**(103), 95–100 (2002)
9. Goldberg, D.: Genetic algorithms in search, optimization, and machine learning. Addison-wesley (1989)
10. Zitzler, E., Laumanns, M., Bleuler, S.: A tutorial on evolutionary multiobjective optimization. In: Gandibleux, X., (ed.) Metaheuristics for Multiobjective Optimisation. Lecture Notes in Economics and Mathematical Systems, vol. 535. Springer (2004)
11. Silverman, B.: Density estimation for statistics and data analysis. Chapman and Hall, London (1986)
12. Zhang, Q., Zhou, A., Zhao, S., Suganthan, P. N., Liu, W., Tiwari, S.: Multi-objective optimization test instances for the CEC 2009 special session and competition. University of Essex and Nanyang Technological University, Tech. Rep. CES-487 (2008)

Multi-Objective Particle Swarm Optimization Algorithm Based on Comprehensive Optimization Strategies

Huan Luo$^{(\boxtimes)}$, Minyou Chen, and Tingjing Ke

State Key Laboratory of Power Transmission Equipment & System Security
and New Technology, Chongqing University, Chongqing, China
yolanda_1989@163.com, minyouchen@cqu.edu.cn, ketingjing@126.com

Abstract. Multi-objective particle swarm optimization algorithm based on comprehensive optimization strategies (MOPSO-COS) is proposed in this paper to deal with the problems of premature convergence and poor diversity. The velocity updating mode is modified by incorporating the information of the global second best particle to promote information flowing among particles. In order to improve the convergence accuracy and diversity, some effective strategies, such as chaotic mutation, external archiving with dynamic grid method, selection strategy based on a temporary population and so on, are introduced into MOPSO-COS. Theoretical analysis of MOPSO-COS is carried out including convergence and time complexity. Performance tests are conducted with ZDT test functions. Simulation results show that MOPSO-COS can improve the convergence accuracy and diversity of Pareto optimal solutions simultaneously, and particles can escape from local optimum point effectively.

Keywords: MOPSO · Comprehensive optimization · The global second best particle · External archiving strategy · Chaotic mutation

1 Introduction

Particle swarm optimization (PSO) algorithm solves complex optimization problems by simulating foraging of birds, fish and other groups. PSO is widely applied because it's simple, easy to realize and has less parameters. The velocity v_i and position x_i of i-th particle in standard PSO are updated respectively according to Eq.1 and Eq.2.

$$v_i(t+1) = \omega v_i(t) + c_1 r_1 \left(p_i - x_i(t) \right) + c_2 r_2 \left(p_{gi} - x_i(t) \right). \tag{1}$$

$$x_i(t+1) = v_i(t+1) + x_i(t). \tag{2}$$

where $x_i = [x_{i1}, x_{i2}, \ldots, x_{id}]$ represents a candidate solution, and d is the total dimensions; t is the current iteration times; p_i, which is called personal best, is the previous best location of i-th particle; p_{gi}, which is called global best, is the location of the particle with best fitness; ω is inertia weight; c_1 and c_2 are acceleration constants which show the contributions of p_i and p_{gi}; r_1 and r_2 are independent random numbers within [0, 1]. According to Eq.1, each particle adjusts its velocity and track according

© Springer International Publishing Switzerland 2015
Y. Tan et al. (Eds.): ICSI-CCI 2015, Part I, LNCS 9140, pp. 479–486, 2015.
DOI: 10.1007/978-3-319-20466-6_50

to the flying experience from itself and the whole group. Therefore they have the capacity to search for better position in the search space.

Now it is common to solve multi-objective optimization problems by PSO algorithm. Multi-objective particle swarm optimization (MOPSO) algorithm inherits the advantages of PSO, but it also has some shortcomings, such as premature, low convergence accuracy and poor diversity. Therefore, MOPSO has been improved at different points in recent years, including population initialization [1], the setting of inertia weight [2] and acceleration constants [3], selection methods for the global best particle [4], modification of the position and velocity updating equation [5], and co-evolution of multi-population [6].

The performance of modified MOPSO is better. But there are still some problems found by simulation and experiments. When the population falls into the area around local optimum point, it is difficult for non-convex and multimodal problems to get rid of it effectively. The diversity of non-dominated solutions needs to be further improved. And the convergence and diversity indices are seriously fluctuant among different runs. To solve these problems, this paper proposes multi-objective particle swarm optimization algorithm based on comprehensive optimization strategies (MOPSO-COS). In MOPSO-COS, velocity updating equation is modified by introducing the global second best particle, and chaotic mutation, external archiving strategy based on dynamic grid method and so on are incorporated. All the strategies work simultaneously. Simulation is carried out with ZDT test functions. Results show that good performance can be obtained.

2   MOPSO Based on Comprehensive Optimization Strategies

Traditional velocity equation only involves personal best and global best. Information from other particles in the population hasn't been utilized effectively. It results in low information sharing rate, poor diversity and slow convergence. To handle these problems, this paper changes velocity equation as Eq. 3.

$$v_i(t+1) = \omega v_i(t) + c_1 r_1 (p_i - x_i(t)) + c_2 r_2 (p_{gi} - x_i(t)) + c_3 r_3 (\sec p_{gi} - x_i(t)) . \tag{3}$$

where $\sec p_{gi}$ is the position of the global second best particle, whose fitness is only worse than p_{gi}'s. c_3 is a coefficient like c_1 and c_2. r_3 is a random number within [0,1]. Therefore, the whole population will move towards the personal best particle, the global best particle and the global second best particle at the same time. Compared with Eq.1, Eq.3 can promote information sharing among particles in theory, enhance information flowing within the population, and avoid the population gathering excessively at the global best point.

To escape from local optimum point effectively and break highly aggregated state, this paper introduces chaotic mutation [7]. If the population has a highly aggregation, even overlap, in the target space, the evolution is marked as stagnation once. When the evolution stagnates K times consecutively, the algorithm relying on current strategies is deemed to fail to escape from local optimum point, and chaotic mutation starts to work. Aggregation index, a, is introduced to quantize the aggregation degree of the population, and it can be expressed as Eq. 4. The closer a is to 1, the more seriously the population gathered and the worse the diversity is.

$$a = \frac{f_{1best} \cdot f_{2best} \cdots f_{Mbest}}{f_{1mean} \cdot f_{2mean} \cdots f_{Mmean}} . \tag{4}$$

where f_{jbest} is the best value of the j-th $(j = 1,2, ..., M)$ objective function, and f_{jmean} is the mean value; M is the total number of objective functions.

To make the Pareto optimal solutions distribute more uniformly in target space, external archiving strategy [8] is adopted to store non-dominated particles gained in each calculation. When external archive overflows, dynamic grid method [9] is employed to maintain archive. In addition, selection strategy based on a temporary population [10] is adopted to choose particles in the next population. In order to enhance the ability of global search and improve convergence speed, random mutation works when the flight speed of the population is less than the threshold value [10].

3   Theoretical Analysis of MOPSO-COS Algorithm

3.1   Convergence Analysis of MOPSO-COS Algorithm

Compared with Eq. 1, Eq.3 has a new part including the global second best particle. How is the convergence of MOPSO-COS? How will the parameters be set? These problems will be discussed below. v_i and x_i are independent on each dimension. For simplicity, the following analysis is based on one-dimensional space and all the random values are ignored.

$$
\begin{aligned}
x(t+2) &= x(t+1)+v(t+2) \\
&= x(t+1)+\omega(x(t+1)-x(t))+c_1(p_i-x(t+1))+c_2(p_{gi}-x(t+1))+c_3(\sec p_{gi}-x(t+1)) .
\end{aligned} \tag{5}
$$

$$x(t+2)+(c-\omega-1)x(t+1)+\omega x(t) = c_1 p_i +c_2 p_{gi} +c_3 \sec p_{gi} . \tag{6}$$

Supposing that $c_1 +c_2 +c_3 = c$, Eq. 6 is available. Ignoring the change of p_i, p_{gi} and $\sec p_{gi}$, Eq. 6 is a non-homogeneous second-order differential equation with constant coefficients. The characteristic equation is expressed as Eq. 7.

$$s^2 +(c-\omega-1)s+\omega=0 . \tag{7}$$

Let $\Delta = (c-\omega-1)^2 -4\omega$, so the solution of Eq.6 is $x(t) = As_1^{t} + Bts_2^{t} +C$, which can be divided into the following three cases: 1) $\Delta=0$, $s_1 = s_2 =-0.5(c-\omega-1)$; 2)$\Delta>0$, $s_1 = -0.5((c-\omega-1)+\sqrt{\Delta})$; $s_2 = -0.5((c-\omega-1)-\sqrt{\Delta})$; 3)$\Delta<0$, $s_1 =-0.5((c-\omega-1)+i\sqrt{-\Delta})$; $s_2 =-0.5((c-\omega-1)-i\sqrt{-\Delta})$. Where A, B and C are uncertain coefficients determined by x (0) and v (0).

If MOPSO-COS converges, when $t \to \infty$, $x(t)$ is finite, namely, $|s_1|<1$ and $|s_2|<1$. Considering three cases comprehensively, feasible domain of parameters in MOPSO-COS algorithm can be described as: $c>0$、$-1<\omega<1$ and $2\omega-c+2>0$.

The parameter values have significant influence on the performance of MOPSO-COS. The convergence analysis provides reference for the setting of some parameters.

3.2   Time Complexity Analysis of MOPSO-COS Algorithm

Time complexity is an important index that weighs the performance of modified algorithm. The strategies used most frequently and holding higher degree of time complexity are external archiving strategy and selection strategy based on a temporary population. According to the research in [11], time complexity of selection strategy based on a temporary population can be expressed as: $O(\sum_{i=N}^{2N} Mi \log(i))$, where N is the population size. Set archive size to Ne. There are some assumptions for the worst case: 1) the current archive is full; 2) all the N particles in the current population are non-dominated, further when they're added into archive, there are no new dominated particles nor overlap in the target space. Hence dynamic grid method needs to remove N particles. By calculation, time complexity for removing the first particle is $O((N + Ne)^2)$, and that for removing the second particle is $O((N + Ne - 1)^2)$ ...and by this analogy, that for removing the N-th particle is $O((Ne+1)^2)$. So the total time complexity for archive updating with the worst case is $O\left[\sum_{i=0}^{N-1} (N + Ne - i)^2 \right]$.

According to the relationship of time complexity, the total time complexity for MOPSO-COS is $O\left[\sum_{i=0}^{N-1} (N + Ne - i)^2 \right]$. Therefore, MOPSO-COS increases operation time. However, the efficiency of MOPSO-COS is still high. When N is bigger, fast convergence is available .

4   Performance Tests of MOPSO-COS Algorithm

Performance tests are based on ZDT test functions. Generation distance (GD) [4] and diversity index (Δ) [12] are used to evaluate the convergence accuracy and distribution properties of Pareto solutions. The smaller GD is, the higher convergence accuracy is. The smaller Δ is, the more evenly Pareto optimal solutions distribute.

Comparisons will be made between MOPSO-COS and some other similar typical algorithms, such as SPEA2, NSGA2 and MOPSO [13], to evaluate the performance of MOPSO-COS more objectively and comprehensively. For all the algorithms, set N=100, Ne=100, G_{max}=250 (maximum iteration times). Only for MOPSO-COS, set $\omega = \omega_1 + (1-\omega_1) \times r$, ω_1 =0.5, c_1=0.7(2.5-2t/G_{max}), c_2=0.5+2t/G_{max}, c_3=0.3(2.5-2t/G_{max}), T=3, m=5, h=20, where r is a random number. The value of ω in MOPSO is set the same as MOPSO-COS, while c_1=2.5-2t/G_{max}, c_2=0.5+2t/G_{max}. The probabilities of crossover and mutation in NSGA2 are respectively set p_c=0.9 and p_m=0.1.

Table 1. Comparison of Convergence Index-GD

		SPEA2	NSGA2	MOPSO	MOPSO-COS
ZDT1	meanG	6.2205e-	2.9997e-	1.6537e-4	2.1638e-4
	varGD	1.3218e-	3.1154e-	2.1596e-	2.7426e-9
ZDT2	meanG	2.3288e-	2.2078e-	3.1045e-2	1.0597e-4
	varGD	2.1089e-	7.6374e-	8.8444e-3	3.0936e-11
ZDT3	meanG	2.6240e-	6.2196e-	1.5469e-1	6.1302e-4
	varGD	7.0359e-	3.3547e-	4.1214e-2	2.8784e-9
ZDT4	meanG	8.8982e-	0.12657	3.1461e-4	6.5150e-4
	varGD	4.5395e-	1.0551e-	8.2928e-	7.7298e-10
ZDT6	meanG	3.3990e-	0.57436	2.7691e-2	7.2604e-3
	varGD	4.7913e-	6.4193e-	2.2267e-3	2.8939e-4

Table 2. Comparison of Diversity Index- Δ

		SPEA2	NSGA2	MOPSO	MOPSO-COS
ZDT1	mean Δ	0.69577	0.38010	0.52879	0.43395
	var Δ	1.2828e-2	1.0412e-3	6.9777e-4	1.3164e-3
ZDT2	mean Δ	0.81973	0.48544	0.90792	0.44556
	var Δ	1.9481e-2	8.2685e-3	1.3987e-2	1.3223e-3
ZDT3	mean Δ	0.91386	0.75506	0.72845	0.68625
	var Δ	1.3084e-2	5.9355e-2	2.9714e-2	1.4590e-3
ZDT4	mean Δ	0.80941	0.69995	0.49158	0.37226
	var Δ	3.0119e-2	2.8908e-2	2.7190e-3	6.3067e-4
ZDT6	mean Δ	0.77787	1.0299	1.1245	0.85959
	var Δ	2.3101e-3	2.7091e-2	8.1845e-2	3.7125e-3

Optimization for each algorithm is done 50 times independently. The results are shown in Table 1 and 2, where *mean* represents the mean value and *var* represents the variance. Fig. 1 shows the difference between optimal front obtained from MOPSO-COS and the true Pareto front. In fact, it is difficult for many improved MOPSO algorithms to escape from local optimal point effectively, particularly for non-convex function ZDT2. Besides, many MOPSO algorithms are difficult to converge to the true Pareto front for ZDT6 because ZDT6 is multimodal function. It is easy to find in Table 1 that convergence of MOPSO-COS is obvious improved for ZDT2 and ZDT6, and variance of GD is smaller than other algorithms. According to Table 2, diversity of MOPSO-COS is best, except that it is worse only than NSGA2 for ZDT1 and worse than SPEA2 for ZDT6. Combing Table 1, Table 2 and Fig.1, it can be drawn that MOPSO-COS is able to converge to the true Pareto front with high accuracy for ZDT test functions. The optimal front of MOPSO-COS distributes uniformly and diversity is good. Although its performance is not the best at certain test function, order of magnitude is similar. Therefore the improved MOPSO-COS is effective.

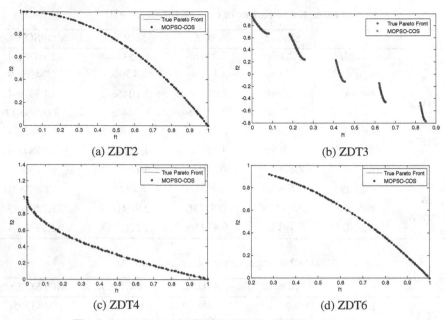

(a) ZDT2

(b) ZDT3

(c) ZDT4

(d) ZDT6

Fig. 1. Pareto Front of MOPSO-COS for ZDT Test Functions

Fig. 2 shows the difference between MOPSO-COS and MOPSO for ZDT2 when the population gets rid of local optimal point. Results for 20 times selected randomly are adopted for comparison. In Fig. 2, the number '0' indicates that final Pareto optimal solutions converge to the true front and distribute uniformly; '1' indicates that the population fails to get out of the local optimal point, and particles overlap at the local extremum finally; '2' indicates that the population aggregates seriously, so the diversity is poor. Through experiments and observation, it is obvious that MOPSO can hardly escape from local optimal point for ZDT2. Even if MOPSO can, aggregation degree of the population is high at last, and solutions on the Pareto front distribute extremely unevenly. On the contrary, MOPSO-COS is able to avoid seriously aggregating and it has a larger probability to escape from the local optimal point.

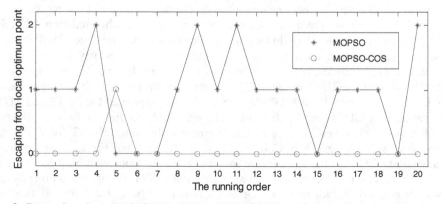

Fig. 2. Comparison between MOPSO-COS and MOPSO When Escaping from Local Optimum

5   Conclusions

This paper presents a multi-objective particle swarm optimization algorithm based on comprehensive optimization strategies. The concept of the global second best particle is proposed and further velocity updating equation is modified, which has improved the utilization of population information and convergence speed. Premature convergence and serious aggregation are avoided by chaotic mutation. External archive strategy, which maintains the archive by dynamic gird method, has increased the diversity of Pareto optimal solutions. However, the analysis of time complexity shows that MOPSO-COS requires a little more time to search for the optimal solutions. Results of simulation based on ZDT test functions show that, compared with several other algorithms, MOPSO-COS is able to improve the convergence accuracy and diversity and escape from local optimum point effectively.

References

1. Liu, B., Zhang, W., Li, G., Nie, R.: Improved Multi-Objective Particle Swarm Optimization Algorithm. Journal of Beijing University of Aeronautics and Astronautics **39**(4), 458–462 (2013). (in Chinese)
2. Chen, M., Wu, C., and Fleming, P.J.: An Evolutionary particle swarm algorithm for multi-objective optimization. In: The 7th World Congress on Intelligent Control and Automation, pp. 3269–3274. IEEE Press, Chongqing (2008)
3. Ratnaweera, A., Halgamuge, S., Watson, H.C.: Self-Organizing Hierarchical Particle Swarm Optimizer with Time-Varying Acceleration Coefficients. IEEE Transactions on Evolutionary Computation **8**(3), 240–255 (2004)
4. Pang, S., Zou, H., Yang, W., et al.: An Adaptive Mutated Multi-objective Particle Swarm Optimization with an Entropy-based Density Assessment Scheme. Information & Computational Science **4**, 1065–1074 (2013)
5. Sun, C., Zeng, J., Chu, S., et al.: Solving constrained optimization problems by an improved particle swarm optimization. In: 2011 Second International Conference on Innovations in Bio-inspired Computing and Applications (IBICA), pp. 124–128. IEEE Press, Shen Zhen (2011)
6. Hu, C., Yao, H., Yan, X.: Multiple Particle Swarms Co-evolutional Algorithm for Dynamic Multi-objective Optimization Problems and Its Application. Journal of computer research and development **50**(6), 1313–1323 (2013). (in Chinese)
7. Zhou, X., Chen, C., Yang, F., Chen, M.: Optimal Coordinated HVDC Modulation Based on Adaptive Chaos Particle Swarm Optimization Algorithm in Multi-Infeed HVDC Transmission System. Transactions of China Electrotechnical Society **24**(50), 193–201 (2009). (in Chinese)
8. Zheng, X., Liu, H.: Progress of Research on Multi Objective Evolutionary Algorithms. Computer Science **34**(7), 187–191 (2007). (in Chinese)
9. Wu, X., Xu, Q.: Optimization Model of Multi-Objective Distribution Based on Adaptive Grid Particle Swarm Optimization Algorithm. Journal of Highway and Transportation Research and Development **27**(5), 132–136 (2010). (in Chinese)

10. Luo, H., Chen, M., Cheng, T.: Adaptive Time-Intervalled Reactive Power Optimization for Distribution Network Containing Wind Power Generation. Power System Technology **38**(8), 2207–2212 (2014). (in Chinese)
11. Chen, M., Cheng, S.: Multi-Objective Particle Swarm Optimization Algorithm Based on Random Black Hole Mechanism and Step-by-Step Elimination Strategy. Control and Decision **28**(11), 1729–1734 (2013). (in Chinese)
12. Li, Y.: Model-Based Multi-objective Constellation Optimization Algorithm Design. China University of Geosciences, May 2010. (in Chinese)
13. Chen, M., Zhang, C., Luo, C.: Adaptive Evolution Multi-Objective Particle Swarm Optimization Algorithm. Control and Decision **24**(12), 1851–1855 (2009). (in Chinese)

Co-Operation of Biology-Related Algorithms
for Constrained Multiobjective Optimization

Shakhnaz Akhmedova[✉] and Eugene Semenkin

Siberian State Aerospace University, Krasnoyarsk, Russia
shahnaz@inbox.ru, eugenesemenkin@yandex.ru

Abstract. A modification of the self-tuning meta-heuristic, called Co-Operation of Biology Related Algorithms for multiobjective optimization problems (COBRA-m) is introduced. Its basic idea consists of a cooperative work of five well-known bionic algorithms such as Particle Swarm Optimization, Wolf Pack Search, the Firefly Algorithm, the Cuckoo Search Algorithm and the Bat Algorithm with the use of Pareto optimality theory. The performance of the mentioned algorithms as well as COBRA-m on the set of benchmark functions is reported. It was established that the proposed approach COBRA-m has performed either comparably or better than its component bionic algorithms. Then the method COBRA-m is modified for solving constrained multiobjective optimization problems. The proposed algorithm is first validated against a subset of test functions, and then applied to known multiobjective design problems such as welded beam design and disc brake design. Simulation results suggest that the proposed algorithm works effectively.

Keywords: Biologically inspired algorithms · Cooperation · Multiobjective optimization · Design problems

1 Introduction

There are numerous industrial and engineering problems whose solving requires the simultaneous optimizing of several objective functions which are in conflict with each other. This kind of problem is known as a multiobjective optimization problem.

Although there are many approaches and methods developed in multiobjective optimization generally [1] and in evolutionary multiobjective optimization [2] in particular, the scientific community continues generating new ideas and algorithms and this area of research is permanently expanding. It is usually in the area of the development of nature-inspired optimization algorithms that every new algorithm is suggested first for conventional one-objective unconstrained optimization problems and later, if it demonstrates high performance and reliability, further modifications for constrained and multiobjective optimization are fulfilled for this algorithm.

In [3], the new bionic self-tuning meta-heuristic called Co-Operation of Biology Related Algorithms (COBRA), based on the collective work of five biology-inspired optimization methods such as particle swarm optimization (PSO), wolf pack search (WPS), the firefly algorithm (FFA), the cuckoo search algorithm (CSA) and the bat

© Springer International Publishing Switzerland 2015
Y. Tan et al. (Eds.): ICSI-CCI 2015, Part I, LNCS 9140, pp. 487–494, 2015.
DOI: 10.1007/978-3-319-20466-6_51

algorithm (BA), for real-valued one-objective unconstrained optimization has been introduced and investigated. Numerical experiments from the CEC 2013 continuous optimization competition benchmark showed that this metaheuristic works better than its component algorithms and can be used instead of them [3].

Having in mind that all component algorithms used in the COBRA metaheuristic have their own modifications for multiobjective optimization problems [4-7], it was natural to assume that a cooperation of these algorithms could give again an improvement in effectiveness. To evaluate the effectiveness of the proposed approach the algorithm COBRA-m (for unconstrained multiobjective optimization) was empirically compared with its above-listed component algorithms on a set of benchmark functions which are test problems for multiobjective unconstrained optimization. Then the heuristic COBRA-cm (for constrained multiobjective optimization) was validated against a subset of corresponding test functions. And finally the algorithms were applied to solving known multiobjective design problems such as welded beam design and disc brake design. The results obtained showed that the COBRA approach to multiobjective optimization is also effective and reliable, i.e., it finds a good approximation of the Pareto front in all runs.

The remainder of this paper is organized as follows. Section 2 explains the original COBRA algorithm. Further, in Section 3, the proposed meta-heuristics COBRA-m and COBRA-cm for multiobjective unconstrained and constrained optimization problems, respectively, are introduced. In Section 4, the results of testing are discussed and in Section 5 the results of solving the design optimization problems are reported. Finally, the conclusion follows in Section 6.

2   Co-operation of Biology Related Algorithms

A new method for solving one-criterion unconstrained real-parameter optimization problems based on the cooperation of five nature-inspired algorithms and called Co-Operation of Biology Related Algorithms (COBRA) was introduced in [3]. The basic idea of this approach consists in generating five populations (one population for each above mentioned algorithm) which are then executed in parallel cooperating with each other. The choice of these five algorithms was explained through their likeness in ideology and a similarity in behaviour that brings troubles for end users when choice of an appropriate tool for solving the problem in hand.

Proposed in [3] algorithm is a self-tuning meta-heuristic so there is no need to choose the population size for algorithms. The number of individuals in the population of each algorithm can increase or decrease depending on whether the fitness value is improving or not. If the fitness value was not improved during a given number of generations, then the size of all populations increases by adding randomly generated individuals. And vice versa, if the fitness value was constantly improved, then the size of all populations decreases by removing the worst population members. Additionally, each population can "grow" by accepting individuals removed from another population. A population "grows" only if its average fitness is better than the average fitness of all other populations. Besides, all populations communicate with each other: they exchange individuals in such a way that a part of the worst individuals of each population is replaced by the best individuals of other populations.

The performance of the proposed algorithm was evaluated on the set of benchmark problems from the CEC'2013 competition [3]. This set of 28 unconstrained real-parameter optimization problems was given in [8]. Experiments showed that COBRA works successfully and is reliable on this benchmark. Results also showed that COBRA outperforms its component algorithms when the dimension grows and more complicated problems are solved. It means that COBRA can be used instead of any component algorithm.

3   Multiobjective Modifications COBRA-m and COBRA-cm

At the beginning, we will assume that the following multiobjective unconstrained optimization problem should be solved:

$$f(x) = (f_1(x), f_2(x), ..., f_K(x)) \rightarrow \min \qquad (1)$$

So, there are K objectives which should be minimized. And later we will assume that there are also M constraint functions $g_i(x) \leq 0$, $i = 1, ..., M$.

For development of the multiobjective version of COBRA, its component algorithms have to be also modified. Therefore, all these techniques were extended to produce a Pareto optimal front directly: PSO and WPS by using the σ-method [4] and the FFA, CSA and BA as suggested in [5-7] correspondingly. Thus, for each component algorithm an external archive S_i ($i = 1, ..., 5$) of nondominated solutions was generated and a general external archive S was created. Solutions in all archives S_i were compared and solutions which were nondominated among all of them were placed in the archive S.

The development of the multiobjective modification of optimization tool COBRA requires changes in the procedure of selecting the winning algorithm. That is why on the next stage of the COBRA-m execution K weight coefficients whose sum is equal to 1 are initialized randomly. Then all objectives are combined into a single objective (weighted sum of K objectives). We will call this single objective "fitness" on the current stage. For each component algorithm its average fitness value is calculated; the one with the best fitness value is the "winner-algorithm".

A migration operator should also be modified. Again, K weight coefficients whose sum is equal to 1 are initialized randomly; and a single objective as a weighted sum of K objectives is generated. For each component algorithm, its individuals are sorted according to this single objective. Finally, the worst solutions of one component algorithm are replaced by the best solutions of others; and archive S is updated.

The next step in the approach development and main point of this study is about the development and investigation of COBRA-cm, i.e., the modification of COBRA-m that can be used for solving constrained multiobjective optimization problems. For this purpose a well-known constraint handling technique, namely Deb's rule [9], is used.

4   Numerical Results

4.1   Comparison Study for Algorithm COBRA-m

There are various test functions for the multiobjective optimization, but to validate the proposed algorithm COBRA-m, a subset of these functions with convex, non-convex and discontinuous Pareto fronts was selected [11]. To be more specific, in this study the following four functions were used for the testing of COBRA-m: ZDT1 function with a convex Pareto front, ZDT2 function with a non-convex Pareto front, ZDT3 function with a discontinuous Pareto front and Schaffer's Min-Min (SCH) test function with convex Pareto front.

The dimension of the first three optimization problems was equal to 30. In papers [5-7] the results for multiobjective FFA, CSA and BA algorithms are given where the number of individuals for each algorithm was equal to 50; and the number of iterations was equal to 1000 for BA and 2000 for FFA and CSA. So, the maximum number of function evaluations for COBRA-m was established to be equal to 50000.

After generating Pareto points by COBRA-m, the corresponding Pareto front was compared with the true front. We define the distance between the estimated Pareto front PF^e and its corresponding true front PF^t as follows:

$$E = \| PF^e - PF^t \|^2 \tag{2}$$

The results for all the problems are summarized in Table 1; the results for CSA and FFA were taken from [5, 6], and the results for all test problems except SCH obtained by method BA were taken from [7].

Table 1. Summary of results for unconstrained problems

Func	PSO	WPS	FFA	CSA	BA	COBRA-m
ZDT1	7.2E-05	4.8E-06	2.3E-06	1.2E-06	3.7E-04	1.1E-06
ZDT2	6.8E-07	3.2E-07	8.9E-06	7.3E-06	2.4E-04	2.6E-07
ZDT3	2.9E-05	1.9E-05	3.7E-05	2.2E-05	5.2E-05	1.4E-05
SCH	2.5E-06	1.7E-06	5.5E-09	4.9E-09	5.1E-07	2.7E-09

As one can see, the simulations for these test problems show that the proposed approach COBRA-m is an efficient algorithm for solving multiobjective optimization problems. It can deal with highly non-linear problems with diverse Pareto optimal sets. Also COBRA-m outperforms its component algorithms. It means that COBRA-m could be recommended for use instead of all of its components.

4.2   Comparison Study for Algorithm COBRA-cm

The subset of test multiobjective optimization problems taken from the CEC 2009 special session on Performance Assessment of Constrained / Bound Constrained Multiobjective Optimization Algorithms [10] was used to validate the proposed algorithm COBRA-cm. All test problems were treated as black-box problems. The results obtained were used for calculating the inverted generational distance (IGD) values for different problems which can be described as follows. Let P^* be a set of uniformly distributed points along the Pareto front in the objective space (for all test problems from mentioned competition it was given). Let A be an approximate set to the Pareto front. The average distance from P^* to A is defined as:

$$IGD(A, P^*) = \sum d(v, A), \qquad (3)$$

where v is a point from P^*, $d(v, A)$ is the minimum Euclidean distance between v and all points in A. For each constraint $g_i(x) \geq 0$ all the solutions in the approximate set for computing the IGD should satisfy $g_i(x) \geq -10^{-6}$.

The maximum number of solutions in the approximate set produced by COBRA-cm for computing the IGD was equal to 100. All problems were two-objective optimization problems. The maximum number of function evaluations was equal to 300000. The problems were solved 30 times. The results achieved and also the results obtained by the other methods which participated in the mentioned special session are presented in Table 2 and Table 3.

Table 2. Summary of results for constrained problems

	CF1		CF2		CF3	
1	LuiLiAlgorithm	0.00085	DMOEADD	0.0021	DMOEADD	0.05630
2	NSGAIILS	0.00692	LuiLiAlg	0.0042	MTS	0.10446
3	MOEDGM	0.0108	MOEDGM	0.008	GDE3	0.12506
4	DMOEADD	0.01131	NSGAIILS	0.01183	LuiLiAlg	0.18290
5	COBRA-cm	0.01775	GDE3	0.01597	NSGAIILS	0.02399
6	MTS	0.01918	COBRA-cm	0.01949	COBRA-cm	0.24441
7	GDE3	0.0294	MTS	0.02677	MOEDGM	0.5131
8	DECMOSA-SQP	0.10773	DECMOSA-SQP	0.0946	DECMOSA-SQP	10^6

Table 3. Summary of results for constrained problems

	CF4		CF5		CF6	
1	DMOEADD	0.00699	DMOEADD	0.01577	LiuLiAlg	0.01395
2	GDE3	0.00799	MTS	0.02077	DMOEADD	0.01502
3	MTS	0.01109	GDE3	0.06799	MTS	0.01616
4	LuiLiAlg	0.01423	COBRA-cm	0.07934	NSGAIILS	0.02013
5	COBRA-cm	0.01542	LiuLiAlg	0.10973	COBRA-cm	0.03125
6	NSGAIILS	0.01576	NSGAIILS	0.1842	GDE3	0.06199
7	MOEDGM	0.0707	DECMOSA-SQP	0.41275	DECMOSA-SQP	0.14782
8	DECMOSA-SQP	0.15265	MOEADGM	0.5446	MOEADGM	0.2071

In Table 2 and Table 3 results were ranked from the best to the worst. In those tables "CF" means constrained multiobjective optimization problems.

The results obtained show that the proposed optimization tool COBRA-cm is efficient enough taking 5th place among 8 winners of competition. Its usefulness is also established as COBRA-cm outperforms its component-algorithm on this test set.

5   Design Optimization Problems

There are various design optimization problems, namely the design of the structure of some kind of objects, which have applications in engineering. In consequence of this fact, nowadays numerous studies with detailed description of the solving of some real-world design problems can be found in literature, for example [12] or [13]. For this study two well-known design optimization problems were chosen: the welded beam design problem [14] and the disc brake design problem [15]. This choice was conditioned by the circumstance that the mentioned problems were solved by other researchers multiple times with a variety of tools. Hence there is a variety of results obtained by alternative optimization tools that can be used for the comparison.

5.1   Welded Beam Design Problem

The multiobjective design problem of a welded beam is a well-known benchmark [14] which has four design variables: the width x_1 and length x_2 of the welded area, the depth x_3 and thickness x_4 of the main beam. The objective is to minimize both the overall fabrication cost and the end deflection. The welded beam structure consists of a beam and the weld required to hold it to the member. Constraints for this problem are related to the shear stress, bending stress in the beam and buckling load on the bar. The detailed formulation of the problem can be found in [14].

The welded beam design problem was solved by using the proposed approach COBRA-cm. The approximate Pareto front was generated by the 50 nondominated solutions. The maximum number of function evaluations was equal to 50000.

Estimated Pareto fronts were consistent with the results obtained by alternative methods [5-7], [14]. Besides, the decreasing of the maximum number of evaluations changed results unessentially.

5.2   Disc Brake Design Problem

The design of a multiple disc brake is another well-known benchmark for multiobjective optimization [15] which was solved by the proposed algorithm COBRA-cm. The objectives are to minimize the overall mass and the braking time by choosing optimal design variables. There are four design variables for this problem: the inner radius and outer radius of the discs, the engaging force and the number of friction surfaces. This is under the design constraints such as the torque, pressure, temperature and length of the brake. The detailed formulation of problem can be found in [15].

The approximate Pareto front for the disc brake design problem was generated by the 50 nondominated solutions. Again the maximum number of function evaluations was equal to 50000. The estimated Pareto fronts were consistent with the results obtained by alternative methods [11-12], [20]. Besides, as for the welded beam design problem, the decreasing of the maximum number of evaluations changed the obtained results unessentially.

6   Conclusions

In this paper a new algorithm for solving multiobjective unconstrained problems, COBRA-m, based on the recently developed method COBRA was described and investigated. The proposed approach COBRA-m has been tested against a subset of well-known test functions and compared with its component algorithms. Thus, the usefulness and workability of the optimization tool COBRA-m were established. Besides, comparison showed that the method COBRA-m has performed either comparably or better than its component bionic algorithms.

Then the proposed approach COBRA-m was modified for solving multiobjective constrained optimization problems. This algorithm (COBRA-cm) has also been tested. A subset of test multiobjective optimization problems taken from CEC 2009 special session was used to validate the algorithm. Eventually, the heuristic COBRA-cm demonstrated competitive results on that subset of test problems.

Finally, two design optimization problems were solved by the COBRA-cm. The simulations for these benchmarks suggest that the COBRA-cm is an efficient algorithm for solving real-world multiobjective constrained optimization problems. It can deal with highly nonlinear problems with complex constraints and diverse Pareto optimal sets.

It should be mentioned that on this stage of research we do not try to develop the best algorithm for the constrained multiobjective optimization. It means that the intention of this study is the development of the optimization tool that could be used instead of five similar algorithms in order to help an end user in choosing an appropriate optimization technique without need to become an expert in swarm intelligence computations. Further studies can focus on parametric studies of component algorithms.

Acknowledgement. Research is fulfilled within governmental assignment of the Ministry of Education and Science of the Russian Federation for the Siberian State Aerospace University, project 140/14.

References

1. Collette, Y., Siarry, P.: Multiobjective Optimization, p. 293. Springer-Verlag, Berlin Heidelberg (2004)
2. Coello Coello, C.A., Van Veldhuizen, D.A., Lamont, G.B.: Evolutionary Algorithms for Solving Multi-Objective Problems. Kluwer Academic Publishers, New York (2002)

3. Akhmedova, Sh., Semenkin, E.: Co-operation of biology related algorithms. In: IEEE Congress on Evolutionary Computation (CEC 2013), pp. 2207–2214. Cancún (México) (2013)
4. Mostaghim, S., Teich, J.: Strategies for finding good local guides in multi-objective particle swarm optimization (MOPSO). In: IEEE Swarm Intelligence Symposium, pp. 26–33. IEEE Service Center (2003)
5. Yang, X.S.: Multi-objective firefly algorithm for continuous optimization. Engineering with Computers **29**(2), 175–184 (2013)
6. Yang, X.S., Deb, S.: Multi-objective cuckoo search for design optimization. Computers & Operations Research **40**, 1616–1624 (2011)
7. Yang, X.S.: Bat Algorithm for Multiobjective Optimization. Bio-Inspired Computation **3**(5), 267–274 (2012)
8. Liang, J.J., Qu, B. Y., Suganthan P.N., Hernandez-Diaz, A.G.: Problem Definitions and Evaluation Criteria for the CEC 2013 Special Session on Real-Parameter Optimization. Technical Report, Computational Intelligence Laboratory, Zhengzhou University, Zhengzhou China, and Technical Report, Nanyang Technological University, Singapore (2012)
9. Deb, K.: An efficient constraint handling method for genetic algorithms. Computer methods in applied mechanics and engineering **186**(2–4), 311–338 (2000)
10. Zhang, Q., Zhou, A., Zhao, S., Suganthan, P.N., Lui, W., Tiwari, S.: Multiobjective optimization Test Instances for the CEC 2009 Special Session and Competition. Technical Report CES, University of Essex, UK, Nanyang Technological University, Singapore, Clemson University, USA (2009)
11. Zitzler, E., Deb, K., Thiele, L.: Comparison of multiobjective evolutionary algorithms: empirical results. Evolutionary Computation **8**(2), 173–195 (2000)
12. Gandomi, A.H., Yang, X.S.: Benchmark problems in structural engineering. In: Koziel, S., Yang, X.S. (eds.) Computational optimization methods and algorithms. SCI, vol. 356, pp. 256–281. Springer, Heidelberg (2010)
13. Ray, T., Liew, K.M.: A swarm methaphor for multiobjective design optimization. Engineering Optimization **34**(2), 141–153 (2002)
14. Gong, W.Y., Cai, Z.H., Zhu, L.: An effective multiobjective differential evolution algorithm for engineering design. Structural Multidisciplinary Optimization **38**, 137–157 (2009)
15. Cagnina, L.C., Esquivel, S.C., Coello Coello, C.A.: Solving engineering optimization problems with the simple constrained Particle Swarm Optimizer. Informatica **32**, 319–326 (2008)

The Initialization of Evolutionary Multi-objective Optimization Algorithms

Mohammad Hamdan[1,2(✉)] and Osamah Qudah[1]

[1] Department of Computer Science, Faculty of IT, Yarmouk University, Irbid, Jordan
m.hamdan@hw.ac.uk, hamdan@yu.edu.jo, osamah_cs@yahoo.com
[2] MACS, Heriot-Watt University DUBAI, Dubai, UAE

Abstract. Evolutionary algorithms are the most widely used meta-heuristics for solving multi objective optimization problems, and since all of these algorithms are population based, such as NSGAII, there are a set of factors that affect the final outcomes of these algorithms such as selection criteria, crossover, mutation and fitness evaluation. Unfortunately, little research sheds light at how to generate the initial population. The common method is to generate the initial population randomly. In this work, a set of initialization methods were examined such as, Latin hypercube sampling (LHS), Quasi-Random sampling and stratified sampling. Nonetheless, we also propose a modified version of Latin Hypercube sampling method called (Quasi_LHS) that uses Quasi random numbers as a backbone in its body. Furthermore, we propose a modified version of Stratified sampling method that uses Quasi-Random numbers to represent the intervals. For our research, a set of well known multi objective optimization problems were used in order to evaluate our initial population strategies using NSGAII algorithm. The results show that the proposed initialization methods (Quasi_LHS) and Quasi-based Stratified improved to some extent the quality of final results of the experiments.

Keywords: Evolutionary algorithms · Initial population · Random numbers · Quasi random numbers · Latin hypercube sampling · Stratified sampling · Quality indicators · Pareto set · NSGAII algorithm

1 Introduction

Multi objective optimization as mentioned in [7] a complex task, since there are a set of objectives that have to be done simultaneously as a result of the optimization process. Moreover, the difficulty will rise when there is a conflict between those objectives in nature.

Based on [10], this type of optimization problems cannot be solved in a reasonable time by using traditional and exact methods such as linear programming and gradient search. Therefore, Evolutionary Algorithms (EAs) are an aspect to be considered when solving this type of problems in a practical manner in terms of solution quality and computation time. For a given optimization problem, the EAs starting from a set of possible solutions called initial population or individuals generated randomly, and the process continues by selecting a subset from those individuals according to the

© Springer International Publishing Switzerland 2015
Y. Tan et al. (Eds.): ICSI-CCI 2015, Part I, LNCS 9140, pp. 495–504, 2015.
DOI: 10.1007/978-3-319-20466-6_52

selection criteria. Then, the crossover and mutation are performed according to a specific criteria and probability as in the case of genetic algorithms (GAs). Finally the function evaluation and fitness assistance take place in order to produce the fittest individuals for the next generation.

The domain of decision variables that form the initial population may be continuous or discrete based on the nature of the problem being formulated. In this research all problems which are undertaken have continuous domain decision variables. Based on [5], the initial values for this kind of variables will be generated randomly in order to create the initial population. For example, if the domain of variable xi $\in$ [0,1] for all x $\in$ X then we can generate a set of random values between lower bound (0) and upper bound (1) in order to represent the value of xi at any given point, i.e. $0 \leq x_i \leq 1$ at every time during the evolution process. However, in the large scale problems the misuse of randomness may worsen the final solution of the problem undertaken in term of quality and evolution speed, i.e. the result of (25000) function evaluation for bad initial population may be achieved using (10000) function evaluation of a good initial population. More precisely, as mentioned in [1], if the initial population to evolutionary algorithms is good, then the possibility of finding a good solution will be increased. Moreover, based on [5] it is a hard problem to find a good initial population for a given problem because in the most cases of optimization we do not have a priori knowledge about the location and the number of local optima in the fitness space.

In the case of multi objective optimization, the overall process is more complex than it appears in single objective family, since the goal of any working evolutionary algorithm is to find the set of non dominated optimal solutions that satisfy the problem objectives. Based on [8, 9], those non dominated solutions are called Pareto set. They are somehow affected by the supplied possible solutions, i.e. population. Furthermore, the complexity of the problem is increased when the number of objectives is increased. Therefore, it is no longer feasible to let the current working algorithm spend more time in the complex function evaluation.

For this reason, in this research we will look at the initial population in order to increase the representativeness of the initial population hoping to increase the evolution speed.   Kalyanmoy et al. in [8], outline the main challenge of the optimization algorithms, such that, the Pareto-optimal solutions cannot be said to be better than the other in the absence of further information. This leads us to let the optimization algorithm find Pareto-optimal solutions as large as possible i.e. diversity must be increased, and based on the results in [5]; the diversity of the current solutions is somehow related to the diversity of the initial population. This was achieved by letting the generated decision variables cover the feasible regions in the problem domain as large as possible while preserving the randomness nature of the initial population, and obviously, this leads us to the fact that not only the representativeness in decision variables is required, but also the randomness nature must be preserved in order to preserve the stochastic nature of evolutionary algorithms because of being the main advantage of this type of optimization algorithms.

The rest of the paper is organized as follows: Section 2 presents literature review. The initialization methods are described in Section 3. Section 4 outlines the proposed initialization method. The experimental environment is outlined in Section 5. In Sections 6 and 7 we show the results and how to obtain them. Finally, we conclude in Section 8.

2   Related Work

Up to our knowledge, little research has been done in the field of initial population. Pedro et al. in [1], focused on the population size in order to understand the trade-offs of population size issue such that, the smaller population size could guide the algorithm to poor solutions. On the other hand, large population size could guide the algorithm to a poor performance and makes the algorithm spend more time in computation i.e. slow evolution. But the issue of how these populations are generated was completely ignored.

However, in Mckay and Beckman [2], a comparison study for single objective problems was made between three methods for generating input variables for a specific complex computer code, such as, simple random sampling, stratified sampling and Latin hypercube sampling (LHS) and the results shown LHS method improves upon simple random and stratified sampling methods when it is adopted, and it appears to be a good method to be used for selecting values of input variables. In Hekki et al. [5], after estimating the initial population of GAs for a single objective family of optimization functions, four methods were adopted in order to generate the initial population such as, pseudo random numbers, quasi random sequence generators and another two statistical methods such as SSI (simple sequential inhibition process) and Nonaligned Systematic sampling (the details of these methods are discussed in [5]).

Unfortunately, there was no explicit conclusion could be made of what is the best method to be considered because of the presence of trade-offs between them in term of uniform coverage, genetic diversity and the evolution lifetime before the premature converge is happened, except the quasi random method shown a significant improvement in many test cases.

3   Initialization Methods

In this Section, we will discuss three initialization methods related to the standard random approach that are commonly used by previous researches, and they were considered to be good methods for generating initial population and performed well in the experiments.

3.1   Stratified Sampling Method

Based on [2, 6], it is a sampling method used in statistical applications in order to generate a set of samples from a given continuous variable, such that, the range of continuous variable is divided into n-equal width intervals, each one of them called strata, and a simple random sampling method or systematic sampling method is used in order to select a random value within each interval often with equal probability for selection. Finally, those collected samples with length n (one sample per strata) are used to represent the continuous variables in experiments and simulations. But in our scope, the sample is used to generate the initial population for the evolutionary algorithms.

For example, consider we have a problem P with 4 continuous input variables x1,x2 x3, and x4, each xi in the range of [0,1], i.e. $0 \leq xi \leq 1$ for all xi $\in$ X, then the range of xi is divided into 4 intervals, each interval length is proportional to the number of variables. In our example, the interval length L is equal to (0.25), i.e. L= (upper bound (1)-lower bound (0)) /4 = (0.25), and this produces the following:

Interval_1 $\in$ [0, 0.25] and x1 value= random number (r1) such that: $0 \leq r1 \leq 0.25$.
Interval_2 $\in$ [0.25, 0.5] and x2 value= random number (r2) such that: $0.25 \leq r2 \leq 0.5$.
Interval_3 $\in$ [0.5, 0.75] and x3 value= random number (r3) such that: $0.5 \leq r3 \leq 0.75$.
Interval_4 $\in$ [0.75, 1] and x4 value= random number (r4) such that: $0.75 \leq r4 \leq 1$.

Hence, the input initial population for the problem P is the sample (r1, r2, r3 and r4) and so on. Actually based on [6], stratified sampling has the advantage of forcing the inclusion of specified subsets of variable range, while maintaining the probabilistic character of random sampling, i.e. the representativeness of the sample is increased while preventing the desired randomness in the sample, but, it is a challenge to define the number of intervals and calculating their interval probabilities, but in our scope this will not be a significant problem because the number of intervals must be defined before the individuals are generated, and in our research, the number of intervals will be equal to the number of input variables of the NSGAII algorithm.

3.2   Latin Hypercube Sampling Method (LHS)

Based on [2, 3, 6, 11 and 12], it is a statistical method that combines the desirable features found in random and stratified sampling, such that, the randomness issue is increased when it is compared to the stratified sampling. But in the other hand, it provides more stable samples when it is compared to the simple random sampling by ensuring that all portions of the continuous variable were sampled.

Table 1. N x K matrix of LHS samples $\in$ [0, 1]

interval number \ variables	x1	x2	x3	x4	x5
1	0.007	0.038	0.039	0.114	0.033
2	0.379	0.256	0.321	0.224	0.263
3	0.491	0.436	0.504	0.593	0.490
4	0.670	0.706	0.699	0.742	0.702
5	0.948	0.950	0.968	0.841	0.803

Latin hypercube sampling works as follows, consider we have K input variables for a citrine optimization algorithm $(x_1, x_2 \ldots, x_k)$, each input variable is divided into N non overlapping intervals on the basis of equal probability, and one value is selected from each interval based on the probability density in the interval using density function, each value in the x_1 intervals is paired at random with N values of x_2 intervals without replacement, and the pair (x_1, x_2) is paired at random with values from x_3

intervals and so on until we have a matrix with N x K input values from K variables. For example, consider that N=5 and there are 5 input variables (x_1, x_2, x_3, x_4 and x_5), such that $0 \leq x_i \leq 1$ for all x_i, then each x_i variable will be divided into 5 intervals as the case of stratified sampling, then we have : $x_{i,1} \in [0,0.2]$, $x_{i,2} \in [0.2,0.4]$, $x_{i,3} \in [0.4,0.6]$, $x_{i,4} \in [0.6,0.8]$, and $x_{i,5} \in [0.8,1]$, and as a result, Table 1 is a possible N x K matrix of our input variables.

Obviously, after random pairing between (x_1, x_2) we may have a possible pair (0.007, 0.436), and the resulted pair is paired at random with x_3 intervals to have (0.007, 0.436,0.504) and so on. Actually, it is possible for intervals to be repeated within the samples in contrast with stratified sampling approach that discussed in Section 3.1. Furthermore, all samples will share the same values generated in the NxK matrix, i.e. the NxK matrix will act as a repository for all input samples. Samples 1 and 2 below are two possible instances from the NxK matrix in Table 1:

Sample 1: (0.007, 0.436, 0.504, 0.224, 0.803).
Sample 2: (0.491, 0.038, 0.699, 0.968, 0.803).

3.3 Niederreiter Quasi Random Method

For more convenience, the details of this method will not be discussed here and available at appendix A. Therefore, we will shed light into the concepts that are related to our interest. Based on [5,13], Niederreiter quasi random sequences are a part of a low-discrepancy sequences with the property that for all values of N, its subsequence x_1, ..., x_N has a low discrepancy. In other words, the values generated in a given range for a continuous variable are less random and cover more regions than strictly random approach.

Actually, Niederreiter quasi random method is useful for global optimization. This is because the low discrepancy property of this method lets us to sample the space of continuous variables more uniformly than random approach. In fact, this is desirable for exploring the space of the problem variables in order to generate the initial population for evolutionary algorithms, since, the generated samples will be more representative and the randomness still preserved. Notice that in our work, Niederreiter quasi random numbers is generated using (martignal.jar) library from open source, and this library is available at [15].

4   Proposed Initialization Method

4.1   Merging the Stratified Sampling with Quasi-random Method

In this approach, when generating random values within such interval in Stratified sampling, these values will be generated using Niederreiter quasi random method rather than classical strictly random. In more detail, once the number of intervals for the given variable X has been determined, we can define the lower and upper bond for each interval as discussed before in section 3.1. These bounds are given to the Quasi-Random generator to generate a value r within strata boundaries such that:

$$\text{lower-bound} \leq r \leq \text{upper-bound}$$

Actually, we can have the properties of the two methods Quasi-Random and Strati-fied sampling at the same time in one step using this approach, but that is not enough to make this method more attractive than stratified sampling. For this reason, the output sample will be randomly shuffled as a finalizing step before producing the output sam-ples to prevent our initial populations from falling in the same regions in the problem domain space. For example, consider we have an output sample $S \in [0,1]$ generated using the modified version of stratified sampling as discussed above say ($r1,r2,r3$ and $r4$), each ri in S falls in a different strata, the sample S for example may be converted to ($r2,r3,r1$ and $r4$) or ($r1,r3,r4$ and $r2$) after random shuffling, this strategy i.e. gener-ate and shuffle will continue for all generated samples in advance in order to preserve randomness beside representativeness in our initial populations.

4.2   Merging LHS with Quasi-random Method

In this approach, all properties of LHS sampling method remain unchanged except those that tell how the value was generated within each interval. This initialization method is similar to the modified version of stratified samplin in the concept of how the values are generated within each interval. In other words, Quasi-Random generator were used to generate the values from each interval while constructing the NxK matrix as discussed in section 3.2 rather than generate those values using strictly random method, and the issues regards to random pairing will remain unchanged to preserve the randomness along with representativeness. For more convenience, this new method will be called QuasiLHS in our experiments. The motivation behind these two approaches is to set the advantages of random, Quasi-random and LHS beside each other's.

5   Experimental Environment

In our experiments, we have used the NSGAII algorithm as a multi objective optimi-zation algorithm and a set of state-of-the-art multi objective optimization problems as shown in Table 2. The algorithm and optimization problems are implemented in jMetal platform [14]. This platform allows user to design an experiment by selecting the algorithm, setting the algorithm parameters and finally choosing the optimization problems and setting the corresponding parameters. Moreover, there are a set of quality indicators have be set before running such experiment, the average of final results of all independent runs of the experiments would be stored in latex files, each one of them contains the mean and standard deviation for all quality indicators that have been set before running the experiments, see [4] for more details, Actually. We have made the required modifications in jMetal platform in order to apply our initiali-zation method in all experiments instead of the default java random number, and also, there were a set of general setting have to be set for all experiments in order to isolate the effects of our initialization method as the following:

- Population size equal to 100 input sample.
- Crossover probability equal to 100%.
- Mutation probability equal to (1 / number of variables).
- Maximum number of function evaluations equal to 25000.
- Number of independent runs for each experiment equal to 30.

Table 2. Testing Problems

Optimization problem	Number of objectives
DTLZ1, DTLZ2, DTLZ3, DTLZ4 DTLZ5, DTLZ6, DTLZ7	2D and 3D
LZ09_F1, LZ09_F2, LZ09_F3, LZ09_F4, LZ09_F5, LZ09_F6, LZ09_F7, LZ09_F8, LZ09_F9	2D
WFG1, WFG2, WFG3, WFG4, WFG5, WFG6, WFG7, WFG8, WFG9	2D and 3D
CEC2009_UF1, CEC2009_UF2, CEC2009_UF3, CEC2009_UF4, CEC2009_UF5, CEC2009_UF6, CEC2009_UF7	2D
CEC2009_UF8, CEC2009_UF9, CEC2009_UF10	3D

6   Quality Indicators

There are a set of quality indicators that have to be used in order to compare the quality of obtained solution. We are using Hyper Volume (HV) and Inverted Generational Distance (IGD) [16]. HV is used to measure the volume of the dominated portion of the objective space and the fitness of Pareto sets in evolutionary multi-objective optimization, whereas, IGD is used to measure how far the elements are in the Pareto optimal set from those in the set of non-dominated solutions found.

Table 3. HyperVolume quality indicator for LZ09, WFG, DTLZ   and CEC2009 Problems (2D then 3D)

	Random	Quasi	Stratified	LHS	QuasiLHS
LZ09_F1	$6.18e-01_{7.0e-03}$	$6.23e-01_{6.0e-03}$	$6.15e-01_{1.1e-02}$	$6.20e-01_{9.1e-03}$	$6.07e-01_{1.2e-02}$
LZ09_F2	$5.16e-01_{2.9e-02}$	$4.76e-01_{2.4e-02}$	$5.91e-01_{3.2e-02}$	$5.11e-01_{4.6e-02}$	$5.22e-01_{3.7e-02}$
LZ09_F3	$5.93e-01_{6.7e-03}$	$5.91e-01_{8.5e-03}$	$5.89e-01_{1.1e-02}$	$5.91e-01_{6.1e-03}$	$5.97e-01_{4.2e-03}$
LZ09_F4	$6.03e-01_{4.6e-03}$	$5.98e-01_{2.1e-02}$	$6.04e-01_{3.3e-03}$	$6.06e-01_{3.1e-03}$	$6.06e-01_{2.4e-03}$
LZ09_F5	$6.06e-01_{8.5e-03}$	$6.06e-01_{7.0e-03}$	$6.09e-01_{3.9e-03}$	$6.07e-01_{5.3e-03}$	$6.10e-01_{4.1e-03}$
LZ09_F6	$1.23e-01_{3.9e-02}$	$1.04e-01_{5.1e-02}$	$1.35e-01_{2.9e-02}$	$1.30e-01_{3.8e-02}$	$1.52e-01_{3.1e-02}$
LZ09_F7	$3.05e-01_{5.1e-02}$	$3.65e-01_{3.0e-02}$	$2.43e-01_{9.0e-02}$	$3.09e-01_{4.5e-02}$	$2.83e-01_{5.1e-02}$
LZ09_F8	$2.43e-01_{5.1e-02}$	$2.85e-01_{5.5e-02}$	$2.28e-01_{4.2e-02}$	$2.44e-01_{4.5e-02}$	$2.41e-01_{3.5e-02}$
LZ09_F9	$1.63e-01_{4.7e-02}$	$2.06e-01_{2.1e-02}$	$1.76e-01_{4.8e-02}$	$1.65e-01_{4.0e-02}$	$1.96e-01_{3.5e-02}$
	Random	Quasi	Stratified	LHS	QuasiLHS
WFG1	$1.84e-01_{3.8e-02}$	$1.60e-01_{4.5e-02}$	$1.79e-01_{5.4e-02}$	$1.80e-01_{3.2e-02}$	$1.75e-01_{4.3e-02}$
WFG2	$5.05e-01_{2.8e-02}$	$5.13e-01_{3.1e-02}$	$4.98e-01_{2.2e-02}$	$5.12e-01_{2.2e-03}$	$4.91e-01_{2.2e-03}$
WFG3	$4.33e-01_{2.8e-03}$	$4.32e-01_{2.4e-03}$	$4.32e-01_{2.6e-03}$	$4.33e-01_{1.9e-03}$	$4.31e-01_{4.3e-03}$
WFG4	$2.14e-01_{4.1e-04}$	$2.15e-01_{7.5e-04}$	$2.15e-01_{4.6e-04}$	$2.14e-01_{8.5e-04}$	$2.15e-01_{8.9e-04}$
WFG5	$1.94e-01_{4.1e-04}$	$1.94e-01_{3.8e-04}$	$1.94e-01_{3.4e-04}$	$1.94e-01_{4.8e-04}$	$1.94e-01_{6.4e-04}$
WFG6	$1.71e-01_{1.1e-02}$	$1.63e-01_{1.3e-02}$	$1.64e-01_{1.1e-02}$	$1.67e-01_{1.1e-02}$	$1.73e-01_{1.1e-02}$
WFG7	$2.01e-01_{6.2e-03}$	$1.89e-01_{5.6e-03}$	$2.00e-01_{7.1e-03}$	$2.00e-01_{6.6e-03}$	$2.03e-01_{1.1e-03}$
WFG8	$1.38e-01_{2.2e-03}$	$1.37e-01_{1.6e-03}$	$1.40e-01_{6.1e-03}$	$1.38e-01_{6.6e-03}$	$1.40e-01_{1.3e-03}$
WFG9	$2.22e-01_{3.1e-02}$	$2.16e-01_{3.5e-02}$	$1.72e-01_{5.0e-02}$	$2.17e-01_{3.5e-02}$	$2.31e-01_{1.9e-03}$
	Random	Quasi	Stratified	LHS	QuasiLHS
WFG1	$2.67e-01_{4.6e-02}$	$2.96e-01_{4.7e-02}$	$2.65e-01_{2.7e-02}$	$2.89e-01_{3.9e-02}$	$2.84e-01_{4.7e-02}$
WFG2	$7.75e-01_{3.4e-02}$	$7.73e-01_{5.5e-03}$	$7.63e-01_{3.1e-02}$	$7.76e-01_{2.9e-02}$	$7.40e-01_{2.9e-02}$
WFG3	$3.01e-01_{4.3e-03}$	$2.98e-01_{1.1e-02}$	$3.00e-01_{6.1e-03}$	$3.02e-01_{6.4e-03}$	$3.07e-01_{4.1e-03}$
WFG4	$3.19e-01_{5.5e-03}$	$3.16e-01_{7.9e-02}$	$3.20e-01_{9.3e-03}$	$3.20e-01_{3.3e-03}$	$3.20e-01_{4.4e-03}$
WFG5	$3.19e-01_{5.9e-03}$	$3.19e-01_{7.8e-03}$	$3.22e-01_{6.1e-03}$	$3.23e-01_{5.5e-03}$	$3.20e-01_{7.0e-03}$
WFG6	$2.93e-01_{1.5e-02}$	$2.86e-01_{1.6e-02}$	$2.85e-01_{1.2e-02}$	$2.89e-01_{1.5e-02}$	$2.90e-01_{1.3e-02}$
WFG7	$3.21e-01_{1.3e-02}$	$3.16e-01_{1.5e-02}$	$3.18e-01_{1.7e-02}$	$3.19e-01_{1.8e-02}$	$3.28e-01_{1.6e-02}$
WFG8	$1.95e-01_{8.7e-03}$	$1.96e-01_{7.4e-02}$	$1.96e-01_{1.1e-02}$	$1.99e-01_{4.4e-02}$	$1.91e-01_{9.1e-03}$
WFG9	$2.35e-01_{1.5e-02}$	$2.44e-01_{6.6e-02}$	$2.33e-01_{5.3e-02}$	$2.43e-01_{3.8e-02}$	$2.50e-01_{1.1e-02}$

Table 3. (*Continued*)

	Random	Quasi	Stratified	LHS	QuasiLHS
DTLZ1	$7.62e-01_{4.5e-03}$	$7.61e-01_{4.6e-03}$	$7.62e-01_{4.6e-03}$	$7.62e-01_{4.3e-03}$	$7.61e-01_{5.2e-03}$
DTLZ2	$3.75e-01_{7.7e-03}$	$3.77e-01_{5.3e-03}$	$3.74e-01_{6.2e-03}$	$3.75e-01_{4.9e-03}$	$3.74e-01_{5.1e-03}$
DTLZ3	$3.75e-01_{9.0e-03}$	$3.73e-01_{7.3e-03}$	$3.75e-01_{7.2e-03}$	$3.78e-01_{8.7e-03}$	$3.74e-01_{1.2e-02}$
DTLZ4	$3.74e-01_{6.0e-03}$	$3.52e-01_{9.4e-02}$	$3.75e-01_{6.6e-03}$	$3.74e-01_{4.4e-03}$	$3.01e-01_{1.5e-01}$
DTLZ5	$9.29e-02_{1.9e-04}$	$9.30e-02_{1.8e-04}$	$9.31e-02_{1.7e-04}$	$9.29e-02_{1.5e-04}$	$9.29e-02_{1.5e-04}$
DTLZ6	$8.42e-02_{1.1e-02}$	$8.36e-02_{1.0e-02}$	$8.09e-02_{1.2e-02}$	$8.43e-02_{1.0e-02}$	$8.73e-02_{1.0e-02}$
DTLZ7	$2.85e-01_{4.0e-03}$	$2.84e-01_{3.6e-03}$	$2.85e-01_{3.8e-03}$	$2.85e-01_{4.2e-03}$	$2.84e-01_{3.7e-03}$

	Random	Quasi	Stratified	LHS	QuasiLHS
CEC2009$_U$F1	$5.30e-01_{3.5e-02}$	$4.88e-01_{2.9e-02}$	$5.24e-01_{3.4e-02}$	$5.31e-01_{4.0e-02}$	$5.55e-01_{2.8e-02}$
CEC2009$_U$F2	$6.26e-01_{7.4e-03}$	$6.21e-01_{8.0e-03}$	$6.29e-01_{4.3e-03}$	$6.27e-01_{5.0e-03}$	$6.29e-01_{4.4e-03}$
CEC2009$_U$F3	$5.25e-01_{6.4e-02}$	$5.26e-01_{5.6e-02}$	$5.02e-01_{6.0e-02}$	$5.25e-01_{6.5e-02}$	$5.44e-01_{6.0e-02}$
CEC2009$_U$F4	$3.33e-01_{2.8e-03}$	$3.24e-01_{5.6e-03}$	$3.31e-01_{2.5e-03}$	$3.32e-01_{3.4e-03}$	$3.24e-01_{3.3e-03}$
CEC2009$_U$F5	$1.16e-01_{8.5e-02}$	$1.11e-01_{7.3e-02}$	$1.54e-01_{9.5e-02}$	$7.92e-02_{7.5e-02}$	$1.40e-01_{9.0e-02}$
CEC2009$_U$F6	$4.14e-01_{6.8e-02}$	$4.13e-01_{6.4e-02}$	$4.62e-01_{6.3e-02}$	$4.11e-01_{9.6e-02}$	$4.46e-01_{6.4e-02}$
CEC2009$_U$F7	$3.80e-01_{6.9e-02}$	$1.69e-01_{5.9e-02}$	$3.68e-01_{8.1e-02}$	$3.31e-01_{8.6e-02}$	$2.54e-01_{4.1e-02}$
CEC2009$_U$F8	$7.81e-01_{3.1e-02}$	$7.27e-01_{7.3e-02}$	$7.58e-01_{2.8e-02}$	$7.83e-01_{4.2e-02}$	$7.57e-01_{2.5e-02}$
CEC2009$_U$F9	$8.40e-01_{6.4e-02}$	$8.42e-01_{3.9e-02}$	$8.72e-01_{6.4e-02}$	$8.53e-01_{5.6e-02}$	$8.31e-01_{4.4e-02}$
CEC2009$_U$F10	$7.28e-01_{1.5e-01}$	$5.22e-01_{6.6e-02}$	$7.38e-01_{1.2e-01}$	$6.84e-01_{1.4e-01}$	$6.76e-01_{1.2e-01}$

Table 4. IGD quality indicator for LZ09, WFG, DTLZ Problems (2D then 3D for any given problem)

	Random	Quasi	Stratified	LHS	QuasiLHS
LZ09$_F$1	$3.61e-03_{1.3e-03}$	$3.27e-03_{1.2e-03}$	$3.85e-03_{1.7e-03}$	$3.04e-03_{1.4e-03}$	$5.11e-03_{1.8e-03}$
LZ09$_F$2	$5.81e-03_{9.0e-04}$	$8.34e-03_{1.5e-03}$	$6.41e-03_{1.3e-03}$	$6.05e-03_{1.7e-03}$	$5.94e-03_{1.5e-03}$
LZ09$_F$3	$3.67e-03_{1.0e-03}$	$4.52e-03_{1.2e-03}$	$3.94e-03_{2.0e-03}$	$3.54e-03_{1.2e-03}$	$3.35e-03_{7.2e-04}$
LZ09$_F$4	$3.35e-03_{5.1e-04}$	$3.80e-03_{9.8e-04}$	$3.11e-03_{4.0e-04}$	$3.40e-03_{6.3e-04}$	$3.40e-03_{5.2e-04}$
LZ09$_F$5	$2.60e-03_{5.3e-04}$	$3.37e-03_{1.3e-03}$	$2.45e-03_{7.7e-04}$	$2.74e-03_{5.2e-04}$	$2.79e-03_{5.9e-04}$
LZ09$_F$6	$1.12e-02_{1.5e-03}$	$1.19e-02_{1.9e-03}$	$1.12e-02_{1.0e-03}$	$1.04e-02_{1.5e-03}$	$1.10e-02_{1.4e-03}$
LZ09$_F$7	$1.69e-02_{4.1e-03}$	$1.36e-02_{3.3e-03}$	$1.88e-02_{6.8e-03}$	$1.79e-02_{5.2e-03}$	$2.04e-02_{5.6e-03}$
LZ09$_F$8	$1.73e-02_{1.6e-02}$	$1.45e-02_{1.9e-02}$	$1.77e-02_{2.1e-02}$	$1.67e-02_{2.1e-02}$	$1.68e-02_{1.5e-02}$
LZ09$_F$9	$8.10e-03_{2.7e-03}$	$8.00e-03_{1.1e-03}$	$7.63e-03_{2.6e-03}$	$7.99e-03_{2.1e-03}$	$6.77e-03_{1.9e-03}$

	Random	Quasi	Stratified	LHS	QuasiLHS
WFG1	$6.22e-03_{2.2e-04}$	$6.34e-03_{2.0e-04}$	$6.22e-03_{3.3e-04}$	$6.19e-03_{2.1e-04}$	$6.24e-03_{2.4e-04}$
WFG2	$1.49e-03_{6.3e-04}$	$1.32e-03_{7.0e-04}$	$1.64e-03_{5.0e-04}$	$1.35e-03_{7.9e-04}$	$1.84e-03_{6.4e-06}$
WFG3	$7.49e-04_{3.4e-05}$	$7.52e-04_{2.7e-05}$	$7.48e-04_{9.0e-05}$	$7.44e-04_{2.7e-05}$	$7.63e-04_{5.2e-05}$
WFG4	$1.48e-04_{1.3e-05}$	$1.59e-04_{2.0e-04}$	$1.49e-04_{1.5e-05}$	$1.57e-04_{2.7e-05}$	$1.49e-04_{1.9e-05}$
WFG5	$5.63e-04_{5.5e-06}$	$5.63e-04_{4.7e-06}$	$5.63e-04_{4.6e-06}$	$5.63e-04_{5.8e-06}$	$5.59e-04_{9.9e-06}$
WFG6	$6.80e-04_{1.8e-04}$	$8.97e-04_{2.2e-04}$	$7.96e-04_{2.0e-04}$	$7.60e-04_{1.8e-04}$	$6.28e-04_{1.8e-04}$
WFG7	$3.56e-04_{9.8e-05}$	$5.31e-04_{1.1e-04}$	$3.86e-04_{1.3e-04}$	$3.60e-04_{9.8e-04}$	$3.37e-04_{9.1e-05}$
WFG8	$1.18e-03_{1.7e-04}$	$1.27e-03_{1.2e-04}$	$1.07e-03_{2.2e-04}$	$1.16e-03_{1.9e-04}$	$1.45e-03_{3.4e-05}$
WFG9	$4.04e-04_{5.4e-04}$	$4.92e-04_{6.1e-04}$	$1.26e-03_{8.7e-04}$	$4.77e-04_{6.1e-04}$	$2.34e-04_{3.5e-05}$

	Random	Quasi	Stratified	LHS	QuasiLHS
WFG1	$1.34e-02_{8.0e-04}$	$1.32e-02_{7.8e-04}$	$1.34e-02_{5.4e-04}$	$1.32e-02_{4.0e-04}$	$1.31e-02_{7.1e-04}$
WFG2	$1.29e-03_{1.3e-04}$	$1.36e-03_{1.1e-04}$	$1.31e-03_{2.8e-04}$	$1.26e-03_{1.2e-04}$	$1.55e-03_{6.0e-04}$
WFG3	$4.70e-04_{5.4e-05}$	$4.78e-04_{1.2e-04}$	$4.84e-04_{9.3e-05}$	$4.70e-04_{8.6e-05}$	$4.03e-04_{7.1e-05}$
WFG4	$7.96e-04_{8.6e-05}$	$7.84e-04_{6.0e-05}$	$7.95e-04_{6.8e-05}$	$7.70e-04_{7.2e-05}$	$7.77e-04_{8.2e-05}$
WFG5	$8.31e-04_{3.4e-05}$	$8.22e-04_{4.4e-05}$	$8.14e-04_{3.1e-05}$	$7.98e-04_{2.9e-05}$	$8.19e-04_{2.8e-05}$
WFG6	$9.55e-04_{7.3e-05}$	$1.32e-03_{2.2e-04}$	$9.87e-04_{6.5e-05}$	$9.90e-04_{7.4e-05}$	$9.74e-04_{7.5e-05}$
WFG7	$9.80e-04_{1.2e-04}$	$1.13e-03_{1.6e-04}$	$9.77e-04_{1.6e-04}$	$1.05e-03_{1.5e-04}$	$9.87e-04_{1.3e-04}$
WFG8	$2.03e-03_{1.0e-04}$	$2.08e-03_{9.7e-05}$	$1.95e-03_{1.8e-04}$	$1.96e-03_{3.1e-04}$	$2.13e-03_{5.1e-05}$
WFG9	$1.10e-03_{7.9e-05}$	$1.08e-03_{1.1e-04}$	$1.13e-03_{4.4e-05}$	$1.09e-03_{1.1e-04}$	$1.03e-03_{3.8e-04}$

	Random	Quasi	Stratified	LHS	QuasiLHS
DTLZ1	$3.84e-04_{1.4e-05}$	$3.84e-04_{1.2e-05}$	$3.87e-04_{1.3e-05}$	$3.84e-04_{1.6e-05}$	$4.20e-04_{1.7e-05}$
DTLZ2	$4.40e-04_{2.6e-05}$	$4.35e-04_{1.7e-05}$	$4.45e-04_{2.2e-05}$	$4.40e-04_{2.7e-05}$	$4.49e-04_{2.6e-05}$
DTLZ3	$4.53e-04_{2.9e-05}$	$4.49e-04_{2.2e-05}$	$4.41e-04_{2.1e-05}$	$4.39e-04_{2.2e-05}$	$4.73e-04_{3.2e-05}$
DTLZ4	$1.64e-03_{2.8e-03}$	$5.60e-03_{2.5e-03}$	$2.52e-03_{3.3e-03}$	$1.42e-03_{2.6e-03}$	$4.06e-03_{3.2e-03}$
DTLZ5	$4.39e-04_{2.3e-05}$	$4.36e-04_{2.1e-05}$	$4.42e-04_{2.6e-05}$	$4.36e-04_{2.5e-05}$	$4.45e-04_{2.4e-05}$
DTLZ6	$1.52e-03_{9.7e-04}$	$1.36e-03_{6.4e-04}$	$1.91e-03_{1.5e-03}$	$1.79e-03_{1.1e-03}$	$1.66e-03_{1.2e-03}$
DTLZ7	$2.09e-04_{9.1e-06}$	$2.13e-04_{1.4e-05}$	$2.08e-04_{8.6e-06}$	$2.11e-04_{6.8e-06}$	$2.18e-04_{1.3e-05}$

	Random	Quasi	Stratified	LHS	QuasiLHS
DTLZ1	$6.00e-04_{3.6e-05}$	$5.85e-04_{3.8e-05}$	$5.95e-04_{3.8e-05}$	$5.84e-04_{3.8e-05}$	$5.99e-04_{3.5e-05}$
DTLZ2	$7.77e-04_{4.9e-05}$	$7.58e-04_{4.3e-05}$	$7.80e-04_{4.0e-05}$	$7.71e-04_{4.5e-05}$	$7.89e-04_{6.1e-05}$
DTLZ3	$1.24e-03_{5.9e-05}$	$1.23e-03_{6.4e-05}$	$1.25e-03_{3.4e-05}$	$1.21e-03_{5.5e-05}$	$1.24e-03_{7.4e-05}$
DTLZ4	$1.23e-03_{1.1e-04}$	$1.77e-03_{5.6e-03}$	$1.22e-03_{1.4e-04}$	$1.21e-03_{1.0e-04}$	$2.84e-03_{3.3e-03}$
DTLZ5	$1.97e-05_{1.2e-04}$	$1.99e-05_{1.5e-06}$	$1.95e-05_{5.9e-07}$	$1.98e-05_{5.5e-07}$	$2.01e-05_{1.2e-06}$
DTLZ6	$1.29e-04_{9.5e-04}$	$1.30e-04_{8.1e-05}$	$1.55e-04_{1.1e-04}$	$1.24e-04_{3.3e-05}$	$1.02e-04_{3.2e-05}$
DTLZ7	$2.21e-03_{1.5e-04}$	$2.27e-03_{1.6e-04}$	$2.21e-03_{1.4e-04}$	$2.22e-03_{1.4e-04}$	$2.24e-03_{1.1e-04}$

Table 4. (*Continued*)

	Random	Quasi	Stratified	LHS	QuasiLHS
CEC2009$_U$ F1	$2.53e-03_{6.6e-04}$	$3.83e-03_{8.1e-04}$	$2.55e-03_{6.2e-04}$	$2.54e-03_{6.6e-04}$	$2.16e-03_{4.4e-04}$
CEC2009$_U$ F2	$1.27e-03_{4.4e-04}$	$2.01e-03_{8.4e-04}$	$1.16e-03_{1.6e-04}$	$1.27e-03_{2.5e-04}$	$1.31e-03_{2.2e-04}$
CEC2009$_U$ F3	$1.91e-02_{4.4e-03}$	$1.82e-02_{3.5e-03}$	$1.99e-02_{4.1e-03}$	$1.86e-02_{4.4e-03}$	$1.76e-02_{3.8e-03}$
CEC2009$_U$ F4	$8.55e-04_{1.6e-04}$	$1.12e-03_{1.3e-04}$	$9.45e-04_{1.9e-04}$	$8.59e-04_{1.7e-04}$	$1.37e-03_{1.8e-04}$
CEC2009$_U$ F5	$3.59e-02_{1.2e-02}$	$4.50e-02_{6.1e-03}$	$3.12e-02_{9.8e-03}$	$4.05e-02_{1.5e-02}$	$3.21e-02_{1.0e-02}$
CEC2009$_U$ F6	$1.26e-02_{4.3e-03}$	$1.44e-02_{5.7e-03}$	$1.03e-02_{9.9e-03}$	$1.36e-02_{6.3e-03}$	$1.15e-02_{5.1e-03}$
CEC2009$_U$ F7	$2.32e-03_{1.4e-03}$	$6.29e-03_{1.0e-03}$	$2.43e-03_{1.6e-03}$	$3.22e-03_{1.6e-03}$	$4.68e-03_{8.8e-04}$
CEC2009$_U$ F8	$3.45e-03_{2.8e-04}$	$4.17e-03_{1.1e-03}$	$3.65e-03_{2.8e-04}$	$3.44e-03_{5.5e-04}$	$3.74e-03_{2.2e-04}$
CEC2009$_U$ F9	$5.63e-03_{1.1e-03}$	$5.70e-03_{8.9e-03}$	$5.42e-03_{1.1e-03}$	$5.59e-03_{8.8e-04}$	$5.83e-03_{6.6e-04}$
CEC2009$_U$ F10	$9.30e-03_{4.0e-03}$	$1.32e-02_{3.1e-03}$	$8.97e-03_{2.9e-03}$	$1.02e-02_{3.8e-03}$	$9.66e-03_{3.0e-03}$

7   Experimental Results

Tables 3 and 4 show the detailed results for all test problems that were used in our experiments using the HV and IGD quality metrics. The columns in tables 3 and 4 were titled by the names of our initialization methods such that: Random represents standard random method, Quasi represents quasi random method, Stratified represents the modified version of stratified sampling, LHS represents the LHS method, QuasiLHS represents the modified version of LHS method.

Actually, the best results were highlighted by a dark gray color, and Light gray for the second best for each problem. The main conclusion that can be made from our results is: it is no longer feasible to use Random initialization, because it has not performed well in our experiments. Different problems prefer different initialization techniques. This is clear from the following: regarding HV. LZ09, WFG (2D & 3D) preferred QuasiLHS, while DTLZ (2D) preferred LHS the DTLZ (3D) preferred Stratified. Also, regarding IGD. LZ09 preferred all except Random, WFG (2D) preferred QuasiLHS, and WFG (3D) preferred both QuasiLHS and LHS. Finally DTLZ (2D and 3D) preferred LHS.

8   Conclusion and Future Work

In this work, we evaluated different initialization methods using NSGAII algorithm such as Random, Quasi, Stratified, LHS and QuasiLHS using different multi objectives optimization problems.   Based on our experimental results, these initialization methods have shown different reactions depending on the optimization problem characteristics.

We have shown through our experiments that the use of traditional Random initialization method would not perform well when it is compared with other initialization methods. Furthermore, our new initialization methods such as Stratified and QuasiLHS have performed well beside LHS in the experiments and they were better than Random and Quasi initialization methods in term of (HV) and (IGD)   quality indicators.

Actually, we have shown in our research that the nature and number of objectives of an optimization problem may affect the output quality of an initialization method; therefore, further researches are required in order to find a possible connection between the optimization problem characteristics and the initialization method being used. Also, the number of intervals inside Stratified, LHS and QuasiLHS initialization methods require further experiments in order to estimate the best number of intervals that maximize the quality of final solution.

References

1. Diaz-Gomez, P.A., Hougen, D.F.: Initial Population for Genetic Algorithms: A Metric Approach. In: Proceedings of the International Conference on Genetic and Evoltionary Methods (2007)
2. Mckay, M.D., Beckman, R.J., Conover, W.J.: A Comparison of three methods for selecting values of input variables in the analysis of output from a computer code. Technometrics **21**(2), 239–245 (1979)
3. Sallaberry, C.J., Helton, J.C., Hora, S.C.: Extension of Latin Hypercube Samples with Correlated Variables. Reliability Engineering & System Safety **93**(7), 1047–1059 (2008)
4. Nebro, A.J., Durillo, J.J.: jMetal 4.3 User Manual, January 3 2013
5. Maaranen, H., Miettinen, K., Penttinen, A.: On initial populations of a Genetic algorithm for continuous optimization problems. Journal of Global Optimization **37**(3), 405–436 (2007)
6. Helton, J.C., Davis, F.J.: Latin hypercube sampling and the propagation of uncertainty in analyses of complex systems. Reliability Engineering & System Safety **81**(1), 23–69 (2003)
7. Timothy, M.R., Arora, J.S.: Survey of multi-objective optimization methods for engineering. Structural and multidisciplinary optimization **26**(6), 369–395 (2004)
8. Deb, K., et al.: A fast and elitist multiobjective genetic algorithm: NSGA-II. IEEE Transactions on Evolutionary Computation **6**(2), 182–197 (2002)
9. Fonseca, C.M., Fleming, P.J.: An overview of evolutionary algorithms in multi-objective optimization. Evolutionary computation **3**(1), 1–16 (1995)
10. Zitzler, E., Deb, K., Thiele, L.: Comparison of multi objective evolutionary algorithms: Empirical results. Evolutionary computation **8**(2), 173–195 (2000)
11. Iman, R.L., Conover, W.J.: A distribution-free approach to inducing rank correlation among input variables. Communications in Statistics-Simulation and Computation **11**(3), 311–334 (1982)
12. Wyss, Gregory D., and Kelly H. Jorgensen. A user's guide to LHS: Sandia's Latin hypercube sampling software. SAND98-0210, Sandia National Laboratories, Albuquerque, NM (1998)
13. Levy, G.: An introduction to quasi-random numbers. Numerical Algorithms Group Ltd (2002). introduction_to_quasi_random_numbers
14. http://jmetal.sourceforge.net, Last visited: March, 2013
15. http://people.sc.fsu.edu/~jburkardt/f_src/niederreiter/niederreiter.html, Last visited: April, 2014
16. Radziukynienė, I., Žilinskas, A.: Evolutionary methods for multi-objective portfolio optimization. In: Proceedings of the World Congress on Engineering, vol. 2 (2008)

Cultural Particle Swarm Optimization Algorithms for Interval Multi-Objective Problems

Yi-nan Guo$^{(\boxtimes)}$, Zhen Yang, Chun Wang, and Dunwei Gong

China University of Mining and Technology, Xuzhou 221116, Jiangsu, China
guoyinan@cumt.edu.cn

Abstract. Traditional dominant comparison never fits for the interval multi-objective optimization problems. The particle swarm optimization for solving these problems cannot adaptively adjust the key parameters and easily falls into premature. So a novel multi-objective cultural particle optimization algorithm is proposed. Its strength are: (i)The possibility degree is introduced to construct a novel dominant relationship so as to rationally measure the uncertainty of particles; (ii)The grid's coverage degree is defined based on topological knowledge and used to measure the uniformity of non-dominant solutions instead of the crowding distance. (iii)The key flight parameters are adaptively adjusted and the local or global best are selected in terms of the knowledge. Simulation results indicate that the proposed algorithms coverage to the Pareto front uniformly and the uncertainty of non-dominant solutions is less. Furthermore, the knowledge plays a rational impact on balancing exploration and exploitation.

Keywords: Interval multi-objective optimization · Cultural particle swarm · Possibility-dominant · The coverage degree of grid · Parameters adjustment

1 Introduction

For most of practical engineering optimization problems, the environmental noise and other uncertain factors make the parameters in optimization models uncertain and dynamic. To define the uncertain information and optimize corresponding problems, stochastic programming, fuzzy programming and interval programming[1] are gradually formed. In the former two methods, the uncertainty of parameters obeys the known probability distribution functions or fuzzy membership functions. But incomplete information and limited cognition make the exact functions matching the actual situation difficultly constructed. In contrast, it is easy to achieve the possible interval of the parameters' values. Consequently, we focus on interval multi-objective optimization problem(IMOP)[2] and their problem-solving methods. Suppose $x \in [\underline{x}, \bar{x}]$ is decision variable. $\alpha \in [\underline{\alpha}, \bar{\alpha}]$ is interval parameter. M is the number of objective functions. IMOP is expressed by P: $\max F(x, a) = \{f_1(x, a), f_2(x, a), \cdots, f_M(x, a)\}$.

Recently, population-based intelligent optimization methods were introduced to solve IMOPs. Philipp[3]proposed imprecise-propagation multi-object evolutionary algorithm. The improved domination sort method from NSGA-II[4] made non-dominant solutions easily lost and the selection pressure higher. Interval robust multi-objective evolutionary algorithm[5] adopted interval analysis techniques to deal with

© Springer International Publishing Switzerland 2015
Y. Tan et al. (Eds.): ICSI-CCI 2015, Part I, LNCS 9140, pp. 505–512, 2015.
DOI: 10.1007/978-3-319-20466-6_53

the uncertainties in a deterministic way. A novel crowding distance based on hyper-volume[4] was proposed for IMOP. The solutions with small volumes and far away from others' midpoint have less crowding distances and more chance to be survived in next generation. Particle swarm optimization(PSO)[2] was also incorporated into solving IMOPs. Though the uniform pareto-optimal set approximates the true pareto front, the key parameters couldn't adaptively adjust and the algorithm easily got premature. It had been proved that PSO with dynamically adjusted parameters, including cognitive coefficient, social coefficient and inertia weight, was capable of handling optimization problems with different characteristics[6]. In addition, more than one hyper-volume solutions compose of the local best set or global best set. How to choose a rational one to direct a particle's flight is a difficult issue. They were commonly selected from the archive randomly or in terms of the distance[7]. Cultural algorithm is introduced to extract the knowledge from the non-dominated individuals and utilize it to adjust the particles' flight parameters and the extremes' selection[8]. However, the adaptive strategies only deal with MOPs with certain parameters.

To alleviate the weakness of existing PSO for IMOPs, we propose a novel multi-objective cultural particle swarm optimization algorithm(MOCPSO). Firstly, a novel possibility-dominant relationship is constructed so as to rationally measure the uncertainty of particles. Secondly, the grid's coverage degree is defined based on topological knowledge and used to measure the crowding degree of optimal non-dominant set instead of the crowding distance. Thirdly, the key flight parameters are adaptively adjusted and the local or global best are selected in terms of the knowledge.

2　The Novel Possibility Dominant Relationship

In IMOP, the individuals' objective values are all intervals, denoted by $f_m(x_i, \alpha) = \left[\underline{f_m(x_i)}, \overline{f_m(x_i)}\right], x_i \in S_P$. They compose of the hyper-volume, limited by $\underline{f_m(x_i)} = \min_{\alpha} f_m(x_i, \alpha)$ and $\overline{f_m(x_i)} = \max_{\alpha} f_m(x_i, \alpha)$. To compare these individuals, the order-dominant comparison based on order relationship between intervals was given[3], which does not fit for the situation that two intervals are embodied each other. In interval probability dominant relationship[5], the individual's rank is the number of individuals who probably dominates it. However, the individuals with same rank may have different chance to be reserved in next generation. In order to fully measure the difference among individuals, we proposed a novel possibility-dominant comparison method by introducing the possibility degree[9]. Let $L\big(f_m(x_i, \alpha)\big) = \overline{f_m(x_i)} - \underline{f_m(x_i)}$ be the span of mth objective function. Defined $\sigma_m(x_i, x_j)$ is the possibility that x_i is superior to x_j.

$$\sigma_m(x_i, x_j) = \frac{\max\left\{0, \left(L\big(f_m(x_i, \alpha)\big) + L\big(f_m(x_j, \alpha)\big) - \max\big(\overline{f_m(x_i)} - \underline{f_m(x_j)}, 0\big)\right)\right\}}{L\big(f_m(x_i, \alpha)\big) + L\big(f_m(x_j, \alpha)\big)}. \tag{1}$$

Suppose $\sigma(x_i \succ x_j)$ denotes the possibility that x_i dominating x_j. If x_i dominats x_j in terms of certain possibility, we describe it as $x_i: \xrightarrow{\sigma(x_i \succ x_j)} x_j$. For ith individual, its rank is decides by the possibility of all individuals dominated by it, defined as

$rank_j(t) = \sum_{\neq i} \sigma\left(x_i(t) > x_j(t)\right), \forall x_i(t) \in X, x_i: \xrightarrow{\sigma(x_i > x_j)} x_j$,   The optimal non-dominant particles satisfy $S_P(t) = \{x_i(t) | rank_i(t) = 0, \forall x_i(t) \in X\}$. Though the particle's rank may be not an integer and the non-dominant particles' ranks may be different, the particles with larger rank must be dominated in larger possibility by more particles.

$$\begin{cases} \sigma(x_i > x_j) = \prod_m \sigma_m(x_i, x_j) & \begin{aligned} &\forall m, \sigma_m(x_i, x_j) \geq \sigma_m(x_j, x_i) \\ &\exists k, \sigma_k(x_i, x_j) > \sigma_k(x_j, x_i) \end{aligned} \\ \sigma(x_j > x_i) = \prod_m \sigma_m(x_j, x_i) & \begin{aligned} &\forall m, \sigma_m(x_j, x_i) \geq \sigma_m(x_i, x_j) \\ &\exists k, \sigma_k(x_j, x_i) > \sigma_k(x_i, x_j) \end{aligned} \\ \sigma(x_i \| x_j) = 1 - \prod_m \sigma_m(x_i, x_j) - \prod_k \sigma_k(x_j, x_i) & \begin{aligned} &\forall m, \sigma_m(x_i, x_j) \geq \sigma_m(x_j, x_i) \\ &\exists k, \sigma_k(x_j, x_i) \geq \sigma_k(x_i, x_j) \end{aligned} \end{cases} \quad (2)$$

3   Multi-objective Cultural Particle Swarm Optimization for IMOP

MOCPSO consists of population space and belief space. In population space, multi-objective PSO realizes the particles' operators and the possibility-dominant comparison. In belief space, all non-dominant particles are preserved and used to extract useful knowledge so as to adjust the flight parameters and direct the extreme selection.

In particle swarm optimization, each particle flies towards optimum along itself trajectory after initialization. The flight direction and distance depend on its velocity. In each generation, particles update their velocity by tracking two extreme values. One is the local best, which records this particle's optimal location. The other is global best, which memories the optimal particle in whole population. Suppose $x_i(t)$ and $v_i(t)$ are ith particle's location and velocity in tth generation. Let $p_i(t)$ be ith particle's local best. $g_i(t)$ is jth global best. c_1 and c_2 are cognitive coefficient and social coefficient. w is inertia weight.

$$x_i(t + 1) = x_i(t) + v_i(t + 1). \tag{3}$$
$$v_i(t + 1) = wv_i(t) + c_1 r_1 (p_i(t) - x_i(t)) + c_2 r_2 \left(g_j(t) - x_i(t)\right). \tag{4}$$

In belief space, three kinds of knowledge are constructed in MOCPSO.

Normative Knowledge. It is defined as $K_{NOM} = \{U(t), L(t), UV(t), LV(t), UF(t), LF(t), W, C_1, C_2\}$. The former four parts record the boundary of each decision variable and particles' extreme velocity. The following two parts memories the extreme objective values of non-dominant particles. The latter three parts save the limits of key parameters. The inertia weight is adjusted in terms of the extreme value of particle's velocity. Let $\Delta w \in (0,1]$ be the incremental of w.

$$w_i^j(t + 1) = \begin{cases} w_i^j(t) + \Delta w & v_i^j(t) < lv^j(t) \\ w_i^j(t) - \Delta w & v_i^j(t) > uv^j(t) \\ w_i^j(t) & other \end{cases} \tag{5}$$

$$w_i(t+1) = \begin{cases} \overline{w} & w_i(t+1) > \overline{w} \\ \underline{w} & w_i(t+1) < \underline{w} \\ w_i(t+1) & other \end{cases} \quad (6)$$

Situational Knowledge. Situational knowledge consists of local optimal set $SP_i(t)$ and global optimal set $SG(t)$. Let $pb_i^j(t)$ be jth non-dominated solution in $SP_i(t)$ and $gb_j(t)$ be jth non-dominated solution in $SG(t)$. They updated in terms of the possibility-dominant relationship between them. $gb_k(t)$ is a particle which have a smallest contribution to hyper-volume in $SG(t+1)$. $\eta \in [0,1]$ is the probability threshold.

$$SP_i(t+1) = \begin{cases} SP_i(t) & rank(x_i(t+1)) > rank\left(pb_i^j(t)\right) \\ SP_i(t) \cup x_i(t+1) & rank(x_i(t+1)) = 0 \\ SP_i(t) \cup x_i(t+1) \backslash pb_i^j(t) & \sigma\left(x_i(t+1), pb_i^j(t)\right) > \eta \end{cases} \quad . \ (7)$$

$$SG(t+1) = \begin{cases} SG(t) & rank\left(pb_i^j(t+1)\right) > rank\left(gb_j(t)\right) \\ SG(t) \cup pb_i^j(t+1) & rank\left(pb_i^j(t+1)\right) = 0 \\ SG(t) \cup pb_i^j(t+1) \backslash gb_j(t) & \sigma\left(pb_i^j(t+1), gb_j(t)\right) > \eta \end{cases} \quad .$$

$$(8)$$

Topographic Knowledge. Topographic knowledge memories the hyper-volumes' distribution about non-dominant particles' objective vector based on the grids. The object space recorded by normative knowledge is evenly partitioned into many grids. Suppose $A_q(x_i)$ is the hyper-volume's acreage of ith particle's objective hyper-volume overlapped by G_q. If $F(x_i, \alpha) \cap G_q \neq \emptyset$, $A_q(x_i) = \prod_{m=1}^{M} |\min(ug_m^{Cm}(t),$ $\overline{f_m(x_i)}) - \max\left(lg_m^{Cm}(t), \underline{f_m(x_i)}\right)|$ .Otherwise, $A_q(x_i) = 0$. A_q is the acreage of each grid. Subsequently, the coverage degree is defined as $CG_q(t) = \sum_{i=1}^{N} A_q(x_i(t))/A_q$ to measure the distribution of non-dominant particles' objective hyper-volume. Larger $CG_q(t)$ means qth grid is crowed and covered by more particles or larger hyper-volumes. Let d_m be the division depth for mth objective. Let $ug_m^{Cm}(t) = lg_m^{Cm}(t) + \Delta, c_m = 1,2,\cdots,d_m.\Delta = [uf_m(t) - lf_m(t)]/d_m$ and $lg_m^{Cm}(t) = lf_m(t) + (c_m - 1)\Delta$.

Topographic knowledge not only directs the adjustment of cognitive coefficient and social coefficient, but also helps to choose the extreme. In order to avoid premature and ensure the uniformity of pareto-optimal set, global best shall direct the evolution to exploit the less crowded area. Thus, we choose $gb_j(t)$ that locates in the grid with $_q^{min}CG_q(t)$ and has largest coverage acreage in qth grid as global best. The influences of global best and local best on the particle's velocity are decided by social coefficient and cognitive coefficient. The area containing less non-dominated solutions and having less $CG_q(t)$ is needed to be explored more. Hence, we dynamically adjust $c_2(t)$ and $c_1(t)$ as follows. Let α and β be the adjustment factors.

$$c_{2i}(t+1) = c_{2i}(t) - \alpha(CG_l(t+1) - CG_k(t)) . \quad (9)$$
$$c_{1i}(t+1) = c_{1i}(t) - \beta(CG_l(t+1) - CG_k(t)) . \quad (10)$$

4   Simulation and Analysis

In order to further validate the rationality of MOCPSO, three groups of experiments are designed to analyze the effect of the key parameters on the pareto-optimal set, and then different algorithms are compared. All experiments are done aiming at ZDT_1, ZDT_2, ZDT_4 and ZDT_6 given in[3]. The uncertainty is constructed by adding an imprecision factor ε to the benchmark functions. The variables' dimensions are 30,30,10,10. The main parameters' values are set as follows: $\underline{w}=0.1, \overline{w} = 0.9, \overline{c_1} = \overline{c_2} = 3, \underline{c_1} = \underline{c_2} = 1, T = 100, w(0) = 0.4, c_1(0) = c_2(0) = 2$. Moreover, the convergence, the distribution and the uncertainty of pareto-optimal set are measured by interval hyper-volume(IH), interval spacing(ISP), interval purity(IP) and interval uncertainty(IX)[2,3, 10].

Under different Δw, ISP, IX and IP metrics are compared in Table.1. IH-metric are shown in Fig.1 by box plot. We see that IP-metric is largest and ISP-metric is smallest as $\Delta w = 0.1$. For $\overline{ZDT_6}$, the performances are better as $\Delta w = 0.05$. Under smaller Δw, the inertia weight is not adjusted enough. This makes the distribution of the pareto-optimal set worse. The particles fly over the possible global best with larger Δw. Hence, $\Delta w = 0.1$ has the exploration–exploitation tradeoff.

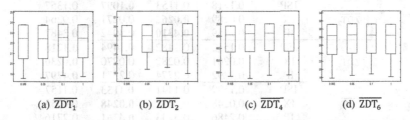

(a) $\overline{ZDT_1}$ (b) $\overline{ZDT_2}$ (c) $\overline{ZDT_4}$ (d) $\overline{ZDT_6}$

Fig. 1. The box plot of IH-metric obtained by different Δw

Table 1. The performances of MOCPSO with different Δw

Functions	measures	Δw			
		0.005	0.05	0.1	1
$\overline{ZDT_1}$	ISP	0.1267	0.1246	**0.1091**	0.1215
	IX	**0.0241**	0.0290	0.0289	0.0299
	IP	0.2800	0.2809	**0.4602**	0.3557
$\overline{ZDT_2}$	ISP	**0.1253**	0.1364	0.1568	0.1495
	IX	0.0293	0.0283	0.0269	**0.0267**
	IP	0.2810	0.3554	0.3064	**0.3829**
$\overline{ZDT_4}$	ISP	0.1843	0.1426	0.1325	**0.1266**
	IX	0.0261	0.0261	**0.0254**	0.0259
	IP	0.1694	0.2628	**0.4809**	0.2776
$\overline{ZDT_6}$	ISP	0.1414	**0.1122**	0.1151	0.1311
	IX	0.0271	**0.0216**	0.0229	0.0224
	IP	0.3687	**0.4272**	0.3462	0.2111

Under different α, the statistical results are listed in Table.2 and Fig.2. We see that the performances of $\overline{ZDT_1}$ and $\overline{ZDT_2}$ are best as $\alpha=0.5$. For $\overline{ZDT_6}$, it is better to chose

α=2. Especially, α=10 fits for $\overline{ZDT_4}$. IH-metric indicates that all functions have a better convergence as α equals to 0.5 or 2, except for $\overline{ZDT_4}$. Thus, α=0.5 is better because less α cannot provide enough adjustment for social coefficient while larger α forms a large disturbance for particles flight.

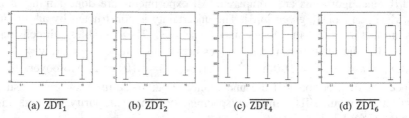

(a) $\overline{ZDT_1}$ (b) $\overline{ZDT_2}$ (c) $\overline{ZDT_4}$ (d) $\overline{ZDT_6}$

Fig. 2. The boxplot of IH-metric obtained by different α

Table 2. The performances of MOCPSO with different α

Function	Measure	α			
		0.1	0.5	2	10
$\overline{ZDT_1}$	ISP	0.1198	**0.1089**	0.1197	0.1064
	IX	0.0273	**0.0262**	0.0268	0.0285
	IP	0.2810	**0.3819**	0.3732	0.3010
$\overline{ZDT_2}$	ISP	0.1418	0.1153	**0.1097**	0.1357
	IX	0.0289	**0.0262**	0.0271	0.0264
	IP	0.2159	**0.4310**	0.3942	0.3487
$\overline{ZDT_4}$	ISP	0.1552	0.1482	0.1494	**0.1218**
	IX	0.0261	0.0282	0.0270	**0.0230**
	IP	0.4292	0.2774	0.2571	**0.4797**
$\overline{ZDT_6}$	ISP	0.1579	0.1441	**0.1153**	0.1157
	IX	0.0248	0.0251	**0.0248**	0.0255
	IP	0.2486	0.3243	**0.4361**	0.2716

Under different β, the statistical results listed in Table.3 and Fig.3 show that the algorithm's performances is better when β=2.

Table 3. The performances of MOCPSO with different β

Function	Measure	β			
		0.1	0.5	2	10
$\overline{ZDT_1}$	ISP	**0.1192**	0.1234	0.1297	0.1298
	IX	0.0291	**0.0269**	0.0288	0.0271
	IP	0.3950	**0.4576**	0.3462	0.2397
$\overline{ZDT_2}$	ISP	0.1203	**0.1157**	0.1361	0.1367
	IX	0.0253	0.0259	**0.0236**	0.0261
	IP	0.4245	**0.4647**	0.1418	0.1345
$\overline{ZDT_4}$	ISP	0.1987	**0.1150**	0.1240	0.1465
	IX	**0.0236**	0.0264	0.0265	0.0247
	IP	0.5025	**0.5103**	0.2109	0.1341
$\overline{ZDT_6}$	ISP	**0.0994**	0.1034	0.1423	0.1228
	IX	0.0264	**0.0242**	0.0255	0.0245
	IP	0.2977	**0.5523**	0.1800	0.1682

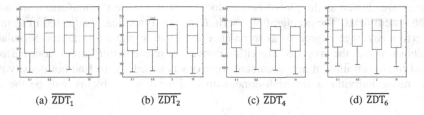

(a) $\overline{ZDT_1}$ (b) $\overline{ZDT_2}$ (c) $\overline{ZDT_4}$ (d) $\overline{ZDT_6}$

Fig. 3. The box plot of IH-metric obtained by different β

The rationality and validity of MOCPSO is further analyzed by compared with multi-objective particle swarm optimization algorithm(MOPSO) and probability dominate multi-objective particle swarm optimization (PD-MOPSO)[2]. The statistical results listed in Table.4 and Fig.4 show that though MOPSO has better performances than others for $\overline{ZDT_2}$, MOCPSO does better in other benchmark functions.

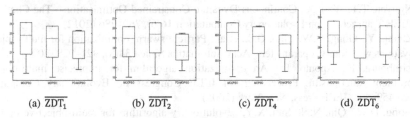

(a) $\overline{ZDT_1}$ (b) $\overline{ZDT_2}$ (c) $\overline{ZDT_4}$ (d) $\overline{ZDT_6}$

Fig. 4. Comparison of the IH-metric among different algorithms

Table 4. Comparison of the performances among different algorithms

Function	algorithm	ISP	IX	IP
$\overline{ZDT_1}$	MOCPSO	**0.1278**	0.0277	**0.5866**
	MOPSO	0.1350	0.0292	0.4741
	PD-MOPSO	0.1403	**0.0271**	0.1238
$\overline{ZDT_2}$	MOCPSO	0.1274	0.0273	0.4081
	MOPSO	**0.1160**	**0.0271**	**0.6279**
	PD-MOPSO	0.1661	0.0304	0.1569
$\overline{ZDT_4}$	MOCPSO	**0.1190**	**0.0230**	**0.7625**
	MOPSO	0.1443	0.0255	0.4621
	PD-MOPSO	0.1706	0.0266	0.1034
$\overline{ZDT_6}$	MOCPSO	**0.0967**	**0.0226**	**0.7368**
	MOPSO	0.1193	0.0260	0.3776
	PD-MOPSO	0.1239	0.0281	0.1249

5   Conclusions

Aiming at interval multi-objective optimization problem, a kind of cultural particle swarm optimization was proposed. In belief space, particles' locations and weight inertia are updated by normative knowledge. The grid's coverage degree is defined in topographic knowledge instead of the crowding distance to measure the distribution of non-dominant solutions in object space. The solutions with smaller coverage degrees are chosen as the global best or local best to ensure the uniform distribution of the pareto front. The accelerating coefficients are dynamically adjusted in terms of the grid's coverage degree so as to balance exploration and exploitation. Simulation results show that the proposed algorithm converges to the better Pareto front uniformly and the uncertainty of non-dominant solutions is less. How to find out density relation among individuals for high-dimension optimization problems will be the next work.

Acknowledgement. This work was supported in part by National Basic Research Program of China(2014CB046300). Thank you for the help from Collaborative Innovation Center of Intelligent Mining Equipment of Jiangsu province.

References

1. Nguyen, T.T., Yao, X.: Continuous Dynamic Constrained Optimisation---The Challenges. IEEE Transactions on Evolutionary Computation **16**(6), 769–786 (2012)
2. Zhang, Y., Gong, D.W., Hao, G.S., et al.: Particle swarm optimization for multi-objective systems with interval parameters. Acta Automatica Sinica **34**(8), 921–928 (2008)
3. Limbourg, P., Aponte, D.: An optimization algorithm for imprecise multi-objective problem functions.In: Proceedings of IEEE Congress on Evolutionary Computation, pp. 459–466. IEEE Press, New York (2005)
4. Gong, D.W., Qin, N.N., Sun, X.Y.: Evolutionary algorithm for multi-objective optimization problems with interval parameters. In: Proc.of 5th IEEE International Conference on Bio-Inspired Computing:Theories and Applications, pp. 411–420. IEEE Press, New York (2010)
5. Soares, G.L., Guimars, F.G., Maia, C.A., et al.: Interval robust multi-objective evolutionary algorithm.In: 2009 IEEE Congress on Evolutionary Computation, pp. 1637–1643. IEEE Press, New York (2009)
6. Tripathi, P.K., Bandyopadhyay, S., Pal, S.K.: Multi-Objective Particle Swarm Optimization with time variant inertia and acceleration coefficients. Information Sciences **177**, 5033–5049 (2007)
7. Branke, J., Mostaghim, S.: About selecting the personal best in multi-objective particle swarm optimization. In: Runarsson, T.P., Beyer, H.-G., Burke, E.K., Merelo-Guervós, J.J., Whitley, L., Yao, X. (eds.) PPSN 2006. LNCS, vol. 4193, pp. 523–532. Springer, Heidelberg (2006)
8. Daneshyari, M., Yen, G.G.: Cultural-based multi-objective particle swarm optimization. IEEE Transactions on Systems, Man, and Cybernetics, Part B: Cybernetics **41**(2), 553–567 (2011)
9. Cheng, Z.Q., Dai, L.K., Sun, Y.X.: Feasibility analysis for optimization of uncertain systems with interval parameters. Acta Automatica Sinica **30**(3), 455–459 (2004)
10. Bandyopadhyay, S., Pal, S.K., Aruna, B.: Multi-objective Gas, Quantitative Indices, and Pattern Classification. IEEE Transactions on Systems, Man, and Cybernetics, Part B: Cybernetics **34**(5), 2088–2099 (2004)

A New Multi-swarm Multi-objective Particle Swarm Optimization Based Power and Supply Voltage Unbalance Optimization of Three-Phase Submerged Arc Furnace

Yanxia Sun[1(✉)] and Zenghui Wang[2]

[1] Department of Electrical Engineering,
Tshwane University of Technology, Pretoria 0001, South Africa
sunyanxia@gmail.com
[2] Department of Electrical and Mining Engineering,
University of South Africa, Florida 1710, South Africa
wangzengh@gmail.com

Abstract. To improve the production ability of a three-phase submerged arc furnace (SAF), it is necessary to maximize the power input; and it needs to minimize the supply voltage unbalances to reduce the side effect to the power grids. In this paper, maximizing the power input and minimizing the supply voltage unbalances based on a proposed multi-swarm multi-objective particle swarm optimization algorithm are the focus. It is necessary to have objective functions when an optimization algorithm is applied. However, it is difficult to get the mathematic model of a three-phase submerged arc furnace according to its mechanisms because the system is complex and there are many disturbances. The neural networks (NN) have been applied since its ability can be used as an arbitrary function approximation mechanism based on the observed data. Based on the Pareto front, a multi-swarm multi-objective particle swarm optimization is described, which can be used to optimize the NN model of the three-phase SAF. The simulation results showed the efficiency of the proposed method.

Keywords: Multi-objective optimization · Particle swarm optimization · Submerged arc furnace · Power optimization · Supply voltage unbalances

1 Introduction

In the past decades there has been a drastic increment in the number and size of Submerged Arc Furnaces (SAF) constructed for the production of Ferro-chromium and Ferro-manganese alloys. The economic benefit caused the use of larger furnaces which are relatively large, e.g. 48 MVA for ferro-chromium, and up to 81 MVA for ferro-manganese, with currents ranging from about 50 to 130KA [1]. With the increment of the furnaces' power, it is very important to consider the side effects on the power grim such as supply voltage unbalances for three-phase submerged arc furnaces. Hence we have to consider the constraints or other objectives when furnaces are optimally controlled. There are many parameters or variables about three-phase submerged arc

© Springer International Publishing Switzerland 2015
Y. Tan et al. (Eds.): ICSI-CCI 2015, Part I, LNCS 9140, pp. 513–522, 2015.
DOI: 10.1007/978-3-319-20466-6_54

furnace and the most important variables are voltages, equivalent resistance and temperature, for determining power of SAFs. However, it is difficult to construct mathematical model SAF according to the mechanisms of the actual furnace plant system due to its complexity and many disturbances; and the neural network is a good option to model SAF as it is easy to use in modeling nonlinear functions based on the observed data. Neural networks have been widely used for modeling complicated systems and achieve good results [2], [3]. Hence, the optimization algorithm can be applied based on the neural network model of for three-phase submerged arc furnaces to get the control signals. Here, a proposed multi-swarm multi-objective particle swarm optimization algorithm was used to optimize the power and supply voltage unbalances.

The rest parts of the paper are organized as follows. Section 2 the multi-objective particle swarm optimization was reviewed and a multi-swarm multi-objective particle swarm optimization (MSMOPSO) method was proposed. Section 3 investigated the three-phase SAF. Three-phase SAF was modeled by BP neural network in Section 4. Section 5 presents MSMOPSO based power and voltage unbalances optimization. The concluding remarks were given in the last section.

2   Multi-swarm Multi-objective Particle Swarm Optimization

In general the single objective optimization algorithms will terminate when an optimal solution is obtained. But for most MOO problems, there can be a number of optimal solutions. A multi-objective optimization (MOO) problem can be described by

$$\text{Min } F(x) = \text{Min}(f_1(x), \cdots, f_m(x)) \tag{1}$$

$$\text{Subject to } x \in \Omega .$$

Here Ω is the variable space, R^m is the objective space, and $F : \Omega \to R^m$ consists of m real-valued objective functions.

If there is no information regarding the preference of objectives, a ranking scheme based upon the Pareto optimality is regarded as an appropriate for MOO [4]. The solution to the MOO problem is described by a Pareto front set. For the more details related to Pareto front set, please refer to reference [5].

A good MOO algorithm should guarantee a high probability of finding the Pareto optimal set. Among the MOO algorithms, the multi-objective particle swarm optimization algorithm has been proven to be a promising algorithm [6]. To achieve good optimization performance, the particles can be divided to several swarms. If a multiple-swarm PSO employs an over large number of swarms, it will have a better chance of obtaining possible good solutions that lead to the optimal Pareto set, but it can also suffer from an undesirable computational cost. There are some multiple-swarm PSO algorithms, such as reference [6] [7], which used the adaptive swarm size methods. However, the existing MSMOPSOs do not use the information of the found Pareto front set to allocate the swarms. For most of the continuous optimization problems, the good results may be discovered if the particles search around the Pareto front. Based on this finding, we propose an MSMO optimization method. Several swarms are used to search regions around certain points of the Pareto front set. These swarms are called Pareto front swarms. There is still another swarm, which is called spare swarm and searches other spaces far away from the Pareto front to ensure all the particles are spread around the whole objective space. The main contributions of the proposed algorithm are:

1)   Pareto front swarms are used to search different regions around some points of Pareto front, and the velocity update equation is

$$V_i(t+1) = \omega V_i(t) + c_1 R_1 (P_i - X_i(t)) + c_2 R_2 (P_g - X_i(t)) + c_3 R_3 (Core(m) - X_i(t)) \quad (2)$$

$$X_i(t+1) = X_i(t) + V_i(t+1) \quad (3)$$

Here, $Core(m)$ is central point of the mth swarm and is chosen dynamically, the relationship between m and i is $m = floor(\dfrac{i}{num_g}) + 1$, num_g is the particle number of the mth swarm and $floor(A)$ rounds the elements of A to the nearest integers less than or equal to A. The number of the cores equals the number of the Pareto front swarms. The cores are from the Pareto front set and using the same way as choosing the Pareto front set.

2) The particles of the spare swarm are updated using

$$V_i(t+1) = \omega V_i(t) + c_1 R_1 (P_i - X_i(t)) + c_2 R_2 (P_g - X_i(t)) - c_4 R_4 (Core(m) - X_i(t)) \quad (4)$$

$$X_i(t+1) = X_i(t) + V_i(t+1) \quad (5)$$

Here, c_4 is determined by the sharing function [8] according to the distance between particle i and core particles,

$$R_4 = \frac{1}{m_g} rand(\cdot) \quad (6)$$

and m_g is the number of Pareto front swarms.

3) To avoid the premature of PSO, small disturbance is added, that is,

$$V_i(t+1, irand) = V_i(t+1, irand) + \frac{R_v}{m_g} \quad (7)$$

Here, R_v is a random number within an interval of $[-1,1]$.

The method of choosing P_i and P_g is described in ref. [9].

The following procedure can be used for the proposed particle swarm algorithm:

1) Initialize the parameters of particles.

2) Evaluate the fitness functions for each particle.

3) Find the non-dominated Pareto front particles and store them in the repository set.

4) Determine the cores of Pareto front swarms and dynamically set up the relationship among the swarms and the cores.

5) Using (2) and (3); or (4), (7) and (5) to update particles.

6) Repeat steps (2)-(6) until a stopping criterion is met (e.g., maximum number of iterations or a sufficiently good fitness value).

3   Three-Phase Submerged Arc Furnaces

A typical three-phase SAF consists of a fixed circular bath and three electrodes sub-merged in a charge of raw materials projected into it. The operation of the SAF involves trying to maintain the maximum real power input to the furnace within the constraints or limits of the associated equipment of the furnace [10]. To control the input power, the input voltage can be changed by the transformers. The transformers for the furnaces are different from the standard power system transformers in that the secondary winding has to supply very high currents at low voltages as shown in Fig. 1. Furnace transformers are used to step down from voltages between 11KV and 33 kV to levels of several hundred volts and control the input voltage of the furnace.

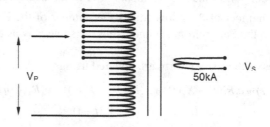

Fig. 1. A single phase furnace transformer [12]

There are also constraints and limits for the operation of the three-phase SAF, for example, the voltage unbalance must be considered. The voltage unbalance occurs in SAFs, when current consumption is not balanced during the 3 cycle of processing (operation) or during a faulty condition before tripping. They impact negatively on three phase asynchronous motors by causing overheating and a tripping of protective devices. A voltage unbalance is a ratio of the negative sequence component to the positive sequence component and it can be determined by the following formulas [11] and the unbalance voltage, u_u, is given as:

$$u_u = \max_i (\frac{U_i - U_{avg}}{U_{avg}})100\%, (i = 1,2,3) \tag{8}$$

where U_i is the phase voltage, and $U_{avg} = \frac{1}{3}\sum_{i=1}^{3} U_i$.

Hence to achieve good control performance of the three-phase SAF, *there are at least two objectives: 1) maximum the power input; 2) minimum the voltage unbalances.* Of course, there are other constraints or limits such as harmonics, which should be considered when the SAF system is optimized/controlled. In this study, we only focus on these two objectives.

4   Neural Network Based Modeling of Three-Phase SAF

One of the most commonly used supervised neural networks is the back-propagation network which uses the back-propagation learning algorithm [13, 14, 15]. It was first proposed by Paul Werbos in 1974, but it wasn't until 1986, through the work of David E. Rumelhart, Geoffrey E. Hinton and Ronald J. Williams [16], that it gained recognition, which led to a "renaissance" in the field of artificial neural network research. The back propagation neural network is essentially a network of simple processing elements working together to produce a complex output. The combination of weights which minimizes the error function is considered to be a solution of the learning problem. Here, the Neural Network Model is of two-layer feed-forward network with the default tan-sigmoid transfer function in the hidden layer with 45 neurons and the linear transfer function in the output layer. The design SAF model is trained using the Levenberg-Marquardt back-propagation method. To test the performance of the proposed SAF neural network model, a set of electro-thermal variables from the 45 MW SAF Wonderkop Chrome Processing Plant (WCP) was used [12]. The input vectors (equivalent resistances, voltages and temperature) and the target vector (power) comprise of 120 samples each. It should be noted that only the first 90 samples are used to train neural network and the last 30 samples are used to validate and test the trained neural network.

The Neural Network Fitting Tool GUI is utilized to construct and train the neural network based on the software MATLAB 2009a. The linear regression performance between the obtained NN model outputs and the corresponding targets (power) shows that the model's output tracks the targets very well for training and validations shown in Fig. 2, which means that the trained neural network model is acceptable.

The output of the SAF NN model and the real power data reveal some similarities explicitly. The real furnace power samples and the trained neural network model output are shown in Fig. 3. As can be seen from Fig. 3, the obtained neural network model showed similar characteristics of the real samples although the last 30 samples were not used to train the BP neural network. Hence this NN model can be used as the representative of the real SAF system.

5   Multi-swarm Multi-objective PSO Based Power and Voltage Unbalances Optimization of Three-Phase SAF

As the data is from the 45 MW SAF Wonderkop Chrome Processing Plant (WCP), the theoretic input power can be 45 MW. However, as can be seen from Fig. 3, the real sample input power is much lower than the theoretic value (the highest input sample power is about 35 MW). Hence, there should be space to improve the input power based on the optimization algorithm although the voltage unbalances to be considered. As mentioned in Section 3, to optimize the performance of this three-phase SAF, the optimization problem can be described by

$$\text{Min } F(x) = (f_1(x), f_2(x)) \qquad (9)$$

$$\text{Subject to } x \in \Omega \qquad (10)$$

where, $f_1(x)$ and $f_2(x)$ are the input power and the voltage unbalance, respectively, and $x = [R_1, R_2, R_3, U_1, U_2, U_3, T]$, R_1, R_2, R_3 are three phase equivalent resistances, U_1, U_2, U_3 are three input phase voltages, and T is the furnace temperature.

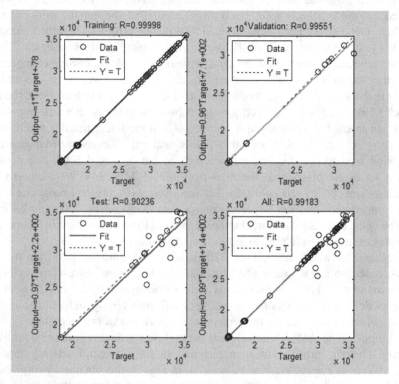

Fig. 2. Simulated Regression Characteristics

Since we cannot get the mathematical model of the input power, the neural network model obtained in Section 4 can be used as $-f_1(x)$. It should be noted that there is a minus sign before $f_1(x)$ since the first objective is to maximum the input power. The second objective is

$$f_2(x) = \min\left(\max_i \left(\frac{U_i - U_{avg}}{U_{avg}}\right) \cdot 100\%\right). \tag{11}$$

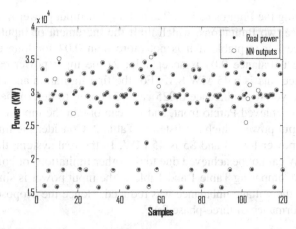

Fig. 3. SAF neural network output versus real output data (power)

Altough there are 7 variables in the NN SAF model, only U_1, U_2, U_3 are looked as control variable and R_1, R_2, R_3, T can be looked as time-variant parameters since this system is a slow response system can R_1, R_2, R_3, T can be measured or calculated in real time. For the condition (10), we can choose $276 < U_1 < 300; 100 < U_2 < 290; 100 < U_3 < 263$ based on the samples. Table 1 gives the first 5 sets of samples and the simulation will be implement to verify the proposed method base on these 5 sets of samples.

Table 1. 5 sets of samples [12]

	RESISTANCE			VOLTAGE (V)			Temp. (°C)	POWER (KW)
	R1	R2	R3	U1	U2	U3	T	P
S1	2.1	2.1	2.77	289	158	188	2349	29417.31
S2	2.73	2.27	4.21	286	187	200	2200	28363.93
S3	2.14	2.12	1.29	299	197	263	2670	33450.08
S4	4.27	4.18	2.09	300	100	100	2717	15914.31
S5	2.14	1.75	1.28	284	290	163	2706	31734.41

In the simulation, the total number of fitness function evaluations was set to 10 000. The particle number is 200. The number of Pareto front swarms is 20 and each swarm has 8 particles. A random initial population was created for each of the 20 runs on each test problem. The maximum number of external repository particles is 100. Parameters are set as $c_1 = c_2 = 2$ and $\omega = 0.5 + rand(.)$.

Using the proposed method, the Pareto fronts were obtained and they are shown in Fig. 4, 5, 6, 7 and 8 for S1, S2, S3, S4 and S5, respectively. Here S1, S2, S3, S4 and S5 are referring to the underlined parameters in Table 1. As can be seen from Fig. 4, 5, 6, 7 and 8, the Pareto front is smooth and uniform which means the proposed multi-swarm multi-objective PSO works well.

Only considering the Figures 4, 5, 6, 7 and 8, higher input power may be obtained. However there are two limitations, which limit the increment of input power: 1) the voltage unbalance is acceptable if it is not more than 0.02 for long time; for short time, the voltage unbalance 0.04 is acceptable; 2) the input power cannot be more than 45 MW since this is a 45 MW SAF. For the first limitation and considering the real situation, the voltage unbalance 0.025 can be chosen to determine the power inputs based on the achieved Pareto front, and we can obtain the input voltages and the corresponding input power which are listed in Table 2. Consider the limitation 2), the maximum input power for S3 and S5 is 45 MW. In the real system, the input power more than 45 MW cannot be achieved due to the other limitations or constraints of the physical system. Comparing Table 1 and Table 2, the input power is similar with each other for S2, but the voltage unbalance was reduced. Hence the proposed method can improve the performance of three-phase SAF.

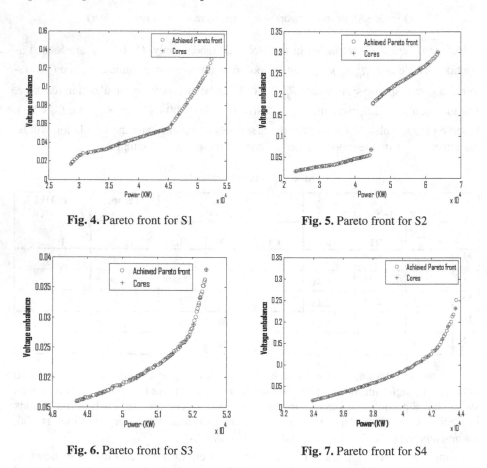

Fig. 4. Pareto front for S1

Fig. 5. Pareto front for S2

Fig. 6. Pareto front for S3

Fig. 7. Pareto front for S4

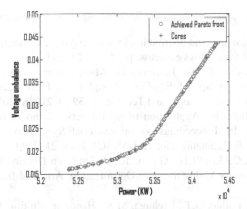

Fig. 8. Pareto front for S5

Table 2. Optimization result based on the five sets of samples

RESISTANCE			VOLTAGE (V)			Temp. (°C)	POWER (KW)
R1	R2	R3	U1	U2	U3	T	P
2.1	2.1	2.77	276	276.5	257.1	2349	29906
2.73	2.27	4.21	283.3	283.1	263	2200	28275
2.14	2.12	1.29	276	276	263	2670	45000
4.27	4.18	2.09	276	276	256	2717	34936
2.14	1.75	1.28	276	276	263	2706	45000

6   Conclusion

The power and voltage unbalance of three-phase SAF were optimized based on a proposed multi-swarm multi-objective particle swarm optimization (MSMOPSO). A back-propagation neural network was used to model the three-phase SAF, and then MSMOPSO was implemented on the obtained neural network model. The achieved Pareto fronts are smooth and uniform which means the proposed multi-swarm multi-objective PSO works well.   Moreover, the simulation result showed the efficiency of the proposed method to improve the performance of three-phase SAF. In our future research, the more constraints such as harmonics, power factor, and so on, will be considered to make the optimization be used in the real SAF plants.

Acknowledgement. This work was supported by China/South Africa Research Cooperation Program (No. 78673), South African National Research Foundation Incentive Grants (No. 81705 and 95687).

References

1. Westly, J.: Resistance and heat distribution in submerged – arc furnace. In: Proceedings of the 1st International ferro-alloys congress, pp. 121–127 (1974)
2. Liao, S., Kabir, H., Cao, Y., Xu, J., Zhang, Q., Ma, J.: Neural-Network Modeling for 3-D Substructures Based on Spatial EM-Field Coupling in Finite-Element Method. IEEE Transactions on Microwave Theory and Techniques 59(1), 21–38 (2011)
3. Moumi, P., Tanushree, B.: Application of neural network model for designing circular monopole antenna. In: Proceedings on International Symposium on Devices MEMS, Intelligent Systems & Communication (ISDMISC), 2, 18–21 (2011)
4. Khor, E.F., Tan, K.C., Lee, T.H., Goh, C.K.: A Study on Distribution Preservation Mechanism in Evolutionary Multi-Objective Optimization. Artificial Intelligence Review 23, 31–56 (2005)
5. Coellocoello, C.A., Pulido, G.T., Lechuga, M.S.: Handling Multiple Objectives With Particle Swarm Optimization. IEEE Transactions on Evolutionary Computation 8, 256–279 (2004)
6. Prasanna, H.A.M., Kumar, M.V.L., Veeresha, A.G., Ananthapadmanabha, T., Kulkarni, A.D.: Multi-objective optimal allocation of a distributed generation unit in distribution network using PSO. Advances in Energy Conversion Technologies (ICAECT). In: 2014 International Conference on Electrical Engineering, Iran, Tehran, pp. 61–66, May 2014
7. Moeini, A.F., Tajvar, P.; Asgharian, R., Yaghoobi, M.: Colonial multi-swarm: a modular approach to administration of particle swarm optimization in large scale problems. In: The 22nd Iranian Conference on Electrical Engineering (ICEE), Jeju, Korea, pp. 986–991 (2014)
8. Goldberg, D.E.: Genetic Algorithms in Search, Optimization and Machine Learning. Addison-Wesley, Reading (1989)
9. Jeong, S., Hasegawa, S., Shimoyama, K., Obayashi, S.: Development and Investigation of Efficient GA/PSO-Hybrid Algorithm Applicable to Real-World Design Optimization. IEEE Computational Intelligence Magazine, 36–44, August 2009
10. Ennie, M.S.: The operation, control and design of submerged-arc ferroalloy furnaces mintek 50. In: Proceedings of International Conference on Mineral Science and Technology, Sandton, South Africa, March 1984
11. Stout, M.B.: Basic electrical measurement. Prentice Hal Inc., Englewood Cliff (1960)
12. Amadi, A.: Measurement and control of electro-thermal variable parameters in three-phase Submerged Arc Furnaces (SAF), Dissertation of MASTERS OF TECHNOLOGY. University of South Africa (2012)
13. Passaro, A., Starita, A.: Clustering particles for multimodal function optimisation. In: Proceedings of ECAI Workshop Evolutionary Computation, Riva del Garda, Italy, pp. 124–131 (2006)
14. Kecman, V.: Learning and Soft Computing, Support Vector machines, Neural Networks and Fuzzy Logic Models, MIT Press (2001)
15. Wang, L.P., Fu, X.J.: Data Mining with Computational Intelligence. Springer Press (2005)
16. Passino, K.M.: Biomimicry of bacterial foraging for distributed optimisation and control. IEEE Control System Magazine, 52–67 (2002)

Handling Multiobjectives with Adaptive Mutation Based ε-Dominance Differential Evolution

Qian Shi[1], Wei Chen[2], Tao Jiang[2], Dongmei Shen[1], and Shangce Gao[1,2(✉)]

[1] College of Information Science and Technology,
Donghua University, Shanghai, China
[2] Faculty of Engineering, University of Toyama, Toyama, Japan
gaosc@eng.u-toyama.ac.jp

Abstract. Differential evolution (DE) is well known as a powerful and efficient population-based stochastic real-parameter optimization algorithms over continuous space. DE is recently shown to outperform several well-known stochastic optimization methods in solving multi-objective problems. Nevertheless, its performance is still limited in finding a uniformly distributed and near optimal Pareto fronts. To alleviate such limitations, this paper introduces an adaptive mutation operator to avoid premature of convergence by adaptively tuning the mutation scale factor F, and adopts ε-dominance strategy to update the archive that stores the nondominated solutions. Experiments based on five widely used multiple objective functions are conducted. Simulation results demonstrate the effectiveness of our proposed approach with respect to the quality of solutions in terms of the convergence and diversity of the Pareto fronts.

1 Introduction

Differential evolution (DE) algorithm [1] is a novel technique that was originally thought to solve the problem of Chebyshev polynomial. It is a population based stochastic meta-heuristic for global optimization on continuous domains which related both with simplex methods and evolutionary algorithms. Due to its simplicity, robustness, and effectiveness, DE is successfully applied in solving optimization problems arising in various practical applications [2], such as data clustering, image processing, etc. DE outperforms many other evolutionary algorithms in terms of convergence speed and the accuracy of solutions. Its performance, however, is still quite dependent on the setting of control parameters such as the mutation factor [3] for complex real-world optimization problems, especially those with multiple objectives [4,5].

In multiple objective problems, several objectives (or criteria) are, not unusually, stay in conflict with each other, thus requiring a set of non-dominated solutions, i.e., Pareto-optimal solutions to be the candidates for decision. The general goals of this requirement are the discovery of solutions as close to the Pareto-optimal as possible, and the distribution of solutions as diverse as

© Springer International Publishing Switzerland 2015
Y. Tan et al. (Eds.): ICSI-CCI 2015, Part I, LNCS 9140, pp. 523–532, 2015.
DOI: 10.1007/978-3-319-20466-6_55

possible in the obtained non-dominated set. Many works have been reported to satisfying these two goals. Wang et al. [6] proposed a crowding entropy-based diversity measure to select the elite solutions into the elitist archive. Zhang et al. [7] utilized the direction information provided by archived inferior solutions to evolve the differential mutations. Gong et al. [8] introduced the ε-dominance and orthogonal design into DE to keep the diversity of the individuals along the trade-off surface. More recently, Chen et al. [9] proposed a cluster degree based individual selection method to maintain the diversity of non-dominated solutions. A hybrid opposition-based DE algorithm was proposed by combining with a multi-objective evolutionary gradient search [10]. Although these variants of multi-objective DE have demonstrated that DE is suitable for handling multiple objectives, rare work, however, is carried out to discuss the setting of control parameters involving the mutation factor in the multi-objective DE.

Based on the above consideration, in this work, we proposed an adaptive mutation operator into DE to avoid the premature convergence of non-dominated solutions. In the former searching phases, the setting of mutation scale factor F remains large enough to explore the search space sounding to the non-dominated solutions, thus maintaining the diversity of the distribution of Pareto set. Along with the lapse of evolution, F is gradually reduced to perform the exploitation around the promising search area, aiming to reserve good information and to avoid the destruction of the optimal solutions. Furthermore, as noticed by Zitzler et al. [11] that elitism helps in achieving better convergence of solutions in multi-objective evolutionary algorithm, an elitist scheme is adopted by maintaining an external archive of nondominated solutions obtained in the evolution process. Moreover, the ε-dominance strategy [12] which can provide a good compromise in terms of convergence near to the Pareto-optimal and the diversity of Pareto fronts is also used in the algorithm. It is expected that, with the utilization of elitist scheme and ε-dominance, the cardinality of Pareto-optimal region can be reduced, and no two obtained solutions are located within relative small regions. To verify the performance of the proposed algorithm, five widely used benchmark multiple objective functions are utilized as the test suit. Experimental results indicate that the proposed adaptive mutation based multi-objective DE outperforms traditional multi-objective evolutionary algorithms in terms of the convergence and diversity of the Pareto fronts.

2   Brief Introduction to DE

The standard DE is essentially a kind of special genetic algorithm based on real parameter and greedy strategy for ensuring quality. An iteration of the classical DE algorithm consists of the four basic steps: initialization of a population of search variable vectors, mutation, crossover or recombination, and finally selection. DE begins its search with a randomly initiated population for a global optimum point in a D-dimensional real parameter space. We denote subsequent generations in DE by $G = \{0, 1, 2, \cdots, G_{max}\}$ and the i-th ($i = 1, 2, ..., NP$) individual of the current population is denoted as $X_{i,G} = (x_{i,G}^1, x_{i,G}^2, ... x_{i,G}^j, ..., x_{i,G}^D)$.

The initial population is randomly generated by:

$$x_{j,i,0} = x_{j,min} + rand_{i,j}[0,1] * (x_{j,max} - x_{j,min}) \tag{1}$$

where $rand_{i,j}[0,1]$ is a uniformly distributed random number in $[0,1]$, $x_{j,min}$ and $x_{j,max}$ represents the boundary values of the search space. For each individual vector $X_{i,G}$ (target vector), differential evolution algorithm uses mutation operator to generate a new individual $V_{i,G}$ (variation vector), which is generated according to Eq. (2).

$$V_{i,G} = X_{r1,G} + F * (X_{r2,G} - X_{r3,G}) \tag{2}$$

where three individuals vectors $X_{r1,G}$, $X_{r2,G}$ and $X_{r3,G}$ are selected randomly from the current populations. $r1, r2, r3 \in \{1, 2, \cdots, NP\}$ are random indexes. F is a real and constant scale factor $\in [0,2]$ which controls the amplification of the differential variation ($X_{r2,G}$ - $X_{r3,G}$). In order to increase the potential diversity of the perturbed parameter vectors, a crossover operation comes into play after generating the donor vector through mutation. The binomial crossover operation was shown in the following.

$$u_{i,G} = \begin{cases} v_{i,G}^j, & \text{if } rand_{i,j}[0,1] \leqslant C_r \ \ or \ \ j = j_{rand} \\ X_{i,G}^j, & \text{otherwise} \end{cases} \tag{3}$$

where C_r is called the crossover rate. $rand_{i,j} \in [0,1]$. After DE generates offspring through mutation and crossover operation, the one-to-one greedy selection operator is performed as:

$$u_{i,G+1} = \begin{cases} U_{i,G}^j, & \text{if } f(U_{i,G}) \leqslant f(X_{i,G}) \\ X_{i,G}^j, & \text{otherwise} \end{cases} \tag{4}$$

3   Design of Multi-objective Differential Evolution Algorithm

For solving multiple objective problems, the general requirements of the approximation of the Pareto optimal set are two-fold: (1) minimize the distance to the true pareto optimal fronts, and (2) the distribution of the obtained non-dominated solutions are located as diverse as possible [13]. The purpose of this research is aimed to address the above two requirements, and the processes of the proposed adaptive mutation based ε-dominance differential evolution (IDE) are summarized in Fig. 1.

To generate initial solutions evenly located over the whole decision space, the orthogonal experimental design method [14] is adopted in IDE. Refer to [15] for detailed description of the orthogonal experimental design in population-based evolutionary algorithm. After generating the orthogonal population (denoted

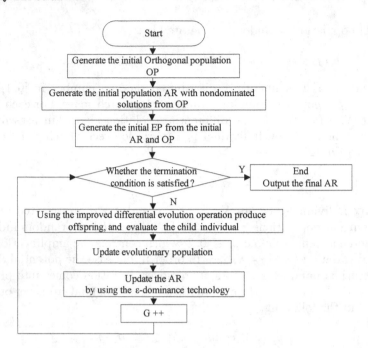

Fig. 1. The general flow chart of the proposed adaptive mutation based multi-objective differential evolution (IDE)

as *OP*), an initial archive with the nondominated individuals extracted from *OP* through the traditional Pareto dominance method [16] is created. Then the initial evolutionary population (*EP*) which is responsible for finding new non-dominated solutions is generated from the initial archive and *OP*. If the size of initial archive is larger than *NP*, we select *NP* solutions from the initial archive randomly, or all of the *ar_size* (which is the size of the initial archive) solutions in the initial archive are inserted into *EP*, and the remainder *NP - ar_size* solutions are selected from *OP* randomly. In order to accelerate the algorithm convergence and make use of the archive individual to guide the evolution, we adopt a hybrid selection mechanism when selecting the target vector X_{r1} as shown in Eq. (2). At the beginning phase of the evolution, all of the parents for mating are randomly selected from *EP* to generate the offspring. With the lapse of evolution, the elitist selection is used. We randomly choose one solution from the archive as the base parent, and the other two parents are selected from the evolution population *EP* randomly.

In previously reported works [6–10], all those multi-objective DE algorithms set the scaling factor *F* as a constant in the whole process of evolution, which made the search appear precocious phenomenon frequently. It is very sensitive to set scaling factor *F* for traditional differential evolution algorithms. Experimental work in a variety of DE algorithms has provided strong evidence supporting the view that the performance of the algorithm is strongly depending on the setting of F values [17,18]. To be more specifically, if the F value is too large, the DE algorithm approximates for random search, thus the search efficiency and the accuracy of getting the global optimal solution are quite low. On the contrary, if the F value is too small, it can lose the diversity of population into the prematurity. To alleviate this problem, we propose an adaptive mutation operator that can determine the mutation rate adaptively according to the progress of the search of the algorithm, thus enabling the algorithm to possess greater mutation rates in the early search stages to maintain the individuals' diversity and to avoid precocious phenomena during the process. Later, the mutation operator was gradually reduced to reserve good information and avoid the destruction of the optimal solution, and meanwhile it increases the probability of searching to the optimal solutions.

To realize the above characteristic of the setting of F, an adaptive setting rule is designed as in Eqs. (6) and (7).

$$t = e^{1 - \frac{G_m}{G_m + 1 - G}} \tag{5}$$

$$F = F_0 * 2^G \tag{6}$$

where F_0 is initial mutation operator. G_m denotes the maximum number of fitness evaluation. G indicates the current evolution number. At the beginning search phase of the algorithm, the adaptive mutation operator is carried out with a probability within $[F_0 - 2F_0]$, which is a relatively large value to maintain the individual diversity. Along with the lapse of evolution, the mutation operator is gradually reduced to reserve good information and expected to well balance the exploration and exploitation of the search.

In addition, as noticed by Zitzler et al. [19] that elitism helps in achieving better convergence in handling multiple objectives. Therefore, in this paper, the elitist scheme is adopted through maintaining an external archive AR of non-dominated solutions found in evolutionary process. In order to achieve faster convergence, we adopted [20] ε-dominance mechanism to update archive population. At each generation, the newly generated non-dominated solution is compared with each other member which is already contained in the archive. The new individual can be saved in the archive only when it meets the requirements that no individuals within a ε distance exist. By doing so, we can ensure both convergence and diversity of the Pareto fronts within reasonable computational times.

4   Simulation and Analysis

Multi-objective optimization problem is also known as multi-criteria optimization problem [21]. In order to evaluate the effectiveness of the proposed IDE and make a comparison with other multi-objective evolutionary algorithms, five widely used benchmark problems [19] involving ZDT1, ZDT2, ZDT3, ZDT4 and ZDT6 are adopted as the test suit. All problems have two objective functions and all objective functions are to be minimized. The parameter settings of IDE are as follows: the maximum number of fitness evaluation $G_m = 5000$, the initial scaling factor value of $F_0 = 0.5$, the crossover probability of $CR = 0.3$, $NP = 100$. For each problem, we run 50 times independently with different random seeds, then compared the performance of IDE with the one of the traditional multi-objective DE variants (MDE) [8]. In addition, we compared the results of IDE algorithm with NSGA-II [16], SPEA2 [22] and MOEO [23]. To assess the performance of the compared algorithms, the convergence metric λ and the diversity metric Δ are used [13]. The first convergence metric λ measures the distance of the obtained non-dominated sets Q and the true Pareto front approximation sets P^* as in Eq. (7).

$$\lambda = \frac{\sum_{i=1}^{|Q|} d_i}{|Q|} \tag{7}$$

where d_i is the Euclidean distance between the solution $i \in Q$ and the nearest member of P^*. It is clear that the lower the λ value, the better convergence of obtained solutions, suggesting that the obtained non-dominated sets are more closer to the true Pareto fronts.

The second diversity metric measures the extent of distribution among the obtained non-dominated sets Q. Δ is defined as in Eq. (8).

$$\Delta = \frac{d_f + d_l + \sum_{i=1}^{|Q|-1} | d_i - \bar{d} |}{d_f + d_l + (|Q| - 1)\bar{d}} \tag{8}$$

where d_i measures the Euclidean distance of each point in Q to its closer point, d_f and d_l denote the Euclidean distance between the extreme points in Q and P^*, respectively. Obviously, the lower the Δ value is, the better distribution of solutions possess.

Table 1 records the convergence metric λ obtained by IDE and the previous MDE algorithm [8]. The diversity metric Δ obtained by IDE and MDE are shown in Table 2. Table 3 shows the convergence metric obtained by IDE and three multi-objective evolutionary algorithms. Table 4 illustrates comparative results in terms of the diversity metric obtained by IDE and its competitors.

Table 1. Comparison of the convergence metric between IDE and MDE

Problem	ZDT1	ZDT2	ZDT3	ZDT4	ZDT6
MDE	0.0028	0.00064	0.0038	0.0026	0.0008
IDE	0.00075	0.00084	0.0030	0.0020	0.00075

Table 2. Comparison of the diversity metric between IDE and MDE

Problem	ZDT1	ZDT2	ZDT3	ZDT4	ZDT6
MDE	0.2536	0.38565	0.40025	0.3850	0.3571
IDE	0.2425	0.2896	0.39575	0.2709	0.2595

Table 3. Comparison of the convergence metric during IDE, NSGA-II, SPEA2, and MOEO

Algorithm	ZDT1	ZDT2	ZDT3	ZDT4	ZDT6
NSGA-II	0.033482	0.072391	0.114500	0.513053	0.296564
SPEA2	0.023285	0.16762	0.018409	4.9271	0.23255
MOEO	0.001277	0.001355	0.004385	0.008145	0.000630
IDE	0.00075	0.00084	0.0030	0.0020	0.00075

Table 4. Comparison of the diversity metric during IDE, NSGA-II, SPEA2, and MOEO

Algorithm	ZDT1	ZDT2	ZDT3	ZDT4	ZDT6
NSGA-II	0.390307	0.430776	0.738540	0.702612	0.668025
SPEA2	0.154723	0.33945	0.4691	0.8239	1.04422
MOEO	0.327140	0.285062	0.965236	0.275567	0.225468
IDE	0.2425	0.2896	0.39575	0.2709	0.2595

From Table 1, we can find that IDE performs better results with respect to the convergence on all tested instances, except on ZDT2, which suggested that the incorporated adaptive mutation strategy indeed help the search finding better solutions. On the other hand, the comparative results in Table 2 show that IDE has capacity of finding a better spread of solutions than MDE on all problems except ZDT6. From Table 3, it is clear that IDE produces solutions significantly closer to the true Pareto fronts than NSGA-II, SPEA2, and MOEO on all tested functions. An exception is that MOEO can find slightly better solutions than IDE on ZDT6. With regards to the diversity of obtained non-dominated solutions, as shown in Table 4, an overall improvement can be found on IDE that its non-dominated solutions located more evenly than those obtained by its competitor algorithms, verifying that the proposed adaptive mutation strategy together with the ε-dominance no doubt improve the performance of DE in terms of the diversity.

Furthermore, to further understand the performance of our improved algorithm more intuitively, Fig. 2 draws the Pareto fronts constructed by the obtained non-dominated solutions that obtained by IDE and MDE on all tested functions respectively. From this figure, it is clear that the Pareto fronts obtained by IDE is much better than those by MDE. The performance on ZDT6 is quite illuminating to further elaborate the search characteristics of the compared algorithms. Almost the same number of non-dominated solutions are obtained by both algorithms, and the average distance (measured by λ) to the true Pareto front is also within an acceptable tolerance (0.0008 vs 0.00075). Nevertheless, the distribution of the non-dominated solutions is quite different (0.3571 vs 0.2595). A significantly evenly distributed non-dominated solutions for ZDT6 are obtained by IDE, implying that IDE is capable of finding a well-distributed and near-complete set of non-dominated solutions when handling multiobjectives.

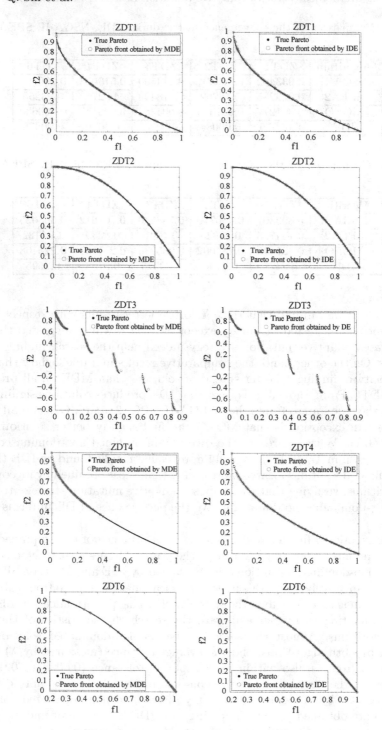

Fig. 2. Pareto fronts obtained by IDE and its competitor algorithm MDE on ZDT1, ZDT2, ZDT3, ZDT4, and ZDT6 respectively

5 Conclusion

This paper proposed an adaptive mutation operator based on the multi-objective differential evolution algorithm. In the beginning of search phase, the algorithm has a relatively large value to maintain the individuals' diversity, and avoid the premature phenomenon of fast convergence. With the lapse of evolution, the mutation operator was gradually reduced to reserve good information and avoid the destruction to the optimal solution. Together with the ε-dominance strategy, we constructed the effective IDE to handling multiple objectives. We test IDE via five standard multi-objective test functions and the performance comparison during MDE, NSGA-II, SPEA2 and MOEO. It can be concluded that IDE is superior to other algorithms on multiple problems, indicating that our approach has ability to obtain effective uniformly distributed and near-optimal Pareto sets.

Acknowledgments. This work is partially supported by the National Natural Science Foundation of China (Grants No. 61203325), Shanghai Rising-Star Program (No. 14QA1400100), "Chen Guang" project supported by Shanghai Municipal Education Commission and Shanghai Education Development Foundation (No. 12CG35), Ph.D. Program Foundation of Ministry of Education of China (No. 20120075120004), the Fundamental Research Funds for the Central Universities and DHU Distinguished Young Professor Program.

References

1. Price, K., Storn, R.M., Lampinen, J.A.: Differential evolution: a practical approach to global optimization. Springer (2006)
2. Das, S., Suganthan, P.N.: Differential evolution: A survey of the state-of-the-art. IEEE Transactions on Evolutionary Computation **99**, 1–28 (2010)
3. Zhang, J., Sanderson, A.C.: Jade: adaptive differential evolution with optional external archive. IEEE Transactions on Evolutionary Computation **13**(5), 945–958 (2009)
4. Wang, J., Liao, J., Zhou, Y., Cai, Y.: Differential evolution enhanced with multiobjective sorting-based mutation operators. IEEE Transactions on Cybernetics **12**(44), 2792–2805 (2014)
5. Santana-Quintero, L.V., Coello, C.A.C.: An algorithm based on differential evolution for multi-objective problems. International Journal of Computational Intelligence Research **1**(1), 151–169 (2005)
6. Wang, Y.N., Wu, L.H., Yuan, X.F.: Multi-objective self-adaptive differential evolution with elitist archive and crowding entropy-based diversity measure. Soft Computing **14**(3), 193–209 (2010)
7. Zhang, J., Sanderson, A.C.: Self-adaptive multi-objective differential evolution with direction information provided by archived inferior solutions. In: IEEE Congress on Evolutionary Computation, pp. 2801–2810 (2008)
8. Gong, W., Cai, Z.: An improved multiobjective differential evoluton based on pareto-adaptive epsilon-dominance and orthogonal design. European Journal of Operational Research **198**(2), 576–601 (2009)

9. Chen, B., Lin, Y., Zeng, W., Zhang, D., Si, Y.W.: Modified differential evolution algorithm using a new diversity maintenance strategy for multi-objective optimization problems. Applied Intelligence, 1–25 (2015)

10. Chong, J.K., Tan, K.C.: An opposition-based self-adaptive hybridized differential evolution algorithm for multi-objective optimization (osade). In: Proceedings of the 18th Asia Pacific Symposium on Intelligent and Evolutionary Systems, pp. 447–461. Springer (2015)

11. Zitzler, E., Thiele, L.: Multiobjective evolutionary algorithms: a comparative case study and the strength pareto approach. IEEE Transactions on Evolutionary Computation 3(4), 257–271 (1999)

12. Laumanns, M., Thiele, L., Deb, K., Zitzler, E.: Combining convergence and diversity in evolutionary multiobjective optimization. Evolutionary computation 10(3), 263–282 (2002)

13. Zitzler, E., Thiele, L., Laumanns, M., Fonseca, C.M., Da Fonseca, V.G.: Performance assessment of multiobjective optimizers: An analysis and review. IEEE Transactions on Evolutionary Computation 7(2), 117–132 (2003)

14. Fang, K., Ma, C.: Orthogonal and uniform experimental design. Science press, Beijing (2001)

15. Leung, Y.W., Wang, Y.: An orthogonal genetic algorithm with quantization for global numerical optimization. IEEE Transactions on Evolutionary Computation 5(1), 41–53 (2001)

16. Deb, K., Pratap, A., Agarwal, S., Meyarivan, T.: A fast and elitist multiobjective genetic algorithm: Nsga-II. IEEE Transactions on Evolutionary Computation 6(2), 182–197 (2002)

17. Dasgupta, S., Das, S., Biswas, A., Abraham, A.: On stability and convergence of the population-dynamics in differential evolution. AI Communications 22(1), 1–20 (2009)

18. Brest, J., Greiner, S., Boskovic, B., Mernik, M., Zumer, V.: Self-adapting control parameters in differential evolution: A comparative study on numerical benchmark problems. IEEE Transactions on Evolutionary Computation 10(6), 646–657 (2006)

19. Zitzler, E., Deb, K., Thiele, L.: Comparison of multiobjective evolutionary algorithms: Empirical results. Evolutionary computation 8(2), 173–195 (2000)

20. Hernández-Díaz, A., Santana-Quintero, L., Coello Coello, C., Molina, J.: Pareto-adaptive ε-dominance. Evolutionary Computation 15(4), 493–517 (2007)

21. Deb, K.: Multi-objective optimization using evolutionary algorithms, vol. 16. John Wiley & Sons (2001)

22. Zitzler, E., Laumanns, M., Thiele, L.: Spea 2: Improving the strength pareto evolutionary algorithm. In: Proc. Evolutionary Methods for Design Optimization and Control with Applications to Industrial Problems, pp. 95–100 (2001)

23. Chen, M.R., Lu, Y.Z.: A novel elitist multiobjective optimization algorithm: Multiobjective extremal optimization. European Journal of Operational Research 188(3), 637–651 (2008)

Multi-Agent Systems
and Swarm Robotics

A Fleet of Chemical Plume Tracers with the Distributed Architecture Built upon DaNI Robots

David Oswald[1], Henry Lin[1], Xiaoqian Mao[2],
Wei Li[1,2]([✉]), Linwei Niu[3], and Xiaosu Chen[4]

[1] Department of Computer and Electrical Engineering and Computer Science,
California State University, Bakersfield, CA 93311, USA
rdoswald@hotmail.com, {hlin5,wli}@csub.edu
[2] School of Electrical Engineering and Automation, Tianjin University, Tianjin 300072, China
mxq613@126.com
[3] Department of Mathematics and Computer Science, West Virginia State University,
Institute, WV 25112, USA
lniu@wvstateu.edu
[4] Environment and Meteorology, Clustertech Ltd., Beijing 100080, China
xschen@clustertech.com

Abstract. This paper presents a fleet of chemical plume tracers with the distributed architecture developed at California State University, Bakersfield (CSUB). Each chemical plume tracer built upon a DaNI robot integrates multiple sensors, including a wind sensor, chemical sensors, a wireless router, and a network camera. The DaNI robot is an advanced platform embedded with a single control board (sbRIO-9632), consisting of a 400 MHz industrial processor, a 2M gate Xilinx Spartan FPGA, and a variety of I/Os, In order demonstrate the feasibility of the designed chemical plume tracers, the experiments on moth-inspired plume tracing are conducted under the turbulent airflow environment. This fleet of chemical plume tracers is a powerful tool for investigating algorithms for the tracking and mapping of chemical plumes via swarm intelligence.

Keywords: Chemical plume tracer · Bio-inspired algorithm · Moth behavior · LabVIEW · DaNI robot · Swarm robots

1   Introduction

A potential application of chemical plume tracing (CPT) is searching for sources of hazardous chemicals or pollutants, or victims in earthquake wreckage using a swarm robot system. One of the critical problems in this area is to develop the swarm robot system with an effective navigation mechanism which guides a fleet of plume tracers to track a chemical plume towards its source. Chemical information-based plume tracing phenomena widely exist in a variety of biological swarm behaviors [1].

An initial approach to designing a chemical plume tracer might attempt to calculate a concentration gradient [2], with subsequent plume tracing based on gradient following. This approach suits for a chemical source under diffusive airflow environments. At medium and high Reynolds numbers, however, the evolution of the chemical

© Springer International Publishing Switzerland 2015
Y. Tan et al. (Eds.): ICSI-CCI 2015, Part I, LNCS 9140, pp. 535–542, 2015.
DOI: 10.1007/978-3-319-20466-6_56

distribution in the flow is turbulence dominated. The result of the turbulent diffusion process is a highly discontinuous and intermittent distribution of the chemical. A dense array of sensors distributed over the area of interest and a long time-average of the output of each sensor (i.e., several minutes per sensor) would be required to estimate a smooth (time-averaged) chemical distribution. However, the required dense spatial sampling and long time-averaging makes such an approach ineffective for implementation on an autonomous robot. In addition, even decameters from the chemical source in the direction of the flow the gradient is too shallow to detect in a time-averaged plume.

Over the past ten years, various biomimetic robotic CPT studies were developed. Grasso et al. [3] evaluated biomimetic strategies and challenged theoretical assumptions of the strategies by implementing biomimetic strategies on their robot lobster. Inspired by moth behavior [4], Li et al. [5] developed, optimized, and evaluated counter-turning strategies. Li et al. [6] implemented the moth-inspired plume tracing strategies on autonomous underwater vehicles (REMUS). The in-water tests conducted in near-shore ocean conditions in [6-7] demonstrate that the REMUS-based plume tracers track the chemical plumes over hundred meters and achieve the average source declaration accuracy of approximately 13m. Webster et al. [8] developed a dynamic CPT algorithm inspired by the behaviors of blue crabs in a turbulent flow environment. Success rates and movement patterns compare favorably to that of blue crabs.

This paper introduces a fleet of chemical plume tracers built upon the DaNI robots manufactured by National Instruments (NI), which is embedded with a single control board (sbRIO-9632), consisting of a 400 MHz industrial processor, a user-reconfigurable field-programmable gate array (FPGA) – a 2M gate Xilinx Spartan, and a variety of I/Os. Each plume tracer integrates multiple sensors, including a wind sensor, chemical sensors, a wireless router, and a camera. The DaNI robot platform uses LabVIEW – the visual programming language environment.

2   Integration of Multiple Sensors

2.1   DaNI Robot

A group of DaNI robots are used to build a fleet of the chemical plume tracers. A DaNI robot comes preassembled and has two motors, encoders, and an ultrasonic sensor, as shown in Fig. 1. It is equipped with the sbRIO-9632 embedded control and acquisition device, integrating a 400 MHz industrial processor, a user-reconfigurable field-programmable gate array (a 2M gate Xilinx Spartan FPGA), and I/O on a single printed circuit board. It features 110 3.3V (5V tolerant/TTL compatible) digital I/O lines, 32 single-ended/16 differential 16-bit analog input channels at 250 kS/s, and four 16-bit analog output channels at 100 kS/s. It also has three connectors for expansion I/O using board-level NI C Series I/O modules. The sbRIO-9632 offers a -20 to 55 °C operating temperature range, and includes a 19 to 30 VDC power supply input range, 128 MB of DRAM for embedded operation, and 256 MB of nonvolatile memory for storing programs and data logging. It also features a built-in 10/100 Mbit/s Ethernet port you can use to conduct programmatic communication over the network and host built-in Web (HTTP) and file (FTP) servers. A user also can use the RS232 serial port to control peripheral devices.

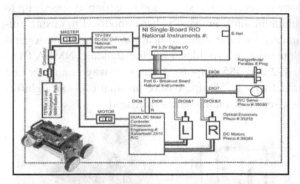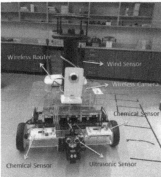

Fig. 1. A DaNI robot is equipped with a sbRIO-9632 embedded control and acquisition device, integrating a 400 MHz industrial processor and a 2M gate Xilinx Spartan FPGA. A chemical plume tracer built upon the DaNI robot integrates multiple sensors, including a wind sensor, chemical sensors, a wireless router, a camera, and an ultrasonic transducer.

2.2   Ultrasonic Sensor and DC Motors

A DaNI robot is equipped with a Parallax ultrasonic sensor. It detects objects by emitting a short ultrasonic burst and then "listening" for the echo. Under the control of a host microcontroller (trigger pulse), the sensor emits a short 40 kHz (ultrasonic) burst. This burst travels through the air at about 1130 feet per second, hits an object, and then bounces back to the sensor. The PING sensor provides an output pulse to the host that terminates when the echo is detected. Meaning the width of this pulse corresponds to the distance to the target. The DC motors use 12V of power and offer 300 oz-in. of torque and 152 RPM. The encoders use 5V of power and offer 100 cycles per revolution and 400 pulses per revolution.

2.3   Chemical Sensor

The chemical sensor made by Figaro uses a metal oxide semiconductor layer on an alumina substrate of a sensing chip with an integrated heater. When the chemical is detectable, the sensor's conductivity increases depending on the gas concentration in the air. The electrical circuit board that converts chemical concentrations into electrical output signals is shown in Fig. 2 (left).

2.4   Wind Sensor

The Gill WindSonic wind sensor is a low-cost anemometer, which utilizes the Gill's ultrasonic technology to provide wind speed and direction data via one serial or two analogue outputs, as shown in Fig. 2 (middle). To confirm correct operation, outputs are transmitted together with an instrument status code. It features a robust, corrosion-free polycarbonate housing which makes it very lightweight. The WindSonic anemometer is a robust ultrasonic wind speed and direction sensor with aluminum alloy construction. The WindSonic sensor is solid-state with no moving parts and uses the ultrasonic measurement technology to detect wind speed and direction at speeds up to 60m/s (134mph). The robust aluminum alloy housing is hard-anodised to ensure suitability in harsh marine environments, and the optional heating system allows operation down to -40°C. The WindSonic sensor provides a marine-standard NMEA 0183 output, with options for RS232, 422 and 485 outputs. A single 9-way connector and three mounting holes for attachment to a 1.75" pipe ensure installation is straightforward. The WindSonic sensor has a very low maintenance overhead.

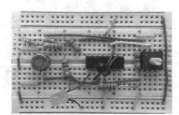  

Fig. 2. The electrical circuit board for the Figaro chemical sensor converts chemical concentrations into electrical output signals (left). The Gill WindSonic sensor is a low-cost anemometer, which provides wind speed and direction data via one serial or two analogue outputs (middle). A wireless router integrated on the DaNI robot is for wireless communication between a computer and the DaNI robot (right).

2.5   Wireless Router and Camera

Since the DaNI robot is open architecture it is easy to add a wireless router to establish wireless communication with the internet, as shown in Fig. 2 (right). Next, an AXIS M1011 network camera is mounted on the DaNI robot to view surroundings from the robot's perspective, as shown in Fig. 1. The camera provides multiple, individually configurable video streams in H.264 as well as Motion JPEG and MPEG-4. It also offers video quality at 30 frames per second in VGA resolution. The NI Measurement and Automation software can configure the camera format and image resolution.

2.6   Swarm Robot System for Chemical Plume Tracing

A swarm robot system with a distributed structure must be powerful to trace chemical plumes and to build plume maps in the real world. However, most of the existing studies on the tracking and mapping of chemical plumes via swarm robots are based on simulation evaluations, e.g., [9].

In order to validate the strategy for plume tracing via the swarm robots, it is essential to develop a fleet of chemical plume tracers with distributed structure, as shown in Fig. 3. Because each robot is equipped with multiple sensors, including a wireless router and two chemical sensor boards, the plume tracers in the swarm robot system are able to talk each other in LabVIEW Robotics projects. Each plume tracer performs its maneuver and shares the collected information with the others, such as, chemical concentrations, airflow orientations and magnitudes, and its current state.

The swarm robots are able to dynamically form their shape to perform CPT missions according to the formation control mechanism [9]. Conducting CPT missions via the swarm robots under formation control in an open and large-scale operation area is on schedule.

Fig. 3. The swarm robot system for plume mapping and tracing is developed in Robotics Lab, at California State University, Bakersfield

3 Algorithms Implementation in Labview

The sbRIO-9632 device can be programmed in the LabVIEW graphical development environment. The real-time processor runs the LabVIEW Real-Time Module on the Wind River VxWorks real-time operating system (RTOS) for extreme reliability and determinism. With the addition of the LabVIEW MathScript RT Module, custom.m file are easily deployed to NI real-time hardware while combining both graphical and textual syntax. The onboard reconfigurable FPGA can be quickly programmed using the LabVIEW FPGA Module for high-speed control, custom I/O timing, and inline signal processing. LabVIEW contains built-in drivers and APIs for handling data transfer between the FPGA and real-time processor.

The moth-inspired chemical plume tracing algorithms proposed in [5], including the four plume tracing behaviors, including Find-Plume, Maintain-Plume, Reacquire-Plume, and Declare-Source, are implemented on the plume tracers using the Lab-VIEW language, as shown in Fig. 4. The program needs the Initialize Starter Kit 2.0 (sbRIO) VI. This VI begins a communication session with the FPGA on the plume tracer and returns a reference to read from or write to the FPGA. This VI must be called before accessing I/O with the FPGA. The next VI in the program is the Create Starter Kit 2.0 Steering Frame VI. This VI generates a steering frame object for the robot and the steering frame object can be used with the other Steering VIs to implement steering for the plume tracer.

The plume tracer needs the VISA Configure Serial Port VI and it initializes the serial port specified by "VISA resource name" to the specified settings. This VI writes data from the write buffer to the device or interface specified by "VISA resource name." This VI reads the specified number of bytes from the device or interface specified by "VISA resource name" and returns the data in "read buffer." The

VISA Close VI closes a device session or event object specified by "VISA resource name." The Write DC motor Velocity Setpoints VI applies velocity values to the drive motors on the robot. The left and right motors are defined by their positions. The maximum motor velocity is 15.7 rad/s.

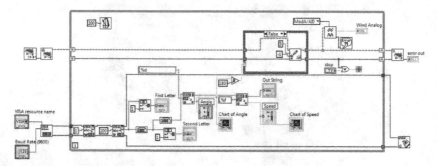

Fig. 4. The moth-inspired chemical plume tracing algorithms are implemented using LabVIEW block diagram which includes Initialize Starter Kit 2.0 (sbRIO) VI and The Close Starter kit VI

The moth-inspired plume tracing algorithm controls the plume tracer to follow the wind. When the wind is coming from the left, the robot turns left. When the wind is from the right, it turns right. Since the front of the sensor is 0°/360° and counts from 0 to 360 clockwise, if the wind is coming from somewhere between 180° to 360° the robot turns left and if the wind is coming from somewhere between 0° to 180° the robot turns right. The case statement block in Fig. 4 checks if the value from the WindSonic sensor for wind direction is greater than 180, the statement is true and it uses the true block. If the value is less than 180 it is false and uses the false block. Both the cases use the Write DC Motor Velocity Setpoints VI. If the statement is true the wind is coming from the left and speeds up the right motor in order to turn left and for false speeds up the left motor to turn right. For this VI, a negative value in the right motor makes it go forward.

For the chemical sensor, a LabVIEW block diagram designed for the program reads the analog inputs converted from chemical. The block is called Read AI Line VI and it reads the voltage value from an analog input line on the Starter Kit FPGA. The Close Starter kit VI terminates a communication session with the FPGA. When the communication session ends, the drive motors, distance sensor, and sensor servo stop operating.

4   Moth-Inspired Plume Tracing Experiments

In order demonstrate the feasibility of the designed chemical plume tracers, the experiments on moth-inspired plume tracing are conducted under the turbulent airflow environment. A humidifier pumps alcohol as the chemical source and a fan with a fix heading or a varying heading are used to generate a chemical plume, as shown in Fig. 5. The first step for the experiments is to calibrate the performance of following the wind by turning off the chemical source. A true-false case structure indicates whether the angle read from the wind sensor is less than or greater than 180 degrees. If the wind is to the left, the tracer will turn left. If the wind is to the right, the tracer will turn right. The tracer keeps turning until the angle approaches zero when the tracer is facing the wind. This behavior is responsible for its zigzag movement, which moths behave when tracking odors.

Fig. 5. The experiments on moth-inspired plume tracing under turbulent airflow environments are conducted by the plume tracer. A humidifier pumps alcohol as the chemical source and a fan with a fix heading or a varying heading is used to generate turbulent airflow.

Next, the plume tracer is controlled to track air-borne chemicals under turbulent airflow environments by using chemical concentrations in conjunction with wind information. The chemical detection is created on two breadboards with both having the same circuit to monitor both sides of the tracer. The chemical sensor circuit uses one of two Figaro sensors and outputs a voltage value corresponding to the detected chemical concentration. This voltage value is fed into the analog input. The plume tracer needs to detect above a chemical threshold before it starts tracking. The experiment settings were given as follows: The experiments show that nearly 100% success rate is achieved using the moth-inspired plume tracing algorithms. Fig. 6 shows chemical concentrations, wind speeds, wind angles measured during a CPT mission under the turbulent airflow environment.

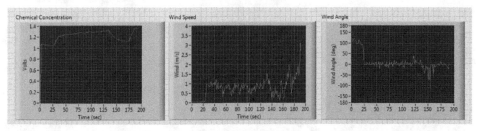

Fig. 6. Chemical concentrations (left), wind speeds (middle), and wind angles (right) were measured during a CPT mission

5   Conclusions

This paper presents a fleet of the chemical plume tracers with the distributed architecture built by the DaNI robots. The experiments on the moth-inspired plume tracing under the turbulent airflow environment demonstrate the feasibility of the designed plume tracers. In our further research, we will validate the plume mapping and plume tracing algorithm via swarm robots proposed in [9].

References

1. Dusenbery, D.B.: Sensory Ecology: How Organisms Acquire and Respond to Information. W.H. Freeman, New York (1992)
2. Sandini, G., Lucarini, G., Varoli, M.: Gradient driven self-organizing systems. In: IEEE/RSJ Int. Conf. on Intelligent Robots and Systems, pp. 429–432. IEEE Press, New York (2001)
3. Grasso, F.W., Consi, T.R., Mountain, D.C., Atema, J.: Biomimetic Robot Lobster Performs Chemo-Orientation in Turbulence Using a Pair of Spatially Separated Sensors: Progress and Challenges. Robotics and Autonomous Systems 30, 115–131 (2000)
4. Cardé, R.T.: Odour plumes and odour-mediated flight in insects. In: Olfaction in Mosquito-host Interactions, CIBA Found. Symp., pp. 54–70. John Wiley & Sons (1996)
5. Li, W., Farrell, J.A., Cardé, R.T.: Tracking of Fluid-advected Chemical Plumes: Strategies Inspired by Insect Orientation to Pheromone. Adaptive Behavior 9, 143–170 (2001)
6. Li, W., Farrell, J.A., Pang, S.: Moth-inspired Chemical Plume Tracing on an Autonomous Underwater Vehicle. IEEE T. Robot. 22, 292–307 (2006)
7. Farrell, J.A., Pang, S., Li, W.: Chemical plume tracing via an autonomous underwater vehicle. IEEE J. Oceanic Eng. 30, 428–442 (2005)
8. Webster, D.R., Volyanskyy, K.Y., Weissburg, M.J.: Bioinspired Algorithm for Autonomous Sensor-driven Guidance in Turbulent Chemical Plumes. Bioi. Biomim. 7, 23–34 (2012)
9. Kang, X.D., Li, W.: Moth-inspired Plume Tracing via Multiple Autonomous Vehicles under Formation Control. Adaptive Behavior 20, 131–142 (2012)

Hierarchical Self-organization for Task-Oriented Swarm Robotics

Yuquan Leng[1,2], Cen Yu[3], Wei Zhang[1], Yang Zhang[1(✉)], Xu He[1], and Weijia Zhou[1]

[1] State Key Laboratory of Robotics, Shenyang Institute of Automation,
University of Chinese Academy of Science, Shenyang, China
zhangyang@sia.cn
[2] University of Chinese Academy of Sciences, Beijing, China
[3] Anhui Xinhe Defense Equipment Technology Corporation Limited, Hefei, China

Abstract. The problems of diversity of tasks and non-structural environment have been put in front of robotic development, on the other hand, we urgently hope they consume low cost and have high reliability, so the method of multi-cooperation is wildly used. Then we would get the swarm robotics social system with the individual growing. In this paper, we proposed hierarchical organizational model to definite social order during task decomposition; then, we design the method of behavior generation based on proposition/transition Petri networks, which would assist the system to construct combined behavior using the sample individual behavior to solve a variety of tasks.

Keywords: Hierarchical self-organization · Task-oriented · Swarm robotics

1 Introduction

Three characters of robotic development are summarized as following: 1) Integrate and improve the function of individual, such as stability and load capacity; 2) Increase of intelligence of individual, which makes robots have autonomy; 3) Number of robots in whole system, which spurs collaboration and solves more tasks. The research contents of this paper relate to the last two aspects, hoping to make swarm system work autonomously.

Swarm robotics occurs from artificial swarm intelligence, the biological studies of insects, ants and other fields in nature, so swarm robotics is defined as a new approach to the coordination of multi-robot systems which consist of large numbers of most simple physical robots and which is supposed that a desired collective behavior emerges from the interactions among the robots and interactions between the robots and environment [1, 2]. Considering swarm robotic application, we focus on the characteristic of organization structure and propose that:

Swarm robotics is a robotics system with a special organization structure, which has the flexibility, unpredictability and infinite increase or decrease mechanism, and consisted of any form of robots, except for interactions with robots or environment [3].

© Springer International Publishing Switzerland 2015
Y. Tan et al. (Eds.): ICSI-CCI 2015, Part I, LNCS 9140, pp. 543–550, 2015.
DOI: 10.1007/978-3-319-20466-6_57

Task-oriented robot means that operators do not need to understand "How to do it", and just need to know "what to do"[4,5,6]. For example, if you want a cup of coffee, you just need to click the button identifying the kind of coffee you like, and you do not need to know how to do it. It is simple to realize for one machine, especially which uses switch control. But the statuses will shapely change with the number of robots increase and complex function of individual. To solve this problem, we put forward some ways to realize task-oriented in swarm system.

In recent years, interest of swarm robotics has been greatly increasing, for the following characteristics: 1) Robustness; 2) Scalability; 3) Flexibility; 4) Economy [1,2]. At the end of 2014, the journal of Science choose top 10 breakthrough of the whole year, which includes the Kilobots [7, 8], a swarm system made by Self-organization system research group in Harvard University. They could organize themselves into stars and other two-dimensional shapes. This news fully affirms the scientific significance of swarm robotics.

In addition, other swarm robotics platforms are created to verify swarm theory, such as Pheromone robotics [9], Swarm-bots[10], Swarmanoid [11] and Termites robotics [12] etc. In practical application, Amazon has hired about 15000 Kiva robots to help company arrange millions of stock. In the foreseeable future, a lot of swarm systems will get into our life such as driverless cars, intelligent manufacturing factory and etc.

In order to make system astronomical, we general adopt MAS (Multi-Agent System) technology to formal description and design robotic society. In the society, individual is abstracted as agent with the ability of perception, decision and action. All agents construct social organization with special order to realize common task. In the MAS technology, robotic society requires to abstract many virtual agents, which are on hierarchical system for task decomposition, action planning, self-organization and etc.

For the research of organizational model, some typical methods have been proposed, such as AGR (Agent/Group/Role) model [13], electronic institution model [14], HARMONIA model [15] and so on, but they have some disadvantages as following:

- The models are built just for static organizational structure, not for dynamic;
- There are no unified standard, so they could just be used on someone system and have no good generality.
- Although there are series of theory of organization design, most of them adopt complex logical tools to describe and infer rules, which are not adaptive to engineering application.

The rest of this paper is organized as follows: Section 2 describes a hierarchical organizational model for task-oriented swarm system. Method of behavior generation in hierarchical organizational model is provided in Section 3. Finally, the concluding remarks follow in Section 5.

2   Task-Oriented Hierarchical Organizational Model

The organizational model defines a series of autonomous agent and the social order among them, intending to decompose social target and define cooperative demand. In the robotic society, cooperation means that many agents in one organization complete social target together.

2.1   Cooperation Types

From social aspect, cooperation is a management mode of the organization, managing the interaction among agents and the dependencies in activities. According to different methods for realizing social target, cooperation can be divided into three basic types: market cooperation, network cooperation and hierarchical cooperation.

Market cooperation is committed to promote the exchange between different agents. Agents provide services and express their capacities, and contest for the executive priority with each other. The common aim of market mechanism is to complete the most tasks with the least cost of the least resources. In the network cooperation, there is a common aim of the agents to make them build a steady trust relationship. Hierarchical cooperation guides resources and information flow from the centralized management perspective of one hierarchy, and determine the affiliation of members by the predefined structure but not by negotiation or communication.

Table 1. Specific characteristics of each coordination model

Cooperation model	market	network	hierarchy
Society type Membership	Open Completely selfish	mutual trust common interest	Close subordination
Social purpose	Exchange resource and information	Cooperate to achieve common purpose	Generate activities from lower to upper
Interactive mode	Predefined standard interaction	Negotiate interactive process and content	Be confirmed when the organization is designed

2.2   Self-configuration/Self-reconfiguration of Organizational Model

In the robotic society, in order to complete one task together, the robots interact with others through respective communication channel and interface. It's a natural network cooperative relationship. But network cooperation problems in space application are as follows:

(1) It needs to make a plan for each robots designating how to do when the task explicit what to do. With the expanding number and kind of robots in society, it will confront the state space explosion problem to get a complete plan directly.

(2) The executive process involving too many individualities and large time and spatial span is highly complicated, which is difficult to control and manage.

(3) Because of higher communication cost, the limitation of the bandwidth, more unpredictable and indefinite events, the organizational structure is inclined to choose the hierarchical cooperation model to reduce the interaction cost.

We can solve this problem by adding task plan agent to organization and making the original agent as the functional agent in bottom layer. Task plan agent is running on robots controller or computer with calculating ability in the form of a program. It decomposes the task layer by layer, till the agent could understand it and execute it. The closeness of hierarchical cooperation determines its limitation in one task or one kind of tasks, so we need to design different hierarchical cooperation model to satisfy different task designation. In another word, task-oriented self-configuration/self-reconfiguration is an indispensable ability for the robotic society.

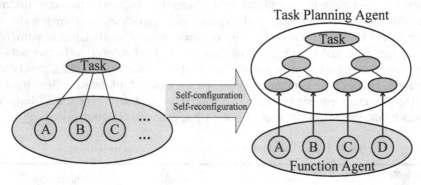

Fig. 1. Self-configuration/self-reconfiguration of an organization

In the task-oriented hierarchical organizational model, task planning agent has determined affiliation and interaction of common goals and specific organizational relationships, and knows how to interact and decompose task through communication.

3   Method of Behavior Generation

Hierarchical organizational model can get the complicated behavior of upper hierarchy through the lower hierarchy. Researches on behavior generation methods, such as CAMPOUT, encapsulate the behavior of lower agent as a series of functional interface, which can be used as the basic unit to build the behavior of upper agent. But this method is generally used in mobile robot platform because it is only active in the atomic behavior choices and is unable to monitor or manage complicated concurrent behavior.

Behavior generation method is asked to monitor and manage concurrent activities of many agents in an organization in order to realize the generality of it.

3.1   Control Structure

Formal description and analysis of the behavior needs to be defined in a control structure, and we use classical Kripke structure model.

Kripke structure makes $M = (S, R, L, S_0)$, in which S means nonempty finite set of states, R means state transition set, $L : S \mapsto 2^{AP}$ means every state corresponds to an atomic proposition, and $S_0 \subseteq S$ means initial set of states.

In the process of monitoring and management of multi-agents action, we use state set $S_{mark,t} \subseteq S$ to describe all effective states at time t, and use sequence $\pi = S_{mark,0}, S_{mark,1} \cdots$ to represent implementation process.

3.2   Definite of Behavior

The lower layer agent provides a series of function interface for the agent, which is belonged. In essence, the behavior is a strategy to operate this functional interface, according to the current events. Next, we definite Event, Action (function interface) and Behavior in formal ways.

(1) Event

$E(x_1, \ldots x_n) = \{Assertion_1, \cdots Assertion_m\}$ is a set of assertions, in which, E means name of event, and $x_1, \ldots x_n$ are parameters of event. When all assertions are true, we could admit that event occurs. Ranges of parameters of event are described as $Parameter_E = \{C_{E,1}(x_1), \ldots C_{E,n}(x_n)\}$.

(2) Action

$A(x_1, \ldots x_n) = (Enable_A, Ready_A, Disable_A, Result_A)$ is one function interface, in which, A means the name of action, $x_1, \ldots x_n$ are parameters of action, and $Enable_A$, $Ready_A$, $Disable_A$, and $Result_A$ represent enable event, ready event, forbid event and result event. Ranges of parameters of action are described as $Parameter_A = \{C_{A,1}(x_1), \ldots C_{A,n}(x_n)\}$.

(3) Behavior

$B(x_1, \ldots x_n) = (P_B, Terminate_B, Input_B, Output_B, E_B)$ is one strategy of action control based on event, in which, B means the name of behavior, $x_1, \ldots x_n$ are parameters of behavior, P_B means pre-event, $Terminate_B$ means end time, $Input_B$ and $Output_B$ means a set of event about action input and output, and E_B represents a set of post event. Ranges of parameters of behavior are described as $Parameter_B = \{C_{B,1}(x_1), \ldots C_{B,n}(x_n)\}$.

3.3   Behavior Generation

Behavior is a motion control strategy that means series of action to realize one behavior. When one behavior just contains one action, we call it atomic behavior, otherwise we call it combined behavior.

In this paper, we adopt proposition/transition Prtri network with enable arc and suppression arc to describe inner structure of combined behavior. Every library

Place$_i$ of Prtri network is given a set of propositions Assertion_Set$_i$. When Place$_i$ contains Token, which means all propositions in this set are true.

Combined behavior is composed by two parts: structure module and control module as shown in Fig. 2. Structure module encapsulates one behavior and provides interface for receiving and sending event. Control module links a plurality of interfaces of structure modules and controls the execution process.

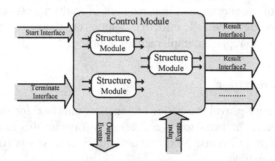

Fig. 2. Components of combined behavior

To simplify the description, existing compound behavior could be set as structure modules when building a more compound behavior.

(1)   Structure Modules

Structure modules of primitive behavior is shown in Fig.3.

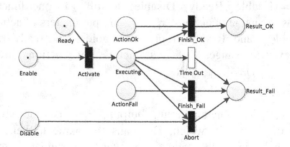

Fig. 3. Structure modules of primitive behaviour

Enable、Disable, Result_OK and Result_Fail are external interface, and Ready, ActionOK and ActionFail are provided by sub layer agent.

As shown in Fig.2, we assume that there are two action results of primitive behavior: Action OK and Action Fail. In the real application, there may be three or even more action results, and we should adjust accordingly.

Structure modules of combined behavior are shown in Fig.4.

P_Block, Terminate_Block, Input, Output, Result1, Result2 are external interface of this structure module.

(2) Control module

Control module controls event flow of each structure modules, and can be defined flexibly according to actual need. Take sequence control module as an example: Sequence(A, B) controls structure module A and B, module B will be skipped if module A fail in execution, as shown in Fig. 5.

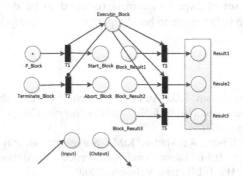

Fig. 4. Structure modules combined behaviour

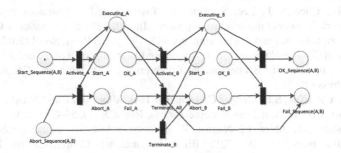

Fig. 5. Sequence control module

The advantage of this generation method of organization behavior method lies in: the relationship between compound behavior and its structure module is operation of activation or ban, but not invoking, which is pretty important because it can avoid using invoking relationship effectively when dealing with concurrent activities of many agents thereby avoiding error-prone problems in the executive process. A task plan is a living example of replacing all variables in one behavior with constant, and it can be considered as an executive strategy that can activate or end the action of robot according to occurred event.

4   Conclusion

In order to avoid the state explosion problem in the decomposition process of robot and reduce the interaction cost in executive process, first, we added task plan agent with explicit affiliation and interaction to space robotic society, and formed a hierarchical organizational model, both of which can definitely understand the common aim and the relationship of organization, and know how to realize interaction through communicating; second, we designed a method of behavior generation based on proposition/transition Petri network. we can see that method of behavior generation could produce compound behavior responding to all kind of events to satisfy demand of users who is not familiar with technical details and it also has a good generality. A set of expert system is required to be developed for automatic generation or used to help people to build a plan of task description in future research.

References

1. Mohan, Y., Ponnambalam, S.G.: An extensive review of research in swarm robotics. In: IEEE World Congress on Nature & Biologically Inspired Computing, pp. 140–145. IEEE Press, Coimbatore (2009)
2. Higgins, F., Tomlinson, A., Martin, K.M.: Survey on security challenges for swarm robotics. In: 2009 IEEE International Conference on Autonomic and Autonomous Systems, pp. 307–312. IEEE Press, Valencia (2009)
3. Leng, Y.Q., Zhang, Y., He, X., Zhou, W.J.: SociBuilder System: A Novel Task-oriented Swarm Robotic System (submitted)
4. Kim, H., Cheong, J., Lee, S., Kim, J.: Task-oriented synchronous error monitoring framework in robotic manufacturing process. In: 2012 IEEE International Conference on Automation Science and Engineering, pp. 485–490. IEEE Press, Seoul (2012)
5. Loizou, S.G., Kyriakopoulos, K.J.: Automated planning of motion tasks for multi-robot systems. In: 44th IEEE International Conference on Decision and Control, pp. 78–83. IEEE Press (2005)
6. Zhang, Y.Z., Xue, S.D., Zeng, J.C.: Dynamic Task Allocation with Closed-Loop Adjusting in Swarm Robotic Search for Multiple Targets. Robot 36(1), 57–67 (2014)
7. Rubenstein, M., Ahler, C., Nagpal, R.: Kilobot: a low cost scalable robot system for collective behaviors. In: 2012 IEEE International Conference on Robotics and Automation, pp. 3293–3298. IEEE Press, Saint Paul (2012)
8. Pennisi, E.: Cooperative 'bots' don't need a boss. Science 346(6216), 1444 (2014)
9. Purnamadjaja, A.H., Russell, R.A.: Robotic pheromones: using temperature modulation in tin oxide gas sensor to differentiate swarm's behaviours. In: 2006 IEEE International Conference on Control, Automation, Robotics and Vision, pp. 1–6. IEEE Press (2006)
10. Dorigo, M.: SWARM-BOT: an experiment in swarm robotics. In: 2005 IEEE Swarm Intelligence Symposium, pp. 192–200. IEEE Press (2005)
11. Dorigo, M., Floreano, D., Gambardella, L.M., et al.: Swarmanoid: A novel concept for the study of heterogeneous robotic swarms. IEEE Transactions on Robotics & Automation Magazine 20, 60–71 (2013)
12. Durrant-Whyte, H., Roy, N., Abbeel, P.: TERMES: An autonomous robotic system for three-dimensional collective construction, pp. 257–264. MIT press (2012)
13. Ferber, J., Gutknecht, O.: A meta-model for the analysis and design of organizations in multi-agent systems. In: 1998 IEEE International Conference on Multi Agent Systems, pp. 128–135. IEEE Press (1998)
14. Cardoso, H.L., Leitão, P., Oliveira, E.: An approach to inter-organizational workflow management in an electronic institution. In: 11th IFAC Symposium on Information Control Problems (2006)
15. Kim, Y., Oral, S., Shipman, G.M., et al.: Harmonia: a globally coordinated garbage collector for arrays of solid-state drives. In: 27th IEEE Symposium on Mass Storage Systems and Technologies, pp. 1–12. IEEE Press, Denver (2011)

Power-Law Distribution of Long-Term Experimental Data in Swarm Robotics

Farshad Arvin[1]([✉]), Abdolrahman Attar[1], Ali Emre Turgut[2], and Shigang Yue[1]

[1] School of Computer Science, University of Lincoln, Lincoln LN6 7TS, UK
{farvin,syue}@lincoln.ac.uk
[2] Mechanical Engineering Department,
Middle East Technical University, 06800 Ankara, Turkey

Abstract. Bio-inspired aggregation is one of the most fundamental behaviours that has been studied in swarm robotic for more than two decades. Biology revealed that the environmental characteristics are very important factors in aggregation of social insects and other animals. In this paper, we study the effects of different environmental factors such as size and texture of aggregation cues using real robots. In addition, we propose a mathematical model to predict the behaviour of the aggregation during an experiment.

Keywords: Power-law distribution · Swarm robotics · Aggregation · Modelling

1 Introduction

Aggregation is a common phenomenon in social behaviour of animals which can be observed from microscopic amoeba to insects and other animals [8]. A cue-based aggregation helps to gather a group of animals at the optimal zones with following the environmental cue. In swarm robotics [16], aggregation is defined as gathering of randomly distributed robots into a single aggregate. It is one of the fundamental behaviours in swarm robotics which helps the robots to get closer to each other and interact in order to perform other behaviours such as flocking and collective transport.

BEECLUST aggregation method proposed in [19] is inspired from simple behaviours in honeybees aggregation. The aggregation method is based on collisions between robots. A gradient light source in the arena is used as the aggregation cue. Each robot moves randomly and stops when it meets another robot. The waiting time depends on the intensity of the light at the particular location where the robot collied. The more the intensity, the longer it waits. After the waiting time is over, the robot turns randomly and moves forward. Results of the performed experiments showed that robots are able to aggregate on the optimal zone where the intensity of the light is the highest. Schmickl et al. [18] proposed two types of experiments: (i) *static* experiments in which there is a single light source and (ii) *dynamic* experiments in which there are two

© Springer International Publishing Switzerland 2015
Y. Tan et al. (Eds.): ICSI-CCI 2015, Part I, LNCS 9140, pp. 551–559, 2015.
DOI: 10.1007/978-3-319-20466-6_58

light sources with different intensities. The intensities of the sources are changed during an experiment. They showed that, robots aggregated on the optimal zone in static experiments. Whereas, in dynamic experiments, robots are able to aggregate under the highest intensity source. To improve the performance of the cue-based aggregation, two modifications on BEECLUST were proposed in [4]. One is the *dynamic velocity* in which robots are allowed to select three different speeds based on intensity of light; higher intensity results in slower speed. The second modification is the *comparative waiting time* in which the waiting time of a robot increases in the presence of the other robots. The results showed that both methods improve aggregation performance. In addition, the effects of turning angle has been studied in [6]. In this study, the performance of two proposed aggregation algorithms – *vector averaging* and *naïve* – was compared with BEECLUST. The results showed that the proposed strategies outperform BEECLUST method. Fuzzy-based aggregation method has been introduced in [5]. The results showed that the proposed fuzzy decisioning method improves the performance of BEECLUST especially in the presence of noise.

In order to analyse a collective behaviour in swarm robotics, macroscopic modelling is considered to be a more comprehensible approach to analysis different effective parameters of the behaviour. Stochastic characteristic of swarm algorithms leads to use a probabilistic modelling to depict the collective behaviour of the swarm systems [20]. To that end, various models from macroscopic behaviours of swarms have been proposed in [10,13,14]. The swarm scenarios are mostly the result of the inter-agent interactions which can be modelled by chemical reaction network model [15]. Macroscopic model of an aggregation behaviour must be able to predict the final distribution of the cluster [21]. Bayindir and Şahin [7] proposed a macroscopic model for a self-organized aggregation using probabilistic finite state automata. Schmickl et al. [17] proposed *Stock & Flow* model to model the macroscopic behaviour of a cue-based aggregation.

In this paper, we analyse a cue-based aggregation scenario based on the state-of-the-art BEECLUST method. In particular, we investigate different characteristics of a cue in the environment (size and texture) to check the influence of the changes on the swarm performance with a real robot, *Colias*.

2   Power-Law Distribution

The power-law distribution is a mathematical model that depicts a dynamic and functional relationship between two variables [9]. This is an important approach to model performance pattern of a long-term activity in an experiment and can reveal the reliability of a system. Mathematically, it is said that two quantities are related by power-law relationship, when one quantity varies as a power of another one (Eq. 1).

$$y = \alpha x^k , \tag{1}$$

where y and x are variables of interest, k is called the power-law exponent, and α is a constant amplitude. Distributions of the form Eq. 1 are said to follow a power-law. Power-law is very important because it reveals reliability in the properties of a system. Therefore, the result we get at one level would be very similar to the obtained result at former levels. This self-similar property makes the system predictable. A simple way to test whether an activity follows the power-low is construct a histogram representing the frequency distribution and re-plot the data on a log-log scaled graph. Hence, if we take the logarithms of both sides of Eq. 1, we get $\log y = k \log x + \log \alpha$. This says that if we have a power-law relationship, and we plot $\log y$ as a function of $\log x$, then we should see a straight line. Such a plot thus provides a quick way to see if one's data exhibits an approximate power-law. Assuming $\log y = V$ and $\log x = U$, simply k and α can be obtained from: $V = kU + \log \alpha$, where k is slope of the straight line and $\log \alpha$ is the value of the intercept when $U = 0$. Practically, some empirical phenomena completely comply with power-law for all values. Therefore, we can hardly ever be certain that an observed quantity is drawn from a power-law distribution. The most we can say is that our observations are consistent with a form of probability distribution like Eq. 1. Usually power-law applies only for values greater than a lower bound. In such cases, we say that the tail of the distribution follows a power-law. Therefore, a probability distribution that follows the power-law is possible:

$$P(x) = Cx^{-k} \qquad \text{for} \qquad x > x_{min} \qquad (2)$$

Generally, if distribution of variables follows a strict/pure power-law, then:

$$P(x) = \frac{k-1}{x_{min}} \left(\frac{x}{x_{min}} \right)^{-k}, \qquad (3)$$

where x_{min} is the smallest value for which the power-law exist. Assuming $x_{min} = 1$ simplifies the distribution form to Eq. 2. We rewrite the power-law for our purpose as following equation where $D(t)$ is the size of aggregation at time t.

$$D(t) = \alpha\, t^k \qquad (4)$$

Examining the size of aggregate on our method outputs, parameters α and k for different experiments are extracted and listed in the results section.

3   Swarm Scenario

We implement a cue-based aggregation scenario with two different configurations: i) effects of different sizes of cue and ii) effects of the cue's texture on the performance of the swarm.

3.1   Aggregation Method

We use BEECLUST method [18] as the aggregation scenario. In BEECLUST, robots move randomly in the environment. When they detect an object, they check whether it is an obstacle or another robot. If it is an obstacle, the robot avoids the obstacle. If not, it stops and waits for a particular amount of time, $w(t)$, depends on the intensity of the light.

3.2   Size of Cue

In this setup, we study the effects of the different cue sizes on the performance of the aggregation with the simulated gradient light. We assume $A_r = \pi R_s^2$ is the area which a robot covers using its sensory system with radius of R_s. Therefore, the total area which can be covered by radial arrangement of the robots is $A_{sw} = NA_r$, where N is the number of the robots in an experiment. In this phase of the experiments, we use three different sizes of cue for each population, $A_c = \beta NA_r$, $\beta \in \{2, 2.5, 3\}$. Therefore, with an increase in population size we increase the size of the cue relatively.

3.3   Texture of Cue

In this experiment, we study on effects of two types of cues with different lighting which are gradient and non-gradient. In the gradient cue, the luminance reduces gradually from the center to the edge of the cue; however the non-gradient cue has similar luminance at every part of the cue. We study effects of the texture on the performance of the aggregation at two fixed sizes of cues, a small cue with radius of R_c =16 cm and a big cue with radius of R_c =20 cm. We then extract the model parameters (Eq. 4) from the observed results to investigate the effects of the different texture of cue on the model.

4   Experiments

4.1   Experimental Setup

We use Colias [2] as the robot platform in our experiments. It is specially designed for swarm robotics research with a very compact size of 4 cm. Fig. 1(a) shows a Colias robot and its different modules. Two micro DC gearhead motors each connected to a wheel with diameter of 2.2 cm actuate Colias attaining a maximum speed of 35 cm/s. The rotational speed for each motor is controlled individually using pulse-width modulation [1]. The basic Colias uses only IR proximity sensors to avoid obstacles as well as the collision with the other robots [3], and a light sensor to read intensity of the ambient light. Colias is a modular robot which supports extension modules such as bio-inspired vision board developed in [11].

We used a horizontally placed 42" LCD flat screen as the ground that the robots move on as shown in Fig. 1(b). A very useful feature of Colias is the

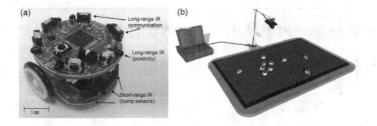

Fig. 1. (a) Colias micro robot. (b) Experimental setup.

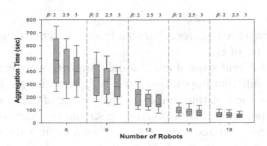

Fig. 2. Aggregation time in different population sizes at different cue sizes $\beta \in \{2, 2.5, 3\}$

light (illuminace) sensor face to the bottom side of the robot which gives an opportunity to use a LCD screen. In our experiments, all the aggregation cues are circular light spots with maximum illuminance of 420 lux. We use visual localisation software [12] to track the robots.

4.2   Results

Aggregation time, T_a, and size of the aggregate, D_a, are two metrics used in this study. Aggregation zone is defined as the area at the cue zone and set the robots within that area as an aggregated robot. Therefore, the aggregation time is defined as the time that the aggregate size reaches at 70% of the total number of robots.

The results of the aggregation at different cue sizes are shown in Fig. 2. In general, an increase in population size reduces the aggregation time. In the experiments with the same number of robots, aggregation at a big cue accomplishes faster than at a small size cue. It is because of the increase in probability of the successful collisions which result in a longer resting time for the robot at the high luminance spots. The reduction in aggregation time also depends on the population size, which in the big population the reduction is less than the small populations.

Fig. 3 shows size of the aggregate and the best fitted model. The model fitting is investigated in three population sizes of minimum, middle and maximum number of robots ($N = \{6, 12, 18\}$). The results show that, the proposed

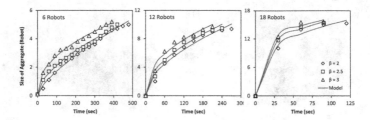

Fig. 3. Median of size of the aggregate during aggregation process with different β values

model meets the captured aggregation size from the experiments with different populations and sizes of cue.

In case of the different sizes of cue, model parameters are extracted from the recorded results in different population sizes. In general, for all β, with increasing the population size, the constant amplitude parameter of the model (α) increases and the exponent parameter (k) decreases. In similar populations, an increase in the size of cue, increases α and reduces k. As shown in the extracted parameters, the changes on environments have clear influence on the model parameters. In addition, all the results are fitted to the model with high coefficient of determination ($R^2 > 0.97$).

Table 1. Extracted model parameters for different cue sizes

Population	$\beta = 2$			$\beta = 2.5$			$\beta = 3$		
	α	k	R^2	α	k	R^2	α	k	R^2
6 Robots	0.063	0.714	0.99	0.145	0.580	0.99	0.353	0.447	0.99
12 Robots	0.470	0.547	0.97	0.626	0.505	0.98	0.994	0.431	0.98
18 Robots	3.999	0.288	0.98	5.072	0.249	0.99	6.407	0.203	0.99

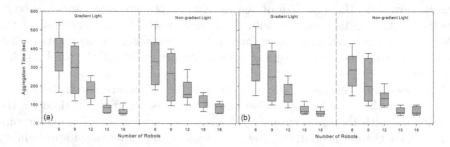

Fig. 4. Aggregation time with gradient and non-gradient lights in different population sizes at (a) a small size cue (with radius of 16 cm) and (b) a big size cue (with radius of 20 cm)

In the second configuration, we study the effects of different methods of lighting on swarm performance. Fig. 4 reveals the results of four different configurations with different number of robots. As shown in the all experiments, an increase in number of robots reduces the aggregation time. In addition, it is observed that, in all runs the non-gradient cue reduces the aggregation time slightly due to its higher average luminance which resulted in longer resting time. However, in higher populations (12 and 15 robots) the aggregation time increased. Since cue has same luminance, in high populations the aggregate formed nearby the edges hence the way to reach the centre of the cue by other robots is blocked. However, the swarm performance in the big size cue was less affected by the phenomenon.

In addition, we modelled the recorded data from aggregation experiment using Eq. 4 and extracted the model parameters for the different population sizes in the small size cue. Median of the size of aggregate during an aggregation process and the predicted model are shown in Fig. 5. We stop the experiments when the aggregate is formed ($t = T_a$).

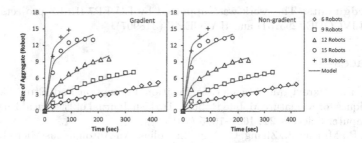

Fig. 5. Median of size of the aggregate in gradient and non-gradient environments

Table 2 shows the model parameters in different populations and the coefficient of determinations for each configuration. All the results are fitted in the model with high R^2 values. The results of the modelling reveal that, an increase in the population size increases parameter α and reduces parameter k. Moreover, α and k in a similar population size are different for gradient and non-gradient cues. In non-gradient cue, α is higher than the gradient cue, however, k is less than the gradient cue, except in the case of 6 robots which both α and k showed an opposite behaviour than the higher populations which could be due to lower population size in a large swarm arena.

The anticipated changes on the model parameters due to the physical changes on the swarm configuration demonstrate that, the environmental changes can also be predicted by the proposed model.

5   Conclusion

In this paper we analysed the effects of two environmental factors in a cue-based aggregation method called BEECLUST. We investigate two metrics, namely, aggregation time and size of the aggregate and evaluated the performance of

Table 2. Extracted model parameters for small cue

Population	Gradient light			Non-gradient light		
	α	k	R^2	α	k	R^2
6 Robots	0.134	0.576	0.95	0.099	0.641	0.98
9 Robots	0.431	0.479	0.96	0.515	0.449	0.98
12 Robots	0.811	0.460	0.98	1.272	0.372	0.99
15 Robots	3.113	0.289	0.96	3.854	0.249	0.99
18 Robots	4.555	0.253	0.99	5.870	0.198	0.99

the swarm aggregation using real mobile robots. We also modelled the experimental data with the simplified Power-Law distribution. The model parameters were extracted from the results observed from the experiments in different configurations.

Acknowledgments. This work was supported by EU FP7-IRSES projects EYE2E (269118), LIVCODE (295151) and HAZCEPT (318907).

References

1. Arvin, F., Bekravi, M.: Encoderless Position Estimation and Error Correction Techniques for Miniature Mobile Robots. Turkish Journal of Electrical Engineering & Computer Sciences **21**, 1631–1645 (2013)
2. Arvin, F., Murray, J., Zhang, C., Yue, S.: Colias: An Autonomous Micro Robot for Swarm Robotic Applications. International Journal of Advanced Robotic Systems **11**(113), 1–10 (2014)
3. Arvin, F., Samsudin, K., Ramli, A.R.: Development of IR-Based Short-Range Communication Techniques for Swarm Robot Applications. Advances in Electrical and Computer Engineering **10**(4), 61–68 (2010)
4. Arvin, F., Samsudin, K., Ramli, A.R., Bekravi, M.: Imitation of Honeybee Aggregation with Collective Behavior of Swarm Robots. International Journal of Computational Intelligence Systems **4**(4), 739–748 (2011)
5. Arvin, F., Turgut, A.E., Bazyari, F., Arikan, K.B., Bellotto, N., Yue, S.: Cue-based aggregation with a mobile robot swarm: a novel fuzzy-based method. Adaptive Behavior **22**, 189–206 (2014)
6. Arvin, F., Turgut, A.E., Bellotto, N., Yue, S.: Comparison of different cue-based swarm aggregation strategies. In: Tan, Y., Shi, Y., Coello, C.A.C. (eds.) ICSI 2014, Part I. LNCS, vol. 8794, pp. 1–8. Springer, Heidelberg (2014)
7. Bayindir, L., Şahin, E.: Modeling self-organized aggregation in swarm robotic systems. In: Swarm Intelligence Symposium, pp. 88–95 (2009)
8. Camazine, S., Franks, N., Sneyd, J., Bonabeau, E., Deneubourg, J.L., Theraulaz, G.: Self-organization in Biological Systems. Princeton University Press (2001)
9. Clauset, A., Shalizi, C.R., Newman, M.E.: Power-law distributions in empirical data. SIAM Review **51**(4), 661–703 (2009)
10. Correll, N., Martinoli, A.: Modeling self-organized aggregation in a swarm of miniature robots. In: IEEE International Conference on Robotics and Automation, Workshop on Collective Behaviors Inspired by Biological and Biochemical Systems (2007)

11. Hu, C., Arvin, F., Yue, S.: Development of a bio-inspired vision system for mobile micro-robots. In: 4th International Conference on Development and Learning and on Epigenetic Robotics, pp. 137–142 (2014)
12. Krajník, T., Nitsche, M., Faigl, J., Vaněk, P., Saska, M., Přeučil, L., Duckett, T., Mejail, M.: A Practical Multirobot Localization System. Journal of Intelligent & Robotic Systems **76**(3–4), 539–562 (2014)
13. Lerman, K., Galstyan, A., Martinoli, A., Ijspeert, A.: A macroscopic analytical model of collaboration in distributed robotic systems. Artificial Life **7**(4), 375–393 (2001)
14. Martinoli, A., Ijspeert, A., Mondada, F.: Understanding collective aggregation mechanisms: From probabilistic modelling to experiments with real robots. Robotics and Autonomous Systems **29**(1), 51–63 (1999)
15. Mermoud, G., Matthey, L., Evans, W., Martinoli, A.: Aggregation-mediated collective perception and action in a group of miniature robots. In: International Conference on Autonomous Agents and Multiagent Systems, pp. 599–606 (2010)
16. Şahin, E., Girgin, S., Bayındır, L., Turgut, A.E.: Swarm robotics. In: Swarm Intelligence, vol. 1, pp. 87–100 (2008)
17. Schmickl, T., Hamann, H., Worn, H., Crailsheim, K.: Two different approaches to a macroscopic model of a bio-inspired robotic swarm. Robotics and Autonomous Systems **57**(9), 913–921 (2009)
18. Schmickl, T., Thenius, R., Moeslinger, C., Radspieler, G., Kernbach, S., Szymanski, M., Crailsheim, K.: Get in touch: cooperative decision making based on robot-to-robot collisions. Autonomous Agents and Multi-Agent Systems **18**(1), 133–155 (2009)
19. Schmickl, T., Hamann, H.: BEECLUST: A Swarm Algorithm Derived from Honeybees. Bio-inspired Computing and Communication Networks. CRC Press (2011)
20. Soysal, O., Şahin, E.: Probabilistic aggregation strategies in swarm robotic systems. In: Swarm Intelligence Symposium, pp. 325–332 (2005)
21. Soysal, O., Şahin, E.: A macroscopic model for self-organized aggregation in swarm robotic systems. In: Şahin, E., Spears, W.M., Winfield, A.F.T. (eds.) Swarm Robotics Ws. LNCS, vol. 4433, pp. 27–42. Springer, Heidelberg (2007)

Effect of the Emergent Structures in the Improvement of the Performance of the Cognitive Agents

Abdelhak Chatty[1,2]([✉]), Philippe Gaussier[2], Ilhem Kallel[1], and Adel M. Alimi[1]

[1] ReGIM-lab: REsearch Groups on Intelligent Machine,
National School of Engineers (ENIS), Sfax University, Sfax, Tunisia
[2] ETIS: Neuro-cybernetic Team, Image and Signal Processing,
National School of Electronics and Its Applications (ENSEA),
Cergy-Pontoise University, Paris, France
abdelhak_chatty@ieee.org

Abstract. This paper tries to analyze the positive effect of the emergent structures in the objects' aggregation task which is performed by a cognitive multi-agent system (CMAS). Indeed, these structures allow improving overall performance of the system by the optimization of the planning time and satisfaction level of the cognitive agents. A series of simulations enables us to discuss our system.

Keywords: Emergent structures · Cognitive multi-agent system · Optimization · Cognitive maps

1 Introduction

In the field of swarm intelligence [1–7], a number of successful experiments in the objects' aggregation task was performed based on simple local rules. A model relying on biologically plausible assumptions was proposed in [8] to analyse the phenomenon of dead bodies' aggregation by ants. The authors in [9] showed that interacting directly with objects simplifies the reasoning needed by multi-agent system and allows the aggregation of scattered objects. The aggregation and the sorting of colored frisbees by a multi-agent system was studied in [10]. In the experiments cited above, the agents do not have a navigation strategy; they move randomly using only local rules. Thus, it's interesting to know what will happen if the agents were able to use beside the local rules a bio-inspired navigation system which allows them to learn the objects' positions in their environment. Our case study is part of the objects' aggregation task. In fact, the simulated environment contains three infinite resources (A, B and E). All resources are composed of a set of objects. The agent life cycles are linked to their supply levels from each resource type (which is already discovered and learned). Each agent possesses drives which corresponds to a resource type. The level of each supply levels is internally represented by an essential variable $ei(t)$ whose value

© Springer International Publishing Switzerland 2015
Y. Tan et al. (Eds.): ICSI-CCI 2015, Part I, LNCS 9140, pp. 560–569, 2015.
DOI: 10.1007/978-3-319-20466-6_59

is in $[0; 1]$ and varies with time as in Equation 1.

$$\frac{de_i}{dt} = -\alpha_n e_i(t) \tag{1}$$

In the equation, α_n represents the decreasing rate of the essential variable. When the level of one supply decreases to a critical threshold level (TL), the drive related to that type of supply triggers and the agent starts to reach the resource that allows the satisfaction of that need. If the agent fails to go back to that resource before the corresponding satisfaction level reaches a very low level, it dies. To maintain the satisfaction level of the agents, instead of navigation between the three learned original resources, it's interesting if the accumulation of the local rules and the learning ability of agents give birth of the creation of relevant warehouses in the environment. Warehouse creation is possible only if agents are able to carry and deposit a quantity of products taken from resources. The localisation of warehouses is important because when agents reach them easily, they can increase their average satisfaction level and optimize the planning time spent looking for warehouses. So what is the effect of the localisation of warehouses in the improvement of the performance of the cognitive agents? The aim of this work is to show how the emergent structures (the warehouses) are able to improve the overall performance of the cognitive multi-agent system (CMAS) through the optimization of the planning time and satisfaction level of the cognitive agents. This paper is organized as follow: in section 2 the internal behavior of a cognitive agent is presented. Section 3 describes the emerging of relevant warehouses. Before concluding, section 4 and 5 are devoted to show the positive effect's of the emergent warehouses to improve the performance of the cognitive multi-agent system (CMAS)

2   The Internal Behavior of a Cognitive Agent

An agent is able to perceive objects only in a local neighborhood. Thanks to the perception-action loop, which implies sensing from the environment, an agent is able to learn and build its own cognitive map.

2.1   The Agent's Neural Networks Architecture

Starting from neurobiological hypotheses on the role of hippocampus in the spatial navigation, several works [11,12] revealed special cells in the rats hippocampus that becomes active whenever the animal passes through a given place which it already visited. These neurons have been called place cells (PCs). We do not use PCs directly to navigate, plan or build a map, we rather use neurons called transition cells (TC) [13]. These cells represent the basis of the neurobiological model of temporal learning sequences in the hippocampus. A transition cell encodes a spatio-temporal transition between two PCs consecutively winning the place recognition competition, respectively at time t and δt. The set of PCs and TCs constitutes a non-cartesian cognitive map. To develop the bio-inspired cognitive map, we took inspiration from the model presented in [14] which describes the role of the hippocampus. In practice, to create the PC the agent takes a visual panorama of the surrounding environment. The views are processed to extract visual landmarks. After learning these landmarks, a visual code is created by combining the landmarks of a panorama with their azimuth. This configuration serves as a code for PCs. We suppose the signals provided by the EC (the entorhinal cortex:an input structure to the hippocampus) are solely spatial and consistent with spatial cells? activities. Spatial cells activities are submitted to a Winner-Take-All competition in order to only select the cell with the strongest response at a specific location. We will subsequently talk about the current location by indicating the spatial cell which has the highest activity at a given location. Thus, the temporal function at the level of the DG (dentate gyrus) is reduced to the memorization of past location. The acquired association at the level of CA3 (the pyramidal cells) is then the transition from a location to another. Once the association from the past location to the new one is learned, every new entry will reactivate the corresponding memory in the DG. A schematic view of our architecture is shown in Fig. 1.

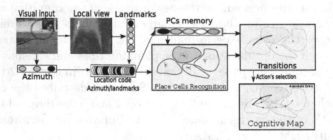

Fig. 1. From the construction of the visual code of place cells to the creation of the cognitive map [15]

During the exploration of the environment each agent is able, independently of the other agents, to navigate, learn and create its own cognitive map on-line whose structure depends on the agent's own experience and discovery of the environment in which it lives [16]. After having explored the environment, the agents are able to predict, in each position the locations directly reachable. The equations 2 and 3 govern the learning in the cognitive map, where $T(t)$ is a binary signal (0 or 1) which is activated when a transition is made (moving from one place to another). This signal controls the learning of recurrent connections W^{CC}. γ is a parameter less than 1 which regulates the distribution of the motivation activity on the map. $\lambda1$ and $\lambda2$ are parameters of respectively active and passive forgotten on the recurrent connections. $S(t)$ is a signal marking the satisfaction of an objective. This signal controls the learning of synaptic connections between neurons in W^{MC} motivations activity X^M and neurons of the cognitive map of activity X^C.

$$\frac{dW_{ij}^{CC}(t)}{dt} = T(t).((\gamma - W_{ij}^{CC}).X_i^C(t).X_j^C(t) - W_{ij}^{CC}(t).(\lambda1.X_j^C(t) - \lambda2)) \tag{2}$$

$$\frac{dW_{ij}^{MC}(t)}{dt} = S(t) \, for \, i, j = argmax_{k,l}(X_l^C(t).X_M^K(t)) \tag{3}$$

2.2   Agent's Local Rules

In the context of situated cognition, the local rules can lead to create emergent structures allowing the creation of relevant warehouses as shown in the sorting strategy used by the authors in [8]. In the simulated environment, the agents have two behaviors. The first one is the exploration mode which allows them to discover the environment without the need to satisfy their drives. The resources and the new warehouses discovered will be learned and added in the cognitive maps of the agents in order to allow them to satisfy their needs. The second behavior is manifested when the need arises and the drives trigger. Indeed, the agents switch to the planning mode using their cognitive maps to reach a resource or a warehouse. Thereby, the agents are able to return to these resources or warehouses in order to pick up an object and deposit elsewhere in the environment. The pick up and the diposit local rules are functions of the number of agents perceived. The agent can indeed, tend to favor the location which contains other agents rather than empty regions in order to deposit the object. The pick up condition follows equation 4, which means, that the probability to take an object from a resource or a warehouse is inversely proportional to the number of agents surrounding the resource, the more isolated is the resource, the higer picking probability is.

$$Pr_{(pick\_up)} = \exp^{-\lambda N_R} \tag{4}$$

where N_R is the number of agents in the neighborhood, λ is a positive constant. Equation 5 describes the deposit conditions. The probability of deposit increases with time from the original resource or the warehouse from where the agent took the objects. It also depends on the number of the agents in the neighborhood: the agent wants more to deposit when the place is frequented by other agents. The deposit operation is also built on the concept of refueling : the agent puts objects in the warehouses that already exist.

$$Pr_{(Deposits)} = (1 - \exp^{-\alpha N_R}) * (1 - \exp^{-\beta t}) \tag{5}$$

where α, β are environmental factors, N_R is the number of agents in the neighborhood and t is the time since the taking.

3   Emerging of Relevant Warehouses

We placed the three original resources (A, B and E) in the summits of an isosceles triangle knowing that the centroid is the relevant place. The three original resources allow the creation of three different types of warehouses ("a" is from A, "b" from B and "e" is from E). In Fig. 3 (a), the 48 agents start moving randomly in the environment, with a limited field of view that restricts the ability to perceive the entire environment and let the agents detect only the close agents. While passing through a resource or a warehouse, a agent increases its level of satisfaction and applies the local rule to pick up an object ($\lambda = 1$). The probability to pick up an object increases when the agent does not detect other agents next to the resource. Once the pick up is succeeded, the agent tries to find a suitable location to deposit the object and to create thus a new warehouse ($\alpha = \beta = 0.3$). The probability of deposit increases when the agent detects other agents and it is sufficiently away from the original resource as indicated by equation 5. This means that the locations chosen for the deposit are often common to several agents. Fig. 3 (b) shows the creation of warehouses. Agents also have the possibility of refueling warehouses by adding objects to them. This provides stability for the warehouse in relevant locations which are close to several agents. However, warehouses that are abandoned or poorly visited will eventually disappear since the number of objects available will decrease rapidly (see Fig. 3 (c)). When a planning agent ($TL = 70$) tries to reach a previously known warehouse and realizes that it has disappeared, the agent dissociates the current TC from the formerly-corresponding warehouse, and it resets the motivation to 0. Since the TC does not fire any more when the agent feels the need for this warehouse, the transitions leading to this place will be progressively forgotten. Similarly, when a new matching warehouse is discovered, the paths leading to the warehouse are immediately reinforced, making the cognitive map evolving in accordance with the environment (see Figure 2).

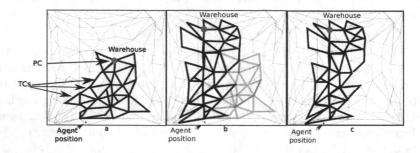

Fig. 2. The evolution of the cognitive map. The disappearance of the warehouse leads to the dissociation of the motivation from the old warehouse place. After discovering a new warehouse, the TCs, leading the old warehouse disappear and the new ones become completely reinforced.

Finally, the cognitive multi-agent system (CMAS) converges to a stable configuration (see Fig. 3 (d)) with a fixed number of warehouses in fixed places at 7587 time steps and remains the same for more than 20000 time steps (see Fig. 3 (e)). We note that the agents were able to create villages of warehouses (composed by a fixed number of warehouses, in fixed positions), which consist of objects from the three different original resources in an appropriate location at the centroid of triangle. Thus, instead of browsing an Euclidean distance between the three original resources which is equal to 59.2 to look for objects, agents can reduce this distance (equal to 12.08) with the creation of near-perfect villages. This shows that emergent warehouses ensure an optimization of the distance traveled by the agents to return to resources, using their cognitive maps (see Fig. 3 (f)).

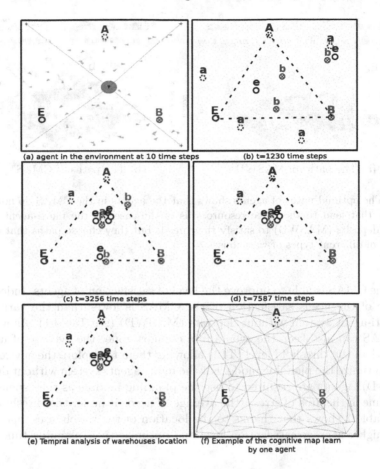

Fig. 3. The cognitive multi-agent system (CMAS) were able to create a village of different kinds of new stable warehouses in relevant places which allowed agents to optimize the distance walked to return to the different kind of resources

4   The Increasing of the Needs Satisfaction

In order to show that the CMAS is able to create emergent warehouses in relevant places frequented by the agents (without having to use thresholds in order to limit the number of warehouses nor to specify their locations), we tried to experimentally count the number of visits of the warehouses created by the CMAS, compared to the number of visits of the resources for 20000 time steps. We noted that the average number of visits to the warehouses (115) is more important than for the resources (27). Indeed, the agents of the CMAS do not have to follow the same paths as the multi-agent system without deposits (MASWD) to satisfy their needs (see Figure 4 (a)). They can use the warehouses placed in relevant locations that form a village of different types of warehouses (see Figure 4 (b)).

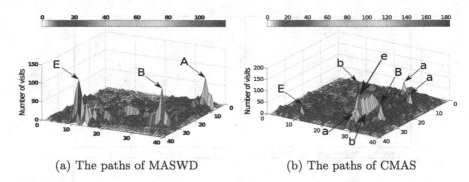

(a) The paths of MASWD   (b) The paths of CMAS

Fig. 4. The optimal paths of agents shows that the agents in the CMAS do not follow the path that lead to the three resources as is the case for the multi-agent system without deposits (MASWD) to satisfy their needs but they choose paths that contain a village of different types of warehouses

Thus, the CMAS is able to improve the level of satisfaction of agents. Indeed, the variance of the needs satisfaction of the CMAS is lower than the variance of the multi-agent system without deposits (MASWD) (see Table 1). As a result, the CMAS can also keep the agents in a comfort zone (the average of needs is greater than the threshold level (TL)) allowing them to explore the environment and to optimize the planning mode. For the multi-agents system without deposits (MASWD), agents are required to use the planning module as they spend most of the time in the area of stress (the average of needs is under the threshold level (TL). Table 1 shows that thanks to the location of the warehouses, agents can keep a higher average satisfaction level and optimize their planning time.

Table 1. Improvement of the satisfaction Level (time steps)

Average	MASWD	CMAS
Planning Time	1600ts	450ts
Satisfaction level	59,89	88.07
Variance of the needs satisfaction	74.272	31.596

5   Optimization of the Planning Time

In order to assess the quality of the solution, we tried to change the environment, since the location of resources and the shape of the environment can affect the quality of the final solution. Thus, we put 10 agents in an environment (in form of "T") consisting by three sides with different length, containing the three resources in the three ends, as shown in Figure 5 (a). Example of the cognitive map of an agent after 20000 time steps is presented in Figure 5 (e). During the experiment, the agents first deposit two warehouses (type a and e) near to the resource B which are placed on the longest side (see Figure 5 (b)) and also frequented by agents (see Figure 7 (d)). Thus, agents do not have to move to other resources to satisfy their needs, they just have to visit the village composed by the resource B and the two warehouses a and e. This solution is interesting, except that agents exploit in full the warehouses without to refuel them, since they are far from the resources (A nd E). Thus, the warehouses (a and e) eventually disappear. At stability, the solution converges to Figure 5 (c): the agents ended up creating a village composed by three different types of warehouses (a, b and e) in the intersection of three paths leading to the different resources (A, B and E). It's the optimal solution since the village is located at the centroid of the resources. The number of visits in this location is the highest and therefore the most frequented by the agents (see Figure 5 (d)).

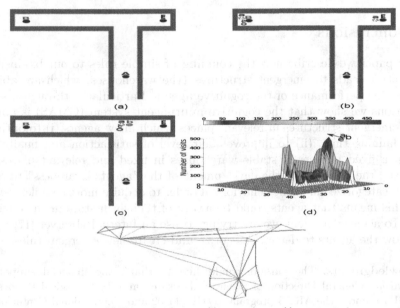

(e) Example of a cognitive map learned by one agent

Fig. 5. Relevance in the location of the warehouses: based on the cognitive process the CMAS was able to create emergent warehouses in relevant places more precisely in the intersection of three paths leading to the different resources (A, B and E)

Thus, this emergent configuration allows the agents to no longer use their cognitive maps and therefore to: optimise their planning time (i) to improve their needs satisfaction and (ii) to refuel the warehouses (see Table 2). It is important to note that it is possible to program a single agent to deposit the warehouses in the centroid of plants once their positions are discovered. However, this solution is not attractive in the case where the environment is constituted by several villages of plants. Indeed, the agent will place the warehouses not in the centroid of each village but in the centroid of all the villages. The CMAS is able to adapt to changes in the configuration of the environment and allows the emergence of a fixed number of warehouses in fixed and appropriate places without having to use thresholds in order to limit the number of warehouses nor to specify their locations. This warehouses allow the agents to optimize the planning time and to improve of their needs satisfaction.

Table 2. Optimization of planning time (time steps)

Average	MASWD	CMAS
Planning Time	1825ts	0ts
Satisfaction level	54,36	95,89
Variance of the needs satisfaction	97,294	13,275

6   Conclusions

In this paper we describe how the coupling of simple rules to our bio-inspired architecture leads to emergent structures (the warehouses) which are able to improve the performance of the cognitive agents. Particularly, through a set of simulations we show that the cognitive multi-agent system (CMAS) is able to create emergent structures in relevant places which allow agents (i) to optimize their planning time, (ii) to improve their level of satisfaction and finally (iii) to keep a fixed number of stable warehouses in fixed and relevant places. As prospects, the challenge is the development of the "agents awareness" to allow them to categorize emergent behaviors in order to acquire more complex behaviors. This means that agents could be aware of their own state in the environment. To achieve our goal, we are trying to use an internal observer [17] which will allow the agents to detect, categorize and create new emergent rules.

Acknowledgments. The authors would like to thank the financial support of the Tunisian General Direction of Scientific Research and Technological Renovation (DGRSRT), under the ARUB program 01/UR/11 02 and The National Center of scientific research (CNRS).

References

1. Mataric, M.J.: Designing emergent behaviors: from local interactions to collective intelligence. In: Meyer, J., Roitblat, H., Wilson, S. (eds.) Proceedings of the Second Conference on Simulation of Adaptive Behavior, pp. 1–6. MIT Press (1992)
2. Kube, C.R., Zhang, H.: Collective robotics from social insects to robots. Adaptive Behavior **2**(2), 189–218 (1993)
3. Holland, O., Melhuish, C.: Stigmergy, self-organization, and sorting in collective robotics. Artif. Life **5**(2), 173–202 (1999)
4. Chatty, A., Kallel, I., Alimi, A.M.: Counter-ant algorithm for evolving multirobot collaboration. In: IEEE Proceedings of the 5th International Conference on Soft Computing as Transdisciplinary Science and Technology, pp. 84–89. CSTST (2008)

5. Chatty, A., Kallel, I., Gaussier, P., Alimi, A.M.: Emergent complex behaviors for swarm robotic systems by local rules. In: IEEE Proceedings of the Symposium Series on Computational Intelligence on Robotic Intelligence in Informationally Structured Space (RiiSS), pp. 69–76 (2011)
6. Chatty, A., Gaussier, P., Kallel, I., Laroque, P., Pirard, F., Alimi, A.M.: Evaluation of emergent structures in a "cognitive" multi-agent system based on on-line building and learning of a cognitive map. In: Proceedings of the 5th International Conference on Agents and Artificial Intelligence (ICAART), pp. 269–275 (2013)
7. Chatty, A., Gaussier, P., Karaouzene, A., Bouzid, M., Kallel, I., Alimi, A.M.: Coupling learning capability and local rules for the improvement of the objects' aggregation task by a cognitive multi-robot system. In: del Pobil, A.P., Chinellato, E., Martinez-Martin, E., Hallam, J., Cervera, E., Morales, A. (eds.) SAB 2014. LNCS (LNAI), vol. 8575, pp. 290–299. Springer, Heidelberg (2014)
8. Deneubourg, J.L., Goss, S., Franks, N., Franks, A.S., Detrain, C., Chrétien, L.: The dynamics of collective sorting robot-like ants and ant-like robots. In: Proceedings of the First International Conference on Simulation of Adaptive Behavior on From Animals to Animats, Cambridge, MA, USA, pp. 356–363. MIT Press (1990)
9. Gaussier, P., Zrehen, S.: Avoiding the world model trap: An acting robot does not need to be so smart!. Robotics and Computer-Integrated Manufacturing 11(4), 279–286 (1994)
10. Beckers, R., Holland, O.E., Deneubourg, J.L.: From local actions to global tasks: stigmergy and collective robotics. In: Articial Life IV. Proc. Fourth International Workshop on the Synthesis and Simulation of Living Systems, Cambridge, Massachusetts, USA, pp. 181–189 (1994)
11. O'Keefe, J., Nadel, L.: The hippocampus as a cognitive map. Clarendon Press, Oxford University Press, Oxford (1978)
12. Milford, M., Wyeth, G.: Mapping a suburb with a single camera using a biologically inspired slam system. IEEE Transactions on Robotics 24(5), 1038–1053 (2008)
13. Gaussier, P., Revel, A., Banquet, J.P., Babeau, V.: From view cells and place cells to cognitive map learning: processing stages of the hippocampal system. Biological Cybernetics 86(1), 15–28 (2002)
14. Banquet, J.P., Gaussier, P., Dreher, J.C., Joulain, C., Revel, A., Gunther, W.: Spacetime, order and hierarchy in fronto-hippocamal system : a neural basis of personality. In: Cognitive Science Perspectives on Personality and Emotion, pp. 123–189. Elsevier Science BV (1997)
15. Chatty, A., Gaussier, P., Hasnain, S.K., Kallel, I., Alimi, A.M.: The effect of learning by imitation on a multi-robot system based on the coupling of low-level imitation strategy and online learning for cognitive map building. Advanced Robotics 28(11), 731–743 (2014)
16. Chatty, A., Gaussier, P., Kallel, I., Laroque, P., Alimi, A.M.: Adaptive capability of the cognitive map to improve behaviors of swarm robotics. In: IEEE Proceedings of the International Conference on Development and Learning and the Epigenetic Robotics (ICDL-EPIROB), pp. 1–7 (2012)
17. Tani, J.: An interpretation of the "self" from the dynamical systems perspective: A constructivist approach. Journal of Consciousness Studies 5, 516–542 (1998)

Multi-agent Organization
for Hiberarchy Dynamic Evolution

Lu Wang, Qingshan Li$^{(\boxtimes)}$, Yishuai Lin, and Hua Chu

Software Engineering Institute, Xidian University, Xi'an 710071, People's Republic of China
qshli@mail.xidian.edu.cn

Abstract. With increasingly dynamic operating environment and user requirements, software adopts a unified strategy to achieve the different levels of evolution, a fact which reduces the flexibility and efficiency. So, in this paper, a method with agent technology is proposed to support the hiberarchy evolution of both the function and service levels. Precisely, a multi-agent organization is proposed to separate the calculation and collaboration logics of software which are corresponding to the different levels of evolution. To achieve the function-level evolution, an adaptive agent model with knowledge reasoning provides the software an ability to dynamically modify the calculation logics. With the adjustment of the collaboration logics, the multi-agent organization can make it convenient for the software to deal with the service-level evolution. Finally, a case study of air defense simulation system and some test metrics indicates that the proposed multi-agent organization can effectively support the hierarchy evolution.

Keywords: Dynamic evolution · Multi-agent system · Adaptive agent

1   Introduction

There is a growing awareness that both users' requirements and the operating environment in which software runs are likely changing. However, the software running in some special fields is unable to stop to update itself. Therefore, the dynamic evolution has become a hot research topic. There are some existing studies of dynamic evolution, such as strategies based on model driven [1] [2], control engineering [3] [4], software architecture [5], programming language, and so on. As the agents have the commendable features such as adaptability, autonomy, initiative and collaboration, so the dynamic evolution strategy based on agents is also becoming a research focus [6].

Some existing studies about the dynamic evolution based on agents mainly pay attentions to build the adaptive agent models [7]. However, these current studies lack of classifying the evolutionary triggered factors by triggered reasons, and ignore the effect of the multi-agent organization on evolutions. Whichever the evolutionary triggered factor is, software adopts a unified strategy based on the adaptive agents to achieve the evolution, undoubtedly reducing the flexibility and efficiency of software.

© Springer International Publishing Switzerland 2015
Y. Tan et al. (Eds.): ICSI-CCI 2015, Part I, LNCS 9140, pp. 570–577, 2015.
DOI: 10.1007/978-3-319-20469-6_60

So this paper proposes an approach that uses agent technology to support the hiberarchy evolution of both the function and service levels. In order to apply this approach for all the software, the first thing is to change the normal software into the multi-agent system. Particularly, the functional units of software are packaged as agents and the relationship between the units are defined as the collaborative relationship between agents. Based on the transformation, the paper proposes a multi-agent organization to support the hiberarchy evolution by separating the different logics.

The paper is organized as follows. The section 2 presents the proposed multi-agent organization for hiberarchy dynamic evolution. Further, the section 3 demonstrates a platform to verify the effectiveness of this multi-agent organization. Finally, the section 4 draws a conclusion.

2   Multi-agent Organization for Hiberarchy Evolution

2.1   Multi-agent Organization and Agent Model

In the proposed evolution, the evolutionary triggered factors are separated into function-level and service-level. So, a multi-agent organization is shown in figure 1.

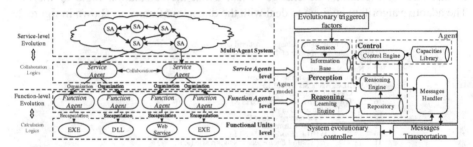

Fig. 1. Multi-Agent Organization and Agent Model

Specially, the factors in function level only influence the local functions of the software. For example, the functional units of software are damaged or lost is a factor of this level. And the calculation logic describes the specific definitions of the software units. So the Function Agent is designed to represent the calculation logics of a software unit. Function Agents are formed by packaging the software units which are in the form of exe, dll, or web service. So the solution of function-level evolution is just to adjust the calculation logics of the units in the corresponding Function Agents.

Furthermore, the service-level triggered factors require the software to provide the new services. For instance, users are likely to require software to provide new services or expand existing services. And when the environment changes, software also needs to dynamically adjust the relationship among functional units. So the *Service Agent* is formed by organizing some *Function Agents*. And there are also collaborations between *Service Agents*. Both the collaborative relationship between *Service*

Agents and the organization of *Function Agents* in a *Service Agent* are the collaboration logics. So the solution of service-level evolution is just to adjust these two kinds
of logics.

Both *Function Agents* and Services Agents are established according to an adaptive
agent model, as shown in figure 1. The Perception Module is designed to use different
Sensors to monitor various triggered factors and use Information Base to store the
collected information. The Control Module uses the Control Engine to make decisions. The capacities stored in the Capacities Library mean the functions or services,
which an agent could provide to external. The Reasoning Module is designed to support the Control Module by using the Learning Engine to learn the knowledge and
using Reasoning Engine to reason and propose the solutions. The evolutionary knowledge stored in the Repository is about the evolution tasks, collaborative relationship,
and other information about the evolution. The Messages Handler is used to package
and submit the messages to the system evolutionary controller through the messages
transportation.

2.2 Function-Level Evolution

In this kind of evolution, the change is just about the calculation logic of the units.
The adaptive algorithm is given to deal with the damages of units, as shown in figure 2.

```
Input:  the name of the damaged functional unit
Output:  the running state of this functional unit
1 BEGIN
2   Name N = Input;
3   Agent Set = all the agents in system
4   WHILE (Agent Set != NULL)
5     Check every agent;
6     IF (agent A has the unit named as N)
7       Find the unit U in A;
8       Check the running state of U;
9       IF (the state = damaged )
10        WHILE (Repository != NULL)
11          Control Engine searches the knowledge K;
12          IF (A has the K)
13            Control Engine sends the operations to Messages Handler;
14            RETURN  "Normal";
15          ELSE
16            Learning Engine learns the K from other agents;
17            IF (other agents have the K)
18              Reasoning Engine gives the suggestions;
19              Control Engine sends the operations to Messages Handler;
20            RETURN  "Normal";
21            ELSE RETURN  "Abnormal" ;
22        END-WHILE;
23        ELSE RETURN  "Normal";
24      ELSE CONTINUE;
25  END-WHILE
26 END
```

Fig. 2. Adaptive Algorithm

The character of this algorithm is the knowledge learning. It means that agent could
learn the knowledge from other agents and deal with the unfamiliar problems, which
enhance the ability of agents to adapt to the different kinds of abnormal situations.

The process of the function-level evolution is shown in figure 3. When the functional unit is damaged or lost is happened, a Function Agent could get this information by its Sensors ①. And the Control Engine queries the Capacities Library to

verify the running state of this unit ②. The agent could adjust itself according to the adaptive algorithm ③. If the agent does not have the appropriate knowledge, it could use the Learning Engine to get the knowledge from other agents ④. With the learned knowledge, the Reasoning Engine can give the preferable suggestions to the Control Engine ⑤.

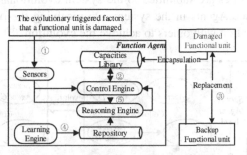

Fig. 3. Process of the Function-Level Evolution

2.3   Service-Level Evolution

The "Events/Conditions/Actions" Rules (ECA Rules) is "ON event of environment information changes; [IF condition on life message or attribute is available]; DO action of switching the integration rules", used as the evolutionary rules to achieve the adjustments. The adjustment algorithm can be described as figure 4.

```
Input:  Evolutional Drivers
Output:  The running state of system
1 BEGIN
2    IF (evolutionary controller finds the triggered factors)
3         EVENT E₁= the corresponding event;
4         Kind K₁= the kind of the E₁;
5         Set S₁= the set of ECA Rules in controller;
6         WHILE (S₁!=NULL)
7              Search the corresponding rules;
8              IF (Rule R₁'s event=E AND kind=K₁)
9                   Send R₁ to all the agents;
10             ELSE CONTINUE;
11        END-WHILE;
12        LABLE:
13        IF (K₁= service)
14             All the Service Agent adjust collaborations;
15             RETURN "Normal";
16        ELSE (K₁= function)
17             Service Agent S₁ adjusts its Function Agents;
18             RETURN "Normal";
19   ELSE (Service Agent S₂ finds the triggered factors)
20        EVENT E₂= the corresponding event;
21        Kind K₂= the kind of the E₂;
22        Set₂= the set of ECA Rules in Repository of S₂;
23        WHILE (Set₂ !=NULL)
24             Search the corresponding rules;
25             IF (Rule R₂'s event=E₂ AND kind=K₂)
26                  Send R₂ to evolutionary controller;
27                  GOTO LABLE;
28             ELSE CONTINUE;
39        END-WHILE;
30   END
```

Fig. 4. Adjustment Algorithm

As shown in figure 5, when some new requirements is put forward by users, these requirements could be sent to the system evolutionary controller as events ①.

It is necessary to trigger the corresponding ECA rules stored in the controller ②. If the events just influence Service Agent, the adjustment algorithm is used to adjust the collaboration among the Function Agents ③. If the events influence all the Service Agents, the collaboration among Service Agents needs to be adjusted ④. When the environment changes, the corresponding ECA rules stored in Service Agents are triggered ⑤. Then the rules are submitted to the system evolutionary controller ⑥ and sent to all the Service Agents in the system. On receiving the rules, Service Agents adjust their collaboration with others to achieve the evolution ③, ④.

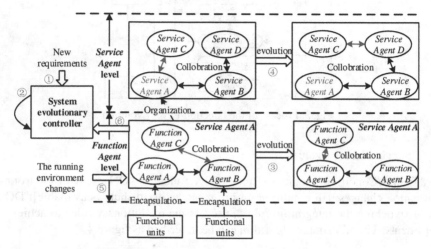

Fig. 5. Process of the Service-Level Evolution

3   Case Study

3.1   Background

To verify whether the proposed multi-agent organization can effectively support the hiberarchy evolution, this paper analyzes the air defense simulation system and proposes a platform to change this software into multi-agent system. The platform packages the functional units as *Function Agents* (such as the Command Agent, Plane Agent). And by organizing the *Function Agent*, a variety of *Service Agents* are formed (such as the Control Agent). In this simulation system, the damage of the functional units is considered as an evolutionary triggered factor of function level. And the facts that users want to enhance the capacity to fight the enemy and the changes of the location of enemies are considered as evolutionary triggered factors of service level.

The experiment conditions are CPU: the Intel (R) Core (TM) 2 Duo the E7500 @ 2.93 GHZ, Memory: 1.87 GB, Operating System: Windows 7 SP1.

3.2   Function-Level Evolution

When the enemies destroy the command, software should dynamically discovery this factor. Based on the analysis of its capacities library, this *Function Agent* can replace the command. As shown in figure 6, when this circumstance is occurred, the platform has the awareness of this factor and the command is replaced.

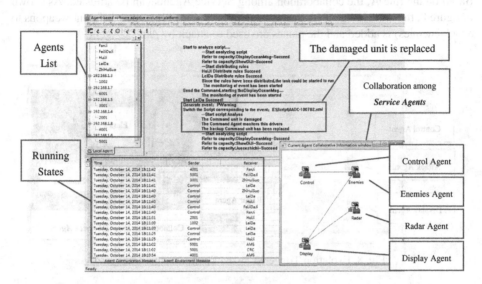

Fig. 6. Evolution Triggered by the Damaged Functional Units

The replacement time is used to test the ability of the multi-agent organization to support the function-layer dynamic evolution. The elapsed time of finding the damaged units to replace the backup unit is called the replacement time. System monitors all the running states of functional units every 5 seconds. So as shown in figure 7, the replacement time remains 6 seconds, a fact indicates that once the triggered factor of units damaged are found, it can be completed within 1 second.

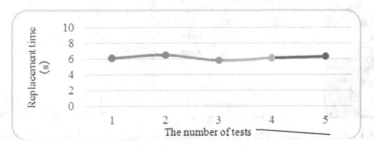

Fig. 7. Replacement Time

3.3   Service-Level Evolution

When the enemies are in the alert area, this evolutionary triggered factor can be perceived by Control Agent (A Service Agent is responsible to monitor the enemies and make operation plans). Then the corresponding evolutionary rule A is submitted to the system evolution controller and then distributed to all the Service Agents in the system. Based on the rule A, the collaboration among Service Agents can be adjusted. As shown in figure 8, the Attack Agent (A Service Agent is responsible to use different weapons to attack enemies) is added and the planes are used to attack by Attack Agent.

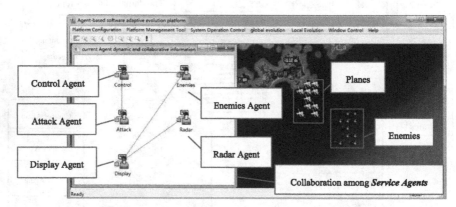

Fig. 8. Evolution Triggered by the Environmental Changes

Following, when users want enhance the ability of this system to attack the enemies, the evolutionary rule B in system evolutionary controller is triggered and distributed to all the Service Agents. Then according to the rule B, the Attack Agent adjusts its internal Function Agents and adds the Missile Agent (A Function Agent is used to control missiles). As shown in figure 9, the missiles are sent to attack the enemies.

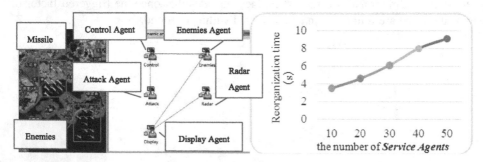

Fig. 9. Evolution Triggered by the Users' Requirement and Test of Reorganization Time

The elapsed time of distributing the rules to finish the adjustment of the collaboration is called the reorganization time, used to test the ability of the multi-agent organization to support the service-layer dynamic evolution. As shown in figure 9, with the number of Service Agents increasing, this time rises proportionally.

4   Conclusion

This paper proposes a multi-agent organization for hiberarchy dynamic evolution by separating calculation logics from collaboration logics. The adaptive agent model is used to support the function-level evolution by dynamically replacing the damaged units. And with the adjustment algorithm modifying the collaborative relationship among agents, the multi-agent organization can supports the service-level evolution. The case study proves that the multi-agent organization is suitable for the hiberarchy dynamic evolution and helpful to decrease the cost and time of evolution. Researchers may mature this multi-agent organization in the future by comparing it with other existing organizations and improving its ability of dealing with more levels evolution.

Acknowledgments. This work is supported by the Projects (61173026, 61373045, 61202039) supported by the National Natural Science Foundation of China; Projects (BDY221411, K5051223008) supported by the Fundamental Re-search Funds for the Central Universities of China; Project (513***103E) supported by the Pre-Research Project of the "Twelfth Five-Year-Plan" of China; Project (2012AA02A603) supported by the National High Technology Research and Development Program of China.

References

1. Vogel, T., Giese, H.: Model-driven engineering of self-adaptive software with eurema. In: Parashar, M., Zambonelli, F. (eds.) ACM Transactions on Autonomous and Adaptive Systems. ACM, vol. 8(4), pp. 18–51. ACM Press, New York (2014)
2. Iftikhar, U.M., Weyns, D.: Assuring system goals under uncertainty with active formal models of self-adaptation. In: 36th International Conference on Software Engineering, pp. 604–605. ACM Press, New York (2014)
3. Perrouin, G., Morin, B., Chauvel, F.: Towards flexible evolution of dynamically adaptive systems. In: 34th International Conference on Software Engineering, pp. 1353–1356. IEEE Press, New York (2012)
4. Brun, Y., et al.: Engineering self-adaptive systems through feedback loops. In: Cheng, B.H.C., de Lemos, R., Inverardi, P., Magee, J. (eds.) Software Engineering for Self-Adaptive Systems. LNCS, vol. 5525, pp. 48–70. Springer, Heidelberg (2009)
5. Souza, S.: A requirements-based approach for the design of adaptive systems. In: 34th International Conference on Software Engineering, pp. 1635–1637. IEEE Press, New York (2012)
6. Liang, X., Dave, R., Madalina, C.: Adaptive agent model: an agent interaction and computation model. In: 31st Annual International Conference on Computer Software and Applications, pp. 153–158. IEEE Press, New York (2007)
7. Leriche, S., Arcangeli, J.: Adaptive autonomous agent models for open distributed systems. In: 2th International Multi-Conference on Computing in the Global Information Technology, p. 4. IEEE Press, New York (2007)

Long Term Electricity Demand Forecasting with Multi-agent-Based Model

Zhang Jian[1](✉), Hu Zhao-guang[1,2], Zhou Yu-hui[1], and Duan Wei[1]

[1] Beijing Jiaotong University, Beijing, China
{12117375,huzhaoguang,yhzhou1,10117336}@bjtu.edu.cn,
huzhaoguang@sgeri.sgcc.com.cn
[2] State Grid Energy Research Institute, Beijing, China

Abstract. Electricity demand and economic growth are closely correlated. Electricity is an important means of production and subsistence and plays an important role in the national economy system. Accurate electricity demand forecasting results could provide the basis for the power grid planning and construction and therefore has important social and economic benefits. In this paper, a long-term electricity demand forecasting model that contains six kinds of Agent is proposed based on multi-agent technology. The model is validated by the electricity consumption data of 2011-2014. Then the industry-wide electricity demand forecasting results from 2015 to 2025 are obtained. Through case study, the results change affected by economic policy is studied. The results show that the electricity demand will increase under loose monetary policy.

Keywords: Multi-agent based model · Electricity demand · Forecasting · Economic policy

1 Introduction

China's electricity demand and economic growth are closely correlated [1]. Electricity is an important means of production and subsistence, the growth of electricity consumption will lead to the growth of GDP, while the shortage of electricity will inhibit the normal development of the economy [2]. Accurate forecasting of electricity demand can provide the basis for the power grid planning and construction. It has important social and economic benefits [3].

Many approaches have been proposed and applied to long term electricity demand forecasting, including regression analysis, time series analysis, gray prediction, expert systems, artificial neural networks and support vector machine method [4,5]. Existing forecasting methods often require large amount of historical data. In these methods the model parameters are set based on historical data and fixed during the forecasting process. It is difficult to evaluate the impact on electricity demand by a variety of factors such as the economic performance and policy changes. As part of the

"The Fundamental Research Funds for the Central Universities" (E14JB00160).

Y. Tan et al. (Eds.): ICSI-CCI 2015, Part I, LNCS 9140, pp. 578–585, 2015.
DOI: 10.1007/978-3-319-20466-6_61

macroeconomic system, power system is inseparable from the operation of economic system are. Considering both the macroeconomic situation and the operation of the power system is the key to get accurate data of electricity demand. Macroeconomic forecasting is a complex economic and social problem. Electricity demand forecasting considering economic factors is a cross-disciplinary problem which is very difficult to solve using traditional tools. With the development of artificial intelligence technology, Agent-based modeling method provides a new way of economic research and also has been widely used in power system [6,7,8]. Ref. [9] studied fluctuations in the development of economy and power system by intelligent engineering methods and obtained the optimal solution of power system development. Ref. [10,11,12] studied the impact of economic policies on electricity consumption based on multi-agent technology.

2   Theory

2.1   Agent

Minsky [13] first proposed Agent in 1986 and believed that Agent is the individual in society who is able to obtain the problem solution through negotiation. There is no unified definition of Agent due to the different backgrounds and views of researchers from different areas. Woodridge and Jennings [14] gave a comprehensive overview of the concept and features of Agent and proposed the most popular definition. They believe that Agent is the independent individual with situatedness, autonomy and flexibility. Holland [15] proposed the concept and theory of complex adaptive systems, and proposed Agent-based simulation (ABS) method, which is a powerful tool to study the socio-economic system. Ref. [16] compared Agent-Based Model and Equation-Based Model applied in different areas. The results show that Agent-Based Model has better performance.

2.2   MAS

Multi-agent system (MAS) is a distributed autonomous system consists of a plurality of mutually interacting Agents [13]. MAS studies Agent behavior coordination. That is how independent autonomous Agents adjust their autonomous intelligent behavior to solve complex problems can not be solved by Agent individuals through interaction and collaboration between each other.

Multi-Agent based modeling method has been widely used in various fields [17]. Tesfatsion [8] proposed Agent-based Computational Economics (ACE) and pointed out the advantages of ACE method to simulate complex adaptive economic system. US Sandia Laboratories developed a simulation model of the US economy ASPEN [18]. ASPEN used multi-agent technology to simulate the decision-making behavior of micro-economic individual, thus, reproduced some of the process of socio-economic system. Kang et al. [7] proposed the theoretical framework of the electricity market simulation based on individual beliefs study of Agent. Yuan et al. [19] added the fuzzy theory into Agent decision and simulated the consultation in electricity contract market.

2.3   ARE

ACE model combines mathematical modeling, game theory and experimental methods, but ignores the human initiative in the economic simulation. Hu [20] proposed Agent Response Equilibrium (ARE) model. ARE model divides the macroeconomic according to the function, organization and structure. Each of the microscopic section can be abstracted as Agent has intelligence and response capabilities.

ARE model sets different Agents and builds the framework of interaction and response to achieve the objective of macroeconomic simulation. Hu et al. [21] derived the input-output table of China 2010 based on ARE model. Duan et al. [12] build a simulation system   using ARE model to study how economic policies affect electricity demand and the validity of the model was proved through data validation.

3   The Model

3.1   Model Design

In the economic system, the changes of the industrial production activities will lead to the electricity consumption change. Through the economic system the individuals will adjust their activity based on their own goals and interests after the implementation of economic policy, thus affecting electricity consumption. Accurately forecasting of electricity demand cannot be separated from the foundation of the economic system. In ARE model, the production activities of power industry and other various industries are fully taken into account thus enabling electricity demand forecasting results more accurate. Given the complexity of the economic system, the economic system can be simplified into five sectors including government, bank, residents, industry and abroad.

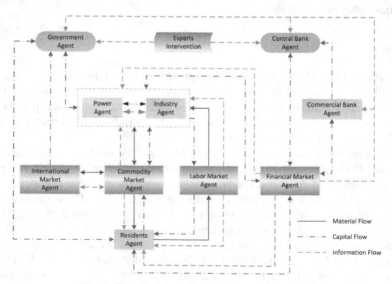

Fig. 1. The electricity demand forecasting model

Based on the simplified macroeconomic system, the electricity demand forecasting model is build using multi-agent technology. The model includes Power Agent, Industry Agent, Residents Agent, Bank Agent, Government Agent, Market Agent and Abroad Agent. In order to unify statistical standard of electricity consumption data and industry production data, the national economic system is divided into 42 sectors based on the input-output table of China 2010. The relationship between the various sectors, as well as electricity consumption of industry production activities is represented by the input-output relationship.

3.2   Model Application

The electricity demand forecasting model is developed using Java programming language on Swarm platform. The flow chart is shown in Fig.2.

4   Results

4.1   Model Validation

The industry-wide electricity consumption data from 2011 to 2014 is predicted and compared with the data published by CEC and NEA [22,23].

Table 1. The forecasting data and the published data 2011-2014

Industry-wide electricity consumption / trillion kWh	2011	2012	2013	2014
Forecasting data	4.1409	4.3375	4.6433	4.8095
Published data	4.1401	4.3429	4.6430	4.8305
Forecasting error	0.02%	-0.12%	0.01%	-0.43%

As can be seen, the forecasting errors of 2011-2014 are: 0.02%, -0.12%, 0.01%, -0.43% . The errors between predicted value and the actual value are small, indicating that the model is valid.

4.2   Scenarios

Steady development of China's "new normal" economy in recent years leads to the smooth power consumption at low growth rate. Total electricity consumption in 2014 reached 5.52 trillion kWh with an increase of 3.8%. To accomplish the industry-wide electricity demand forecasting from 2015 to 2025 and analyze the impact of policy changes on the predicted results, we set two scenarios.

Scenario 1: The initial policy remains unchanged without applying any intervention.

Scenario 2: In 2015-2018, with the implementation of loose monetary policy, the deposit reserve rate is adjusted from 20% to 15% through several adjustments. Meanwhile the deposit rate is adjusted from 3% to 2% and the interest rate is adjusted from 6% to 4%.

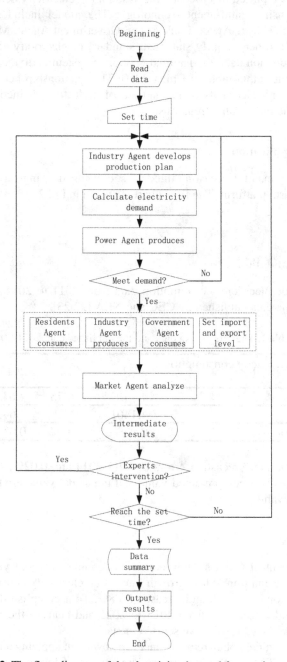

Fig. 2. The flow diagram of the electricity demand forecasting model

4.3   Forecasting Results

According to the forecasting results of scenario 1, the industry-wide electricity consumption of 2015 is 4.86 trillion kWh and grows to 6.03 trillion kWh at 2025 with an average annual growth rate of 2.2%. According to predict two scenarios, the entire industry in 2015 with a capacity of 4.88 trillion kWh, 2025 grew to 6.33 trillion kWh, with an average annual growth rate of 2.6%.

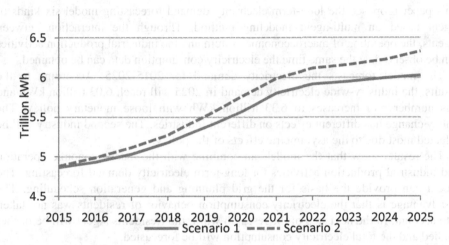

Fig. 3. The industry-wide electricity consumption 2015-2025

Compared to Scenario 1, the industry-wide electricity consumption of 2025 increases by 301.4 billion kWh in Scenario 2. The electricity consumption increment of Scenario 2 rises year by year. The increment changes from 0.3% (2015) to 5.0% (2025).

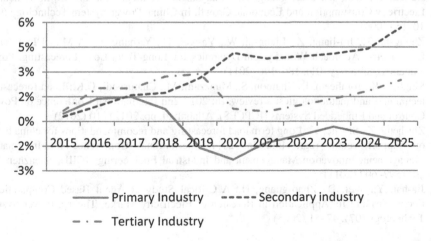

Fig. 4. The increment in forecasting results under different economic policy

From the industry perspective, the increment of primary industry, secondary industry and tertiary industry in Scenario 2 are as follows: -1.2%, 5.6%, 2.5%. The second industry has the biggest increment, mainly due to the asymmetric effects of monetary policy.

5   Conclusion

This paper proposed the long-term electricity demand forecasting model six kinds of Agent based on multi-agent modeling method. Through the interaction between Agents, the operation of macroeconomic system and the industrial production activities can be observed, at the same time the electricity consumption data can be obtained.

The model forecasts the electricity demand for 2015-2025. According to the results, the industry-wide electricity demand in 2025 will reach 6.03 trillion kWh and this number will increases to 6.33 trillion kWh with loose monetary policy. The policy change has different effects on different industries. The second industry will be affected most due to the asymmetric effects of the policy.

The results show that the model can combine with the macroeconomic operation and industrial production activities for long-term electricity demand forecasting. The model can provide the basis for the grid planning and generation scheduling. The disadvantage is that the electricity consumption behavior of residents was not taken into account. In the next step of the research work, the Residents Agent will be further studied and the total electricity consumption will be forecasted.

References

1. Zhao-guang, H., Yan-ping, F.: Analysis on Prospects of Economic Development and Power Demand in China. Electric Power (08), 6–9 (2000)
2. Jia-hai, Y., Wei, D., Zhao-guang, H.: Analysis on Cointegration and Co-movement of Electricity Consumption and Economic Growth in China. Power System Technology (09), 10–15 (2006)
3. Zhong-fu, T., Jin-liang, Z., Liang-qi, W., Ya-wei, D., Yi-hang, S.: A Model Integrating Econometric Approach With System Dynamics for Long-Term Load Forecasting. Power System Technology (01), 186–190 (2011)
4. Singh, A.K., Ibraheem, I., Khatoon, S., Muazzam, M., Chaturvedi, D.K.: Load forecasting techniques and methodologies: a review. In: 2012 2nd International Conference on Power, Control and Embedded Systems (ICPCES), Allahabad, pp. 20121–10 (2012)
5. Zhi-heng, Z., Shi-jie, Y.: Long term load forecasting and recommendations for china based on support vector regression. In: 2011 International Conference on Information Management, Innovation Management and Industrial Engineering (ICIII), Shenzhen, pp. 2011597–602 (2011)
6. Jia-hai, Y., Wei, D., Zhao-guang, H.: A Critical Study of Agent Based Computational Economics and Its Application in Research of Electricity Market Theory. Power System Technology (07), 47–51 (2005)

7. Chong-qing, K., Jian-jian, J., Qing, X.: Theoretical Fundamental and Concepts of Electricity Market Simulation Based on Agents' Belief Learning. Power System Technology (12), 10–15 (2005)
8. Tesfatsion, L.: Agent-based computational economics: Modeling economies as complex adaptive systems. Inform. Sciences **149**(4), 263–269 (2003)
9. Zhao-guang, H.: Study on the baseline space of sustainable power development. Electric Power (04), 6–9 (2004)
10. Min-jie, X., Zhao-guang, H.: Simulating Impact of Macroeconomic Policy on Electricity Consumption Based on Multi-agent. Journal of Systems & Management (05), 539–548 (2011)
11. Jian-wei, T., Zhao-guang, H., Jun-yong, W., Xiao, X., Min-jie, X.: Dynamic Economy and Power Simulation System Based on Multi-agent Modelling. Proceedings of the CSEE (07), 85–91 (2010)
12. Wei, D., Zhao-guang, H., Si-zhu, W., Yu-hui, Z., Ming-tao, Y.: Dynamic Simulation of Economic Policy and Electricity Demand by Agents Response Equilibrium Model. Proceedings of the CSEE (07), 1206–1212 (2014)
13. Hai-gang, L., Qi-di, W.: Summary on Research of Multi-agent System. Journal of Tongji University (06), 728–732 (2003)
14. Woodridge, M., Jennings, N.R.: Intelligent agents theory and practice. Knowl. Eng. Rev. **10**(2), 115 (1995)
15. Holland, J.H.: Hidden order: How adaptation builds complexity. Basic Books (1995)
16. Bollinger, L.A., Davis, C., Nikolic, I., Dijkema, G.P.J.: Modeling Metal Flow Systems: Agents vs. Equations. J. Ind. Ecol. **16**(2), 176–190 (2012)
17. Shao-ping, Z., Feng, D., Cheng-zhi, W., Qin, Z.: Summary on Research of Multi-Agent System. Complex Systems and Complexity Science (04), 1–8 (2011)
18. Laboratory, S.N.: Aspen's information page (2002). http://www.cs.sandia.gov
19. Jia-hai, Y., Zhao-guang, H.: A Multi-agent Based Negotiation Simulation System for Electricity Contract Market. Power System Technology (11), 49–53 (2005)
20. Zhao-guang, H.: Study on Agents Response Equilibrium Models. Energy Technology and Economics (06), 9–15 (2011)
21. Zhao-guang, H., Wei, D., Xiao, X., Jian-wei, T.: Derivation of China's 2010 Input-Output Table Based on Agent Response Equilibrium (ARE) Model. Energy Technology and Economics (11), 8–14 (2011)
22. Council, C.E.: Annual Statistics of China Power Industry (2015). http://www.cec.org.cn/guihuayutongji/tongjxinxi/
23. Administration, N.E.: National Energy Administration released the total electricity consumption in 2014 (2015). http://www.nea.gov.cn/2015-01/16/c_133923477.htm

Ontology Based Fusion Engine for Interaction with an Intelligent Assistance Robot

Nadia Touileb Djaid[1,2(✉)], Nadia Saadia[1], and Amar Ramdane-Cherif[2]

[1] LRPE Laboratory, University of Houari Boumediene, Algiers, Algeria
n-touileb@hotmail.fr, saadia_nadia@hotmail.com
[2] LISV Laboratory, University of Versailles Saint-Quentin-en-Yvelines, Versailles, France
rca@lisv.uvsq.fr

Abstract. Every human continuously interacts with his environment and its entities. To interact with the environment, humans use language and physical expression to understand the events and be understood. These communication methods are natural features acquired at birth, with a few exceptions. Unfortunately, some people face interaction difficulties because of disabilities or illnesses. To remedy to these problems, researchers have been designing assistance robots which can imitate human interaction using multiple modalities. To do so, the robot must be able to interact with humans using natural methods used by people such as speech, gestures, eye movements, etc. The robot must be able to understand and execute the commands issued by the user through the different modalities. To do so, we propose a smart system that will use a knowledge base to achieve the three tasks of "sensing-understanding-acting" in an ambient environment.

Keywords: Assistance robot · Multimodal systems · Ontology · SWRL rules · Fusion engine

1   Introduction

With the increasing emergence of ambient intelligence, sensors and wireless network technologies, robotic assistance becomes a very active area of research in autonomous intelligent systems. The robot, which becomes an intelligent system, will be able to understand an environment in which events are detected by sensors. This system must be able to merge events in order to understand a situation and be able to decide, act and perform different services.

The aim of this research work is to build a multimodal fusion engine using the semantic web. This multimodal system will be applied on a wheelchair with a manipulated arm to help people with disabilities interact with their main tool of movement and their environment. This work focuses on building a multimodal interaction fusion engine to better understand the multimodal inputs using the concept of ontology. Given that the system components are interconnected in a network, our architecture allows a robot to provide services to humans anytime and anywhere.

© Springer International Publishing Switzerland 2015
Y. Tan et al. (Eds.): ICSI-CCI 2015, Part I, LNCS 9140, pp. 586–593, 2015.
DOI: 10.1007/978-3-319-20466-6_62

2   Related Work

In recent years, multimodal fusion is gaining attention of researchers of various domains due to the benefits of using multimodal inputs and outputs. Multimodality provides access to various modalities, and their use based upon accessibility and availability. Since the first multimodal system, the famous Bolt's [1] system "Put that Here", several multimodal system have been proposed. For instance, we can find a review of sensor fusion algorithms for wearable robots [2]. In this paper, the authors highlight that the fusion combines information from different sensors either by using a single fusion algorithm, a unimodal switching, a multimodal switching or a parallel multiple sensor fusion algorithm (mixing) [2]. Another work on mobile robot uses vison and RFID data fusion for tracking and following a person of interest using a mobile robot [3]. For tracking method, the authors use particle filtering framework, and for following the person, they have designed a multi-sensor-based control strategy based on the tracker output and RFID data. Furthermore, an example is presented in the concept-based evidential reasoning form multimodal fusion in human-computer interaction [4]. In this work, an approach is proposed for the semantic fusion of different input modalities based on transferable belief models. On the other hand, the Adaptive Resonance Theory (ART)-based fusion of multimodal perception for robots [5] uses ART for multimodal fusion. The ART is an unsupervised neural network which has the ability of fast incremental on-line learning. Furthermore, the work of detection of violence in movies [6] uses ontology for multimodal fusion. To do this, two different fusion approaches are used: "the first one is a multimodal fusion that provides binary decisions on the existence of violence, and the second one is an ontological and reasoning fusion that combines the audio-visual cues with violence and multimedia ontologies" [6]. In the same way, ontology is used in multimodal fusion for interaction systems [7]. In this work, the environment is described in ontology and then used in the fusion engine. It uses semantic web languages based on W3C standards.

Based to the works mentioned above, we conclude that no work has been done using a multimodal ontology-based fusion engine to control a device such as a wheelchair with an embedded manipulation arm. This choice is guided by the fact that the use of ontology allows the full description of the environment of a user and takes into consideration its context. It provides an easy access to information and allows the possibility of reusing it and allows us to introduce fusion rules to facilitate the fusion process according to predefined models in the ontology. We have chosen to use a rule based fusion engine because it allows a good temporal alignment between different modalities and is often used to better estimate the state of a moving object.

3   Proposed Architecture

In representing our system architecture our approach of choice is a multimodal fusion and service composition engine. The fusion engine will be associated with an assistant robot. In this case, the robot is a wheelchair with a manipulator arm used by persons with disabilities to interact with the environment and to provide them services. The user can use multimodal inputs, such as speech, gesture, etc. to request a service. The proposed system will take into account the contextual information and combine it with the inputs modalities. When the fusion is made, the composition service engine will subdivide the fusion results and send it to output modalities to answer the request of the user.

Fig. 1 shows a general view of our architecture. This architecture is composed of four parts: the input, the multimodal system architecture (MSA), the knowledge base and the output. The input part is responsible for input modalities that detect events coming from the environment, such as a user request or contextual information. This event will be sent to the MSA. The MSA is an essential part of our architecture because it is that part where the fusion and the service composition will take place. When an event is detected by the input modalities, it will be understood by the system through the different components of the MSA. First, it will make a modality selection according to the information obtained from the environment. Then, based on the knowledge base, which is the ontology that describes the surroundings of the user, the MSA will merge the information obtained from the environment using the fusion engine. Then, its result will be subdivided into subtasks using the service composition engine which will be sent to the available output modalities to provide the service demanded by the user. In this work, we will focus on the knowledge base, the understanding of the request and the fusion engine. The service composition engine will be presented in a future article.

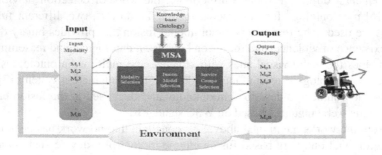

Fig. 1. The multimodal and composition of service engines [8]

We use ontology to describe the environment of a user. The ontology will serve as the knowledge base of the global multimodal system. This concept allows taking into consideration a large number of elements that describes the user's environment. Also, it allows the integration of rules that must be taken into account during the fusion process. These rules will be stored in the ontology using the semantic web rule languages SWRL (for rules) and SQWRL (for queries) based on W3C standards. To achieve this, we use the open source tool ROTEGE (www.protege.stanford.edu).

In our work, the environment of the robotic system is also the user's environment. As such, this environment shall be the interior of the user's home and all the elements that comprise it, and the outdoors which can be a garden or the neighborhood. The environment ontology, represented by a hierarchical graph, will allow us to define in detail all the elements that comprise the environment and provide the relationships among them.

Fig. 2 shows the classes defined in our ontology. The class environment is the super class that contains all the entities present around the user. These entities are described in different classes according to their types. The subclass "Modality" contains all the modalities used in our system; we have chosen to use speech, gesture, eye gaze, the manual modality (the keyboard and the mouse of a computer) and touch. The "Alarm" subclass contains all the alarms used by our system. In our work, we have two types of alarms. The system alarm is triggered when the battery level of the system is low and when one of the four sensors of the wheelchair detects an obstacle. The surroundings subclass is composed of all the elements present in the surrounding of the user. By elements, we mean the persons, the objects, and the places where the user is. The vocabulary subclass contains the words that the user will use when requesting a service. The time subclass defines the maximum time allowed for a command to be accepted and the maximum time between two modalities. The coordinates subclass gives the coordinates of the environment elements. Finally, the command model and the fission model subclasses contain our predefined models of different commands that can be emitted by the user to use in the fusion engine and the composition of service engine, respectively. The system context takes into account the information obtained from the different sensors embedded on the wheelchair. We have chosen to consider the battery level of the system and four sensors (ultrasonic and infrared). These sensors are on the four edges of the wheelchair which detect obstacles when moving. Finally, the environment context takes into account the elements which can affect modalities; we have chosen to consider the lighting level, the noise level, the weather condition, and the ambient temperature.

Fig. 2. The class Environment and its sub classes

After defining the classes, we add individuals (instances) to the ontology. The individuals represent all the entities present in the environment, such as persons, food, objects, etc. This integration makes our use of ontology more efficient because it makes the system reusable and open. For this paper, we will just give some examples of individuals. For instance, the class indoors (subclass of the class surroundings) has as individual: bathroom, bedroom, corridor, kitchen, living room, sitting room and toilet. The class words for tracking (subclass of the class vocabulary), has as individuals: find, locate, return, search and take.

Then, we define the relations between different entities in the environment using properties. We use object properties to link two objects, and data type properties to link an object to XML schema datatype or rdf: literal. For this article, we will just give two examples rather than all the properties used. For instance, Fig. 3 shows two of the properties used in our ontology. The object property "IsActivatedBy" links the individual "Alarm2" to the individual "Battery". Here, alarm2, which is the system alarm, is triggered by the level of the battery of the system. If the level is low, the alarm will be triggered and advises the user about it. Also, we find the datatype property "hasBatteryLevel" which links the individual "Battery" to its value which is 15 % (we provide this value in percentage and is of type "int"). In addition to the properties just cited, we use semantic relations to allow robot to understand some actions that are common knowledge for people. For example, we can define a relationship between liquid drinks and objects used for liquids. This relationship will allow the robot to understand that a liquid needs a container, and when the user requests for a drink, it is implied to use an object to hold a liquid (for example, a glass or a bottle).

Fig. 3. The object property: "IsAffectedBy" and the data type property "hasBatteryLevel"

4   Fusion Engine

The fusion engine is the most important part of our system. It allows the merging of information obtained from the environment, such as the modalities and the contextual information. When the user makes a request, the system detects the events and merges them to offer the service requested by the user. To do so, our system has to go through different stages to complete the fusion.

First, the system will have to make a semantic check. Indeed, by using the concept of ontology, we have to check the consistency of the ontology by checking the relation between its components. This can be done by using the inference engine Pellet and the Jess plug-in of Protégé. The Pellet engine will check the inferences, the taxonomy and the consistency between the classes of the ontology and the Jess engine will check the SQWRL queries defined in our ontology.

Secondly, the system will check the presence of the detected events in the predefined vocabulary in the ontology. This checking will allow the system to reject any undefined event that will not be used in the fusion engine such as the sound of a door or a TV.

Thirdly, the order of the events will be checked. To do so, the events obtained from different modalities will be checked using SQWRL queries. We have defined nineteen models that describe examples of requests made by the user, and each model has its own query defined in the SWRL tab of Protégé.

Finally, the time checking will allow the system to verify that the time between two successive modalities is less than the maximum predefined time allowed (5seconds). As well as the global time of the command which has to be under the maximum allowed predefined command time (15 seconds). We have chosen to introduce this checking so that the system will not wait indefinitely for commands from the user.

We will present an examples of commands made by the user and how the system deals with it and merge the information using the fusion engine. The scenario chosen to test our fusion engine is "Give Mother Some Juice". In this case, the system has to understand the meaning of the phrase, find a model that corresponds to it and merge the information, and take into account the maximum time allowed between the modalities and the full command. We highlight that when asking for juice, the system has to understand that the juice has to be brought within an appropriate object, such as a glass or a bottle.

After launching the reasoning engine Pellet in Protégé and the check is completed and no problem was detected, we start our test of the fusion engine. We assume that the system detected the inputs described in table 1.When detecting inputs, the fusion engine checks their presence in the ontology and gives the classes where they are defined as individuals in the ontology, as shown in Fig.4 .We note that the fusion engine has recognized the words: "give", "juice", and "mother" and gives us their location in different classes of the ontology. The word "ding" which is a sound emitted by the television and detected by the system is rejected because it is not found in any class within the ontology. Furthermore, the fusion engine will find the predefined model of the command "Give Mother Some Juice" that is represented by the following order: words convenient objects → People → Liquid. Also, the fusion engine has to understand that the liquid requested has to be brought in an object used for liquids.

Table 1. Input modalities

Modality	Event	Arrival time (s)
Speech	Give	0
Speech	Mother	3
Sound from TV	Ding	6.2
Speech	Juice	7

The result of the query is presented in Fig.5. The fusion engine has recognized the model as being model 19 of the ontology and that the answer for such a request will be the answer of the model: words convenient objects →People →Object used for liquid →Liquid. In fact, the fusion engine has recognized the model and added the objects: bottle, glass or jug that can be used for the juice.

Finally, the fusion engine will make a time verification. The result is given in the Fig.5, where Txy and Tyz are the times between the successive modalities "Bring", "Mother" and "Juice" respectively, and Full-Time-C is the time of the full command. We notice that the modalities times are less than "5 seconds" which is the maximum allowed, and the full command time is less than "15 seconds" (the maximum command time allowed), so time condition has been satisfied.

SQ SQWRLQueryTab	→ systemOntology:Vocabulary_Verification	
?m	?y
systemOntology:Command_Model	systemOntology:Give
systemOntology:Command_Model	systemOntology:Juice
systemOntology:Command_Model	systemOntology:Mother
systemOntology:Surroundings	systemOntology:Juice
systemOntology:Surroundings	systemOntology:Mother
systemOntology:Vocabulary	systemOntology:Give

Fig. 4. Vocabulary verification result

SQ SQWRLQueryTab	→ systemOntology:Command_selection19			
?m	?x	?y	?a	?z
systemOntology:Command_Model_19	systemOntology:Give	systemOntology:Mother	systemOntology:Bottle	systemOntology:Juice
systemOntology:Command_Model_19	systemOntology:Give	systemOntology:Mother	systemOntology:Glass	systemOntology:Juice
systemOntology:Command_Model_19	systemOntology:Give	systemOntology:Mother	systemOntology:Jug	systemOntology:Juice

SQ SQWRLQueryTab	→ systemOntology:Time_Verification				
?x	?y	?z	?Txy	?Tyz	?Full-Time-C
systemOntology:Bring	systemOntology:Mother	systemOntology:Juice	3	4	7

Fig. 5. Command selection result and Time verification result

To validate our approach, we have chosen to use Colored Petri Nets (CPN). Indeed, the validation of an IT architecture is an important step in the development process of a real time system. Our use of CPN is also justified by the existence of software tools that support the graphical construction and visualization of models. To do that, we have chosen to use the CPN-Tools (University of Aarhus 2012) to explore the behavior of the architecture model.

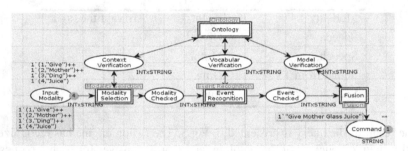

Fig. 6. The colored Petri Net of the proposed architecture

Fig 6 shows the general view of the CPN. This net is composed of all the elements that compose our architecture. The doubled rectangle represents a subnet that deals with specific tasks. For instance, the transition "Modality selection" is composed of a sub net that is responsible of the selection of the modalities according to the state of the context. The verification will be done using the information stored in the ontology. The Fig 6 shows the results after launching the simulation. We used for this simulation the same example presented before "Give mother juice". We notice that the result is "Give mother glass juice", which means that the system has successfully merged the inputs according to the conditions of the fusion engine presented before. Moreover, we notice that the event "Ding" which is the sound issued by a television has been rejected and that the word "Glass" has been added to the request after the fusion process. We conclude that the fusion process has been made correctly and that this net validate our proposed approach.

5   Conclusion

We have described in this paper a multimodal fusion engine used to control a wheelchair with a manipulated arm. We aimed to facilitate the interaction between the user and his main tool of living. Our proposed architecture uses the concept of ontology as a knowledge base for the fusion engine. By using this concept, we have ensured the reusability of the information at any time. Thereby, our proposed architecture is an autonomous system that is able to interact and make decisions for answering the user requests. To validate our approach, we have built a Colored Petri Net of the architecture and simulated it using CPNTools. Indeed, the Petri nets are a graphical and mathematical tool to verify systems and protocols. The model built using the CPNTools allowed us to verify the correctness of our system and the absence of deadlock and bugs.

Our perspective for this work is to build a composition of service engine that will subdivide the result of the fusion engine into unimodal tasks and send them to the available output modality. By doing this, the global system will be a full multimodal system that is able to understand the requests and offer services to the user using a multimodal interaction.

References

1. Bolt, R.A.: "Put-that-there": Voice and gesture at the graphics interface. J. ACM SIGGRAPH Comput. Graph. **14**, 262–270 (1980)
2. Novak, D., Riener, R.: A survey of sensor fusion methods in wearable robotics. J. Robot. Autonom. Syst. (2014)
3. Axenie, C., Conradt, J.: Cortically inspired sensor fusion network for mobile robot egomotion estimation. J. Robot. Autonom. Syst. (2014)
4. Reddy, B.S., Basir, O.A.: Concept-based evidential reasoning for multimodal fusion in human-computer interaction. J. App. Soft. Comp. **10**, 567–577 (2010)
5. Berghöfer, E., Schulze, D., Rauch, C., Tscherepanow, M., Köhler, T., Wachsmuth, S.: ART-based fusion of multi-modal perception for robots. J. Neurocomputing **107**, 11–22 (2013)
6. Perperis, T., Giannakopoulos, T., Makris, A., Kosmopoulos, D.I., Tsekeridou, S., Perantonis, S.J., Theodoridis, S.: Multimodal and ontology-based fusion approaches of audio and visual processing for violence detection in movies. J. Expert syst. with Applic. **38**, 14102–14116 (2011)
7. Wehbi, A., Zaguia, A., Ramdane-Cherif, A., Tadj, C.: Multimodal Fusion for Interaction Systems. J. Emerg. Trends. Comp. Info. Sciences **4**, 445–458 (2013)
8. Arlyn Toolworks. http://www.osbornej.com/ChairBotHome.html

Author Index

Printed in the United States
By Bookmasters